MICROCOMPUTER-BASED NUMERICAL METHODS FOR SCIENCE AND ENGINEERING

Gary J. Lastman
University of Waterloo
Waterloo, Ontario

Naresh K. Sinha
McMaster University
Hamilton, Ontario

Saunders College Publishing
A Division of Holt, Rinehart and Winston, Inc.

New York Chicago San Francisco Philadelphia
Montreal Toronto London Sydney Tokyo

Publisher *Ted Buchholz*
Acquisitions Editor *Robert Argentieri*
Senior Project Manager *Marc Sherman*
Production Manager *Roger Kasunic*
Design Supervisor *Judy Allan*

Library of Congress Cataloging-in-Publication Data
Lastman, Gary J.
Microcomputer-based numerical methods for science and engineering.
Includes bibliographies and index.
1. Numerical analysis—Data processing.
2. Microcomputers. I. Sinha, N. K. II. Title.
QA297.S53 1988 519.4 88-26406
ISBN 0-03-013792-6

Printed in the United States of America

9 0 1 2 039 9 8 7 6 5 4 3 2 1

Preface

This textbook is intended for use by science and engineering students for a first undergraduate course in numerical methods. Many examples and problems with applications in science and engineering have been included. The book is intended for all physical science and engineering students; there is no conscious bias toward any particular discipline. It is not intended to be a completely rigorous introduction to numerical analysis. Instead, we have provided an introduction, with sufficient background material, to some of the *general* numerical methods most likely to be used by a scientist or engineer.

The problems and programs in this book have been developed for use on a personal computer, because we believe that the scientist or engineer of the future is very likely to have a personal computer on his or her desk, which will be used for much of the professional work. These powerful microcomputers can perform much of the computing that was formerly done using mainframe computers. For work that requires a mainframe, it will be possible to save considerable time and effort by communicating with such a machine through a personal computer.

Hence, it will be imperative for all forward-looking universities to have a suitable course in the area of numerical computational methods as a requirement for most undergraduate programs. Many universities have already introduced such courses at either the sophomore or the junior level. Moreover, it is reasonable to expect that emphasis will be on the use of microcomputers. A large number of engineering schools are trying to make small personal computers available to their undergraduate students. This establishes the need for a suitable textbook on the subject, designed particularly for scientists and engineers.

Although at present several books are available on numerical methods for scientists and engineers, this is one of the few books on the subject that emphasizes the microcomputer implementation of numerical methods using the Pascal language. Pascal is now considered a very suitable language for introducing science and engineering students to programming. As far as the authors are aware, most departments of Computer Science, as well as Electrical and Computer Engineering departments, teach Pascal to their freshmen to encourage students to approach programming in a disciplined manner.

This book provides many examples of programs in Pascal. Furthermore, the accompanying floppy disk contains Pascal programs that can be used for solving all the computational problems given in the book. These problems have been chosen from different scientific and engineering areas. Although many of the computational problems are straightforward, we have tried to omit problems with easily obtained closed-form solutions. Most of the problems require numerical computation to obtain a solution; they are not the typical "toy" problems found in many numerical methods books for science and engineering students.

In addition, listings of the Pascal programs, or user's manuals in the case of very long programs, are included as appendices for each chapter. This is especially desirable in view of the fact that although the price of microcomputer hardware has dropped considerably during the past five years, suitable microcomputer software for solving problems in science and engineering is still unavailablc at reasonable prices.

We think that *Microcomputer-Based Numerical Methods for Science and Engineering* will encourage all students of science and engineering to use personal computers effectively for numerically solving a wide variety of problems which cannot normally be solved in closed-form.

Scope of the Book

In Chapter 1 we introduce numerical computing and microcomputers, and discuss some pitfalls of numerical computation. Floating point number systems are introduced.

In Chapter 2 we discuss the familiar problem of solving a set of linear simultaneous algebraic equations. The most common methods, Gaussian elimination and L-U factorization, are presented in detail. The problem of ill-conditioning is introduced and iterative methods are discussed. A Pascal program for solving a set of linear simultaneous equations is given.

In Chapter 3 we first present Horner's rule and synthetic division for efficient evaluation of polynomials for real or complex arguments. This section is followed by a discussion of the problem of finding the roots of a quadratic in a manner which maximizes the accuracy. Cardan's method for the cubic is also discussed. Iterative search methods for higher-order polynomials are then studied, and a Pascal program using Laguerre's method is presented, as well as a Pascal procedure using the Jenkins-Traub algorithm. The problem of sensitivity of the roots to small changes in coefficients is discussed.

Chapter 4 describes methods for solving nonlinear equations. A number of popular iterative methods and their rates of convergence are discussed, and a Pascal version of the routine ZeroIn is presented. The case of a system of simultaneous nonlinear equations is also studied and a program is included.

Chapter 5 presents the topics of interpolation and curve-fitting, which often arise in scientific and engineering problems. Lagrange, Newton, and Hermite interpolating polynomials and cubic splines are discussed, with programs included for the last three. The properties of orthogonal polynomials are discussed with special attention to Chebyshev polynomials. Methods for least-squares fitting of orthogonal polynomials to experimentally obtained data are presented. The singular value decomposition of a real $m \times n$ matrix is introduced and used to solve linear least squares problems. The technique of Richardson extrapolation to the limit is also discussed. Computer programs for curve-fitting with orthogonal polynomials and the singular value decomposition are given.

Chapter 6 describes methods for numerical differentiation and integration. The latter include the Mid-Point and the Trapezoidal rules, Romberg integration and Simpson's rule, as well as Gaussian quadrature. In addition, adaptive quadrature is discussed. Pascal programs for these methods are given.

Chapter 7 describes various methods for the numerical solution of ordinary differential equations. Starting with the simple Euler's method, the reader is led to more sophisticated Runge-Kutta and predictor-corrector algorithms. Practical problems like estimation of the error in approximation and the selection of stepsize are discussed in detail. Higher-order differential equations are presented as a generalization of these methods using the vector differential equation formulation. Two general Pascal procedures are included, RKF45 and ODE.

The solution of a system of first-order linear differential equations leads naturally to the problem of determining the eigenvalues and eigenvectors of a matrix. This is discussed in Chapter 8 in a manner which will motivate engineering and science students and make the theory easier to understand. Although some familiarity with the basic theory of linear algebra is assumed, there is a brief review in Appendix A. The importance of orthogonal transformations which will reduce a matrix to a block triangular form without altering its eigenvalues is discussed, and leads to the QR algorithm. This chapter includes Pascal programs for calculating the eigenvalues and eigenvectors of a real matrix.

Methods for maximizing or minimizing a function of one or more independent variables are discussed in Chapter 9. The Golden Section Search Method, the Fibonacci Search Method, and Brent's Method are examined for the minimization of a function of one variable. The Gradient Method, the Conjugate-Gradient Method, and quasi-Newton methods are also considered for multidimensional minimization.

The appendices include a review of linear algebra, a review of Pascal, a summary of Turbo Pascal, and a listing of some utility Pascal procedures and functions that are used in many of the book's programs. For those persons who do not have access to a Pascal compiler, or who may wish to use FORTRAN, we have provided a short list of some of our major Pascal procedures, along with possible FORTRAN equivalent or substitute programs in Appendix E.

Notation

Throughout this book we use the notation $\ell og(x)$ to denote the logarithm to the base e of the real, positive number x. For logarithms to the base b, $b > 0$, and $b \neq e$, we write $\ell og_b(x)$. The exponential function is written as e^x or $\exp(x)$.

Vectors and matrices are written in bold face; thus, $\mathbf{x}$ denotes a vector and $\mathbf{A}$ denotes a matrix. All vectors are assumed to be column vectors, unless otherwise stated.

Computer Programs

All the programs in the book have been written in Turbo Pascal. The computational results were obtained using version 3 of Turbo Pascal with an 8087 mathematical coprocessor (approximately 16 decimal digits of precision for floating point real numbers). Many programs contain machine-dependent constants such as the machine number base (2.0 for Turbo Pascal) or the effective machine infinity (approximately 10^{308} for Turbo Pascal with the 8087 coprocessor in which REALS are stored in eight bytes). These constants may have to be changed if the programs are to be run using other versions of Pascal. For example, Turbo Pascal without the 8087 coprocessor uses six-byte REALS which have approximately 12 decimal digits of precision for floating point and a machine infinity of about 10^{38}. The programs can be run using Turbo Pascal 4 with some minor modifications; see Appendix C.

The floppy disk accompanying this book contains the Pascal programs, procedures, or functions listed in the appendices for each chapter or given as a user's manual in the case for very long programs or procedures. The files on the disk are named in two different ways. A file named AppX-Y.Pas refers to the program, function or procedure that is in Appendix X.Y of the text. For example, the file App8-3.Pas contains the program that is listed in Appendix 8.3 of the text.

The second way that files are named is by a specific name; these files are generally used as Turbo Pascal include-files. They are referenced in programs by the Turbo Pascal compiler option (*$I filename.Pas *). Most of these files are rather large, and are stored on the disk in a compacted form. This compacted form contains no comment statements or the usual Pascal indentation. The compacted files will compile more quickly than their noncompacted versions. The files stored in the compacted form are

EigenR.Pas	EigenRS.Pas	Files.Pas	Jenkins.Pas
LinearSy.Pas	MatrixIO.Pas	ODE.Pas	Powell.Pas
PrPlot.Pas	QNM.Pas	QUANC8.Pas	ReadNum.Pas
RKF45.Pas	Roots.Pas	SVD.Pas	ZeroIn.Pas
ZeroODE.Pas	ZeroRKF.Pas		

In addition, the Pascal procedure Spline, in the file Spline.Pas, is also stored in the compacted form; however, the Pascal function EvaluateSpline, included in the file Spline.Pas, is stored in the noncompacted form.

Acknowledgments

Preliminary manuscript versions of this book were used in 1986, 1987, and 1988 by J. Reilly, McMaster University, in a numerical methods course for sophomore students in the Department of Electrical and Computer Engineering. We wish to thank him and his students for many suggestions which greatly improved the manuscript.

Also, we express our sincere thanks to the following reviewers, whose helpful suggestions and comments were so useful to us in the preparation of this book: P. Belanger, McGill University; N. K. Bose, University of Pittsburgh; B. Farouk, Drexel University; W. Gray, University of Notre Dame; L. Jehn, University of Dayton; R. Jost, Air Force Institute of Technology; T. Kearns and Z. Motteler, California Polytechnic State University; E. Stuffle, University of Missouri; and J. R. Too, The Catholic University of America.

Legal Matters

Although the Pascal programs, Pascal procedures and Pascal functions (abbreviated as programs) in this book have been tested by the authors, no warranty, express or implied, is made by the authors as to the accuracy and functioning of the programs, nor shall the fact of distribution within the book constitute such warranty, and no responsibility is assumed by the authors in connection therewith.

In particular, these programs should not be relied on for solving a problem whose incorrect solution could result in injury to a person or loss of property. If you do use the program in such a manner, it is at your own risk.

In this book we refer to several registered trademarks. IBM is a trademark of International Business Machines Corporation; Turbo Pascal refers to copyrighted computer software of Borland International; UCSD Pascal and UCSD p-System are trademarks of the Regents of the University of California; Microsoft

G.J. Lastman
N.K. Sinha

October 1988

Contents

Chapter 1

Introduction to Numerical Computation Using Microcomputers

Microcomputers and numerical computations are discussed in this chapter. Floating point number systems are introduced. Such problems in computing with finite precision numbers as roundoff, catastrophic cancellation, and various sources of error in computation are examined.

1.1 MICROCOMPUTERS AS TOOLS FOR PROBLEM SOLVING IN SCIENCE AND ENGINEERING

Science has been defined as an attempt to understand the forces of nature whereas engineering has been defined as the use of the forces of nature for the benefit of mankind. Engineers employ both science and art for the development of a process or product which is beneficial as well as economical. Although both science and engineering have different goals, they both use models to try to understand the basic processes involved in their respective disciplines. Such models are often mathematical; by studying the behavior of the model the scientist or engineer tries to obtain an understanding of the underlying ''real-world'' process. This may require a careful examination of several possible mathematical models in order to determine the one which is the most suitable or best mimics the real-world process. After a model is compared with ''reality'' it may have to be improved in several steps to become more realistic. A great deal of numerical computation may be needed when developing and testing a model. Therefore, all scientists and engineers need to know about the methods and techniques of numerical computation.

In the days when the authors were undergraduate students, the trademark of an engineering or science student was the slide rule that he or she always carried to school. In the early 1970s the slide rule was replaced with an electronic handheld calculator. With further advances in microelectronics, programmable pocket calculators became available in the late 1970s at prices which an undergraduate student could afford. These helped in reducing the drudgery in those situations where a certain set of calculations had to be done repetitively; these calculators were most suitable for problems requiring a small number of inputs and outputs.

Electronic digital computers became available for scientific and engineering problems in the 1950s. These early computers utilized vacuum tubes and were rather bulky as well as very expensive. Also, they consumed large amounts of electric power. The overall reliability was rather poor due to frequent failures in the vacuum tubes. Often it required hours to locate and replace the faulty tubes. Digital computers utilizing transistors became available in the 1960s. Large-scale integration of electronic circuits on a silicon chip in the 1970s led to the development of microcomputers at much more affordable prices. Today it is possible to buy a microcomputer for about $2000, which has more computing power than the IBM 360/50, a popular mainframe computer available in 1965 for approximately $500,000. As a result of this, microcomputers are now being used extensively in science, business, and industry. It is reasonable to expect that microcomputers will be utilized extensively for solving engineering and scientific problems in the near future. With the large amount of computing power associated with microcomputers, much of the numerical computation that was formerly done on a mainframe computer can now be done on a microcomputer. Even for cases where a mainframe is required, it will be possible for the scientist or engineer to save a great deal of time and effort by communicating with such a machine through his or her microcomputer.

In order to use a microcomputer effectively, it is necessary to have a basic understanding of the numerical methods associated with science and engineering problems. Instead of reinventing the wheel, an engineer or scientist should be able to utilize standard mathematical software, and should also be thoroughly familiar with its shortcomings and possible pitfalls. An attempt has been made in this book to describe these basic numerical methods in a form which will motivate science and engineering students to use microcomputers for solving their problems. In the sequel, software has also been developed and programs listed which can be readily used. Some of the possible pitfalls in numerical computation are discussed in the next section.

1.2 SOME PITFALLS IN NUMERICAL COMPUTATION

Many novice computer users have the tendency to believe that the output from a computer is always correct and unquestionable. This is often far from reality.

It is of interest to examine some possible sources of error in numerical computation. These are listed below.

1.2.1 Errors in the Input Data

The input data could be influenced by the following types of errors.

1. The input data is the result of measurements which have been influenced by systematic errors or temporary disturbances.
2. There are roundoff errors: numbers such as $\sqrt{2}$ are "shortened" in order to use them in a computer, which has only a finite precision.
3. There are conversion errors: numbers such as 0.1 require an infinite number of bits to be exactly represented in base 2 arithmetic.

1.2.2 Roundoff Errors During Computations

A basic limitation with all digital computers is that they can use only a finite number of digits for computation. On the other hand, most of the time we come across real numbers, such as π, e, or $\sqrt{2}$, which require an infinite number of digits for exact representation. Hence, these numbers can be represented on a digital computer only approximately. Such numbers often arise in our computations and if we are not careful, these may lead to inexact results. Eventually, the finite precision of the computer will catch up with you, if the computations are sufficiently lengthy. Some care in programming may postpone the buildup of this type of error. It is possible to postpone the problem even further by carrying more digits. However, it must be noted that carrying more significant digits will not necessarily solve all problems.

EXAMPLE 1.1

Consider the following simple computation

$$Q = 10^6 \times \left[\left(\frac{1}{27}\right)^3 - \left(\frac{25}{27} - \frac{8}{9}\right)^3 \right]^{1/3}$$

If we use a calculator with 9-digit accuracy and calculate the value of Q after first evaluating the fractions separately, we get 55.4. On the other hand, if the quantity within the parentheses is simplified before division by 27, we get the correct answer, which is zero. The error is caused by the fact that division by 27 leads to a number with an infinite number of digits. Rounding this number to nine digits causes the difference between the two numbers within the parentheses to be nonzero, whereas if we first reduce the fractions to a common denominator and thus postpone the division by 27, we avoid the error.

1.2.3 Truncation or Discretization Errors

Truncation or discretization errors are due to breaking off an infinite process after a finite number of "terms," or replacing a continuous process by a discrete process. An example would be the replacement of a derivative by a difference quotient.

1.2.4 Simplifications in the Mathematical Model

Certain assumptions may have been made in order to obtain the mathematical model on which the numerical computations are based. These assumptions could give a model that is unrealistic, or a model with solutions that are physically unrealizable, or a model with no solutions.

1.2.5 Errors Due to the Chosen Numerical Method

If we choose a poor numerical method to solve our problem, we should not be surprised if the results are unexpected. The following example illustrates this.

EXAMPLE 1.2

Consider the evaluation of the exponential function e^x by adding the terms in the series

$$e^x = 1 + x + \frac{x^2}{2!} + \frac{x^3}{3!} + \cdots + \frac{x^n}{n!} + \cdots$$

Although this is an infinite series, it might be expected that the result will be fairly accurate if we truncate it at some large n. Define S_n to be the sum of the first n-terms in the series

$$S_n = 1 + x + \frac{x^2}{2!} + \frac{x^3}{3!} + \cdots + \frac{x^{n-1}}{(n-1)!}$$

A Pascal program for adding terms of this series for any given x and a specified n is given below.

```
PROGRAM Exponential;
VAR
  n, i : INTEGER;
  x, t, sum : REAL;
BEGIN
  WRITE('Enter the value of x : ');
  READLN(x);
  WRITE('How many terms in the series are to be added? ');
  READLN(n);
  sum := 1.0;
  t := 1.0;
  FOR i := 1 TO (n-1) DO
    BEGIN
      t := x*t/i;
      sum := sum + t
```

EXAMPLE 1.2 (cont'd)

```
      END;
    WRITELN('exp(',x:6:4,') is ',sum);
    WRITELN('The correct value is ',exp(x))
  END.
```

It can be verified that the program works quite well for small values of $|x|$. For instance, if $x = 2$, we get $S_{22} = 7.38905609893065$, correct to 14 digits after the decimal point by adding only 22 terms in the series. The convergence can also be checked by noting that we get the same answer if we add 23 terms in the series. We can expect some problem when x is larger. For example with $x = 25$, we need 77 terms in the series to obtain convergence to 15 significant digits. For this case, we find that $S_{77} = 72004899337.3859$. However, the result is rather unexpected when we try to evaluate e^{-25} using this series. This time the series converges to $-7.1297804 \times 10^{-7}$ when 94 terms are used. This is not the correct answer! The correct answer should be positive and equal to the reciprocal of $e^{25} \simeq 1.3887944 \times 10^{-11}$.

It is important to note that adding more terms to the series does not improve the result at all. The main reason for inaccuracy in this case is that, for negative x, our algorithm leads to evaluating the difference between two numbers which are nearly equal. This is known as *catastrophic cancellation* and should be avoided in good computational algorithms (see Section 1.4). For another simple example of catastrophic cancellation, the reader is referred to Problem 6 at the end of this chapter.

Another form of computational problem can arise when a user selects a numerical method that is inherently unstable; eventually, any answers produced will be essentially meaningless. Example 1.3 illustrates the effect of an unstable numerical method.

EXAMPLE 1.3

As an example of an unstable numerical method, consider the sequence of numbers, $\{y_n\}$, defined by

$$y_n = -5y_{n-1} + \frac{1}{n}$$

where $y_0 = \log(6/5)$. It can be shown that $y_n > 0$ for $n = 0, 1, 2, \ldots$. In addition, $y_{n+1} \leq y_n$. When this sequence was computed in Turbo Pascal (with an 8087 coprocessor, see Section 1.8), it was found that $y_{22} < 0$, while $y_{27} > 293$. To see why this happens, let y_n denote the exact solution, while z_n denotes the numbers found by our computer. Define d_n as

$$d_n = z_n - y_n$$

EXAMPLE 1.3 (cont'd)

so that $d_n = -5d_{n-1}$. Consequently,

$$| d_n | = 5^n | d_0 |$$

If $| d_0 | > 0$, then $| d_n |$ can become arbitrarily large. Therefore, the numerical scheme is unstable.

1.2.6 Human Errors and/or Machine Errors

Of these two sources of error, the former is very much more likely to occur than the latter. It is a common failing of beginning scientific programmers to be convinced of their own infallibility. Sometimes this is also true of experienced programmers, who should know better.

Of all the types of errors we have discussed in this section, the ones due to roundoff (Sections 1.2.1 and 1.2.2) are due to the representation of numbers within the computer. In the next section we will discuss floating point number systems, the most commonly used method of number representation. In a floating point number system we can represent numbers with greatly different magnitudes.

1.3 REPRESENTATION OF FLOATING POINT NUMBERS

The usual representation of a real number in a digital computer is called *normalized floating point*. A number, x, is represented by a machine number of the form

$$s \times f \times b^k$$

where b is the machine number base (b is a positive integer greater than one), f is a machine-representable number in the interval $[b^{-1}, 1 - b^{-t}]$ if $b > 2$, s is the sign of the number ($s = 1$ if x is positive, $s = -1$ if x is negative), k is the exponent (an integer in the interval $[L, U]$, where L and U are integers), t is the number of significant b-digits carried in f. For example, to represent the number $x = 2138.65$ in base $b = 10$ with $t = 6$ significant decimal digits we would have $s = 1$, $f = 0.213865$, and $k = 4$. Note that f is in the interval $[0.1, 0.999999]$ in this case.

If $b = 2$, then f is a machine-representable number in the interval $[1, 2 - 2^{1-t}]$. To see why $b = 2$ can be different from other values of b consider a computer with q bits physically available to represent f in base $b = 2$. If we take f in the interval $[2^{-1}, 1 - 2^{-q}]$ then f has exactly q significant bits. However, if f is in the interval $[1, 2 - 2^{-q}]$ then f has exactly $(q + 1)$ significant bits, with one bit to the left of the binary point and q bits to the right. Normally, this would require $(q + 1)$ bits to be physically present, but we can represent the number by using only the available q bits to the right of the binary point and remembering that all such numbers are considered to have a bit (with the value 1) to the left of the binary point. Thus, the leading bit need not be physically present. The number of significant bits is $t = q + 1$, so that f is in the interval $[1, 2 - 2^{1-t}]$.

The number 0 usually has a special representation, possibly $f = 0$, $s = 1$, and $k = L$. Most normalized floating point systems have $L < 0 < U$, and $t \leq U$. We shall refer to a normalized floating point system with these properties as the set of numbers M. It is important to realize that there are gaps between the numbers in M. The following example illustrates M for a simple case.

EXAMPLE 1.4

Consider the normalized floating point system with

$$b = 2, \quad t = 3, \quad L = -2, \quad \text{and} \quad U = 3$$

The set M consists of the number 0 and numbers of the form $\pm f \times 2^k$, where f is one of the four numbers 1, $\frac{5}{4}$, $\frac{3}{2}$, or $\frac{7}{4}$, with $-2 \leq k \leq 3$. Consequently, there are exactly 49 numbers in M (zero and 4×6 positive and 4×6 negative numbers).

The floating point number system in Example 1.4 obviously has large gaps between the numbers in M. A typical, realistic floating point system might have $b = 2$, $t = 53$, $L = -1023$, and $U = 1023$, although there still are gaps between the numbers.

A normalized floating point number, $s \times f \times b^k$, may have the following physical representation within a digital computer memory. The number would be stored in several memory-contiguous bytes (a group of 8 binary digits). One or two of the bytes would contain the sign, s, and the exponent k. The sign occupies one bit in this group of bytes, with 0 denoting positive and 1 denoting negative. The remaining bits consist of the *characteristic*, the actual exponent plus a constant. The constant is chosen so that every characteristic is a positive number. An obvious choice for the additive constant is $|L|$. This technique of storing characteristics rather than exponents means that special arrangements do not have to be made for the sign of the exponent. The remaining bytes form f. Figure 1.1 shows a floating point number stored in memory.

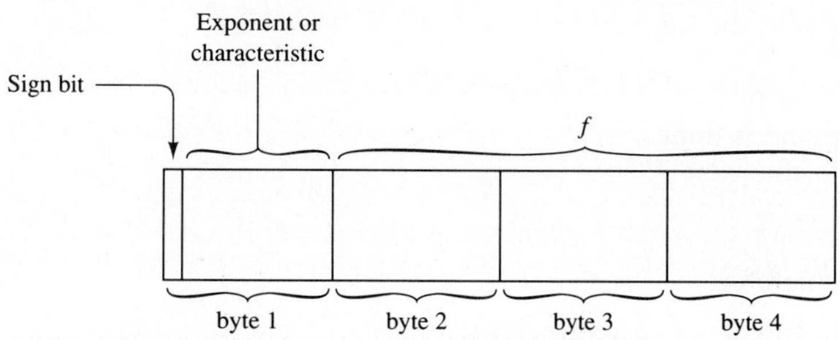

Figure 1.1 A floating point number machine storage.

A normalized floating point system, M, is a subset of the rational numbers; the rational numbers are a subset of the real numbers. Consequently, a real number x is usually represented by its closest neighbor in M, denoted by $fl(x)$. The notation $fl(x)$, an abbreviation for *float*(x), was introduced by Wilkinson; it is standard for denoting the floating point representation of the real number x. Real numbers that have magnitudes less than the smallest positive machine number cannot be represented in M. When such numbers arise in the course of computations an error flag may be set, indicating the condition known as *underflow*, or such numbers may be replaced by 0, without indication of underflow. A similar condition occurs for real numbers with magnitudes exceeding the largest positive machine number; the appearance of such numbers in a computation signifies the condition of *overflow* and the setting of an error indication flag. The operating system usually terminates the program if an overflow is detected.

When one does machine arithmetic (such as $+$, $-$, $\times$, $\div$) on two machine numbers x and y in M, the resulting number $x \circ y$ (where $\circ$ denotes one of the arithmetic operations $+$, $-$, $\times$, $\div$) may not be in M. If we do not have underflow or overflow, then we must ensure that the resultant is changed to a number that is in M. There are two common methods for doing this: *rounding* and *chopping*. Suppose that $|x \circ y| = gb^k$. If rounding is done then $fl(|x \circ y|)$ is the closest machine number in M to gb^k. If chopping is done then $fl(|x \circ y|)$ is the largest machine number in M that is less than or equal to gb^k. Whether rounding or chopping is done, $fl(x \circ y) = \text{sign}(x \circ y) fl(|x \circ y|)$. We refer to either rounding or chopping as *roundoff*.

An important machine number is the *relative machine precision* or *machine epsilon*, the smallest positive machine number, u, such that

$$(1 + u) > 1$$

The value of u is determined in the following example.

EXAMPLE 1.5

Suppose we have a floating point system with parameters b, t, L, and U. Let the nonzero real number x have the machine representation

$$fl(x) = \text{sign}(x) \times f \times b^k$$

where f is in the interval $[b^{-1}, 1 - b^{-t}]$. Then,

$$f \times b^k \leq |x| < (f + b^{-t}) \times b^k$$

a. Chopping is done.

In this case $fl(x) = \text{sign}(x) \times f \times b^k$; hence,

$$|fl(x) - x| = |f \times b^k - |x|| < b^{-t}b^k$$

and, therefore

$$\frac{|fl(x) - x|}{|x|} < \frac{b^{-t}b^k}{|x|} \leq \frac{b^{-t}b^k}{b^{-1}b^k} = b^{1-t}$$

EXAMPLE 1.5 (cont'd)

b. Rounding is done.

In this case the floating point representation of x is given by

$$fl(x) = \begin{cases} \text{sign}(x) \times f \times b^k & \text{if } |x| < (f + b^{-t}/2)b^k \\ \text{sign}(x) \times (f + b^{-t}) \times b^k & \text{if } |x| \geq (f + b^{-t}/2)b^k \end{cases}$$

Therefore

$$|fl(x) - x| = \begin{cases} |f \times b^k - |x|| < b^{-t}b^k/2 & \text{if } |x| < (f + b^{-t}/2)b^k \\ |f \times b^k - |x|| \leq b^{-t}b^k/2 & \text{if } |x| \geq (f + b^{-t}/2)b^k \end{cases}$$

and

$$\frac{|fl(x) - x|}{|x|} < \frac{b^{-t}b^k}{2|x|} \leq \frac{b^{-t}b^k}{2b^{-1}b^k} = b^{1-t}/2$$

Thus, $u = b^{1-t}$ if chopping is done, while $u = b^{1-t}/2$ if rounding is done. Clearly, rounding is the more accurate procedure; however, many floating point arithmetic systems use chopping because it is easier to implement than rounding.

Some programs for calculating the floating-point characteristics of a given computer system are listed in the appendices to this chapter.

EXERCISE 1.1

Verify that a floating point system with $b = 2$ using t binary digits (only $t - 1$ are present as the leading 1 to the left of the binary point is not physically represented) has a relative machine precision of 2^{1-t} if chopping is done and 2^{-t} if rounding is done.

In the next section we shall discuss the effect of the use of floating point numbers in computation.

1.4 COMPUTING WITH FLOATING POINT NUMBERS

One of the first difficulties faced by someone who is relatively inexperienced with computer computations is the loss-of-significance problem. This occurs because the numbers in M only have t b-digits of accuracy. For example: if $b = 10$, $t = 3$ with chopping, suppose we have the numbers $x = 0.1231$ and $y = 0.1224$, then $fl(x) = 0.123$ and $fl(y) = 0.122$. But $fl(fl(x) - fl(y)) = 0.100 \times 10^{-2}$, whereas

$x - y = 0.700 \times 10^{-3}$. This is known as *catastrophic cancellation*; it occurs because of the limited precision available and the subtraction of two quantities of nearly the same magnitude. This type of problem can be alleviated by careful programming.

Another type of computational problem that surprises the novice is the behavior of such standard functions as sines or cosines. We know that $\sin(x + 2n\pi) = \sin(x)$ for all integers n. If we let SIN denote the machine version of the function sine, the corresponding machine computation is for $fl(\text{SIN}(fl(fl(x) + fl(2n\pi))))$. Setting $x = \pi/4$, we tested the sine function in Turbo Pascal (with an 8087 coprocessor) using $n = 10^k$, for $k \geq 0$. The results that were obtained are given in Table 1.1. The digits in agreement with $fl(\text{SIN}(fl(\pi/4)))$ are underlined.

As you can see in Table 1.1, the loss of significance is quite rapid until at $k = 15$, none of the digits is correct. This occurs for two reasons:

1. As k increases, the term $\pi/4$ becomes smaller and smaller relative to the term $2\pi 10^k$, until eventually

$$fl(fl(\pi/4) + fl(2\pi 10^k)) = fl(2\pi 10^k).$$

When this happens, the argument of the SIN function has no correct digits, and therefore, we would not expect SIN to return a meaningful answer. Notice that the sines have lost all significant digits when $k = 15$; $(\pi/4)/(2\pi 10^{15}) = 1.25 \times 10^{-16} < 2.22 \times 10^{-16} = u$ for Turbo Pascal with an 8087 coprocessor.

Table 1.1 NUMERICAL COMPUTATION OF SIN($\pi/4 + 2\pi 10^k$)

k	$fl(\text{SIN}(fl(fl(\pi/4) + fl(2\pi 10^k))))$
0	0.707106781186547
1	0.707106781186544
2	0.707106781186528
3	0.707106781186473
4	0.707106781184954
5	0.707106781199345
6	0.707106781837167
7	0.707106781629947
8	0.707106783509005
9	0.707106960350244
10	0.707110583214003
11	0.707173782772094
12	0.707934802727108
13	0.716862812793496
14	0.752794691590865
15	0.579910047194619
16	0.460199103021975
17	−0.935246268220856

2. The computer algorithms to compute SIN($fl(z)$) or COS($fl(z)$) have their greatest accuracy only for z in the interval $[-2\pi m, 2\pi m]$, for some small positive integer m. For $z = 2\pi 10^{14}$ we see that SIN($fl(2\pi 10^k)$) is not close to zero. This shows that when computing these functions we should first obtain z as $z = 2\pi p + \theta$, where p is an integer and θ is in the interval $[-2\pi, 2\pi]$. Now we can compute SIN($fl(\theta)$) or COS($fl(\theta)$) and be reasonably assured of the accuracy, unless z is so inaccurate that it has no correct digits, as in (1).

Sometimes, the organization of the computations can influence the final result. For example, if we want to obtain the sum

$$1 + \frac{u}{2} + \frac{u}{2}$$

where u is the relative machine precision, we could add from left to right, giving the machine sum 1; however, adding from right to left we obtain the sum $fl(1 + u) > 1$.

1.5 CHOICE OF UNITS

Another source of trouble that often contributes to the loss of significant digits during computations is the choice of units which are to be used in a problem. Problems in engineering and science are usually formulated in some convenient set of units which depend on the physical nature of the problem. However, these units may be totally unsuitable for computational purposes. As an example, consider a problem in which the unit of length has been taken as the meter, but all the lengths that occur in the problem are of the order of 10^{16} meters. If we did the computations using the given unit (using Turbo Pascal with 8087 coprocessor) there is a very good chance that we would lose significance or run into overflow or underflow. The reciprocal of the lengths is of the same order of magnitude as our relative machine precision. For this example, a more natural unit of length would be the distance 10^{16} meters. If all lengths in the problem were divided by the factor 10^{16}, then the new dimensionless lengths are of the order of 1. This illustrates an important concept in the numerical solution of engineering and science problems: each computational variable should be nondimensional (it is the ratio of a physical variable to some chosen scale with the same physical units) and take its values in an interval such as $[-10^p, 10^p]$ or $[0, 10^p]$, where p is a small positive integer; usually $p = 1, 2$, or 3.

1.6 PROGRAMMING ACCURATE EVALUATION OF FUNCTIONS

When one does any scientific computation, either writing one's own programs, or using procedures or functions from some program library, a function or procedure must be supplied so that the numerical computations can be carried out. For example, suppose we want to find

$$\int_{-1}^{2} \left(\frac{\sin (x)}{x}\right) dx$$

We want to be able to evaluate the integrand $f(x) = \sin(x)/x$, for each x in the interval of integration, as accurately as possible, using any given digital computer. The integrand is undefined at $x = 0$; however, elementary calculus tells us that

$$\lim_{x \to 0} \left(\frac{\sin (x)}{x}\right) = 1$$

We could use this fact when writing our function that returns $f(x)$ for each x; if $x = 0$ return $f(0) = 1$, otherwise set $f(x) = \sin(x)/x$. This will certainly work. But for $|x|$ very close to zero, we may have underflow problems and consequently, we could end up trying to compute 0/0. There is a better way. Replace $\sin(x)$ by its infinite series and divide by x to obtain

$$f(x) = \frac{\sin(x)}{x} = 1 - \frac{x^2}{3!} + \frac{x^4}{5!} - \frac{x^6}{7!} + \cdots$$

This series for $\sin(x)/x$ will be accurate only for fairly small $|x|$. This is because large $|x|$ will cause some of the terms to have large magnitudes. The alternating sign will then give us the difference between two large magnitude terms, and thus we have catastrophic cancellation. We use this series to find $f(x)$ if $|x| < 0.1$, because for $|x| < 0.1$ the series will rapidly converge. For $|x| \geq 0.1$ we use the formula $\sin(x)/x$. Our function to evaluate $f(x)$ will now accurately compute the integrand for each x in the given interval $[-1, 2]$.

EXAMPLE 1.6

We want to write a Pascal function to accurately compute

$$f(x) = \frac{\sin(x)}{x}$$

for all values of x in $[-a, a]$, for $a > 0$. To accomplish this we first rewrite the series for $f(x)$ as

$$f(x) = \frac{\sin(x)}{x} = \sum_{k=0}^{\infty} \frac{(-1)^k x^{2k}}{(2k + 1)!}$$

Let t_k denote the kth term in the sum and let s_k denote the partial sum

$$s_k = \sum_{j=0}^{k} t_j$$

We have $s_{k+1} = s_k + t_{k+1}$, and

$$t_{k+1} = \frac{(-1)^{k+1} x^{2k+2}}{(2k + 3)!} = \frac{-x^2}{(2k + 3)(2k + 2)} t_k$$

EXAMPLE 1.6 (cont'd)

Given $s_0 = 1 = t_0$, we can determine t_{k+1} from t_k, and hence find s_{k+1}. Because we are going to be computing the sums s_k on a computer with finite precision, we would eventually reach a value of k (call it n) at which $|t_n| < u|s_{n-1}|$, where u is the relative machine precision. Thus, $s_n = s_{n-1}$ to machine accuracy, and the computations can be stopped. The following Pascal procedure finds $f(x)$, using the series if $|x| < 0.1$ and the formula $\sin(x)/x$ if $|x| \geq 0.1$.

```
FUNCTION SineXbyX( x : REAL ) : REAL;

  FUNCTION Series( x : REAL ) : REAL;
  VAR
    olds, s, t, xsqr2 : REAL;
    k, m : INTEGER;
  BEGIN
    k := 0 ; s := 1.0 ; t := 1.0 ; (* initialize *)
    xsqr2 := Sqr(x)/2.0 ;
    REPEAT
      olds := s ; m := k + 1 ;
      t := -t*xsqr2/(m*(2*k + 3)) ; k := m ;
      s := s + t
    UNTIL  olds = s; (* stop when successive partial sums are equal *)
    Series := s
  END; (* Series *)

BEGIN (* SineXbyX *)
  IF  Abs(x) < 0.1  THEN
    SineXbyX := Series(x)
  ELSE
    SineXbyX := sin(x)/x
END; (* SineXbyX *)
```

Example 1.6 illustrates some of the methodology in designing mathematical software. The software is based on a mathematical formula or technique to find the solution to some problem. The basis of the method must be sound mathematics. Next, the mathematics must be translated into or given in the form of an algorithm, which is a description of the computational process to be followed in applying the mathematical method to obtain a solution as accurately and efficiently as possible. The algorithm then must be converted into mathematical software. Here, one needs to know the details of the computer arithmetic as well as the types and ranges of input data that the program should expect. Once the computer program has been written it must be tested, and, if it is to be part of a larger programming project, documented as well.

1.7 CHOICE OF THE PROGRAMMING LANGUAGE

One of the first problems faced by a new user of computers is the choice of the programming language. Currently, several of these are available for use on microcomputers; the most popular are BASIC, Pascal, FORTRAN, C, and Modula-

2. The choice should be based on considerations of economy, the ease of handling inputs and outputs, the readability of the language and the provision for program editing and debugging.

FORTRAN (FORmula TRANslator) was the first high level programming language to be developed in the 1950s for use on a digital computer. At that time, there was a widespread belief in the programming community that programs written in high level languages utilizing compilers or language translators could not be as efficient as those written in the machine language. It was, therefore, felt that programmers accustomed to the machine language could be attracted to FORTRAN only if the execution time would be of the same order. The main objectives in the development of FORTRAN, therefore, were to make programming faster, cheaper, and more reliable. It was later found that FORTRAN had many disadvantages. The worst aspect was the lack of suitable control structures, which resulted in the programmer resorting to the undisciplined use of the "goto" statement. This had the effect of making such a program difficult to understand, debug, or modify. Many scientific programs tend to be long and complicated, and are based on many different ideas. However, each line of a FORTRAN program typically draws on several of these ideas at once. This makes reading and documentation of the program difficult. It is harder to understand and maintain a FORTRAN program written by another person, unless that program has been extensively interlaced with comments.

In the first version of FORTRAN, developed for the IBM 704, no special provision was made for debugging, as it was naively assumed that the need for debugging would be eliminated. Even later versions of this language can give rise to problems which a compiler will fail to detect. Perrot and Allison, in their book *Pascal for FORTRAN programmers*, give the following example. The first American space probe to the planet Venus was aborted after it went off course. The reason was traced to a FORTRAN program statement which was syntactically correct. It occurred in a statement of the form

```
DO 6 I = 1,25
```

where the comma had been mistakenly punched as a period. This caused the FORTRAN compiler to create a new variable called DO6I and to give this variable the value of 1.25. This error, due to confusion between a comma and a period, went undetected during the reading, testing and debugging of the program. Such an error would have been detected by a Pascal compiler in the very first attempt at compilation. A Pascal compiler subjects a program to a series of rigorous compile-time and optional run-time checks, which are applied to both the program and data structures. In general, this has the desirable effect of substantially reducing the chance of programmer errors by requiring a little more discipline in programming.

FORTRAN was used mostly for scientific computation, but was soon found cumbersome to learn and use for other applications. Need was felt for a programming language which would be easy to learn and could be executed in an interactive or conversational manner. The result was BASIC (Beginners' All-purpose Symbolic Instruction Code). It is easy to implement on a small computer as it requires a very small amount of memory. Because of its simplicity, BASIC has become the most widely used language for small personal computers. However, it suffers from several disadvantages. The most serious problem with BASIC is that a program written in this languages is usually incomprehensible to almost everyone except its author. Often it has been found that in the case of a relatively large program, even its author may have some difficulty in understanding or modifying it after a period of a few months. Quoting David Fox and Mitchell Waite in the book *Pascal Primer*, a complex program written in BASIC often looks like the "rat's nest" of wires that one found in early radio and television sets, which made them rather difficult to repair.

The programming language, Pascal, was developed by Professor Niklaus Wirth of Switzerland in the early 1970s. It was designed to overcome the problems in software engineering. The name of the language is a tribute to the French mathematician Blaise Pascal, who had invented the first mechanical calculating machine in 1640 at the age of eighteen. It is a structured language, rich in expressive power and applicable to a wide variety of uses while being unusually simple as well as consistent in its syntax and semantics. An important feature of the language is its modular structure. The blocks forming the various modules can be read like the paragraphs of a book, since each block is designed to contain just the sufficient amount of information required for the reader to understand its purpose. To continue the analogy of electronic hardware mentioned in the previous paragraph, a well-designed Pascal program is very much like a modern television set built with plug-in printed-circuit boards (or modules). In the event of malfunction, it is much simpler to replace or modify the bad module. Furthermore, it is much easier to test each module separately. Hence, well-written Pascal programs are easier to understand, maintain and modify. These advantages gain more importance as a program grows in size and complexity.

A good strategy for solving a complex problem is to first decompose it into smaller and simpler subproblems. Each of these subproblems is then solved separately. Finally, these are combined together to obtain the solution to the original problem. This is called "top-down design" and can also be applied to computer programming. A large program is often written by a team of programmers working separately, each writing and testing a different module. The Pascal language is ideally suited for this purpose. Pascal goes even further by imposing requirements that make it awkward or difficult to avoid good structure. Hence, the programming process lends itself to the same orderly and systematic approach underlying the design and development of any other product. Because of these features, Pascal is now used by many computer science departments as the main language for teaching programming methodology.

Pascal, as originally developed by Professor Wirth, was not suitable for use on a small computer. Professor Kenneth L. Bowles of the University of California, San Diego, adapted the language for small computers in the late 1970s, and developed the UCSD p-system, the operating system which supports the Pascal programming language on many microcomputers. Subsequently, many other versions of Pascal have been developed for small computers. The most noteworthy are Microsoft Pascal and Turbo Pascal. Each of these compiles a program into code which is almost identical to the native code of the machine.

It is worth noting that the programming language Ada, developed under contract with the United States Department of Defense, can be considered as an enhancement of Pascal. Ada compilers are now available for microcomputers. Furthermore, attempts have been made to imitate the structured nature of Pascal in the newer versions of FORTRAN. Another enhancement of Pascal is the language Modula-2, which has been developed recently by Professor Wirth. At present, several implementations of Modula-2 are available for small computers.

A well-written and well-documented program in a structured language, like Pascal, is easy to understand and modify, if required. Because of this, we include programs written in Turbo Pascal throughout this book. This choice was made not only for the reason that Turbo Pascal is available at a very reasonable price (less than $100, with special discounts for educational purposes), but also because of the very fast compiling of Turbo Pascal. It is this fast-compiling that makes Turbo Pascal most useful for student projects or program debugging. It is felt that the use of Pascal for a solution, rather than FORTRAN, will make a difference to the usefulness of the book. Not only will the data and program structure make the program transparent to the user, but will also develop a better understanding of the design techniques of structured programming and stepwise refinements.

1.8 MICROCOMPUTER AND SOFTWARE REQUIREMENTS

It will be assumed throughout this book that the reader has access to an IBM Personal Computer (either a PC, XT, or AT model) or an IBM PS/2 computer or a compatible machine. These microcomputers use the Intel 8088, or 8086, or 80286, or 80386 microprocessor as their main central processor. It is expected that the reader's microcomputer has at least 256 kilobytes of RAM (random-access memory), but in most cases, one will not be using all of the memory. In addition, it is advisable if the microcomputer used has the Intel 8087, or 80287, or 80387 mathematical coprocessor.

The 8087 mathematical coprocessor is designed to work as a companion to the 8088 or the 8086 microprocessor. Similarly, the 80287 and the 80387 are designed to work with the 80286 or the 80386 microprocessors, respectively. We will mainly refer to the 8088 and the 8087 processors, but the comments apply to the other processors as well. The 8087 coprocessor can perform the common arithmetic operations of addition, subtraction, multiplication, and division, as well as such operations as calculating trigonometric functions and natural logarithms. The 8087 has several 80-bit registers where temporary results can be stored while performing calculations. This reduces memory access and improves the speed of calculations. Since the coprocessor has been designed to allow the 8088 microprocessor to pass work to it, the overall result is an increase in speed. This is

Table 1.2 COMPARISON OF RUNNING TIMES USING TURBO PASCAL ON AN IBM PC FOR FLOATING POINT COMPUTATIONS*

Operation/Function	Without an 8087	With an 8087
addition	0.28	0.17
subtraction	0.28	0.17
multiplication	1.04	0.17
division	1.71	0.22
square root	9.34	0.11
sine	12.42	0.50
cosine	12.58	0.50
arctan	12.31	0.34
exponential	14.34	0.56
logarithm	15.38	0.28

* All times are in seconds for 1000 repetitions of the indicated operations or functions.

very noticeable in the calculation of floating point multiplications and divisions, and the calculation of trigonometric and other common functions such as the logarithm and the square root. The gain in speed is very marked, as the results given in Table 1.2 indicate. Most of the Pascal programs in this book will work better if your computer has a mathematical coprocessor installed. If your microcomputer has an 8087 installed, you should make sure that the Pascal compiler you are using can make use of the coprocessor.

The programs described in this book are written in Turbo Pascal for an IBM PC, but they can also run on other machines if a version of Turbo Pascal is available for that machine. It is also possible to use the programs with versions of Pascal other than Turbo Pascal, with some minor modifications (see Appendix C).

Often the results of numerical computations should be presented in a graphical form, rather than presented as a long list of numbers. Graphical results can be displayed on a microcomputer if some form of graphics adaptor card is installed. Turbo Pascal has some built-in functions and procedures to allow graphics to be displayed on an IBM PC with a graphics card. However, many microcomputers do not have graphics cards installed. Rather than supply a graphics package that requires a graphics card, we have included a Turbo Pascal package of procedures which allows rough graphs to be plotted on the screen and/or sent to a printer (see Appendix D.6 and D.7).

1.9 PROBLEMS

1. Use the programs given in Appendices 1.1, 1.2, and 1.3 to determine the floating point characteristics of the computer with which you plan to work.

2. The exponential series

$$e^x = 1 + x + \frac{x^2}{2!} + \frac{x^3}{3!} + \cdots$$

is known to be uniformly convergent for all x. Hence, in theory, it is possible to compute e^x by truncating the series after a sufficient number of terms. We run into numerical problems, however, when x is negative and large, due to catastrophic cancellation. Use the Pascal program in Example 1.3 to compute (a) e^{20} and (b) e^{-20} by adding 60 terms of the series. Compare the results with those obtained from tables. Will the accuracy be improved by adding more terms of the series? Why does the algorithm calculate e^{20} with a greater accuracy? Suggest a better scheme for calculating e^{-20}.

3. One way to improve the accuracy of the algorithm discussed in the previous problem is to add the positive and negative terms of the series separately, and then combine them to evaluate the exponential when the exponent is negative. Another approach is to sum the series for the case when the exponent is positive and then take the reciprocal. Use both of these methods to evaluate e^{-20}. Which is more accurate?

4. As in Problem 3, one may calculate the sine of an angle x, in radians, using the series

$$\sin(x) = x - \frac{x^3}{3!} + \frac{x^5}{5!} - \frac{x^7}{7!} + \cdots$$

which is also known to be uniformly convergent. Here again we run into computational problems when $|x|$ is much greater than 1. Write a computer program to sum the above series for a specified number of terms and use it for calculating $\sin(8\pi + \pi/6)$ retaining 60 terms in the series. How does it compare with the actual value of

$$\sin\left(8\pi + \frac{\pi}{6}\right)?$$

5. Repeat Problem 4 by adding the positive and negative terms of the series separately and then combining the results to evaluate the sine. How does this compare with the actual value of $\sin(8\pi + \pi/6)$? Suggest a better method for evaluating $\sin(8\pi + \pi/6)$.

6. The roots of a quadratic $ax^2 + bx + c$ can be found by using the well-known formula

$$x = \frac{-b + (b^2 - 4ac)^{1/2}}{2a}$$

Use this formula to determine the roots of the quadratic

$$x^2 - 100000000.00000001x + 1 = 0$$

The correct values of the roots are 10^8 and 10^{-8}. Do you get these values? Explain the reason for the discrepancy.

7. The roots of the cubic

$$x^3 + a_2x^2 + a_1x + a_0 \tag{1.9–1}$$

can be found by using Cardan's method. This requires the calculation of Q and R, given by

$$Q = \left(\frac{a_1}{3}\right) - \left(\frac{a_2}{3}\right)^2, \qquad R = \frac{a_2Q + (a_2/3)^3 - a_0}{2}$$

If $S = (Q^3 + R^2) \leq 0$, then all roots are real and one root is given by

$$x_1 = 2\,|\,Q\,|^{1/2}\cos\left(\frac{\arccos(R/\,|\,Q\,|^{3/2})}{3}\right) - \frac{a_2}{3};$$

otherwise, there is only one real root given by

$$x_1 = (R + \sqrt{S})^{1/3} + (R - \sqrt{S})^{1/3} - \frac{a_2}{3}.$$

After the root x_1 is determined, the cubic is divided by the factor $(x - x_1)$ and the roots of the resulting quadratic are computed. Use Cardan's method to determine the roots of the cubic

$$x^3 - 73x^2 + 1776x - 14400 = 0.$$

Do these agree with the roots 25.0, 24.0 and 24.0? How would you explain the discrepancy?

8. Now consider another version of Cardan's formula, where we define

$$q = 3a_1 - a_2^2 \quad \text{and} \quad r = \tfrac{1}{2}[a_2(3q + a_2^2) - 27a_0]$$

Let $s = (q^3 + r^2)$. If $s \leq 0$ one real root of the cubic (1.9–1) is

$$x_1 = \tfrac{1}{3}[2\,|\,q\,|^{1/2}\cos(\arccos(r/\,|\,q\,|^{3/2})/3) - a_2];$$

otherwise, there is only one real root given by

$$x_1 = \tfrac{1}{3}[(r + \sqrt{s})^{1/3} + (r - \sqrt{s})^{1/3} - a_2].$$

Again, after the root x_1 is determined, the cubic is divided by the factor $(x - x_1)$ and the roots of the resulting quadratic are obtained. Use this method of computing q, r, and s to determine the roots of the cubic

$$x^3 - 73x^2 + 1776x - 14400 = 0$$

Why does this method give better answers than the version in Problem 7?

9. Use a hand-held calculator to determine

$$(257889569)^{1/2} - (257889568)^{1/2}$$

correct to eight significant digits. Compare this with the value obtained by using the following identity which avoids subtraction

$$\frac{(257889569)^{1/2} - (257889568)^{1/2}}{1} = \frac{1}{(257889569)^{1/2} + (257889568)^{1/2}}$$

10. Consider the functions

$$f(x) = \begin{cases} (1 - \cos(x))/\sin(x) & \text{if } 0 < |x| < \pi \\ 0 & \text{if } x = 0 \end{cases}$$

and

$$g(x) = \sin(x)/(1 + \cos(x)) \quad \text{if } |x| < \pi.$$

Show that they are identical on the interval $-\pi < x < \pi$. Which of these is more suitable for computation in the case when $|x|$ is very small? Use a hand-held calculator for comparing the values of $f(x)$ and $g(x)$ if $x = 0.000000055$.

11. Write a Pascal program to accurately compute $f(x) = \sinh(x)/x$ for all x in $[-a, a]$, where $a > 0$.

12. Two mathematical formulas are listed below with a range of argument values x. Formula (b) has a removable singularity. For each of the formulas write two Pascal functions:

(i) a function which evaluates the formula as given;

(ii) a function which accurately produces values defined by the formula for arguments in the given range.

Compare the accuracy of the two functions for each of the given test values.

(a) $\log(1 + x)$, $|x| < 1$
Test values: $x = 7.6 \times 10^{-9}, 7.7 \times 10^{-11}, -0.01, 0.01, 0.673$

(b) $(e^{-x} - 1)/x$, $|x| \leq 1$
Test values: $x = 0.0, -7.6 \times 10^{-9}, 0.0036$

13. Consider the function

$$f(x) = \left[\frac{1 - \sin(x)}{1 + \sin(x)}\right]^{1/2} \quad \text{for } 0 \leq x \leq \frac{\pi}{2}.$$

(a) In what approximate region (or regions) of the given interval do you expect this formula for computing $f(x)$ to have poor accuracy? Why?

(b) Propose a better formula for each region that you identified in part (a).

14. A certain computer is known to represent floating point numbers in 10 bytes: base $b = 2$, precision $t = 65$, the number of bits in the characteristic $= 15$, exponent = characteristic − 2^{14}, one bit for the sign, the leading bit in front of the binary point is not physically present, zero is represented by a characteristic of 0 and a fractional part equal to zero.

(a) What is the machine epsilon for rounding and for chopping?

(b) If the values 0 and $2^{15} - 1$ are reserved in the characteristic, what are the values of U and L, the maximum and minimum exponents?

(c) What are the largest and smallest magnitudes of representable nonzero floating point numbers?

(d) Suppose that a nonzero real number x is written as $|x| = fb^{\ell}$, where $L < \ell < U$. Let p be a given floating point number such that $|p| = qb^{\ell}$. Determine the maximum and minimum values of f that map into the same number q if

(i) chopping is done

(ii) rounding is done.

1.10 REFERENCES

G. E. Forsythe, M. A. Malcolm, and C. B. Moler. *Computer Methods for Mathematical Computations*. Englewood Cliffs, NJ: Prentice-Hall, 1977.

David Fox and Mitchell Waite. *Pascal Primer*. Indianapolis, IN: Howard M. Sams & Co., 1981.

M. A. Malcolm. "Algorithms to Reveal Floating Point Arithmetic." *Communications of the A.C.M.* 11 (Nov. 1972):949–951.

R. H. Perrot and D. C. S. Allison. *Pascal for FORTRAN Programmers*. Rockville, MD: Computer Science Press, 1984.

J. H. Wilkinson. *Rounding Errors in Algebraic Processes*. Englewood Cliffs, NJ: Prentice-Hall, 1963.

K. G. Wilson. "Science, Industry and the New Japanese Challenge." *Proceedings of the Institute of Electrical and Electronic Engineers* 72 (1984):6–18 (Special issue on Supercomputers).

The three programs in Appendices 1.1, 1.2, and 1.3 determine some floating point characteristics. The program in Appendix 1.1 finds the machine number base, b, the number of b-digits of precision, whether rounding or chopping is done, and the relative machine precision. The programs in Appendices 1.2 and 1.3 determine the exponent range, $[b^L, b^U]$. Approximate values of L and U can usually be determined from the reference manuals supplied by the manufacturer, or supplier of the programming language. The Pascal function in Appendix 1.4 can be used to compute an approximation to the relative machine precision.

APPENDIX 1.1

Floating Point Characteristics

```
PROGRAM FloatingPointCharacteristics;
 (* Determines number base, number of "digits" of precision, whether
    proper rounding or chopping is done, and the theoretical machine
    epsilon (relative machine precision).
    Based on the FORTRAN subroutine in
      M.A. Malcolm, 'Algorithms to reveal floating-point arithmetic'
                   Comm. A.C.M.  vol. 11  Nov. 1972   pp. 949-951    *)

VAR
  beta, t : INTEGER;
  big, epsilon : REAL;
  round : BOOLEAN;

PROCEDURE Print;
BEGIN
  WRITELN;
  WRITELN('Machine number base = ', beta:2);
  WRITELN('There are ',t:3,' base ',beta,' "digits" ');
  IF  round  THEN
    WRITELN('The machine does proper rounding')
  ELSE
    WRITELN('The machine does chopping');
  WRITELN('Theoretical machine epsilon = ' ,epsilon);
  WRITELN
END; (* Print *)

PROCEDURE Large;
 (* Find floating point integer, big, such that
       (big + 1.0) - (big)  is not equal to  1.0 *)
VAR
  a, b : REAL;
BEGIN
  big := 1.0;
  REPEAT
    big := 2.0*big ; a := big + 1.0 ; b := a - big
  UNTIL  b <> 1.0
END; (* Large *)

PROCEDURE Precision;
 (* Determine the number of base beta digits and the theoretical machine
     epsilon, depending on whether rounding or chopping is done. *)
VAR
  a, b, c, d : REAL;
BEGIN
  t := 0 ; a := 1.0 ; b := beta ;
  REPEAT
    t := t + 1 ; a := a*b ; c := a + 1.0 ; d := c - a
  UNTIL  d <> 1.0;
  epsilon := Exp((1.0 - t)*Ln(b));
  IF  round  THEN
    epsilon := epsilon/2.0
END; (* Precision *)
```

```
PROCEDURE MachineParameters;
VAR
  a, b, c : REAL;
BEGIN
  a := 1.0;
  REPEAT
    a := 2.0*a ; b := a + big ; c := b - big
  UNTIL  c <> 0.0;
  beta := Trunc(c);
  a := big + (beta - 1) ; b := a - big ;
  IF  b = 0.0  THEN
    round := FALSE
  ELSE
    round := TRUE;
  Precision
END; (* MachineParameters *)

BEGIN (* Main program *)
  Large;
  MachineParameters;
  Print
END.
```

APPENDIX 1.2

Finding a Lower Bound on Exponents in Floating Point

```
PROGRAM ExponentL; (* To find a lower bound, L, of exponents *)
CONST
  base  =  2.0;
   (*  Machine number base.  This is for Turbo Pascal with  8087 *)
VAR
  b, OldValue : REAL;
  i : INTEGER;
BEGIN
  b := 1.0 ;
  FOR  i := 1  TO  1100  DO
    BEGIN
      OldValue := b ; b := b/base ;
      IF  (i > 1019) AND (b <> OldValue)  THEN
        BEGIN  (* print out only those values near  L *)
          WRITE(i,' ',b,'    ');
          IF  (i MOD 2) = 1  THEN
            WRITELN
        END
    END
END.
```

APPENDIX 1.3

Finding an Upper Bound on Exponents in Floating Point

```
PROGRAM ExponentU; (* To find an upper bound, U, of exponents *)
CONST
  base = 2.0;
    (* Machine number base.  This is for Turbo Pascal with 8087 *)
VAR
  b : REAL;
  i : INTEGER;
BEGIN
  b := base;
  FOR  i := 2  TO  2000  DO
    BEGIN
      b := b*base;
      IF  i > 1019  THEN
        BEGIN  (* print out only those values near  U *)
          WRITE(i,' ',b,'    ');
          IF  (i MOD 2) = 1  THEN
            WRITELN
        END
    END
END.
```

APPENDIX 1.4

A Pascal Function to Determine an Approximation to the Relative Machine Precision. This function is in the file Machine. Pas

```
FUNCTION MachineEpsilon( base : REAL ) : REAL;
(*  base  is the machine number base *)
VAR
  a, b : REAL;
BEGIN
  a := 1.0;
  REPEAT
    a := a/base ; b := 1.0 + a
  UNTIL  b = 1.0;
  MachineEpsilon := a*base
END;
```

Chapter 2

Systems of Linear Simultaneous Equations

Numerical solution of systems of linear simultaneous algebraic equations is considered. The methods discussed include Gaussian elimination with partial pivoting, L-U factorization, and iterative methods. Vector and matrix norms are introduced.

2.1 INTRODUCTION

The solution of systems of linear simultaneous equations is one of the most common problems encountered by scientists and engineers. Typical examples are analysis of electrical networks, heat transfer, vibrations, and elasticity. The problem also arises in many other applications of numerical analysis—for example, curve-fitting, approximate numerical integration of partial differential equations and least-squares approximation, as well as many problems related to engineering design.

In the general case, the equations will have the following form,

$$\left.\begin{array}{l} a_{11}x_1 + a_{12}x_2 + \cdots + a_{1n}x_n = b_1 \\ a_{21}x_1 + a_{22}x_2 + \cdots + a_{2n}x_n = b_2 \\ \quad \vdots \qquad\quad \vdots \qquad\qquad\quad \vdots \qquad\; \vdots \\ a_{n1}x_1 + a_{n2}x_2 + \cdots + a_{nn}x_n = b_n \end{array}\right\} \tag{2.1–1}$$

where $x_1, x_2, \ldots, x_n$ are the n unknowns. These equations can be written in the following compact form using matrix notation (see Appendix A for a review of linear algebra)

$$\mathbf{Ax} = \mathbf{b} \tag{2.1–2}$$

where $\mathbf{A}$ is the $n \times n$ matrix of coefficients

$$\mathbf{A} = \begin{bmatrix} a_{11} & a_{12} & \cdots & a_{1n} \\ a_{21} & a_{22} & \cdots & a_{2n} \\ \vdots & \vdots & \ddots & \vdots \\ a_{n1} & a_{n2} & \cdots & a_{nn} \end{bmatrix}$$

and $\mathbf{x}$ and $\mathbf{b}$ are n-dimensional vectors,

$$\mathbf{x} = \begin{bmatrix} x_1 \\ x_2 \\ \vdots \\ x_n \end{bmatrix} \quad \text{and} \quad \mathbf{b} = \begin{bmatrix} b_1 \\ b_2 \\ \vdots \\ b_n \end{bmatrix}$$

Because we want a unique solution to (2.1–2) the matrix $\mathbf{A}$ must be nonsingular ($\mathbf{A}$ is invertible). In addition, we shall assume that the linear system (2.1–2) is *irreducible*; that is, solving (2.1–2) cannot be reduced to obtaining the solution for two or more smaller systems of lower order.

Two methods for the direct solution of these equations are often discussed in high school algebra. These are (1) the method of successive elimination by combining equations and (2) Cramer's rule, which utilizes determinants. (See Appendix A for discussion of Cramer's rule.)

Cramer's rule appears quite efficient while solving two or three simultaneous equations. However, the calculation of determinants with expansion by minors gets very cumbersome as the number of unknowns increases. It can be shown that the evaluation of the determinant of an $n \times n$ matrix by expansion in minors will require $(n - 1)$ multiplications for each of the $n!$ terms. Since solution by Cramer's rule will require the evaluation of $(n + 1)$ determinants for n unknowns, the total number of multiplications will be $(n + 1)!$. When n is 13, this implies 8.7178×10^{10} multiplications. Even with a very fast computer that can perform 10^6 multiplications per second, this would require more than twenty-four hours. One can calculate determinants using better methods and thus reduce the required time, but the method is still prone to excessive roundoff error due to the large number of computations.

Another direct approach is to premultiply equation (2.1–2) on both sides by $\mathbf{A}^{-1}$ so that we get $\mathbf{x} = \mathbf{A}^{-1}\mathbf{b}$. In practice, it is seldom advisable to perform matrix inversion, as it requires more computation and increases the possibility of roundoff error.

We shall discuss better and more efficient direct methods in this chapter. In addition, there are many indirect or iterative methods which are specially suitable when the number of unknowns exceeds 100. These will be discussed later in the chapter. A special case arises when the matrix $\mathbf{A}$ is *sparse*; i.e., very few of its elements are nonzero. For such matrices one must use special techniques for storing the nonzero elements, as well as for solving the equations without requiring an excessive amount of memory. A similar problem occurs if the matrix **A** has a *banded* structure. A very special case of a banded matrix is a tridiagonal matrix which occurs in many scientific and engineering problems. Tridiagonal matrices are discussed in Section 2.4.2. Banded matrices, which are not necessarily tridiagonal, occur frequently in the numerical solution of partial differential equations. The special techniques for solving problems with sparse and general banded matrices will not be described in this book. Interested readers should refer to Golub and Van Loan, *Matrix Computations*.

2.2 GAUSSIAN ELIMINATION

This is one of the most efficient methods for solving a set of n linear simultaneous equations since it requires only $2n^3/3$ arithmetic operations (additions, subtractions, multiplications, divisions). See Appendix 2.3 for a discussion of how this result is obtained. The basic idea behind the method is the elimination of one variable at a time, until we have one equation with only one variable. This is called *forward elimination*. All the variables are then evaluated by *back substitution*. For example, consider the following equations for the case when $n = 4$:

$$\left.\begin{aligned} a_{11}x_1 + a_{12}x_2 + a_{13}x_3 + a_{14}x_4 &= b_1 \\ a_{21}x_1 + a_{22}x_2 + a_{23}x_3 + a_{24}x_4 &= b_2 \\ a_{31}x_1 + a_{32}x_2 + a_{33}x_3 + a_{34}x_4 &= b_3 \\ a_{41}x_1 + a_{42}x_2 + a_{43}x_3 + a_{44}x_4 &= b_4 \end{aligned}\right\} \tag{2.2–1}$$

The first step is the elimination of x_1 from the last three equations. The form of the resulting equations is

$$\left.\begin{aligned} a_{11}x_1 + a_{12}x_2 + a_{13}x_3 + a_{14}x_4 &= b_1 \\ a_{22}^{(2)}x_2 + a_{23}^{(2)}x_3 + a_{24}^{(2)}x_4 &= b_2^{(2)} \\ a_{32}^{(2)}x_2 + a_{33}^{(2)}x_3 + a_{34}^{(2)}x_4 &= b_3^{(2)} \\ a_{42}^{(2)}x_2 + a_{43}^{(2)}x_3 + a_{44}^{(2)}x_4 &= b_4^{(2)} \end{aligned}\right\} \tag{2.2–2}$$

The second equation of the new set is obtained by multiplying the first equation by $-a_{21}/a_{11}$ and adding to the second equation in the old set. The third and the fourth equations are obtained by multiplying the first equation by $-a_{31}/a_{11}$ and $-a_{41}/a_{11}$, respectively, and adding to the corresponding equation in the old set. This process of elimination is equivalent to premultiplying the matrix equation $\mathbf{Ax} = \mathbf{b}$ by the following three elementary matrices in the order indicated:

$$\mathbf{P}_1 = \begin{bmatrix} 1 & 0 & 0 & 0 \\ c_1 & 1 & 0 & 0 \\ 0 & 0 & 1 & 0 \\ 0 & 0 & 0 & 1 \end{bmatrix} \quad \mathbf{P}_2 = \begin{bmatrix} 1 & 0 & 0 & 0 \\ 0 & 1 & 0 & 0 \\ c_2 & 0 & 1 & 0 \\ 0 & 0 & 0 & 1 \end{bmatrix} \quad \mathbf{P}_3 = \begin{bmatrix} 1 & 0 & 0 & 0 \\ 0 & 1 & 0 & 0 \\ 0 & 0 & 1 & 0 \\ c_3 & 0 & 0 & 1 \end{bmatrix}$$

where $c_1 = -a_{21}/a_{11}$, $c_2 = -a_{31}/a_{11}$, and $c_3 = -a_{41}/a_{11}$. It may be noted that this method requires that $a_{11} \neq 0$. The element a_{11} is called the *pivot* for this set of transformations. If the pivot is zero then one must interchange the equations (interchange the rows of **A** as well as the corresponding rows of **b** such that this condition is satisfied). Numerical accuracy is increased if this interchange is carried out in such a manner that the resulting pivot is the element with the largest absolute value in the first column of **A**.

Proceeding in this manner, the second step is to eliminate x_2 from the last two equations in (2.2–2). This is done by multiplying the second equation by $-a_{32}^{(2)}/a_{22}^{(2)}$ and $-a_{42}^{(2)}/a_{22}^{(2)}$, respectively, and adding to the third and fourth equations. If necessary, the rows of equation (2.2–2) must be interchanged so that the new pivot is the element with the largest absolute value. The third step is to eliminate x_3 from the last equation to obtain

$$\begin{aligned} a_{11}x_1 + a_{12}x_2 + a_{13}x_3 + a_{14}x_4 &= b_1 \\ a_{22}^{(2)}x_2 + a_{23}^{(2)}x_3 + a_{24}^{(2)}x_4 &= b_2^{(2)} \\ a_{33}^{(3)}x_3 + a_{34}^{(3)}x_4 &= b_3^{(3)} \\ a_{44}^{(4)}x_4 &= b_4^{(4)} \end{aligned}$$

It may be noted that the equations have been transformed so that the matrix of the coefficients of **x** is of the upper triangular form. We shall denote this matrix by **U**. We are now able to solve for x_4 by simply dividing the last equation by $a_{44}^{(4)}$. This is equivalent to premultiplying by the elementary matrix

$$\mathbf{R}_1 = \begin{bmatrix} 1 & 0 & 0 & 0 \\ 0 & 1 & 0 & 0 \\ 0 & 0 & 1 & 0 \\ 0 & 0 & 0 & 1/a_{44}^{(4)} \end{bmatrix}$$

Hence:

$$\mathbf{R}_1\mathbf{U}\mathbf{x} = \begin{bmatrix} a_{11} & a_{12} & a_{13} & a_{14} \\ 0 & a_{22}^{(2)} & a_{23}^{(2)} & a_{24}^{(2)} \\ 0 & 0 & a_{33}^{(3)} & a_{34}^{(3)} \\ 0 & 0 & 0 & 1 \end{bmatrix} \begin{bmatrix} x_1 \\ x_2 \\ x_3 \\ x_4 \end{bmatrix} = \begin{bmatrix} b_1 \\ b_2^{(2)} \\ b_3^{(3)} \\ b_4^{(5)} \end{bmatrix}$$

The next step in back substitution is premultiplication by $\mathbf{R}_2$ and $\mathbf{R}_3$ where

$$\mathbf{R}_2 = \begin{bmatrix} 1 & 0 & 0 & 0 \\ 0 & 1 & 0 & 0 \\ 0 & 0 & 1 & -a_{34}^{(3)} \\ 0 & 0 & 0 & 1 \end{bmatrix} \quad \text{and} \quad \mathbf{R}_3 = \begin{bmatrix} 1 & 0 & 0 & 0 \\ 0 & 1 & 0 & 0 \\ 0 & 0 & 1/a_{33}^{(2)} & 0 \\ 0 & 0 & 0 & 1 \end{bmatrix}$$

to obtain

$$\mathbf{R}_3\mathbf{R}_2\mathbf{R}_1\mathbf{U}\mathbf{x} = \begin{bmatrix} a_{11} & a_{12} & a_{13} & a_{14} \\ 0 & a_{22}^{(2)} & a_{23}^{(2)} & a_{24}^{(2)} \\ 0 & 0 & 1 & 0 \\ 0 & 0 & 0 & 1 \end{bmatrix} \begin{bmatrix} x_1 \\ x_2 \\ x_3 \\ x_4 \end{bmatrix} = \begin{bmatrix} b_1 \\ b_2^{(2)} \\ b_3^{(6)} \\ b_4^{(5)} \end{bmatrix}$$

Proceeding in this manner, we finally obtain

$$\begin{bmatrix} 1 & 0 & 0 & 0 \\ 0 & 1 & 0 & 0 \\ 0 & 0 & 1 & 0 \\ 0 & 0 & 0 & 1 \end{bmatrix} \begin{bmatrix} x_1 \\ x_2 \\ x_3 \\ x_4 \end{bmatrix} = \begin{bmatrix} b_1^{(8)} \\ b_2^{(7)} \\ b_3^{(6)} \\ b_4^{(5)} \end{bmatrix}$$

In an actual computer program for this algorithm it is only necessary to carry out the elementary row operations, for successive elimination on the matrix **A** and the vector **b**, simultaneously until **A** is reduced to an upper triangular matrix. This is done by augmenting the vector **b** to the matrix **A** and then performing the elementary row operations. Once the transformed **A** is in upper triangular form, back substitution can be used to solve for **x**. The following example will illustrate the procedure.

EXAMPLE 2.1

Consider the set of three linear equations

$$\begin{bmatrix} 10 & 8 & -5 \\ 2 & 1 & 4 \\ 5 & -1 & 2 \end{bmatrix} \begin{bmatrix} x_1 \\ x_2 \\ x_3 \end{bmatrix} = \begin{bmatrix} 31 \\ 0 \\ 1 \end{bmatrix}$$

The first step in solving this equation using Gaussian elimination is to append the vector **b** to the matrix **A**. We use the notation $[\mathbf{A} \mid \mathbf{b}]$ to denote this $n \times (n + 1)$ matrix. Hence we get

$$[\mathbf{A} \mid \mathbf{b}] = \left[\begin{array}{ccc|c} 10 & 8 & -5 & 31 \\ 2 & 1 & 4 & 0 \\ 5 & -1 & 2 & 1 \end{array}\right]$$

Next, we reduce column one in $[\mathbf{A} \mid \mathbf{b}]$. Since the element in the first row of column one is larger than the others, we need not interchange rows. With the element as the pivot, we subtract 0.2 times the first row from the second and 0.5 times the first row from the third to obtain

$$[\mathbf{A} \mid \mathbf{b}]^{(2)} = \left[\begin{array}{ccc|c} 10 & 8 & -5 & 31 \\ 0 & -0.6 & 5 & -6.2 \\ 0 & -5 & 4.5 & -14.5 \end{array}\right]$$

The next step would be to reduce column two in $[\mathbf{A} \mid \mathbf{b}]^{(2)}$. From the second column we see that we must interchange rows two and three in order that the pivot may be -5, the largest magnitude element in the unreduced portion of this column. Following this interchange, we subtract 0.12 times the second row from the third. The result is the matrix

$$[\mathbf{A} \mid \mathbf{b}]^{(3)} = \left[\begin{array}{ccc|c} 10 & 8 & -5 & 31 \\ 0 & -5 & 4.5 & -14.5 \\ 0 & 0 & 4.46 & -4.46 \end{array}\right]$$

EXAMPLE 2.1 (cont'd)

The matrix $\mathbf{A}^{(3)}$ in $[\mathbf{A} \mid \mathbf{b}]^{(3)}$ is upper triangular; hence, the process of successive elimination has been completed and we are now ready for back substitution. To solve for x_3 recall that the last row of $[\mathbf{A} \mid \mathbf{b}]^{(3)}$ represents the equation

$$0x_1 + 0x_2 + 4.46x_3 = -4.46$$

Dividing through by 4.46 gives $x_3 = -1$. Using this value in the second row of $[\mathbf{A} \mid \mathbf{b}]^{(3)}$ we obtain

$$-5x_2 + 4.5(-1) = -14.5$$

so that $x_2 = 2$. Finally, we substitute the computed values of x_2 and x_3 into the first row of $[\mathbf{A} \mid \mathbf{b}]^{(3)}$ to obtain

$$10x_1 + 8(2) - 5(-1) = 31$$

which gives $x_1 = 1$. The final answer is $x_1 = 1$, $x_2 = 2$, and $x_3 = -1$.

EXERCISE 2.1

Solve the following set of four simultaneous equations using the Gaussian elimination algorithm.

$$\begin{bmatrix} 2 & 6 & 10 & -2 \\ 1 & 4 & 11 & 0 \\ 3 & 8 & 11 & 6 \\ -2 & -6 & -9 & 4 \end{bmatrix} \begin{bmatrix} x_1 \\ x_2 \\ x_3 \\ x_4 \end{bmatrix} = \begin{bmatrix} 8 \\ -2 \\ 20 \\ -13 \end{bmatrix}$$

Answer $x_1 = -87$, $x_2 = 46$, $x_3 = -9$, $x_4 = 2$.

It must be emphasized that this algorithm requires that the pivot elements be nonzero. If this is not the case, we must interchange the rows. If the absolute value of the pivot is very small, severe error may result in the computations. Hence it is very good strategy to interchange rows so that the pivot element is the element with the largest absolute value in the unreduced part of the column. This process, called *partial pivoting*, may have to be repeated at each subsequent stage. The following example will show how roundoff errors affect the final answer if partial pivoting is not carried out.

EXAMPLE 2.2

Consider solving the following set of equations on a machine which can carry out floating point arithmetic with only five decimal digits. It will be assumed that numbers with more digits are chopped (see Chapter 1).

$$\mathbf{Ax} = \begin{bmatrix} 20 & 8 & -4 \\ 10 & 3.9999 & 15 \\ 5 & 6 & 12 \end{bmatrix} \begin{bmatrix} x_1 \\ x_2 \\ x_3 \end{bmatrix} = \begin{bmatrix} 52 \\ 8.9999 \\ 4 \end{bmatrix} = \mathbf{b}$$

It can be easily verified that the correct solution to these equations is $x_1 = 2$, $x_2 = 1$ and $x_3 = -1$. We shall now try to solve these equations by the method of Gaussian elimination without partial pivoting. We form the matrix $[\mathbf{A} \mid \mathbf{b}]$. The first step is to subtract 0.5 times the first row of $[\mathbf{A} \mid \mathbf{b}]$ from the second and 0.25 times the first row from the third. This gives us

$$[\mathbf{A} \mid \mathbf{b}]^{(2)} = \left[\begin{array}{ccc|c} 20 & 8 & -4 & 52 \\ 0 & -0.0001 & 17 & -17.0001 \\ 0 & 4 & 13 & -9 \end{array}\right]$$

Since our hypothetical computer cannot carry more than five significant digits, the last entry in the second row of $[\mathbf{A} \mid \mathbf{b}]^{(2)}$ will be chopped to -17.0. Hence, in the next step of Gaussian elimination (without row interchanges) we must add 40,000 times the second row to the third. With our floating point arithmetic limited to only five digits, the resulting matrix will be

$$[\mathbf{A} \mid \mathbf{b}]^{(3)} = \left[\begin{array}{ccc|c} 20 & 8 & -4 & 52 \\ 0 & -0.0001 & 17 & -17 \\ 0 & 0 & 680010 & -680000 \end{array}\right]$$

From the last row of $[\mathbf{A} \mid \mathbf{b}]^{(3)}$ we get $x_3 = -0.99998$, again when the number is chopped after five significant digits. This is not so bad since the correct value is -1.0. However, using this value in the second row for back substitution, we get $x_2 = 10.0$. Finally substituting these values in the first row we get $x_1 = -1.5999$. The last two values are very different from the correct solution. This error has been caused by our selecting a very small pivot at the second step of elimination, which resulted in our having to multiply the second row by 40,000. This, in turn, led to catastrophic cancellation due to roundoff error. On the other hand, if we had interchanged rows two and three of $[\mathbf{A} \mid \mathbf{b}]^{(2)}$, we would have obtained the correct answer, in spite of the fact that our hypothetical computer cannot retain more than five digits in floating point arithmetic. This can be easily verified by the reader.

The method of Gaussian elimination can also be utilized to invert a matrix. In this case, we append the identity matrix, **I** to **A** instead of the vector **b** and carry out the process as described above. After the steps of successive elimination and back substitution are completed, the resulting matrix will be $[\mathbf{I} \mid \mathbf{A}^{-1}]$. This requires approximately n^3 arithmetic operations for an $n \times n$ matrix.

2.3 SCALING

The accuracy of the Gaussian elimination algorithm can be improved by adding scaling to partial pivoting. Two types of scaling are possible. In the first, one can multiply any row of $[\mathbf{A} \mid \mathbf{b}]$ by a nonzero constant. This is equivalent to multiplying both sides of the corresponding equation by that constant. Another kind of scaling can be obtained by multiplying the jth column of **A** by a nonzero constant, k. This amounts to replacing x_j by kx_j in the set of equations. The main idea is to make all the elements of **A** have approximately the same magnitude. Although there is no general agreement on the best method of scaling to minimize the effect of roundoff errors, the most popular approach is that of *equilibrating* a matrix. In this case, the equations are scaled so that the largest elements in the rows and columns of **A** have roughly the same absolute value, which is commonly unity. For example, consider the following equations

$$\begin{aligned} 0.1x_1 + 0.02x_2 &= 4 \\ 0.0001x_1 + 0.01x_2 &= 2 \end{aligned} \tag{2.3–1}$$

We can equilibrate this system of equations by multiplying the first equation by 10 and the second equation by 100, to give the system

$$\begin{aligned} x_1 + 0.2x_2 &= 40 \\ 0.01x_1 + x_2 &= 200 \end{aligned} \tag{2.3–2}$$

Computational results with (2.3–2) are less likely to be corrupted by roundoff errors in floating point arithmetic than results obtained with (2.3–1). The reason why equilibration may improve the computational results has to do with a certain quantity that is a measure of the *computational condition* of a system of linear equations, (2.1–2). This is discussed in Section 2.5. Because equilibration may improve the computational condition it is recommended that the scaling be carried out before starting the process of elimination.

The major objective of scaling is to obtain a system with an improved computational condition at a computational cost that is almost negligible when compared with the cost of solving the system.

2.4 L-U FACTORIZATION

A better insight into the method of Gaussian elimination is obtained by noting that the process of successive elimination is equivalent to premultiplication by a sequence of elementary lower triangular matrices, which leaves us with an upper triangular matrix for back substitution. Thus, in effect, the matrix **A** is factored into a lower triangular matrix **L** and an upper triangular matrix **U** so that **A** = **LU**.

The L-U factorization of **A** can be made unique by requiring that the diagonal elements of either **L** or **U** are unity. We shall discuss Crout's method, in Section 2.4.1, which is based on making the diagonal elements of U equal to one. In order to use this approach, it may be necessary to interchange the rows and/or the columns of **A**, and to equilibrate the matrix so that the magnitude of the diagonal element of each row is larger than that of the other elements in that row. Interchanging the columns of **A** is equivalent to interchanging the order of the elements of the solution vector **x**; that is, we renumber the components of **x**. For example, the system of equations

$$\begin{aligned} x_1 + 10x_2 + 8x_3 &= 5 \\ 11x_1 - 5x_2 - 16x_3 &= 7 \\ 3x_1 + x_2 + x_3 &= 1 \end{aligned}$$

can be rearranged as

$$\begin{aligned} 10x_2 + 8x_3 + x_1 &= 5 \\ -5x_2 - 16x_3 + 11x_1 &= 7 \\ x_2 + x_3 + 3x_1 &= 1 \end{aligned}$$

so that rearranged solution vector is $[x_2\ x_3\ x_1]^T$, and the rearranged system has the largest magnitude entry in each row appearing on the main diagonal of the **A**-matrix.

There are systems of linear simultaneous equations in which it may not be possible to rearrange the order of the unknowns to have the largest magnitude entry in each row of **A** occurring on the main diagonal of **A**. For example, the system

$$\begin{aligned} x_1 + 10x_2 + 8x_3 &= 5 \\ 11x_1 - 20x_2 - 16x_3 &= 7 \\ 3x_1 + x_2 + x_3 &= 1 \end{aligned}$$

cannot be rearranged to have the largest magnitude entry in each row on the main diagonal of **A**. However, many linear systems of simultaneous equations can be arranged so that the largest magnitude entry in each row occurs on the main diagonal of **A**. A special case of this is given in the following definition.

Definition 2.1

An $n \times n$ matrix $\mathbf{A} = (a_{ij})$ is said to have (row) *diagonal dominance* if

$$|a_{ii}| \geq \sum_{\substack{j=1 \\ j \neq i}}^{n} |a_{ij}| \qquad \text{for all } i.$$

A has (row) *strict diagonal dominance* if

$$| a_{ii} | > \sum_{\substack{j=1 \\ j \neq i}}^{n} | a_{ij} | \qquad \text{for all } i.$$

It is clear that column diagonal dominance could also be defined for a matrix **A**; however, we shall use row diagonal dominance.

We need a way to represent the operation of interchanging two rows of a matrix. This is done by multiplication of the matrix by a permutation matrix.

Definition 2.2

A matrix is said to be a *permutation matrix* if it is obtained by interchanging the rows of the identity matrix, **I**.

Premultiplication of the $n \times n$ matrix **A** by an $n \times n$ permutation matrix **P** will alter the order of the rows of **A** in exactly the same way as **P** was obtained by altering the order of the rows of **I**. If more than two rows are interchanged, then the product of the permutation matrices multiplying **A** gives the transformed matrix.

Theorem 2.1

Given any nonsingular matrix, **A**, there exists a permutation matrix **P** such that **PA** = **LU**, where **U** is an upper triangular matrix with unit diagonal elements and **L** is a lower triangular matrix with nonzero diagonal elements. Once **P** is chosen, the factorization is unique.

Proof

Select **P** so that none of the pivot elements is zero during the forward elimination phase; then the Gaussian elimination algorithm can be expressed as

$$\mathbf{E}_m\mathbf{E}_{m-1} \ldots . \mathbf{E}_1\mathbf{PA} = \mathbf{U}$$

where $\mathbf{E}_i$, $i = 1, 2, \ldots, m$ are lower triangular nonsingular elementary matrices. The inverse of a lower triangular matrix is also lower triangular as is the product of lower triangular matrices. It follows that **PA** can be written as the product of a lower triangular matrix, **L**, and an upper triangular matrix, **U**. Because the diagonal elements of **U** are all one, a straightforward calculation of the product **LU** shows that each element in **L** and in **U** can be uniquely determined in terms of the elements of **PA**.

To illustrate Theorem 2.1 consider the following system:

$$x_1 + 10x_2 + 8x_3 = 5$$

$$x_1 + 10x_2 - 16x_3 = 7$$

$$3x_1 + x_2 + x_3 = 1$$

where

$$\mathbf{A} = \begin{bmatrix} 1 & 10 & 8 \\ 1 & 10 & -16 \\ 3 & 1 & 1 \end{bmatrix}$$

If we perform the forward elimination phase of the Gaussian elimination method on **A** we obtain the matrix

$$\begin{bmatrix} 1 & 10 & 8 \\ 0 & 0 & -24 \\ 0 & -29 & -23 \end{bmatrix}$$

At the second stage we have a zero pivot in the second row. The second row corresponds to the second equation in the original set. However, if we had interchanged the second and third equations in the original set, then the matrix would have been

$$\mathbf{PA} = \begin{bmatrix} 1 & 10 & 8 \\ 3 & 1 & 1 \\ 1 & 10 & -16 \end{bmatrix}$$

where **P** is the permutation matrix

$$\mathbf{P} = \begin{bmatrix} 1 & 0 & 0 \\ 0 & 0 & 1 \\ 0 & 1 & 0 \end{bmatrix}$$

Notice that **P** has the effect of interchanging rows two and three in **A**. The forward elimination phase of Gaussian elimination on **PA** gives the matrix

$$\begin{bmatrix} 1 & 10 & 8 \\ 0 & -29 & -23 \\ 0 & 0 & -24 \end{bmatrix}$$

which contains only nonzero pivots.

Using Theorem 2.1 we have

$$\mathbf{PA} = \mathbf{LU} \tag{2.4–1}$$

where **P** represents the product of the permutation matrices required to do the row interchanges. Because we are trying to solve (2.1–2), that equation can now be written as

$$\mathbf{PAx} = \mathbf{LUx} = \mathbf{Pb} \tag{2.4–2}$$

We can readily solve (2.4–2) by introducing the transformation

$$\left.\begin{aligned} \mathbf{Lz} &= \mathbf{Pb} \\ \mathbf{Ux} &= \mathbf{z} \end{aligned}\right\} \tag{2.4–3}$$

We first solve for **z** and then for **x** by back substitution in both cases. This is made possible due to the triangular nature of **L** and **U**.

2.4.1 Crout's Method for L-U Factorization

Crout's algorithm allows the calculation of the elements of **L** and **U** recursively without writing over previous results. The method requires about n^3 arithmetic operations like the Gaussian elimination method. The advantages of Crout's method are: (1) smaller roundoff error in the computations and (2) reduced number of computations if **A** is a sparse matrix. To understand the procedure, let us consider the case when $n = 4$. Here $\mathbf{LU} = \mathbf{A}$ can be written as

$$\begin{bmatrix} \ell_{11} & 0 & 0 & 0 \\ \ell_{21} & \ell_{22} & 0 & 0 \\ \ell_{31} & \ell_{32} & \ell_{33} & 0 \\ \ell_{41} & \ell_{42} & \ell_{43} & \ell_{44} \end{bmatrix} \begin{bmatrix} 1 & u_{12} & u_{13} & u_{14} \\ 0 & 1 & u_{23} & u_{24} \\ 0 & 0 & 1 & u_{34} \\ 0 & 0 & 0 & 1 \end{bmatrix} = \begin{bmatrix} a_{11} & a_{12} & a_{13} & a_{14} \\ a_{21} & a_{22} & a_{23} & a_{24} \\ a_{31} & a_{32} & a_{33} & a_{34} \\ a_{41} & a_{42} & a_{43} & a_{44} \end{bmatrix}$$

By comparing the various elements of the matrix product on the left-hand side with the elements of **A**, the unknown elements of **L** and **U** can be calculated as

$$\ell_{11} = a_{11}, \qquad \ell_{21} = a_{21}, \qquad \ell_{31} = a_{31}, \qquad \ell_{41} = a_{41}$$

$$u_{12} = a_{12}/\ell_{11}, \qquad u_{13} = a_{13}/\ell_{11}, \qquad u_{14} = \ell_{11}$$

$$\ell_{22} = a_{22} - \ell_{21}u_{12}, \qquad \ell_{32} = a_{32} - \ell_{31}u_{12}, \qquad \ell_{42} = a_{42} - \ell_{41}u_{12}$$

$$u_{23} = (a_{23} - \ell_{21}u_{13})/\ell_{22}, \qquad u_{24} = (a_{24} - \ell_{21}u_{14})/\ell_{22}$$

$$\ell_{33} = a_{33} - \ell_{31}u_{13} - \ell_{32}u_{23}, \qquad \ell_{43} = a_{43} - \ell_{41}u_{13} - \ell_{42}u_{23}$$

$$u_{34} = (a_{34} - \ell_{31}u_{14} - \ell_{32}u_{24})/\ell_{33}$$

$$\ell_{44} = a_{44} - \ell_{41}u_{14} - \ell_{42}u_{24} - \ell_{43}u_{34}.$$

Let us define an auxiliary matrix, **Q**, consisting of the unknown elements of **L** and **U** in the following manner

$$\mathbf{Q} = (q_{ij}) = \begin{bmatrix} \ell_{11} & u_{12} & u_{13} & u_{14} \\ \ell_{21} & \ell_{22} & u_{23} & u_{24} \\ \ell_{31} & \ell_{32} & \ell_{33} & u_{34} \\ \ell_{41} & \ell_{42} & \ell_{43} & \ell_{44} \end{bmatrix}$$

It will be seen that the elements of **Q** can be calculated in a straightforward manner if a particular order is followed. For the general case of an $n \times n$ matrix, the elements of **Q** can be calculated using the following algorithm.

(i) Obtain the elements of the first column of **Q**,

$$q_{k1} = a_{k1}, \qquad k = 1, 2, \ldots, n.$$

(ii) Calculate the remaining elements in the first row of **Q**,

$$q_{1k} = a_{1k}/q_{11}, \qquad k = 2, 3, \ldots, n.$$

(iii) Set $j = 2$.

(iv) Complete the elements of the jth column of $\mathbf{Q}$,

$$q_{kj} = a_{kj} - q_{k1}q_{1j} - q_{k2}q_{2j} - \ldots - q_{k,j-1}q_{j-1,j}$$

$$\text{for} \quad k = j, j + 1, \ldots, n.$$

(v) If $j = n$, terminate the algorithm. Otherwise, continue.

(vi) Complete the entries of the jth row of $\mathbf{Q}$

$$q_{jk} = (a_{jk} - q_{j1}q_{1k} - q_{j2}q_{2k} - \ldots - q_{j,j-1}q_{j-1,k})/q_{jj}$$

$$\text{for} \quad k = j + 1, j + 2, \ldots, n.$$

(vii) Increase j by 1 and return to (iv).

It may be noted that the calculation of an element of the auxiliary matrix, $\mathbf{Q}$, which is an element of either $\mathbf{L}$ or $\mathbf{U}$, involves only the element of $\mathbf{A}$ in the same position and some of the elements of $\mathbf{Q}$ which have already been calculated. As the element is obtained, it is recorded in the matrix $\mathbf{Q}$. Thus it is possible to store the elements of $\mathbf{Q}$ in the corresponding position in the $\mathbf{A}$-matrix if there is no need to store the latter. Thus one can reduce the need for storage, an important consideration when the matrix has a large dimension. Another useful feature of this algorithm is that after the L-U factorization has been obtained, one can solve the equation $\mathbf{Ax} = \mathbf{b}$ for many vectors $\mathbf{b}$ without further L-U factorizations. The following example will illustrate the procedure.

EXAMPLE 2.3

Consider the following equation where $\mathbf{Ax} = \mathbf{b}$ is given by

$$\begin{bmatrix} 2 & 6 & 10 & -2 \\ 1 & 4 & 11 & 0 \\ 3 & 8 & 11 & 6 \\ -2 & -6 & -9 & 4 \end{bmatrix} \begin{bmatrix} x_1 \\ x_2 \\ x_3 \\ x_4 \end{bmatrix} = \begin{bmatrix} 8 \\ -2 \\ 20 \\ -13 \end{bmatrix}$$

To obtain the elements of $\mathbf{Q}$, we first note that

$$q_{11} = a_{11} = 2, \quad q_{21} = a_{21} = 1, \quad q_{31} = a_{31} = 3 \quad \text{and} \quad q_{14} = a_{14} = -2.$$

Next we determine the remaining elements of the first row of $\mathbf{Q}$. These are

$$q_{12} = a_{12}/q_{11} = 3, \qquad q_{13} = a_{13}/q_{11} = 5 \quad \text{and} \quad q_{14} = a_{14}/q_{11} = -1$$

This is followed by the calculation of the elements of the second column of $\mathbf{Q}$, starting with the diagonal element q_{22}. These are

$$q_{22} = a_{22} - q_{21}q_{12} = 1, \qquad q_{32} = a_{32} - q_{31}q_{12} = -1 \quad \text{and}$$

$$q_{42} = a_{42} - q_{41}q_{12} = 0$$

The remaining elements of the second row of $\mathbf{Q}$ are then obtained as

$$q_{23} = (a_{23} - q_{21}q_{13})q_{22} = 6 \quad \text{and} \quad q_{24} = (a_{24} - q_{21}q_{14})/q_{22} = 1$$

EXAMPLE 2.3 (cont'd)

The unknown elements of the third column of **Q** are now calculated as

$$q_{33} = a_{33} - q_{31}q_{13} - q_{32}q_{23} = 2 \quad \text{and} \quad q_{43} = a_{43} - q_{41}q_{13} - q_{42}q_{23} = 1$$

The remaining element of the third column of **Q** is

$$q_{34} = (a_{34} - q_{31}q_{14} - q_{32}q_{24})/q_{33} = 5.$$

Finally we have

$$q_{44} = a_{44} - q_{41}q_{14} - q_{42}q_{24} - q_{43}q_{34} = -3.$$

The resulting matrix **Q** is

$$\mathbf{Q} = \begin{bmatrix} 2 & 3 & 5 & -1 \\ 1 & 1 & 6 & 1 \\ 3 & -1 & 2 & 5 \\ -2 & 0 & 1 & -3 \end{bmatrix}$$

We shall now use **Q** to solve the equation $\mathbf{Lz} = \mathbf{b}$. Due to the lower triangular nature of **L**, we readily obtain

$$z_1 = b_1/2 = 4$$

$$z_2 = (b_2 - (1)z_1)/1 = -6$$

$$z_3 = (b_3 - (3)z_1 - (-1)z_2)/2 = 1$$

$$z_4 = (b_4 - (-2)z_1 - (0)z_2 - (1)z_3)/(-3) = 2.$$

Finally, we solve the equation $\mathbf{Ux} = \mathbf{z}$. Due to the upper triangular nature of **U**, we solve the elements of **x** in the reverse order; thus,

$$x_4 = z_4 = 2$$

$$x_3 = z_3 - (5)x_4 = -9$$

$$x_2 = z_2 - (1)x_4 - (6)x_3 = 46$$

$$x_1 = z_1 - (-1)x_4 - (5)x_3 - (3)x_2 = -87.$$

EXAMPLE 2.4

We shall now solve the following system of equations by Gaussian elimination and L-U factorization

$$\begin{bmatrix} 3 & 9 & 6 \\ 4 & 12 & 12 \\ 1 & -1 & 1 \end{bmatrix} \begin{bmatrix} x_1 \\ x_2 \\ x_3 \end{bmatrix} = \begin{bmatrix} 12 \\ 12 \\ 1 \end{bmatrix}$$

EXAMPLE 2.4 (cont'd)

a. *Using Gaussian elimination.*

Step 1: Obtain the augmented matrix

$$[\mathbf{A} \mid \mathbf{b}] = \left[\begin{array}{ccc|c} 3 & 9 & 6 & 12 \\ 4 & 12 & 12 & 12 \\ 1 & -1 & 1 & 1 \end{array}\right]$$

Step 2: Subtract $\frac{4}{3}$ times row one from row two and $\frac{1}{3}$ times row one from row three in $[\mathbf{A} \mid \mathbf{b}]$.

$$[\mathbf{A} \mid \mathbf{b}]^{(2)} = \left[\begin{array}{ccc|c} 3 & 9 & 6 & 12 \\ 0 & 0 & 4 & -4 \\ 0 & -4 & -1 & -3 \end{array}\right]$$

Step 3: Interchange rows two and three in $[\mathbf{A} \mid \mathbf{b}]^{(2)}$ due to the zero pivot to get

$$[\mathbf{A} \mid \mathbf{b}]^{(3)} = \left[\begin{array}{ccc|c} 3 & 9 & 6 & 12 \\ 0 & -4 & -1 & -3 \\ 0 & 0 & 4 & -4 \end{array}\right]$$

Step 4: Divide the last row in $[\mathbf{A} \mid \mathbf{b}]^{(3)}$ by 4.

$$[\mathbf{A} \mid \mathbf{b}]^{(4)} = \left[\begin{array}{ccc|c} 3 & 9 & 6 & 12 \\ 0 & -4 & -1 & -3 \\ 0 & 0 & 1 & -1 \end{array}\right]$$

Step 5: Adding row three to row two and dividing row two by -4 in $[\mathbf{A} \mid \mathbf{b}]^{(4)}$ yields

$$[\mathbf{A} \mid \mathbf{b}]^{(5)} = \left[\begin{array}{ccc|c} 3 & 9 & 6 & 12 \\ 0 & 1 & 0 & 1 \\ 0 & 0 & 1 & -1 \end{array}\right]$$

Step 6: Subtract 9 times row two and 6 times row three from row one in $[\mathbf{A} \mid \mathbf{b}]^{(5)}$, then divide row one by 3.

$$[\mathbf{A} \mid \mathbf{b}]^{(6)} = \left[\begin{array}{ccc|c} 1 & 0 & 0 & 3 \\ 0 & 1 & 0 & 1 \\ 0 & 0 & 1 & -1 \end{array}\right]$$

Hence, we get $x_1 = 3$, $x_2 = 1$, and $x_3 = -1$.

b. Using L-U factorization.

We start by building the first column and row of $\mathbf{Q}$. This gives us $q_{11} = 3$, $q_{21} = 4$, $q_{31} = 1$, $q_{12} = 3$ and $q_{13} = 2$. When we proceed to calculate q_{22}, however, we get a value of zero. Hence, it is necessary to interchange rows two and three. This must be kept in mind and the corresponding rows of vector $\mathbf{b}$ changed accordingly. Following the steps of Crout's algorithm, we get

EXAMPLE 2.4 (cont'd)

$$\mathbf{Q} = \begin{bmatrix} 3 & 3 & 2 \\ 1 & -4 & 0.25 \\ 4 & 0 & 4 \end{bmatrix} \quad \text{and} \quad \mathbf{Pb} = \begin{bmatrix} 12 \\ 1 \\ 12 \end{bmatrix}$$

where **P** is the permutation matrix. We can now solve the equation $\mathbf{Lz} = \mathbf{Pb}$, to get

$$z_1 = \frac{12}{3} = 4, \qquad z_2 = \frac{1 - z_1}{-4} = \frac{3}{4} \quad \text{and}$$

$$z_3 = \frac{12 - 4z_1 - 0z_2}{4} = -1.$$

Finally, we solve the equation $\mathbf{Ux} = \mathbf{z}$ to obtain

$$x_3 = z_1 = -1, \quad x_2 = z_2 - 0.25, \quad x_3 = 1 \quad \text{and} \quad x_1 = z_1 - 2x_3 - 3x_2 = 3.$$

EXERCISE 2.2

Solve the following set of linear simultaneous equations, using L-U factorization, rearranging them if necessary.

$$\begin{bmatrix} 1 & 2 & 40 & -5 \\ 3 & -2 & 1 & -100 \\ 10 & -1 & -1 & 1 \\ 0 & 15 & -2 & -1 \end{bmatrix} \begin{bmatrix} x_1 \\ x_2 \\ x_3 \\ x_4 \end{bmatrix} = \begin{bmatrix} -34 \\ -84 \\ 70 \\ 31 \end{bmatrix}$$

Answer $x_1 = 7, x_2 = 2, x_3 = -1, x_4 = 1.$

In Crout's method, the main diagonal elements of the upper triangular matrix **U** are set to one. Instead, we could have set the main diagonal elements of the lower triangular matrix **L** to one; this is known as *Dolittle's method* (see Problem 4 in Section 2.9).

There is another factorization, the *Cholesky factorization*, that can be used if the $n \times n$ matrix **A** is real, symmetric, and all its eigenvalues are positive. The number λ is an eigenvalue of **A** if there is a nonzero vector **x** such that $\mathbf{Ax} = \lambda\mathbf{x}$ (see Appendix A.4). In the Cholesky factorization $\mathbf{U} = \mathbf{L}^T$ (see Problem 7 in Section 2.9).

2.4.2 L-U Factorization of Tridiagonal Matrices

Tridiagonal matrices, defined as matrices in which all elements except those on the main diagonal and the diagonals just above and below it (the *superdiagonal* and the *subdiagonal*, respectively) are zero, occur in a variety of engineering and science problems. One can take advantage of this structure to obtain a simple L-U decomposition.

Theorem 2.2

If the $n \times n$ matrix $\mathbf{A}$ is tridiagonal it has a unique L-U factorization in which both $\mathbf{L}$ and $\mathbf{U}$ have only two nonzero diagonals: $\mathbf{L}$ has the main diagonal and the subdiagonal while $\mathbf{U}$ has the main diagonal, containing only ones, and the superdiagonal.

Proof

It can be shown that

$$\mathbf{L} = \begin{bmatrix} \ell_{11} & 0 & 0 & \cdot & \cdot & \cdot & \cdot \\ a_{21} & \ell_{22} & 0 & \cdot & \cdot & \cdot & \cdot \\ 0 & a_{32} & \ell_{33} & \cdot & \cdot & \cdot & \cdot \\ \cdot & \cdot & \cdot & \cdot & \cdot & \cdot & \cdot \\ \cdot & \cdot & \cdot & \cdot & \cdot & \cdot & \cdot \\ \cdot & \cdot & \cdot & \cdot & \cdot & a_{n,n-1} & \ell_{nn} \end{bmatrix}$$

and

$$\mathbf{U} = \begin{bmatrix} 1 & u_{12} & 0 & 0 & \cdot & \cdot & \cdot & \cdot \\ 0 & 1 & u_{23} & 0 & \cdot & \cdot & \cdot & \cdot \\ 0 & 0 & 1 & u_{34} & \cdot & \cdot & \cdot & \cdot \\ \cdot & \cdot & \cdot & \cdot & \cdot & \cdot & \cdot & \cdot \\ \cdot & \cdot & \cdot & \cdot & \cdot & \cdot & 1 & u_{n-1,n} \\ \cdot & \cdot & \cdot & \cdot & \cdot & \cdot & 0 & 1 \end{bmatrix}$$

The unknown elements of $\mathbf{L}$ and $\mathbf{U}$ are obtained using the equations

$$\ell_{11} = a_{11}$$

$$u_{k,k+1} = \frac{a_{k,k+1}}{\ell_{kk}}, \qquad k = 1, 2, \ldots, n-1$$

$$\ell_{kk} = a_{kk} - a_{k,k-1}u_{k-1,k}, \qquad k = 2, 3, \ldots, n.$$

EXAMPLE 2.5

Consider the tridiagonal matrix

$$\mathbf{A} = \begin{bmatrix} 10 & 10 & 0 & 0 \\ 1 & 17 & 8 & 0 \\ 0 & 2 & 13 & 6 \\ 0 & 0 & 4 & 12 \end{bmatrix}$$

EXAMPLE 2.5 (cont'd)

Using the algorithm described above, we get

$$\mathbf{L} = \begin{bmatrix} 10 & 0 & 0 & 0 \\ 1 & 16 & 0 & 0 \\ 0 & 2 & 12 & 0 \\ 0 & 0 & 4 & 10 \end{bmatrix}$$

and

$$\mathbf{U} = \begin{bmatrix} 1 & 1 & 0 & 0 \\ 0 & 1 & 0.5 & 0 \\ 0 & 0 & 1 & 0.5 \\ 0 & 0 & 0 & 1 \end{bmatrix}$$

EXERCISE 2.3

The loop equations for an electrical ladder network are given below. Determine the loop currents by L-U factorization.

$$\begin{bmatrix} 100 & -80 & 0 & 0 & 0 \\ -80 & 180 & -80 & 0 & 0 \\ 0 & -80 & 180 & -80 & 0 \\ 0 & 0 & -80 & 180 & -80 \\ 0 & 0 & 0 & -80 & 200 \end{bmatrix} \begin{bmatrix} i_1 \\ i_2 \\ i_3 \\ i_4 \\ i_5 \end{bmatrix} = \begin{bmatrix} 200 \\ 0 \\ 0 \\ 0 \\ 0 \end{bmatrix}$$

Answer $i_1 = 3.8497$, $i_2 = 2.3121$, $i_3 = 1.3525$, $i_4 = 0.7311$, $i_5 = 0.2924$.

2.5 ACCURACY OF THE SOLUTION

The methods described in the previous sections will give exact solutions only if the computer being used can carry an infinite number of digits. Due to the finite word length of all practical computers, the accuracy will be limited by roundoff errors. It is, therefore, necessary to examine the magnitude of the error in the solution. If we denote $\mathbf{x}_*$ as the exact solution of the equation $\mathbf{Ax} = \mathbf{b}$, and $\mathbf{y}$ as the solution that has been obtained on the computer, then the error is given by the vector

$$\mathbf{e} = \mathbf{y} - \mathbf{x}_* \qquad \textbf{(2.5–1)}$$

The size of error is measured using a *norm* of the vector $\mathbf{e}$, which can be regarded as a measure of the length of the vector. A norm is a real-valued function that satisfies Definition 2.3.

Definition 2.3

Let E denote the space consisting of all vectors of dimension n. A *vector norm*, denoted by $\|\cdot\|$, is a non-negative function with the following properties.

a. $\| \mathbf{x} \| > 0$ if $\mathbf{x} \neq \mathbf{0}$, for every $\mathbf{x}$ in E

b. $\| b\mathbf{x} \| = | b |\cdot\| \mathbf{x} \|$ for any scalar b and every $\mathbf{x}$ in E

c. $\| \mathbf{x} + \mathbf{y} \| \leq \| \mathbf{x} \| + \| \mathbf{y} \|$ for every $\mathbf{x}$ and $\mathbf{y}$ in E.

As a consequence of parts (a) and (b) of Definition 2.3 we have $\| \mathbf{x} \| = 0$ if, and only if, $\mathbf{x} = \mathbf{0}$. Part (c) of the definition is the triangle inequality. A vector norm that is used most often is the ℓ_p norm.

Definition 2.4

The ℓ_p norm of a vector $\mathbf{x}$ is given by

$$\| \mathbf{x} \|_p = \left[\sum_{i=1}^{n} | x_i |^p \right]^{1/p} \tag{2.5–2}$$

The two most common examples of ℓ_p norms are the Euclidean norm (for $p = 2$) and the uniform norm for the case when $p = \infty$. For the latter case, we have

$$\| \mathbf{x} \|_\infty = \underset{1 \leq i \leq n}{\text{maximum}} \, [\, | x_i | \,] \tag{2.5–3}$$

The accuracy of the solution of the equation $\mathbf{Ax} = \mathbf{b}$ depends largely on the nature of the matrix $\mathbf{A}$. The solution is prone to large computational error if $\mathbf{A}$ is *ill-conditioned*, or nearly singular. A quantitative measure of the nearness of a matrix to singularity is the condition number of the matrix. In order to define such a number we must first define the norm of a matrix.

Definition 2.5

Let M denote the space consisting of all $n \times n$ matrices. A *matrix norm*, denoted by $\| \cdot \|$, is a non-negative function with the following properties.

a. $\| \mathbf{C} \| > 0$ if $\mathbf{C} \neq \mathbf{0}$, for every $\mathbf{C}$ in M

b. $\| b\mathbf{C} \| = |b |\cdot\| \mathbf{C} \|$ for any scalar b and every $\mathbf{C}$ in M

c. $\| \mathbf{C} + \mathbf{Q} \| \leq \| \mathbf{C} \| + \| \mathbf{Q} \|$ for every $\mathbf{C}$ and $\mathbf{Q}$ in M

d. $\| \mathbf{CQ} \| \leq \| \mathbf{C} \|\cdot\| \mathbf{Q} \|$ for every $\mathbf{C}$ and $\mathbf{Q}$ in M.

As a consequence of Definition 2.5 $\| \mathbf{C} \| = 0$, if, and only if, $\mathbf{C} = \mathbf{0}$. Although we have used the same symbol to denote both a matrix and a vector norm, it should be emphasized that they are different functions. However, it is possible to choose a matrix norm that is *compatible* or *subordinate* to a given vector norm. To do this we require a condition similar to part (d) of Definition 2.5. For a given vector norm we select our matrix norm so that

$$\| \mathbf{Cx} \| \leq \| \mathbf{C} \| \cdot \| \mathbf{x} \| \quad \text{for every } \mathbf{x} \text{ in } E \text{ and every } \mathbf{C} \text{ in } M.$$

To obtain the compatible matrix norm we use the following definition.

Definition 2.6

The norm of the matrix **A**, induced by the vector norm $\| \mathbf{x} \|$, is defined by

$$\| \mathbf{A} \| = \underset{\|\mathbf{x}\|=1}{\text{maximum}}\{ \| \mathbf{Ax} \| \} \tag{2.5–4}$$

The norms of a matrix that are most commonly used are the ℓ_1 norm, the ℓ_∞ norm and the ℓ_2 norm. These matrix norms are compatible with the corresponding vector ℓ_p norms.

$$\| \mathbf{A} \|_1 = \underset{1 \leq j \leq n}{\text{maximum}}\left\{ \sum_{i=1}^{n} | a_{ij} | \right\} \qquad \text{(column sum)} \tag{2.5–5}$$

$$\| \mathbf{A} \|_\infty = \underset{1 \leq i \leq n}{\text{maximum}}\left\{ \sum_{j=1}^{n} | a_{ij} | \right\} \qquad \text{(row sum)} \tag{2.5–6}$$

$$\| \mathbf{A} \|_2 = (\lambda)^{0.5} \tag{2.5–7}$$

where λ is the largest eigenvalue of $\mathbf{A}^T\mathbf{A}$, and the superscript T represents transposition of a matrix. The number λ, real or complex, is an *eigenvalue* of the $n \times n$ matrix **A** if there is a nonzero vector **x**, called an *eigenvector*, such that $\mathbf{Ax} = \lambda\mathbf{x}$. The eigenvalues of A^TA are all non-negative (see Appendix A.4).

Theorem 2.3

If λ is an eigenvalue of the $n \times n$ matrix **A** then $| \lambda | \leq \| \mathbf{A} \|$ for any matrix norm.

As a consequence of Theorem 2.3, there is some norm of the matrix **A** that is arbitrarily close to $\rho(\mathbf{A}) = \max\{ | \lambda_i | \}$, where λ_i is an eigenvalue of **A**. The number $\rho(\mathbf{A})$ is called the *spectral radius* of **A**.

Definition 2.7

The *condition number* of an $n \times n$ matrix **A** is defined as

$$\text{Cond}(\mathbf{A}) = \| \mathbf{A} \| \cdot \| \mathbf{A}^{-1} \| \tag{2.5–8}$$

A good measure of the condition number of a matrix can be obtained in terms of the ratio of the largest and smallest modulus eigenvalues of **A**, that is

$$\text{Cond}(\mathbf{A}) = \left| \frac{\lambda_1}{\lambda_n} \right| \tag{2.5–9}$$

where λ_1 and λ_n are the largest and smallest modulus eigenvalues of **A**, respectively. If Cond(**A**) is very large we say that **A** is *ill-conditioned*. It is evident that Cond(**A**) will be infinite if **A** is singular; that is, when $\lambda_n = 0$. If **A** is nonsingular then $\mathbf{I} = \mathbf{A}\mathbf{A}^{-1}$; taking norms gives

$$1 = \| \mathbf{I} \| \leq \| \mathbf{A} \| \cdot \| \mathbf{A}^{-1} \| = \text{Cond}(\mathbf{A})$$

Therefore, the least value of Cond(**A**) is one, which is attained if

$$\| \mathbf{A}^{-1} \| = \| \mathbf{A} \|^{-1}$$

Example 2.6 considers two systems, one with a large condition number for its system matrix.

EXAMPLE 2.6

Consider the following equation

$$\begin{bmatrix} 2.000112 & 1.414215 \\ 1.414215 & 1.000105 \end{bmatrix} \begin{bmatrix} x_1 \\ x_2 \end{bmatrix} = \begin{bmatrix} 0.521471 \\ 0.232279 \end{bmatrix} \qquad \textbf{(2.5–10)}$$

It can be easily verified that the correct solution is $x_1 = 607.1248$ and $x_2 = -858.2826$. The eigenvalues of **A** are 3.0001 and 0.00011, which indicates that the condition number is very large. If we now solve these equations on a computer which does floating point arithmetic while chopping numbers after six significant figures, we get

$$\begin{bmatrix} 2.00011 & 1.41421 \\ 1.41421 & 1.00010 \end{bmatrix} \begin{bmatrix} x_1 \\ x_2 \end{bmatrix} = \begin{bmatrix} 0.521471 \\ 0.232279 \end{bmatrix}$$

Using Gaussian elimination, we obtain

$$\begin{bmatrix} 2.00011 & 1.41421 \\ 0 & 0.000167263 \end{bmatrix} \begin{bmatrix} x_1 \\ x_2 \end{bmatrix} = \begin{bmatrix} 0.521471 \\ -0.136435 \end{bmatrix}$$

Hence, we obtain $x_1 = 577.008$ and $x_2 = -815.691$. These are seen to be substantially different from the correct solution; x_1 has no correct significant digits, and x_2 has only one correct significant digit.

Now consider the equations

$$\begin{bmatrix} 2.000112 & 1.414215 \\ -1.414215 & 1.000105 \end{bmatrix} \begin{bmatrix} x_1 \\ x_2 \end{bmatrix} = \begin{bmatrix} 0.521471 \\ 0.232279 \end{bmatrix} \qquad \textbf{(2.5–11)}$$

which are identical to equations (2.5–10) except that the sign of the first element in the second row of the matrix **A** has been changed. As a result, the eigenvalues of **A** are $1.5001 \pm j1.3229$, where $j = \sqrt{(-1)}$, and the condition number is now small. Hence equations (2.5–11) do not represent an ill-conditioned set, and we shall be able to obtain a good approximation to the correct solution even when we use our hypothetical computer that chops floating point numbers after six significant digits. In this case, it is easily verified that the correct solution is $x_1 = 0.0482543934$ and $x_2 = 0.300895456$. We shall now solve equations (2.5–11) on our hypothetical computer. As a result of roundoff we get

$$\begin{bmatrix} 2.00011 & 1.41421 \\ -1.41421 & 1.00010 \end{bmatrix} \begin{bmatrix} x_1 \\ x_2 \end{bmatrix} = \begin{bmatrix} 0.521471 \\ 0.232279 \end{bmatrix}$$

EXAMPLE 2.6 (cont'd)

and Gaussian elimination gives us

$$\begin{bmatrix} 2.00011 & 1.41421 \\ 0 & 2.00003 \end{bmatrix} \begin{bmatrix} x_1 \\ x_2 \end{bmatrix} = \begin{bmatrix} 0.521471 \\ 0.600993 \end{bmatrix}$$

whose solution is $x_1 = 0.0482541$ and $x_2 = 0.300491$. In spite of the fact that all computations were carried to only six significant digits, the computed value of x_1 is correct to five significant digits, while x_2 is correct to three significant digits. The greater accuracy in the solution of this linear system is due to the small condition number of the matrix **A**.

To see why the condition number of **A** influences the solution of (2.1–2), let **y** denote the computed solution and $\mathbf{x}_*$ denote the exact solution. A short calculation shows that

$$\| \mathbf{y} - \mathbf{x}_* \| \le \| \mathbf{A}^{-1} \| \cdot \| \mathbf{r} \|$$

where $\mathbf{r} = \mathbf{b} - \mathbf{A}\mathbf{y}$ is the residual error. Assuming that $\mathbf{x}_* \ne \mathbf{0}$, we have

$$\| \mathbf{b} \| \le \| \mathbf{A} \| \cdot \| \mathbf{x}_* \|$$

so that

$$\frac{\| \mathbf{y} - \mathbf{x}_* \|}{\| \mathbf{x}_* \|} \le \frac{\| \mathbf{A}^{-1} \| \cdot \| \mathbf{r} \|}{\| \mathbf{b} \| \cdot (\| \mathbf{A} \|)^{-1}} = \text{Cond}(\mathbf{A}) \frac{\| \mathbf{r} \|}{\| \mathbf{b} \|} \qquad \textbf{(2.5–12)}$$

Equation (2.5–12) shows how the condition number could affect the relative error in the computed solution. Even if one could make $\| \mathbf{r} \| / \| \mathbf{b} \| = u$, the relative machine precision, a large value of Cond(**A**) could easily leave us with a computed solution that is essentially meaningless. We consider a condition number to be large if Cond(**A**) $\simeq 1/u$, so that the right-hand side of (2.5–12) is approximately 1. The important point to remember is that if Cond(**A**) is small (we consider Cond(**A**) to be small if Cond(**A**) $\le 10^{t/2}$, where t is the number of decimal digits of precision) then small perturbations in **A** or **b** introduces only small perturbations in the solution of (2.1–2); however, if Cond(**A**) is large, there is a chance that small perturbations in **A** or **b** might lead to large perturbations in the computed solution. Recall that (2.5–10) had a large condition number for the system **A**-matrix; this led to large errors in the computed solution on the hypothetical computer used. The two equations (2.5–10) represent straight lines in the (x_1, x_2) plane. Notice that the lines are nearly parallel, so that the small changes in **A** and

b, introduced when converting to 6-digit arithmetic, caused large changes in the computed solution.

The condition number of the matrix **A**, in (2.1–2), may be reduced by scaling (see Section 2.3). For example, the **A**-matrix in (2.3–1) and its equilibrated form in (2.3–2) have condition numbers of 10.02 and 1.094, respectively. We would expect the computational results obtained from (2.3–2) to be somewhat better than those obtained from (2.3–1). Of course, the scaling should be done before the elimination process.

The condition number of a matrix is a better indicator of the nearness of a matrix to singularity than the size of pivots obtained during Gaussian elimination. Hence, if we know the condition number, we can estimate the accuracy of the solutions obtained using a finite number of digits in floating point arithmetic. In particular, if the reciprocal of the condition number is of the same order as the machine epsilon, the results of direct computer solution using either Gaussian elimination or L-U factorization will be completely unreliable.

The calculation of the eigenvalues of a matrix is usually a difficult problem. It will be discussed in detail in Chapter 8. Usually it is more convenient to estimate the condition number using other approaches. The Pascal function FindConditionNumber in the linear systems package in Appendix 2.1 utilizes a method proposed by Forsythe, Malcolm and Moler to estimate the condition number based on a norm of the matrix while it is being triangularized using Gaussian elimination.

Although the condition number of a matrix warns us about the potential difficulty caused by ill-conditioning of the matrix, it does not take into account all the errors in computation. The following theorem from a book of Wilkinson (see References at end of chapter) gives a good estimate of the accuracy of the solution. In fact, it estimates the perturbation **E** for which the computed solution **y** satisfies the equation $(\mathbf{A} + \mathbf{E})\mathbf{y} = \mathbf{b}$.

Theorem 2.4

If t-digit arithmetic is used then the perturbation, **E**, induced by Gaussian elimination with pivoting satisfies the inequality

$$\| \mathbf{E} \|_\infty \leq f(n)\ 10^{1-t}(\underset{i,j,k}{\text{maximum}}[\ |\ a_{ij}^{(k)}\ |\ ])$$

where $a_{ij}^{(k)}$ is the (i, j)th element of the reduced form of **A** at the kth stage in the elimination and $f(n) \simeq n$, but in the worst case

$$f(n) \leq 1.01(n^3 + 3n^2)$$

Theorem 2.4 tells us that if **A** is well-conditioned so that $a_{ij}^{(k)}$ do not become large at any stage of Gaussian elimination and n is not large then the accumulated rounding errors are of the same order as those caused by the floating point representation of the elements of **A**.

2.6 ITERATIVE METHODS

In many engineering and scientific problems, one has to solve a large system of linear equations in which most of the coefficients are zero. Such systems of equations are often obtained while solving partial differential equations. Furthermore, the nonzero coefficients may be related through known formulas. In such cases, it may not be possible to store the entire matrix into the computer. It is much more convenient to generate the nonzero elements when needed. A good approach to solve such equations is to use a method which does not require altering the matrix **A**. We start with a trial solution $\mathbf{x}^{(0)}$ and try to improve it by obtaining the successive iterates $\mathbf{x}^{(k)}$. The process is repeated until the norm of the residuals $\| \mathbf{A}\mathbf{x}^{(k)} - \mathbf{b} \|$ becomes sufficiently small or until no change in $\mathbf{x}^{(k)}$ is observed. Such methods are called *iterative*, and are also useful when the condition number of **A** is large.

Several iterative methods are available. We shall discuss three of the most popular iterative methods. These are

1. Jacobi's method
2. Gauss-Seidel method
3. Successive Over-Relaxation, or SOR.

In each of these methods, one starts by scaling the equations so that all the diagonal elements of the resulting matrix **A** are equal to one; we replace **A** by the matrix $\mathbf{D}^{-1}\mathbf{A}$, where **D** is a diagonal matrix consisting of the main diagonal of **A**. This assumes that none of the diagonal entries is zero.

The following two theorems will be required for our discussion of iterative methods.

Theorem 2.5 (Gerschgorin disk theorem)

Let $\mathbf{A} = (a_{ij})$ be an $n \times n$ matrix. The set of points $\{z\}$, in the complex plane, defined by

$$| z - a_{ii} | \leq \sum_{\substack{j=1 \\ j \neq i}}^{n} | a_{ij} |$$

forms a disk with center at a_{ii} and radius $\sum_{\substack{j=1 \\ j \neq i}}^{n} | a_{ij} |$. The eigenvalues of **A** all lie in the set union of the n disks.

Proof

Let λ be any eigenvalue of **A** and $\mathbf{x} = [x_1, \ldots, x_n]^T$ be an associated eigenvector. There is an integer k, $1 \leq k \leq n$, such that

$$| x_k | = \underset{1 \leq i \leq n}{\text{maximum}}\{ | x_i | \}$$

$|x_k| > 0$ because $\mathbf{x}$ is an eigenvector and cannot be the zero vector. Using the equation $\mathbf{Ax} = \lambda\mathbf{x}$, we can write

$$\lambda x_k = a_{kk}x_k + \sum_{\substack{j=1 \\ j\neq k}}^{n} a_{kj}x_j$$

so that

$$(\lambda - a_{kk})x_k = \sum_{\substack{j=1 \\ j\neq k}}^{n} a_{kj}x_j$$

Take absolute values to get

$$|\lambda - a_{kk}|\cdot|x_k| \leq \sum_{\substack{j=1 \\ j\neq k}}^{n} |a_{kj}|\cdot|x_j|.$$

Because $|x_k| \geq |x_j|$ for $1 \leq j \leq n$, we have

$$|\lambda - a_{kk}|\cdot|x_k| \leq \sum_{\substack{j=1 \\ j\neq k}}^{n} |a_{kj}|\cdot|x_k|$$

and therefore

$$|\lambda - a_{kk}| \leq \sum_{\substack{j=1 \\ j\neq k}}^{n} |a_{kj}|$$

This last inequality defines a disk, in the complex plane, centered at a_{kk} and radius $\sum_{\substack{j=1 \\ j\neq k}}^{n} |a_{kj}|$. For each eigenvalue of $\mathbf{A}$ we will obtain a similar disk. Consequently, all the eigenvalues of $\mathbf{A}$ must lie in the set union of the disks defined by

$$|z - a_{ii}| \leq \sum_{\substack{j=1 \\ j\neq i}}^{n} |a_{ij}|, \qquad 1 \leq i \leq n.$$

The following theorem uses the Gerschgorin disk theorem for a special matrix related to a matrix with strict diagonal dominance.

Theorem 2.6

If $\mathbf{A}$, a matrix with only ones on the main diagonal, has strict diagonal dominance then the matrix $\mathbf{J} = (\mathbf{I} - \mathbf{A})$, where $\mathbf{I}$ is the identity matrix, has all of its eigenvalues inside the unit circle in the complex plane.

Proof

If λ is any eigenvalue of $\mathbf{J}$ then, by Theorem 2.5 and the strict diagonal dominance of $\mathbf{A}$,

$$|\lambda| \leq \sum_{\substack{j=1 \\ j\neq i}}^{n} |a_{ij}| < 1.$$

2.6.1 Jacobi's Iteration Method

In this method, we start by decomposing $\mathbf{A}$ as

$$\mathbf{A} = \mathbf{I} - \mathbf{U} - \mathbf{L} \tag{2.6–1}$$

where it is assumed that $\mathbf{A}$ has ones on its main diagonal and $\mathbf{A}$ has the property of strict diagonal dominance, $\mathbf{U}$ is an upper triangular matrix with zeros on its main diagonal and $\mathbf{L}$ is a lower triangular matrix with zeros on its main diagonal. We may now write the equation $\mathbf{Ax} = \mathbf{b}$ as

$$\mathbf{b} = \mathbf{Ax} = (\mathbf{I} - \mathbf{U} - \mathbf{L})\mathbf{x} = \mathbf{x} - (\mathbf{U} + \mathbf{L})\mathbf{x}. \tag{2.6–2}$$

This may be rearranged as

$$\mathbf{x} = \mathbf{Jx} + \mathbf{b}, \quad \text{where} \quad \mathbf{J} = \mathbf{U} + \mathbf{L}. \tag{2.6–3}$$

Since $\mathbf{x}$ is not known, we start by making an initial guess and improving it using the iterative algorithm

$$\mathbf{x}^{(k+1)} = \mathbf{Jx}^{(k)} + \mathbf{b}. \tag{2.6–4}$$

Note that equation (2.6–4) amounts to solving the ith equation in the matrix equation $\mathbf{Ax} = \mathbf{b}$ by moving every term, except the one which contains x_i, to the right-hand side. This is facilitated by the fact that the every element on the main diagonal is equal to one. In practice, one may have to scale the equations as well as perform some row interchanges in order to get $\mathbf{A}$ into the desired form. It should be noted that there are matrices that cannot be re-ordered to obtain a diagonally dominant matrix.

Theorem 2.7

The Jacobi algorithm, as given in (2.6–4), will always converge if the matrix $\mathbf{A}$ has ones on its main diagonal and has strict diagonal dominance.

Proof

If λ is any eigenvalue of $\mathbf{J} = \mathbf{I} - \mathbf{A} = \mathbf{U} + \mathbf{L}$, then, by Theorem 2.6, $|\lambda| < 1$. As a consequence of this it can be shown that

$$\lim_{k\to\infty} \mathbf{J}^k = \mathbf{0}$$

Now

$$\mathbf{x}^{(k+1)} = \mathbf{J}\mathbf{x}^{(k)} + \mathbf{b}$$

$$\mathbf{x}_* = \mathbf{J}\mathbf{x}_* + \mathbf{b}$$

where $\mathbf{x}_*$ is the solution of the linear system. Subtracting these equations gives the error equation

$$\mathbf{e}^{(k+1)} = \mathbf{J}\mathbf{e}^{(k)}, \qquad k = 0, 1, 2, \ldots$$

Consequently, the error at iteration k is

$$\mathbf{e}^{(k)} = \mathbf{J}^k\mathbf{e}^{(0)}$$

As $k \to \infty$, $\mathbf{e}^{(k)} \to \mathbf{0}$.

The conditions of Theorem 2.7 can be weakened if $\mathbf{A}$ is irreducible and diagonally dominant, with strict inequality holding for at least one row (see Definition 2.1).

EXAMPLE 2.7

The following set of linear equations represents an electrical ladder network. We want to use Jacobi's method to solve the system

$$\begin{bmatrix} 4 & 3 & 0 \\ 3 & 4 & -1 \\ 0 & -1 & 4 \end{bmatrix} \begin{bmatrix} x_1 \\ x_2 \\ x_3 \end{bmatrix} = \begin{bmatrix} 24 \\ 30 \\ -24 \end{bmatrix}$$

The first step is to scale the matrix so that we have ones on the main diagonal. The equations

$$\begin{bmatrix} 1 & 0.75 & 0 \\ 0.75 & 1 & -0.25 \\ 0 & -0.25 & 1 \end{bmatrix} \begin{bmatrix} x_1 \\ x_2 \\ x_3 \end{bmatrix} = \begin{bmatrix} 6 \\ 7.5 \\ -6 \end{bmatrix}$$

are obtained. The new matrix $\mathbf{A}$ is irreducible and has diagonal dominance with strict inequality holding for rows one and three, so that Jacobi's method will converge. The eigenvalue of $\mathbf{J} = \mathbf{I} - \mathbf{A}$ of maximum modulus is $(\frac{5}{8})^{1/2}$. Using equation (2.6–3) we obtain

$$x_1^{(k+1)} = -0.75x_2^{(k)} + 6$$

$$x_2^{(k+1)} = -0.75x_1^{(k)} + 0.25x_3^{(k)} + 7.5$$

$$x_3^{(k+1)} = 0.25x_2^{(k)} - 6$$

If we let

$$\mathbf{x}^{(0)} = \begin{bmatrix} 1 \\ 1 \\ 1 \end{bmatrix}$$

EXAMPLE 2.7 (cont'd)

then we get the sequence of vectors

$$\mathbf{x}^{(1)} = \begin{bmatrix} 5.25 \\ 7 \\ -5.75 \end{bmatrix}, \quad \mathbf{x}^{(2)} = \begin{bmatrix} 0.75 \\ 2.125 \\ -4.25 \end{bmatrix}, \quad \mathbf{x}^{(3)} = \begin{bmatrix} 4.4063 \\ 5.8750 \\ -5.4688 \end{bmatrix}, \ldots,$$

$$\mathbf{x}^{(15)} = \begin{bmatrix} 3.0838 \\ 4.1118 \\ -5.0279 \end{bmatrix}, \ldots, \mathbf{x}^{(25)} = \begin{bmatrix} 3.0080 \\ 4.0107 \\ -5.0027 \end{bmatrix}.$$

It will be seen that the sequence of iterates $\mathbf{x}^{(k)}$ appears to be converging because the norm of the difference between the successive vectors is decreasing. It can be verified that the correct solution is $\mathbf{x} = [3 \quad 4 \quad -5]^T$.

2.6.2 Gauss-Seidel Method

Although the convergence of the Jacobi iteration algorithm is guaranteed it is often slow. An improvement is obtained if we try to use all the available information as we proceed. For instance, while we are calculating the ith element of $\mathbf{x}^{(k+1)}$ we use the first $(i - 1)$ elements of $\mathbf{x}^{(k)}$ in Jacobi's method, although we have already calculated these elements of $\mathbf{x}^{(k+1)}$. In the Gauss-Seidel method we use the updated values as soon as they are available:

$$x_i^{(k+1)} = b_i - \sum_{j=1}^{i-1} a_{ij}x_j^{(k+1)} - \sum_{j=i+1}^{n} a_{ij}x_j^{(k)} \tag{2.6–5}$$

This algorithm may, therefore, be written as

$$\mathbf{x}^{(k+1)} = \mathbf{L}\mathbf{x}^{(k+1)} + \mathbf{U}\mathbf{x}^{(k)} + \mathbf{b} \tag{2.6–6}$$

or as

$$(\mathbf{I} - \mathbf{L})\mathbf{x}^{(k+1)} = \mathbf{U}\mathbf{x}^{(k)} + \mathbf{b} \tag{2.6–7}$$

Equation (2.6–7) is readily solved, without any matrix inversion, due to the lower triangular nature of **L**. We have assumed that **A** has ones on its main diagonal. It is not necessary to store two vectors $\mathbf{x}^{(k+1)}$ and $\mathbf{x}^{(k)}$; each updated component can replace the old component in the solution vector.

EXAMPLE 2.8

We shall solve Example 2.7 using the Gauss-Seidel method. Using the algorithm given in equation (2.6–6) we obtain the following relationships

$$x_1^{(k+1)} = -0.75x_2^{(k)} + 6$$

$$x_2^{(k+1)} = -0.75x_1^{(k+1)} + 0.25x_3^{(k)} + 7.5$$

$$x_3^{(k+1)} = 0.25x_2^{(k+1)} - 6$$

EXAMPLE 2.8 (cont'd)

It is interesting to compare these equations with the similar equations for the Jacobi method. Starting with the same initial guess as in Example 2.7 we obtain the sequence of vectors

$$\mathbf{x}^{(0)} = \begin{bmatrix} 1 \\ 1 \\ 1 \end{bmatrix}, \quad \mathbf{x}^{(1)} = \begin{bmatrix} 5.25 \\ 3.8125 \\ -5.0469 \end{bmatrix}, \quad \mathbf{x}^{(2)} = \begin{bmatrix} 3.1406 \\ 3.8828 \\ -5.0293 \end{bmatrix},$$

$$\mathbf{x}^{(3)} = \begin{bmatrix} 3.0879 \\ 3.9268 \\ -5.0183 \end{bmatrix}, \ldots, \quad \mathbf{x}^{(10)} = \begin{bmatrix} 3.0033 \\ 3.9973 \\ -5.0007 \end{bmatrix}.$$

Comparison with Example 2.7 shows that this method converges faster than the Jacobi iteration method. The Gauss-Seidel iteration matrix for this example problem

$$\mathbf{A}_{GS} = (\mathbf{I} - \mathbf{L})^{-1}\mathbf{U}$$

has a maximum modulus eigenvalue of $\frac{5}{8}$. Compare this value with the maximum modulus eigenvalue $(\frac{5}{8})^{1/2}$ obtained for the Jacobi method in Example 2.7. The error in the Jacobi method (Theorem 2.7) is given by

$$\| \mathbf{e}^{(k)} \| \leq \| \mathbf{J} \|^k \| \mathbf{e}^{(0)} \| = \left(\frac{5}{8}\right)^{k/2} \| \mathbf{e}^{(0)} \|.$$

The error in the Gauss-Seidel method is

$$\| \mathbf{e}^{(m)} \| \leq \| \mathbf{A}_{GS} \|^m \| \mathbf{e}^{(0)} \| = \left(\frac{5}{8}\right)^{m} \| \mathbf{e}^{(0)} \|.$$

Therefore, for this example the Gauss-Seidel method converges at twice the rate of the Jacobi method.

In general, the Gauss-Seidel method converges if **A** has strict diagonal dominance, or if **A** is a real symmetric positive definite (all eigenvalues are positive) matrix. In the latter case the spectral radii of the Jacobi and the Gauss-Seidel iteration matrices are related: $\rho(\mathbf{A}_{GS}) = [\rho(\mathbf{J})]^2$.

2.6.3 Successive Over-Relaxation

This method, commonly known as SOR, attempts to increase the rate of convergence of the Gauss-Seidel iteration. The main idea behind the method is that if $\hat{x}_i^{(k+1)}$ is the value of x_i obtained at the kth iteration then an even better value may be obtained by forming a weighted average of the old and new values, given by

$$x_i^{(k+1)} = x_i^{(k)} + \omega[\hat{x}_i^{(k+1)} - x_i^{(k)}] \qquad \textbf{(2.6–8)}$$

where ω is a positive number, the *relaxation parameter*. This leads to the algorithm

$$x^{(k+1)} = \omega[\mathbf{L}\mathbf{x}^{(k+1)} + \mathbf{U}\mathbf{x}^{(k)} + \mathbf{b}] + (1 - \omega)\mathbf{x}^{(k)} \tag{2.6–9}$$

For $\omega > 1$, we over-relax, while for $\omega < 1$, we under-relax. Over-relaxation is a form of extrapolation; we over-correct so as to make the residual error change sign. In this way, the convergence of the method can be increased. It will be seen that for ω equal to one, we get the Gauss-Seidel method. The rate of convergence will depend on the choice of ω. Although the optimum value of ω is difficult to find, it can be shown that it lies between 1 and 2 if **A** is positive definite.

EXAMPLE 2.9

We shall solve the problem in Example 2.8 by using successive over-relaxation with $\omega = 1.25$. With the algorithm shown in equation (2.6–9) the equations

$$x_1^{(k+1)} = -0.9375x_2^{(k)} - 0.25x_1^{(k)} + 7.5$$

$$x_2^{(k+1)} = -0.9375x_1^{(k+1)} + 0.3125x_3^{(k)} - 0.25x_2^{(k)} + 9.375$$

$$x_3^{(k+1)} = 0.3125x_2^{(k+1)} - 0.25x_3^{(k)} - 7.5$$

are obtained. Starting with the same initial guess as in Examples 2.7 and 2.8 the sequence of vectors

$$\mathbf{x}^{(0)} = \begin{bmatrix} 1 \\ 1 \\ 1 \end{bmatrix}, \quad \mathbf{x}^{(1)} = \begin{bmatrix} 6.3125 \\ 3.5195 \\ -6.6501 \end{bmatrix}, \quad \mathbf{x}^{(2)} = \begin{bmatrix} 2.6223 \\ 3.9585 \\ -4.6004 \end{bmatrix}$$

$$\mathbf{x}^{(3)} = \begin{bmatrix} 3.1333 \\ 4.0103 \\ -5.0967 \end{bmatrix}, \ldots, \quad \mathbf{x}^{(7)} = \begin{bmatrix} 3.0000 \\ 4.0003 \\ -5.0003 \end{bmatrix}$$

is obtained. It will be seen that the rate of convergence is faster than that for the Gauss-Seidel method. The main problem is the choice of the optimum value of the relaxation factor ω. It can be shown that for a positive definite (all eigenvalues are positive) tridiagonal matrix **A**, the optimum value of ω is given by

$$\omega_{\text{opt}} = \frac{2}{1 + (1 - \lambda^2)^{1/2}}$$

where λ is the magnitude of the largest eigenvalue of **J**, the Jacobi method iteration matrix for **A**.

2.7 A GENERAL LINEAR SYSTEM SOLVER

The Pascal procedure LinearSystemSolver, whose user's manual is in Appendix 2.1, is a good general purpose procedure for solving linear systems. It can handle systems of the form

$$\mathbf{AX} = \mathbf{B} \tag{2.7–1}$$

where $\mathbf{A}$ and $\mathbf{B}$ are $n \times n$ and $n \times m$ real matrices, respectively. The procedure can be used to efficiently solve many different systems of the form $\mathbf{Ax}_{k+1} = \mathbf{x}_k$ because the L-U factorization of $\mathbf{A}$ is retained. The procedure returns an estimate of the condition number of $\mathbf{A}$. LinearSystemSolver should do well provided Cond($\mathbf{A}$) is not too large (less than $10^{t/2}$, where t is the number of decimal digits of precision) and $n < 100$.

LinearSystemSolver performs *iterative improvement* on each computed solution vector $\mathbf{y}$, provided Cond($\mathbf{A}$) is not too large. To use iterative improvement we compute the residual error vector for each solution,

$$\mathbf{r} = \mathbf{b} - \mathbf{Ay}$$

We now solve the equation $\mathbf{Ac} = \mathbf{r}$ to obtain the correction $\mathbf{c}$, which is then added to $\mathbf{y}$. Generally, $\mathbf{r}$ should be computed in multiple precision, if it is available. LinearSystemSolver returns the ℓ_1 norm of each residual error vector.

LinearSystemSolver has an input/output parameter called "digits." On input "digits" is the number of correct digits in the entries of $\mathbf{A}$ and $\mathbf{B}$ in (2.7–1). If "digits" is positive an accuracy test is performed; if "digits" is zero the accuracy test is ignored. On output "digits" gives the number of digits in the answer which were unchanged after iterative improvement. If "digits" is to be positive on input, then it probably should not be larger than $|\ \ell og_{10}(u)\ |/2$, where u is the relative machine precision. LinearSystemSolver returns an indicator "flag," which should be examined to see if any problems were encountered in solving (2.7–1).

2.8 ADDITIONAL TOPICS IN SOLVING SYSTEMS OF SIMULTANEOUS EQUATIONS

In this chapter we have studied several methods for solving sets of linear simultaneous equations. Two types of methods were discussed, those of the direct type and those of the iterative type. The methods of Gaussian elimination and L-U factorization are of the former type and are quite efficient for solving well-conditioned sets of equations. On the other hand, indirect or iterative methods are more suitable for solving large sets of equations ($n > 1000$).

We use the condition number of the system matrix as a measure of well-conditioning. A large condition number implies that the matrix is nearly singular, whereas a condition number near one shows that the matrix is well-conditioned. A large condition number should serve as a warning that the computed solution may be in error. One can also use the norm of the residual error as an indicator of the accuracy of the solution of (2.1–2); however, it should be considered in light of the estimated condition number. It is possible to have a small residual error norm and yet have an inaccurate solution. These comments should help to interpret computational results obtained using the Pascal procedure in Appendix

2.1. There is a rule-of-thumb which may help: there is a loss of $\ell og_{10}(\text{Cond}(\mathbf{A}))$ decimal places due to errors in Gaussian elimination.

2.9 PROBLEMS

1. Show that the method of Gaussian elimination, together with backward substitution, requires a total of $(n^3 + 3n^2 - n)/3$ multiplications or divisions and $(2n^3 + 3n^2 - 5n)/6$ additions or subtractions for solving a set of n linear simultaneous equations (2.1–2). Hence, calculate the total number of arithmetic operations of the two types for $n = 20$, 50, and 100.

2. Solve the following set of linear equations by Gaussian elimination.

$$\begin{aligned} x_1 + 2x_2 + 40x_3 - 5x_4 &= 100 \\ 3x_1 - 2x_2 + x_3 - 100x_4 &= 200 \\ 10x_1 - x_2 - x_3 + x_4 &= 0 \\ 15x_2 - 2x_3 - x_4 &= 30 \end{aligned}$$

3. Use the method discussed in Section 2.4 to obtain the L-U factorization of the coefficient matrix in Problem 2, after rearranging the equations for strict diagonal dominance.

4. The L-U factorization method described in Section 2.4 makes the elements on the main diagonal of the upper triangular matrix equal to one. This is also known as *Crout's method*. Another choice is to make the main diagonal elements of the lower triangular matrix equal to one instead of those of the upper triangular matrix. This is known as *Dolittle's method*. Solve Problem 2 by this method.

5. Determine the number of arithmetic operations required for the L-U factorization of an $n \times n$ matrix.

6. Use the Gauss-Seidel iteration method for solving Problem 2.

7. When a matrix **A** is real, symmetric, and positive definite (all eigenvalues are real and positive) the L-U decomposition can be greatly simplified by requiring that $\mathbf{U} = \mathbf{L}^T$. This is called the *Cholesky factorization* of **A**. Determine the number of arithmetic operations required for Cholesky factorization of an $n \times n$ real, symmetric, positive definite matrix **A**.

8. Write a Pascal procedure to compute the Cholesky factorization of a real, symmetric, positive definite matrix.

9. The following tridiagonal matrix occurs in the interpolation of data using cubic spline functions, when the data $\{x_i, y_i\}$ have equal x-spacing (see Chapter 5). Obtain an L-U factorization of the matrix **A**.

$$\mathbf{A} = \begin{bmatrix} -1 & 1 & 0 & 0 & 0 & 0 & 0 \\ 1 & 4 & 1 & 0 & 0 & 0 & 0 \\ 0 & 1 & 4 & 1 & 0 & 0 & 0 \\ 0 & 0 & 1 & 4 & 1 & 0 & 0 \\ 0 & 0 & 0 & 1 & 4 & 1 & 0 \\ 0 & 0 & 0 & 0 & 1 & 4 & 1 \\ 0 & 0 & 0 & 0 & 0 & 1 & -1 \end{bmatrix}$$

10. Consider the tridiagonal matrix

$$\mathbf{A} = \begin{bmatrix} a_1 & c_1 & 0 & 0 & 0 & 0 & 0 \\ b_2 & a_2 & c_2 & 0 & 0 & 0 & 0 \\ 0 & b_3 & a_3 & c_3 & 0 & 0 & 0 \\ 0 & 0 & b_4 & a_4 & c_4 & 0 & 0 \\ 0 & 0 & 0 & b_5 & a_5 & c_5 & 0 \\ 0 & 0 & 0 & 0 & b_6 & a_6 & c_6 \\ 0 & 0 & 0 & 0 & 0 & b_7 & a_7 \end{bmatrix}$$

Write a computer program for an L-U factorization of this matrix.

11. Use your program from Problem 10 to solve the loop equations for an electrical ladder network.

$$\begin{bmatrix} 100 & -50 & 0 & 0 & 0 & 0 & 0 \\ -50 & 150 & -50 & 0 & 0 & 0 & 0 \\ 0 & -50 & 150 & -50 & 0 & 0 & 0 \\ 0 & 0 & -50 & 150 & -50 & 0 & 0 \\ 0 & 0 & 0 & -50 & 150 & -50 & 0 \\ 0 & 0 & 0 & 0 & -50 & 150 & -50 \\ 0 & 0 & 0 & 0 & 0 & -50 & 200 \end{bmatrix} \mathbf{x} = \begin{bmatrix} 200 \\ 0 \\ 0 \\ 0 \\ 0 \\ 0 \\ 0 \end{bmatrix}$$

12. The 5 × 5 Hilbert matrix

$$\mathbf{A} = \begin{bmatrix} 1 & \frac{1}{2} & \frac{1}{3} & \frac{1}{4} & \frac{1}{5} \\ \frac{1}{2} & \frac{1}{3} & \frac{1}{4} & \frac{1}{5} & \frac{1}{6} \\ \frac{1}{3} & \frac{1}{4} & \frac{1}{5} & \frac{1}{6} & \frac{1}{7} \\ \frac{1}{4} & \frac{1}{5} & \frac{1}{6} & \frac{1}{7} & \frac{1}{8} \\ \frac{1}{5} & \frac{1}{6} & \frac{1}{7} & \frac{1}{8} & \frac{1}{9} \end{bmatrix}$$

is known to be rather ill-conditioned. The equation $\mathbf{Ax} = \mathbf{b}$ is to be solved for the case when $\mathbf{b} = [244 \quad 197 \quad 171 \quad 152 \quad 137]^T$.

(a) Solve these equations using Gaussian elimination. How does the solution compare with the correct answer

$$\mathbf{x} = [60 \quad 60 \quad -420 \quad -840 \quad 2520]^T?$$

(b) Presumbly the inaccuracy is caused by roundoff errors in entering the elements of **A**, particularly, $\frac{1}{3}$, $\frac{1}{6}$, $\frac{1}{7}$ and $\frac{1}{9}$. This can be corrected by multiplying each equation with the lowest common multiple of the denominators of the elements in that row. What solution do you get with the scaling as suggested above?

13. In many engineering and scientific problems, the elements of the matrix **A** and the vector **b** are complex numbers. It is possible to replace such a system of equations by another set involving only real numbers. Let

$$\mathbf{A} = \mathbf{P} + j\mathbf{Q}, \qquad \mathbf{x} = \mathbf{y} + j\mathbf{z} \quad \text{and} \quad \mathbf{b} = \mathbf{c} + j\mathbf{d}$$

where j is the square root of -1, and **P**, **Q**, **y**, **z**, **c** and **d** have only real elements. Show that in this case, the original equation $\mathbf{Ax} = \mathbf{b}$ leads to the matrix equation

$$\begin{bmatrix} \mathbf{P} & -\mathbf{Q} \\ \mathbf{Q} & \mathbf{P} \end{bmatrix} \begin{bmatrix} \mathbf{y} \\ \mathbf{z} \end{bmatrix} = \begin{bmatrix} \mathbf{c} \\ \mathbf{d} \end{bmatrix}.$$

14. Use the method discussed in Problem 13 to solve the following set of equations which arose in the steady-state solution of an electrical network subject to a sinusoidal input.

$$\begin{bmatrix} 5 + j2 & -1 + j3 & -j2 \\ -1 - j3 & 7 + j4 & -2 + j1 \\ -j2 & -2 + j1 & 8 - j4 \end{bmatrix} \begin{bmatrix} i_1 \\ i_2 \\ i_3 \end{bmatrix} = \begin{bmatrix} 100 + j0 \\ 50 - j20 \\ 0 \end{bmatrix}$$

15. The 4×4 matrix in the following equation is called *Wilson's matrix*, which is known to be badly ill-conditioned. Try using (a) Gaussian elimination, (b) the Gauss-Seidel iteration method, and (c) the SOR method (with $\omega = 1.25$, 1.5, 1.75) to obtain the solution. For parts (b) and (c), start with $\mathbf{x}^{(0)} = \mathbf{0}$. *Note*: Wilson's matrix does not have diagonal dominance so that the Gauss-Seidel method may have poor convergence properties.

$$\begin{bmatrix} 10 & 7 & 8 & 7 \\ 7 & 5 & 6 & 5 \\ 8 & 6 & 10 & 9 \\ 7 & 5 & 9 & 10 \end{bmatrix} \begin{bmatrix} x_1 \\ x_2 \\ x_3 \\ x_4 \end{bmatrix} = \begin{bmatrix} 32 \\ 23 \\ 33 \\ 31 \end{bmatrix}$$

16. If **A** is an $n \times n$ matrix with ones on the main diagonal, and **A** has strict diagonal dominance, show that $\| \mathbf{I} - \mathbf{A} \|_\infty < 1$.

17. Using a suitable or convenient matrix norm for the matrix

$$\mathbf{A} = \begin{bmatrix} -602.9 & -0.4762 & 301.0 \\ -248.8 & -0.1048 & 124.2 \\ -200.6 & 0.00001367 & 101.7 \end{bmatrix}$$

and given the following estimate for $\mathbf{A}^{-1}$

$$\mathbf{A}^{-1} = \begin{bmatrix} 0.1224 & -0.5563 & 0.3170 \\ -4.462 & 10.73 & 0.09902 \\ 0.2415 & -1.097 & 0.6352 \end{bmatrix}$$

estimate Cond(**A**).

18. Use the Pascal procedure LinearSystemSolver in Appendix 2.1 to solve the system of equations (2.7–1) for

$$\mathbf{A} = \begin{bmatrix} 33.0 & 16.0 & 72.0 \\ -24.0 & 10.0 & -57.0 \\ -8.0 & -4.0 & -17.0 \end{bmatrix}$$

$$\mathbf{B} = \begin{bmatrix} 1.0 & 0.0 & 12.0 & -359.0 \\ 11.0 & 6.0 & -7.0 & 281.0 \\ -5.0 & 9.0 & 10.0 & 85.0 \end{bmatrix}$$

19. Many problems in science and engineering can be reduced to a system of m linear simultaneous equations in n unknowns, where $m > n$. The system can be written as

$$\mathbf{Cx} = \mathbf{d} \tag{2.9–1}$$

where $\mathbf{C}$ is an $m \times n$ matrix, and $\mathbf{x}$ and $\mathbf{d}$ are column vectors of dimensions n and m, respectively. In general, (2.9–1) will be inconsistent. One way to "solve" (2.9–1) is to look for a solution that minimizes the expression

$$(\mathbf{Cx} - \mathbf{d})^T(\mathbf{Cx} - \mathbf{d})$$

Performing the minimization gives the least squares solution to Eq. (2.9–1), when $\mathbf{C}$ has rank n, as the solution $\mathbf{x}_*$ of the linear system

$$(\mathbf{C}^T\mathbf{C})\mathbf{x} = \mathbf{C}^T\mathbf{d}.$$

(a) Obtain the least squares solution, $\mathbf{x}_*$, of (2.9–1) when $\mathbf{C} = (c_{ij})$ and $\mathbf{d} = (d_i)$; c_{ij} and d_i are defined as

$$c_{i1} = w_i, \qquad c_{i2} = w_{i-1}, \qquad c_{i3} = 1, \qquad d_i = w_{i+1} \quad \text{for} \quad i = 1, 2, \ldots, 9.$$

The numbers w_i are

i	w_i
0	0.000000
1	0.004528
2	0.016429
3	0.033587
4	0.054344
5	0.077409
6	0.101790
7	0.126710
8	0.151620
9	0.176280
10	0.199790

(b) Obtain an estimate of Cond($\mathbf{C}^T\mathbf{C}$).

(c) Obtain the error vector $\mathbf{e} = \mathbf{Cx}_* - \mathbf{d}$.

20. The Pascal procedure LinearSystemSolver can be used to invert a nonsingular matrix **A**. Write a Pascal procedure that accepts as input an $n \times n$ real matrix **A** and returns as its output either an $n \times n$ matrix **B**, which is an approximation to the inverse of **A** if **A** is nonsingular, or an error flag if **A** is singular. Test your procedure on the matrices in Problems 9, 12, 15, 17, 18.

2.10 REFERENCES

L. V. Atkinson and P. J. Harley. *An Introduction to Numerical Methods with Pascal.* London: Addison-Wesley, 1983.

R. L. Burden, J. D. Faires, and A. C. Reynolds. *Numerical Analysis.* 2d ed. Boston: Prindle, Weber & Schmidt, 1981.

G. E. Forsythe, M. A. Malcolm, and C. B. Moler. *Computer Methods for Mathematical Computations.* Englewood Cliffs, N.J: Prentice-Hall, 1977.

G. H. Golub and C. F. Van Loan. *Matrix Computations.* Baltimore, MD: The Johns Hopkins University Press, 1983.

G. Strang. *Linear Algebra and Its Applications.* New York: Academic Press, 1976.

J. H. Wilkinson. *The Algebraic Eigenvalue Problem.* Oxford: Oxford University Press, 1965.

D. M. Young. *Iterative Solution of Large Linear Systems.* New York: Academic Press, 1971.

APPENDIX 2.1

User's Manual for the Pascal Procedure LinearSystemSolver, which is in the file LinearSy.Pas

```
PROCEDURE LinearSystemSolver( n : INTEGER ; VAR a : MatrixNN ; m : INTEGER ;
                VAR b : MatrixNM ; VAR digits : INTEGER ; LUform : BOOLEAN ;
                               VAR flag : INTEGER ; VAR Data : MatrixData );
```

The procedure LinearSystemSolver can be used to solve the linear system

AX = B

using Gaussian elimination (Crout algorithm) with equilibration, partial pivoting, and iterative improvement as required.

The following must be declared in the calling program

```
TYPE
  VectorN = ARRAY[ 1..MaxN ] OF REAL;
  VectorM = ARRAY[ 1..MaxM ] OF REAL;
```

```
MatrixNN = ARRAY[ 1..MaxN , 1..MaxN ] OF REAL;
MatrixNM = ARRAY[ 1..MaxN , 1..MaxM ] OF REAL;
IntVectorN = ARRAY[ 1..MaxN ] OF INTEGER;
MatrixData = RECORD
               LU : MatrixNN;
               pivot : IntVectorN;
               ConditionNumber : REAL;
               residuals : VectorM
             END;
```

where $n \le$ MaxN and $m \le$ MaxM

Input:

n	The order of the matrix **A**.
a	The matrix **A**.
m	The number of solutions, the column dimension of the matrix **B**.
b	The matrix **B**.
digits	If positive, the elements of **A** and **B** are assumed to be correct to *digits* decimal digits, and the routine performs an accuracy test on the L–U decomposition of **A**. If zero, the accuracy test is bypassed.
LUform	Set to TRUE if **A** is already in L–U form, which is stored in Data.LU . The vector Data.pivot must also be known, and the appropriate output value of flag, either 0 or 1. Set to FALSE if either the L–U form of **A** or Data.pivot is unknown.

Output:

b	Contains the solution **X**.
digits	Contains the approximate number of decimal digits in the answer which were unchanged after iterative improvement.
flag	An error indicator. 0 All ok 1 Accuracy test failed 2 Singular matrix 3 Matrix is too ill-conditioned for iterative improvement to be effective
Data.LU	The L–U decomposition of **A**.
Data.pivot	The vector of pivot information obtained during the L–U decomposition of **A**.
Data.ConditionNumber	Contains an estimate of the condition number of **A**. Computed only if the L–U decomposition of **A** is done.
Data.residuals	The ℓ_1 norm (1–norm) of the residual of each solution.

Procedure LinearSystemSolver contains the following machine dependent constants which are set for Turbo Pascal with 8087 support.

```
infinity = 1.0E+307
MaxDigits = 16
```

Procedure LinearSystemSolver is based on the IMSL FORTRAN subroutine LEQT2F

APPENDIX 2.2

Listing of a Pascal Program for Solving Systems of Linear Equations

```
PROGRAM LinearSystem;

CONST
  device = 'c';
  MaxN = 7 ; MaxM = 7 ; Field = 10 ; Decimal = 6 ;

TYPE
  String14 = String[14];
  String80 = String[80];
  VectorN = ARRAY[ 1..MaxN ] OF REAL;
  VectorM = ARRAY[ 1..MaxM ] OF REAL;
  MatrixNN = ARRAY[ 1..MaxN , 1..MaxN ] OF REAL;
  MatrixNM = MatrixNN;
  IntVectorN = ARRAY[ 1..MaxN ] OF INTEGER;
  MatrixData = RECORD
                 LU : MatrixNN;
                 pivot : IntVectorN;
                 ConditionNumber : REAL;
                 residuals : VectorM
               END;

VAR
  i, n, m, digits, flag : INTEGER;
  a : MatrixNN;
  b : MatrixNM;
  Data : MatrixData;
  Title : String80;
  OutFile : TEXT;

(*$I Response.Pas *)
(*$I Files.Pas *)
(*$I ReadNum.Pas *)
(*$I MatrixIO.Pas *)
(*$I LinearSy.Pas *)

BEGIN (* Main program *)
  IF device IN [ 'C', 'c' ] THEN
    ASSIGN(OutFile,'Con')
  ELSE
    ASSIGN(OutFile,'Prn');
  REWRITE(OutFile) ;
  ClrScr ; LowVideo ;
  WRITE('Solution of A*X = B. Get matrix A.');
  GetMatrix(a,n,n,TRUE,MaxN,MaxN,Field,Decimal );
  WRITELN;
  WRITE('Solution of A*X = B. Get matrix B.');
  GetMatrix(b,n,m,FALSE,MaxN,MaxM,Field,Decimal);
  PutMatrix(a,n,n,Field,Decimal,'A:',device);
  WRITELN(OutFile);
  PutMatrix(b,n,m,Field,Decimal,'B:',device);
  WRITELN(OutFile);
  digits := 10;
```

```
  LinearSystemSolver(n,a,m,b,digits,FALSE,flag,Data);
  IF  flag <> 2  THEN
    BEGIN
      IF  m = 1  THEN
        Title := 'Solution of the linear system:'
      ELSE
        Title := 'Solutions of the linear system:';
      PutMatrix(b,n,m,Field,Decimal,Title,device);
      WRITELN(OutFile);
      IF  m = 1  THEN
        WRITELN(OutFile,'The 1-norm of the residual is ',
                                             Data.residuals[1]:10)
      ELSE
        BEGIN
          WRITELN(OutFile,'The 1-norm of the residuals are:');
          WRITELN(OutFile);
          FOR  i := 1  TO  m  DO
            WRITELN(OutFile,'    Residual[',i,'] = ',Data.residuals[i]:10)
        END;
      WRITELN(OutFile);
      WRITELN(OutFile,'The condition number of A is ',
                               Data.ConditionNumber:12);
      IF  flag > 0  THEN
        BEGIN
          WRITELN(OutFile);
          CASE  flag  OF
                 1 : WRITELN(OutFile,'The accuracy test failed');
                 3 : WRITELN(OutFile,'The matrix A is too ill-conditioned',
                                ' for iterative improvement to be effective')
          END
        END;
      WRITELN(OutFile);
      WRITELN(OutFile,'digits = ',digits)
    END
  ELSE
    WRITELN(OutFile,'The matrix A is singular');
  CLOSE(OutFile)
END.
```

APPENDIX 2.3

Computational Cost of the Gaussian Elimination Method

When the linear system (2.1–2) is solved by the Gaussian elimination method, we can estimate the computational cost of the calculations by counting the number of floating point arithmetic operations (additions, subtractions, multiplications, and divisions). We want to consider the Gaussian elimination method for a system (2.1–2), with n unknowns, n equations and one right-hand side vector.

First, we consider the forward elimination phase. In order to eliminate the variable x_1 from the second equation we must multiply the first equation by a_{21}/a_{11} and then subtract the result from the second equation. When performing the calculations we do not have to do the multiplication of the coefficient of x_1 in the first equation. This requires one division and $(n - 1)$ multiplications for the left-hand side of the first equation, plus one multiplication for the right-hand side. When the equations are subtracted from one another, n subtractions are required. This gives a total of $2n + 1$ operations to eliminate x_1 from the second equation. But, we must also eliminate x_1 from equations 3, 4, . . . , n; consequently, a total of $(n - 1)(2n + 1)$ operations are needed to eliminate x_1 from equations 2, 3, . . . , n. To eliminate x_2 from equations 3, 4, . . . , n requires $(n - 2)(2n - 1)$ operations. In general, to eliminate x_k from equations $k + 1$, $k + 2$, . . . , n requires $(n - k)[2(n - k) + 3]$ operations, for $k = 1, 2, \ldots, n - 1$. Therefore, the total number of operations required in the forward elimination phase is

$$\sum_{k=1}^{n-1} (n - k)[2(n - k) + 3] = \frac{n(n - 1)(4n + 7)}{6}$$

For the back substitution phase we start at the last equation of the system obtained after the forward elimination phase has been completed. In equation n we solve for x_n; this requires one division. Using the value of x_n in equation $(n - 1)$ we solve for x_{n-1}, which requires one division, one multiplication and one subtraction. At equation $(n - k + 1)$ we find x_{n-k+1}; this requires one division, $k - 1$ multiplications and $k - 1$ subtractions, for $k = 1, 2, \ldots, n$. The total number of operations is

$$\sum_{k=1}^{n} (2k - 1) = n^2$$

Finally, the total number of arithmetic operations for the Gaussian elimination method, with one right-hand side vector, is

$$\frac{n(n - 1)(4n + 7)}{6} + n^2 = \frac{2}{3}n^3 + \frac{3}{2}n^2 - \frac{7}{6}n$$

When $n \geq 2$ the term $2n^3/3$ dominants the term $(3n^2/2 - 7n/6)$. Thus the Gaussian elimination method requires approximately $2n^3/3$ operations. Notice that the major contribution to the total number of operations occurs in the forward elimination phase.

Chapter

3

Roots of Polynomials

Efficient evaluation of polynomials, for real or complex arguments, by Horner's rule and synthetic division is discussed. Roots of a quadratic are determined in a manner that maximizes accuracy. Cardan's method for the roots of a cubic is given. Iterative methods for finding the roots of higher degree polynomials are introduced. These methods include Newton's method, Laguerre's method, and the Jenkins-Traub method.

3.1 INTRODUCTION

Algebraic functions which contain a weighted sum of the non-negative integral powers of the variable s are called polynomials in s. The following general form will be used for a polynomial of degree n ($a_n \neq 0$)

$$P(s) = a_n s^n + a_{n-1} s^{n-1} + \ldots + a_1 s + a_0 \tag{3.1–1}$$

The constant a_n is called the *leading coefficient* of the polynomial; if $a_n = 1$, then the polynomial is called *monic*. The coefficients of $P(s)$ can be either real or complex numbers. If the coefficients are all real we say that $P(s)$ is a real polynomial, even though it could have complex roots. In this chapter we are mainly interested in real polynomials.

In a number of scientific and engineering problems it is necessary to determine the zeros of the equation

$$P(s) = 0 \tag{3.1–2}$$

These zeros are called the *roots* of the polynomial. For example, in order to determine the natural frequencies of vibration of a mechanical system, which is described by a linear constant-coefficient differential equation, one must determine the roots of the characteristic polynomial of the differential equation. Similarly, one must find the roots of the numerator and denominator polynomials of a transfer function (Laplace transform) of a linear system in order to determine its zeros and poles.

The following properties of the roots of a polynomial follow from the fundamental theorem of algebra.

1. A polynomial of degree n will have n roots. These may be either real or complex. A root may be repeated m times, in which case we say that the root has *multiplicity* m; if $m = 1$ we have a *simple* root.

2. If all the coefficients, a_i, of $P(s)$ are real, then the complex roots, if any, will occur in conjugate pairs.

In general, direct methods (using only a finite number of algebraic operations) for the computation of roots of polynomials exist only if n is less than five. There are certain polynomials, such as $s^n + a$, whose roots can be easily found; however, no general polynomial equation of degree greater than four can be solved by using a finite number of algebraic operations. Generally, then, we must use an iterative method to determine the roots. After a real root or a pair of complex conjugate roots are obtained by this iterative process, the polynomial is divided by either a linear factor for a real root or a quadratic factor for complex roots. This process is called *deflation*, as it results in a polynomial of reduced degree for which the iterative process may be repeated. To carry out this process efficiently we need a good method for calculating $P(s)$ for a given s, which may be either real or complex. Such methods will be discussed in the next section. Before proceeding to that discussion, we briefly consider the problem of root sensitivity.

It is well known that one, or more, roots of a polynomial can be very sensitive to small changes in the polynomial's coefficients. There is a famous example credited to J. H. Wilkinson in which a real polynomial, with roots $1, 2, \ldots, 20$, has the coefficient of s^{19}, which is 210, changed by adding $2^{-23} \simeq 1.2 \times 10^{-7}$ to it; the resulting polynomial now has five pairs of complex conjugate roots which are at least 1.039 units from the real axis in the complex plane. Such a root is called *ill-conditioned*; roots that are relatively insensitive to small changes in the coefficients are called *well-conditioned*. This example illustrates the extreme sensitivity of the roots; the character of roots may change (real to complex) by quite small changes in the coefficients. Consequently, roundoff errors introduced by the deflation process may lead to grossly incorrect roots for high-degree polynomials. Approximations to roots that have been determined through a deflation process should, therefore, be checked with the original polynomial.

3.2 EVALUATION OF POLYNOMIALS

It would appear that for a given value of s, say α, we can calculate $P(\alpha)$ by first calculating $\alpha^2, \alpha^3, \ldots, \alpha^n$, multiplying each of these by the corresponding coefficient and then adding these terms. In general, this is rather laborious and requires $(2n - 1)$ multiplications and n additions. It is possible to reduce the required number of arithmetic operations by using other methods. The two most common methods will be discussed in this section. These are Horner's rule and synthetic division, which require the same number of arithmetic operations.

3.2.1 Horner's Rule

This is based on writing the polynomial in a nested form. For example, we may express $P(s)$, given in equation (3.1–1), as

$$P(s) = \{\ldots((a_n s + a_{n-1})s + a_{n-2})s + \ldots\}s + a_0 \qquad \textbf{(3.2–1)}$$

It follows that evaluation of $P(s)$ using equation (3.2–1) requires only n multiplications and n additions.

EXAMPLE 3.1

Consider the polynomial

$$P(s) = 2s^3 + 3s^2 - 4s + 5.$$

For $s = 2$, we have $P(2) = [(2 \times 2 + 3)2 - 4]2 + 5 = 25$. It will be seen that only three multiplications and three additions are required in this case.

EXERCISE 3.1

Calculate $P(2.5)$ and $P(4)$ for the polynomial

$$P(s) = 3s^5 + 7s^4 - 12s^3 + 20s^2 - 33s + 50.$$

Answer $P(2.5) = 471.40625$, $P(4) = 4334$.

3.2.2 Synthetic Division

Another way to calculate $P(\alpha)$ is to divide $P(s)$ by the factor $s - \alpha$. The remainder is then the value of $P(\alpha)$. This follows from the fact that we can write

$$P(s) = Q(s)(s - \alpha) + r \qquad \textbf{(3.2–2)}$$

In equation (3.2–2), $Q(s)$ is the quotient and r is the remainder obtained as a result of the division of $P(s)$ by $(s - \alpha)$. Since this is valid for all values of α, by setting $s = \alpha$ in (3.2–2) we see that $P(\alpha) = r$.

The process of division of $P(s)$ by $(s - \alpha)$ can be carried out most efficiently by using synthetic division. This will be explained through the following example.

EXAMPLE 3.2

We shall calculate $P(2)$ for the polynomial in Example 3.1. This will be done by the synthetic division of $P(s)$ by $(s - 2)$. It involves writing the coefficients of $P(s)$ in the descending order in the top line, as shown in the tableau given below. The leftmost coefficient is then written in the third line, but in the same column. This is followed by multiplying the leftmost coefficient in the last line by -2 and writing the product in the second line below the next coefficient. The corresponding element in the last line is obtained by subtracting the element in the second line from the one above it. Note that every element in the second line is obtained by multiplying the element to the left of it in the last line, by -2.

$$\begin{array}{rrrr} 2 & 3 & -4 & 5 \\ & -4 & -14 & -20 \\ \hline 2 & 7 & 10 & 25 \end{array}$$

The value of $P(2)$ is obtained as the rightmost element in the last line of the tableau. The remaining elements in the last line are also useful, as they are the coefficients of the quotient obtained by the division; thus, $P(s) = (2s^2 + 7s + 10)(s - 2) + 25$, as can be easily verified. It will also be seen that this procedure has required exactly three multiplications and three additions, as was the case with Horner's rule.

To illustrate the method further, we shall obtain $P(3)$ for the polynomial in Example 3.1. The tableau is

$$\begin{array}{rrrr} 2 & 3 & -4 & 5 \\ & -6 & -27 & -69 \\ \hline 2 & 9 & 23 & 74 \end{array}$$

Hence, $P(3) = 74$.

EXERCISE 3.2

Calculate $P(2.5)$ and $P(4)$ for the polynomial given in Exercise 3.1 by using synthetic division.

Answer $P(2.5) = 471.40625$, $P(4) = 4334$.

3.2.3 Evaluation of a Polynomial at Complex Arguments

The evaluation of $P(s)$ for complex s, say $s = \sigma + j\omega$, $\{j = \sqrt{(-1)}\}$, can be carried out by using either Horner's rule or synthetic division as discussed above. However, both of these methods will require arithmetic using complex numbers. As this is often slower on computers, a better approach is to divide $P(s)$ by a quadratic whose roots are $\sigma \pm j\omega$. This division does not require complex arithmetic. As a result, we may write

$$P(s) = (s^2 - 2\sigma s + \sigma^2 + \omega^2)Q(s) + as + b \tag{3.2–3}$$

Hence, for $s = \sigma + j\omega$ we get

$$P(\sigma + j\omega) = [as + b]_{s=\sigma+j\omega} = (a\sigma + b) + ja\omega. \tag{3.2–4}$$

Synthetic division by a quadratic factor is a little more complicated than that by a linear factor. The procedure is shown in the following examples.

EXAMPLE 3.3

We shall compute $P(-1 + j2)$ for $P(s) = 3s^4 + 14s^3 + 30s^2 + 50s + 68$. In this case, we divide $P(s)$ by $s^2 + 2s + 5$. The process of synthetic division gives

3	14	30	50	68
	6	15	40	−5
		15	10	
		16	−2	
3	8	−1	12	73

Note that in this case again, we have first written the coefficients of $P(s)$ in the top line and copied the leftmost element in the botttom line. This is followed by multiplying the two coefficients, 2 and 5 of the monic polynomial $s^2 + 2s + 5$, by the element in the bottom row and writing these just below the second and third entries in the top line. These entries are then subtracted from the corresponding entries in the top line of the tableau, but whereas the number in the second column is written in the bottom line, that in the third column is written in an intermediate line (third from the top). We then repeat the procedure, using the element in the second column of the last line for multiplication with 2 and 5 and writing the products in the third and fourth columns, respectively. The process terminates after $(n - 1)$ steps, where n is the degree of the polynomial. From the last two entries we get

$$P(-1 + j2) = [12s + 73]_{s=-1+j2} = 61 + j24.$$

EXAMPLE 3.4

We compute $P(-2 + j3)$ for the polynomial in Example 3.3. In this case, we must divide by the quadratic $s^2 + 4s + 13$. Synthetic division gives the tableau

3	14	30	50	68
	12	39	26	−221
		−9	24	
		16	−68	
3	2	−17	92	289

The last two entries give $P(-2 + j3) = [92s + 289]_{s=-2+j3} = 105 + j276.$

EXERCISE 3.3

Calculate $P(1 + j2)$ and $P(3 + j4)$ for the polynomial

$$P(s) = 3s^5 + 7s^4 - 12s^3 + 20s^2 - 33s + 50.$$

Answer $P(1 + j2) = 173 - j224$, $P(3 + j4) = 14635 - j11800$.

3.3 ROOTS OF QUADRATICS

If the degree of $P(s)$ is two, the resulting quadratic polynomial has the following form (setting $a_2 = 1$)

$$P(s) = s^2 + a_1 s + a_0 \tag{3.3–1}$$

The roots of $P(s)$ can be calculated directly using the well-known formula

$$s = \frac{-a_1 \pm (a_1^2 - 4a_0)^{1/2}}{2} \tag{3.3–2}$$

The roots are real if the discriminant, $a_1^2 - 4a_0$, is non-negative and complex if the discriminant is negative. It usually gives satisfactory results except that numerical problems arise due to catastrophic cancellation when $a_1^2 \gg 4 \mid a_0 \mid$. For example, consider the quadratic

$$s^2 - 10000000.0000001s + 1 \tag{3.3–3}$$

The correct roots of (3.3–3) are 10^7 and 10^{-7}, but some computer floating point systems will give the values of the roots as 10^7 and 0. The error is due to the fact that a_1 and $(a_1^2 - 4a_0)^{1/2}$ are nearly equal so that, with the limited number of digits that can be accommodated on a computer, the difference is shown as zero. A better numerical algorithm for calculating real roots is based on first evaluating the root with the larger magnitude; we consider the sum of the two numbers instead of their difference. The second root is then obtained by using the fact that a_0 is the product of the roots. This algorithm is

$$x_1 = \frac{-a_1 - (a_1^2 - 4a_0)^{1/2}\, sgn(a_1)}{2} \tag{3.3–4}$$

$$x_2 = \frac{a_0}{x_1} \tag{3.3–5}$$

where the function $sgn(z)$ is defined as

$$sgn(z) = \begin{cases} 1 & \text{if } z > 0 \\ 0 & \text{if } z = 0 \\ -1 & \text{if } z < 0 \end{cases}$$

Using this method with (3.3–3), the two roots are found to be 10^7 and 10^{-7}.

3.4 ROOTS OF CUBICS

If we set $n = 3$ in equation (3.1–1) we get a polynomial of degree 3, or a cubic. It will be of the form, with $a_3 = 1$,

$$P(s) = s^3 + a_2 s^2 + a_1 s + a_0 \tag{3.4–1}$$

For real s, with $|s|$ very large, the sign of $P(s)$ is completely determined by the term s^3; hence, $P(x) > 0$ for x large and positive, and $P(x) < 0$ for $|x|$ large and x negative. Because $P(s)$ is a continuous function, there must be at least one real x such that $P(x) = 0$. Similarly, it can be shown that any polynomial of odd degree has at least one real root.

We shall now present *Cardan's method* to find a real root of the cubic (3.4–1). Let

$$Q = 3a_1 - a_2^2 \tag{3.4–2}$$

$$R = \tfrac{1}{2}(9a_1a_2 - 27a_0) - a_2^3 \tag{3.4–3}$$

If $V = (Q^3 + R^2) \geq 0$ then

$$x_1 = \tfrac{1}{3}[(R + \sqrt{V})^{1/3} + (R - \sqrt{V})^{1/3} - a_2]; \tag{3.4–4}$$

otherwise,

$$x_1 = \tfrac{1}{3}[2\,|Q|^{1/2} \cos(\arccos(R/|Q|^{3/2})/3) - a_2]. \tag{3.4–5}$$

The remaining roots are obtained by first dividing the cubic by $(s - x_1)$ and then solving the resulting quadratic.

EXAMPLE 3.5

Consider the cubic $P(s) = s^3 + 8s^2 + 25s + 26$. Using (3.4–2) and (3.4–3) gives $Q = 3 \times 25 - 8^2 = 11$ and $R = 0.5(9 \times 8 \times 25 - 27 \times 26) - 8^3 = 37$. Because $V = (Q^3 + R^2) = 2700$, we have

$$x_1 = \tfrac{1}{3}[(37 + \sqrt{2700})^{1/3} + (37 - \sqrt{2700})^{1/3} - 8] = -2.$$

Dividing $P(s)$ by the factor $(s + 2)$, we get the quadratic $s^2 + 6s + 13$, the roots of which are readily obtained as $-3 \pm j2$.

3.5 ROOTS OF A POLYNOMIAL OF DEGREE *n*

Although a quartic polynomial has roots that can be found by formulas similar to those of Cardan's method for a cubic polynomial, we shall not list them here. Instead, we include degree four polynomials with higher order polynomials.

When the degree of a polynomial $P(s)$ is greater than four there are, in general, no direct methods for determining the roots. The roots must be found by iterative search techniques. Several methods of this type are presented in Chapter 4 in connection with the solution of nonlinear equations. In this section, we shall examine several methods that are used for polynomials. For example, one may use the popular Newton's method for determining the real roots of any polynomial. Complex roots require some modification. Another popular approach is to search for the coefficients p and q of the quadratic $(s^2 + ps + q)$ which will give a pair of complex conjugate roots of the polynomial; this means that a zero remainder will be obtained when $P(s)$ is divided by this quadratic. This is the basis of the Bairstow method. In this section, we shall discuss these two methods along with Laguerre's method which can find both real and complex roots even if we start with a real initial guess. Other methods that are used for polynomial root-finding include Bernoulli's method, Muller's method, and Lehmer's method; the interested reader should consult texts on numerical analysis.

3.5.1 Newton's Method

Newton's method is described in greater detail in Section 4.6. The basic idea is to make an initial guess, for a real root of the polynomial $P(s)$, and then improve it by calculating the values of $P(s)$ and $P'(s)$ for this value of s, where $P'(s)$ is the derivative of $P(s)$ with respect to s. The expression for obtaining the new guess s_{k+1}, given the estimate s_k of the root at the kth iteration, is

$$s_{k+1} = s_k - \frac{P(s_k)}{P'(s_k)} \tag{3.5–1}$$

The values of $P(s_k)$ and $P'(s_k)$ can be readily calculated by using Horner's rule, discussed in Section 3.2.1. The procedure will be explained through the following example.

EXAMPLE 3.6

Consider the cubic $P(s) = s^3 + 7s^2 + 20s + 24$. Differentiating $P(s)$ with respect to s gives us $P'(s) = 3s^2 + 14s + 20$. Let us start with the initial guess $s_0 = 0$. In this case, we have $P(s_0) = 24$ and $P'(s_0) = 20$. Hence, we obtain $s_1 = 0 - 24/20 = -1.2$. Continuing in this manner, we can obtain s_2, s_3, etc. These are shown in Table 3.1.

Table 3.1 EXAMPLE 3.6 ITERATIONS FOR NEWTON'S METHOD

k	s_k	$P(s_k)$	$P'(s_k)$
0	0	24	20
1	−1.2	8.352	7.52
2	−2.310638	2.823969	3.668212
3	−3.080487	−0.415913	5.341382
4	−3.002621	−0.013119	5.010525
5	−3.000003	−0.000015	5.000012
6	−3.000000	0.000000	5.000000

The main drawback with Newton's method is that convergence may be rather slow, especially if the initial guess is far from the actual root. The convergence is fast, as Example 3.6 shows, if the initial guess s_0 is sufficiently close to the root, s_*. Also, we shall never be able to get to the roots if they are complex and our initial guess is real. The reader is referred to Section 4.6 for more discussion of the method.

3.5.2 Bairstow's Method

In Bairstow's method, we start by making an initial guess of a quadratic factor $s^2 + ps + q$. If we divide $P(s)$ by this factor, the remainder will be zero only if we have located the roots correctly. If the remainder is not zero it will be a polynomial of the first degree, which can be represented by $rs + v$. Thus, our problem is to select p and q in such a manner that r and v approach zero simultaneously. Methods for solving systems of nonlinear equations are discussed in detail in Section 4.10. We shall not pursue this method further because it is not as efficient as Laguerre's method.

3.5.3 Laguerre's Method

Laguerre's method has the property of finding complex roots even if we start with a real initial guess. It also converges globally to roots irrespective of the initial guess. Furthermore, the method converges to simple roots (either real or complex) at a cubic rate as compared with quadratic convergence for the more popular Newton's method. The rate of convergence of an iterative method, to a root s_*, is defined as

$$t = \lim_{k \to \infty} \left[\frac{\ell\text{og}(|s_{k+1} - s_*|)}{\ell\text{og}(|s_k - s_*|)} \right]$$

where s_m is the approximation to s_* obtained at the mth iteration; thus, $t = 2$ for Newton's method, and $t = 3$ for Laguerre's method.

Laguerre's method, for a polynomial $P(s)$ of degree n, is

$$s_{k+1} = s_k - \frac{nP(s_k)}{P'(s_k) \pm [H(s_k)]^{1/2}} \tag{3.5–2}$$

$$H(s) = (n - 1)[(n - 1)(P'(s))^2 - nP(s)P''(s)] \tag{3.5–3}$$

where s_k is the kth approximation to the sought-for root of $P(s)$, $P'(s)$ and $P''(s)$ are, respectively, the first and second derivatives of $P(s)$. In equation (3.5–2), the sign in the denominator is chosen so that $|s_{k+1} - s_k|$ is as small as possible.

Starting with any initial guess, repeated applications of (3.5–2) will lead us closer to a root, as tested by the value of $|P(s_{k+1})|$. In general, this value will seldom be exactly zero due to the finite number of digits that can be used on any computer. Hence, we must specify the value of a tolerance, ϵ, which should be utilized in a computer program to terminate the iterations. Thus we get a close approximation to the root when the tolerance is small. After the root (or a close approximation to the root) has been obtained, we deflate the original polynomial. If a real root has been found, we divide by a linear factor and we obtain a polynomial of degree $n - 1$. On the other hand, if a complex root has been found, we divide a real polynomial $P(s)$ by the quadratic whose roots are this root and its complex conjugate. As a result, we obtain a polynomial of degree $n - 2$. In both cases, we may continue the process of root estimation and deflation until the degree is two or less. The situation for complex polynomials is slightly different because complex polynomials do not necessarily have roots occurring in complex conjugate pairs; we can only deflate by a linear factor. It should be noted that the deflation process works best when the smaller magnitude roots are found first. Most polynomial root-finding computer programs try to do this.

Since the process of deflation often leaves a small residual due to the fact that the root may not be exactly determined, it is desirable to "refine" a root with one or two iterations of the algorithm using the full polynomial with the computed "root" as the starting point. The Pascal program in Appendix 3.1 uses Laguerre's method for real polynomials; in addition, it allows the refinement of the roots as suggested above. Furthermore, the program is written so that it runs in an interactive mode; the user can use his or her intuition at various places where the program requests input data during the iterative search for the roots. The program can be adapted for complex polynomials.

EXAMPLE 3.7

Consider the polynomial $P(s) = s^4 + 4s^3 + 14s^2 + 20s + 25$. We shall obtain its roots using Laguerre's method. The first and second derivatives of $P(s)$ are $P'(s) = 4s^3 + 12s^2 + 28s + 20$ and $P''(s) = 12s^2 + 24s + 28$. If we set $s_0 = 0$ then $P(0) = 25$, $P'(0) = 20$, $P''(0) = 28$, and $H(0) = 3[3 \times 20^2 - 4 \times 25 \times 28] = -4800$. Hence,

$$s_1 = s_0 - \frac{nP(s_0)}{P'(s_0) \pm [H(s_0)]^{1/2}} = \frac{-100}{20 \pm [-4800]^{1/2}} = -0.3846 \pm j1.3323.$$

We select the plus sign so that $s_1 = -0.3846 + j1.3323$. Proceeding in this manner, we can calculate s_2, s_3, etc. These are shown in Table 3.2. After nine iterations the method converged to s_9, which is very close to the root $-1 + j2$. Deflating $P(s)$ by dividing by $s^2 + 2s + 5$ we get the polynomial $s^2 + 2s + 5$. Hence, $P(s) = (s^2 + 2s + 5)(s^2 + 2s + 5)$.

3.5.4 The Jenkins-Traub Method

The Jenkins-Traub algorithm for real polynomials is based on generating a sequence of special polynomials, called *K-polynomials*. The algorithm calculates the roots in roughly increasing order of magnitude. The mathematical method on which the algorithm is based is beyond the scope of this book; for further details the reader should examine the paper by Jenkins and Traub. However, Appendix

Table 3.2 EXAMPLE 3.7 ITERATIONS FOR LAGUERRE'S METHOD

k	s_k	$\lvert P(s_k) \rvert$
0	$0.0 + j0.0$	25.00
1	$-0.38462 + j1.33235$	9.467
2	$-0.78947 + j1.82321$	1.108
3	$-0.93985 + j1.95344$	0.09045
4	$-0.98361 + j1.98760$	0.006719
5	$-0.99559 + j1.99668$	0.0004868
6	$-0.99882 + j1.99911$	0.00003503
7	$-0.99968 + j1.99976$	0.000002517
8	$-0.99991 + j1.99994$	0.0000001807
9	$-0.99997 + j1.99998$	0.00000001298

3.2 contains a user's manual for a Pascal procedure called JenkinsTraub, which implements the algorithm. Appendix 3.3 contains a listing of a main Pascal program using the procedure JenkinsTraub. The JenkinsTraub procedure runs in a non-interactive, or batch, mode whereas the Laguerre method program in Appendix 3.1 is interactive. The JenkinsTraub procedure requires the polynomial $P(s)$ to be written in the form

$$P(s) = \sum_{k=0}^{n} a_{n-k}s^k,$$

which is different from that in Eq. (3.1–1). The main program in Appendix 3.3 reads in the coefficients according to (3.1–1) and rearranges them in the correct order for the procedure JenkinsTraub. For a general real polynomial it might be advantageous to compare the roots determined by the Laguerre's method program and the Jenkins-Traub algorithm program.

EXAMPLE 3.8

The stresses at a point in an elastic body are completely determined by six stress components (σ_{11}, σ_{12}, σ_{13}, σ_{22}, σ_{23}, σ_{33}) measured with respect to a fixed axis system. It is possible to do a rigid rotation of the axis system about its origin so that these six stress components resolve into the three principal stress components σ_1, σ_2, σ_3, which are the roots of the polynomial

$$P(s) = s^3 + a_2s^2 + a_1s + a_0$$

where

$$a_2 = -(\sigma_{11} + \sigma_{22} + \sigma_{33})$$

$$a_1 = \sigma_{11}\sigma_{22} + \sigma_{11}\sigma_{33} + \sigma_{22}\sigma_{33} - (\sigma_{12})^2 - (\sigma_{13})^2 - (\sigma_{23})^2$$

$$a_0 = \sigma_{11}(\sigma_{23})^2 + \sigma_{22}(\sigma_{13})^2 + \sigma_{33}(\sigma_{12})^2 - 2\sigma_{12}\sigma_{13}\sigma_{23} - \sigma_{11}\sigma_{22}\sigma_{33}$$

EXAMPLE 3.8 (cont'd)

The following stresses (in units of 10^9 dyn/cm^2) were measured at a point in an elastic body:

$$\sigma_{11} = 3.6, \quad \sigma_{12} = 2.5, \quad \sigma_{13} = -1.5,$$
$$\sigma_{22} = 5.7, \quad \sigma_{23} = 3.9, \quad \sigma_{33} = -1.0$$

With these values of the stress components, the principal stresses are the roots of the polynomial

$$P(s) = s^3 - 8.3s^2 - 12.49s + 111.101$$

The roots, as found by the procedure JenkinsTraub, were

$$x_1 = 3.758581303010\,91, \quad x_2 = -3.621269296153\,44,$$
$$\text{and} \quad x_3 = 8.162687993142\,53.$$

The principal stress components are approximately $\sigma_1 = 3.759$, $\sigma_2 = -3.621$, and $\sigma_3 = 8.163$ in units of 10^9 dyn/cm^2.

3.6 ADDITIONAL TOPICS IN FINDING ROOTS OF POLYNOMIALS

In this chapter we have discussed methods for obtaining the roots of polynomials. For polynomials of degree more than four, direct methods do not exist, and we have to try iterative methods. Both the Laguerre and the Jenkins-Traub methods are quite efficient, but there may be some difficulty due to roundoff errors in evaluating polynomials of large degree. Furthermore, in some cases a small change in the value of a coefficient of the polynomial may result in a large change in the roots. In Chapter 8 we shall discuss an approach which may be more robust.

Sometimes it may be necessary to change variables (for example, replacing s by $10s$ and then dividing out a constant factor) in order that the values of a polynomial $P(s)$ may be calculated with a reasonable accuracy. If a polynomial contains roots lying some distance outside the unit circle in the complex plane, then these roots might be more accurately computed by using the change-of-variable $x = 1/s$ to get $P(s) = x^{-n}Q(x)$, where

$$Q(x) = \sum_{k=0}^{n} a_{n-k}x^k$$

and $P(s)$ is given by (3.1–1). The roots of $Q(x)$ are the reciprocals of the roots of $P(s)$.

We often encounter polynomials with roots that are close together, forming a *root cluster*. It may be very difficult to distinguish a root cluster containing k

roots from a single root of multiplicity k. Roundoff errors may turn a root of multiplicity k into a root cluster, and vice versa. Therefore, the calculated roots should be examined carefully to see if a root cluster is really a multiple root. Another type of situation arises when we compute a root whose real (imaginary) part is very much smaller than its imaginary (real) part, in magnitude; the root may actually lie on the imaginary (real) axis of the complex plane. These cases should be verified.

3.7 PROBLEMS

1. Evaluate the following polynomials for $s = -2, 4, j3$, and $-1 + j2$.

(a) $s^3 + 4s^2 + 5s + 24$

(b) $5s^4 - 12s^3 + 15s^2 + 23s + 45$.

2. The transfer function of a linear system is usually in the form of the ratio of two polynomials in s, the complex frequency variable. The frequency response for a sinusoidal input at the frequency ω can be calculated by evaluating the transfer function for $s = j\omega$. Use synthetic division to calculate the frequency response for a third-order Butterworth filter for ω equal to 0.5, 1, 2, and 5 if the transfer function is given by

$$G(s) = \frac{1}{s^3 + 2s^2 + 2s + 1}$$

3. Use the Pascal program given in the Appendix 3.1 to determine all the roots of each of the following polynomials.

(a) $s^3 + 2s^2 + 2s + 1$

(b) $s^4 + 6s^3 + 13s^2 + 24s + 36$

(c) $s^6 + 6s^5 - 56s^3 - 97s^2 + 26s + 120$

(d) $s^{12} - 0.5s^2 + 1$.

4. Let a real monic polynomial $P(s)$ be given by

$$P(s) = s^n + a_{n-1}s^{n-1} + \cdots + a_1 s + a_0,$$

where $a_0 \neq 0$ and $n > 1$. Show that $P(s)$ must have at least one positive real root, or at least one pair of complex conjugate roots with positive real part, if at least one of the coefficients, a_i, $i > 0$, is zero or negative.

5. For the polynomial $P(s)$ given in the previous problem, show that

(a) the sum of its roots is given by $-a_{n-1}$.

(b) the sum of the reciprocals of its roots is given by $-a_1/a_0$.

6. Let s_0 be a real root of the real polynomial $P(s)$ of multiplicity two (a double root).

(a) Show that s_0 will also be a root of $P'(s)$.

(b) How will the graph of $P(s)$ look near $s = s_0$?

(c) Verify (a) and (b) for the case when $P(s) = s^3 + 7s^2 + 16s + 12$, which has a double root at $s = -2$.

7. The "poles" of a transfer function (see Problem 2), defined as the roots of the denominator polynomial, play an important role in the computation of the transient response of a system. Determine the poles of the transfer functions

(a) $\dfrac{3s + 2}{s^3 + 5s^2 + 11s + 15}$, **(b)** $\dfrac{5s^2 + 14s + 20}{s^4 + 6s^3 + 42s^2 + 78s + 145}$.

8. The eigenvalues of an $n \times n$ matrix, $\mathbf{A}$, are the roots of its characteristic polynomial, defined as $P(s) = \det(s\mathbf{I} - \mathbf{A})$, where $\det(\mathbf{Q})$ denotes the determinant of the $n \times n$ matrix $\mathbf{Q}$. See Appendix A.4 for a discussion of eigenvalues. Find the eigenvalues of the matrix

$$\mathbf{A} = \begin{bmatrix} -7 & -20 & -24 \\ 1 & 0 & 0 \\ 0 & 1 & 0 \end{bmatrix}$$

9. Repeat Problem 8 for

$$\mathbf{A} = \begin{bmatrix} -10 & -53 & -150 & -180 \\ 1 & 0 & 0 & 0 \\ 0 & 1 & 0 & 0 \\ 0 & 0 & 1 & 0 \end{bmatrix}$$

10. A polynomial is said to be *ill-conditioned* when a small change in the value of a coefficient causes a large change in some roots and may even change the nature of some roots (e.g., from real to complex). For example, consider the polynomial

$$P(s) = s^3 + 61s^2 + 1240s + 8400.$$

The roots of $P(s)$ are known to be -20, -20, and -21. Consider a small change in the coefficient 8400 in $P(s)$; make it 8400.1 instead of 8400. What are the roots in this case?

11. A common problem in vibration analysis of a linear system is to determine the natural frequencies from the differential equation describing the system. The following differential equation was obtained while performing the dynamic analysis of the passenger compartment of an automobile,

$$\frac{d^3x}{dt^3} + 0.45\frac{d^2x}{dt^2} + 0.15\frac{dx}{dt} + 0.025x = 0.$$

Calculate the natural frequencies, which will be the roots of the polynomial $P(s) = s^3 + 0.45s^2 + 0.15s + 0.025$.

12. The polynomial

$$P(s) = s^5 + 18s^4 + 176.75s^3 + 481.5s^2 + 2006.5s + 1692$$

was obtained from the dynamic analysis of a beam-block transducer system. Calculate the natural frequencies.

13. Find the roots of each of the following polynomials.

(a) $\sum_{k=0}^{7} s^k$

(b) $s^{12} - 11s^{11} + 54s^{10} - 160s^9 + 330s^8 - 518s^7 + 648s^6 - 652s^5 + 525s^4 - 335s^3 + 162s^2 - 52s + 8$.

(c) $s^8 - 20s^7 + 200s^6 - 1200s^5 + 4708s^4 - 12080s^3 + 20000s^2 - 19200s + 9216$.

(d) $-0.7s^{39} - 3s^{38} + 3.5s^{37} - 1.3s^{36} + s^{27} + 1.3s^{26} - 2s^{25} + 0.6s^{24} + 3s^{15} - 2.2s^{14} + 1.4s^{13} - 0.5s^{12} - 2.5s^3 + 1.8s^2 - s + 0.3$.

(e) $0.55s^6 + 57.18s^5 + 82.15s^4 + 306.23s^3 + 796.76s^2 + 1135.9s + 418.68$.

(f) $s^{10} + (1 - 10^{-10})s^5 - 10^{-10}$.

(g) $1250162561s^{16} + 385455882s^{15} + 845947696s^{14} + 240775148s^{13} + 247926664s^{12} + 64249356s^{11} + 41018752s^{10} + 9490840s^9 + 4178260s^8 + 837860s^7 + 267232s^6 + 44184s^5 + 10416s^4 + 1288s^3 + 224s^2 + 16s + 2$.

(h) $s^{20} + 20s^{19} + 190s^{18} + 1140s^{17} + 4845s^{16} + 15504s^{15} + 38760s^{14} + 77520s^{13} + 125970s^{12} + 167960s^{11} + 184756s^{10} + 167960s^9 + 125970s^8 + 77520s^7 + 38760s^6 + 15504s^5 + 4845s^4 + 1140s^3 + 190s^2 + 20s + 1$.

(i) $\sum_{k=0}^{50} s^k$

3.8 REFERENCES

F. S. Acton. *Numerical Methods that Work*. New York: Harper & Row, 1970.

R. L. Burden, J. D. Faires, and A. C. Reynolds. *Numerical Analysis*. 2d Ed. Boston: Prindle, Weber & Schmidt, 1981.

S. D. Conte and C. de Boor. *Elementary Numerical Analysis*. 3d Ed. New York: McGraw-Hill, 1980.

G. Dahlquist, A. Bjorck, and N. Anderson. *Numerical Methods* Englewood Cliffs, NJ: Prentice-Hall, 1974.

G. E. Forsythe, M. A. Malcolm, and C. B. Moler. *Computer Methods for Mathematical Computations*. Englewood Cliffs, NJ: Prentice-Hall, 1977.

M. A. Jenkins. "Algorithm 493 Zeros of a Real Polynomial [C2]." *A.C.M. Transactions on Mathematical Software* 1 (June 1975):178–189.

M. A. Jenkins and J. F. Traub. "A Three-Stage Algorithm for Real Polynomials Using Quadratic Iteration." *S.I.A.M. Journal on Numerical Analysis* 7 (Dec. 1970):545–566.

J. H. Wilkinson. *Rounding Errors in Algebraic Processes*. Englewood Cliffs, NJ: Prentice-Hall, 1963.

APPENDIX 3.1

User's Manual for the Pascal Program RootsOfRealPolynomials, which is in the file Roots.Pas

This program uses Laguerre's method for finding the roots of a real polynomial of degree higher than two and the quadratic formula if the degree is two. The polynomial is in the form

$$P(s) = \sum_{k=0}^{n} a_k s^k$$

Although the program normally starts with the initial guess of $0 + j0$ for a root, the user may select any other initial guess as the computation proceeds in an interactive manner. The absolute value of the (deflated) polynomial and the current approximation to a root are displayed as the iterations proceed. To avoid errors due to deflation, the roots may be refined using the full polynomial.

The program is interactive so that the user is prompted for any required inputs.

The program uses the files Response.Pas, Files.Pas, and ReadNum.Pas as Include files.

The program contains the following machine dependent constant which is set for Turbo Pascal with 8087 support.

Infinity $= 1.0E+300$

APPENDIX 3.2

User's Manual for the Pascal Procedure JenkinsTraub, which is in the file Jenkins.Pas

This procedure uses the Jenkins-Traub algorithm for finding the roots of a real polynomial. It requires the polynomial to be written as

$$P(s) = \sum_{k=0}^{n} a_{n-k} s^k$$

The ordering of the coefficients is different from that in Eq. (3.1–1). The main program in Appendix 3.3 reads in the coefficients according to (3.1–1) and then rearranges them according to the ordering required by the procedure JenkinsTraub.

```
PROCEDURE JenkinsTraub( degree : INTEGER ; p : Coefficient ;
                          PrintFlag : BOOLEAN ; VAR Re, Im : Coefficient ;
                             VAR count : INTEGER ; VAR OutFile : TEXT );

The following must be declared in the calling program

        TYPE
          Coefficient = ARRAY[ 0..MaxN ] OF REAL;

        where  MaxN  is the maximum order of the polynomials whose roots
        are to be found.

    Input:
      degree     The degree of the polynomial
       p[i]      The coefficient of the (degree - i) power of  s
     PrintFlag   Set to TRUE if the roots are to be sent to the file OutFile
                  as they are found; otherwise, set to FALSE

   Output:
       Re[i]     The real part of the i-th root found
       Im[i]     The imaginary part of the i-th root found
       count     The number of roots found
      OutFile    A file to which the roots can be written as they are found.
                  The user must open OutFile in the calling program, and then
                  close OutFile after the JenkinsTraub procedure is exited.
                  OutFile can be the console, the printer, or a disk file.


Procedure JenkinsTraub contains the following machine dependent constants
which are set for Turbo Pascal with 8087 support.
                    infinity = 1.0E+80
                    smallest = 1.0E-80


Procedure JenkinsTraub is  based  on  the  FORTRAN  subroutine  RPOLY  that
appeared in
 M.A. Jenkins,
   "Algorithm 493  Zeros of a Real Polynomial [C2]"
   ACM Trans. Math. Software  vol. 1  June 1975, pp. 178-189.

The Jenkins-Traub method is described in
  M.A. Jenkins and J.F. Traub,
   "A three-stage algorithm for real polynomials using quadratic iteration"
   S.I.A.M. Journal Numerical Analysis  vol. 7  Dec. 1970, pp. 545-566.
```

APPENDIX 3.3

Listing of a Pascal Program using the Procedure JenkinsTraub

```
PROGRAM RootsOfRealPolynomials;

CONST
  F = 20; (* Fieldwidth *)
  D = 15; (* Number of digits after the decimal point *)
  order = 50;

TYPE
  String80 = String[80];
  String14 = String[14];
  Coefficient = ARRAY[ 0..order ] OF REAL;

VAR
  count, degree : INTEGER;
  p, Re, Im : Coefficient;
  OutFile : TEXT;
  PrintFlag, changed : BOOLEAN;
  answer : CHAR;

(*$I Response.Pas *)
(*$I Files.Pas *)
(*$I ReadNum.Pas *)
(*$I Jenkins.Pas *)

(***************************************************************************)
(*      Module 0   Utility functions and procedures                        *)
(***************************************************************************)

PROCEDURE Greeting;
BEGIN
  ClrScr ; LowVideo ;
  WRITELN('         This program finds the roots of a polynomial');
  WRITELN;
  WRITELN('                           n           n-1   ');
  WRITELN('         P(x) = a[n]*x  + a[n-1]*x    + ... + a[1]*x + a[0]');
  WRITELN;
  WRITELN('  with real coefficients, of degree up to ',order:3,'.',
          '   The roots are');
  WRITELN('  determined by the Jenkins-Traub method.');
  WRITELN;
  WRITELN('  If you want results printed, please turn on the printer.');
  WRITELN;
  WRITE(  '               ');
  Pause
END; (* Greeting *)


PROCEDURE RequestToPrint;
BEGIN
  ASSIGN(OutFile,'Prn') ; REWRITE(OutFile)
END; (* RequestToPrint *)
```

```
(*****************************************************************************)
(*      Module 1   Obtain a polynomial                                       *)
(*****************************************************************************)

PROCEDURE Verify; (* the polynomial's coefficients *)
VAR
  n, k, top, bottom : INTEGER;

PROCEDURE DisplayOneScreen;
VAR
  i : INTEGER;
BEGIN
  ClrScr;
  WRITELN('The coefficients are:');
  WRITELN;
  FOR  i := top  DOWNTO  bottom  DO
    WRITELN('   a[',i,'] = ',p[degree-i]:F:D);
  WRITELN;
  IF  bottom > 0  THEN
    WRITELN('The remaining coefficients will also be verified.')
END; (* DisplayOneScreen *)

PROCEDURE VerifyOneScreen;
VAR
  i, x, y : INTEGER;
  answer : CHAR;
BEGIN
  REPEAT
    WRITE('Are these entries correct?  (Y or N):  ');
    Echo(answer);
    IF  NOT (answer IN ['Y','y'])  THEN
      BEGIN
        changed := TRUE;
        WRITELN('Which entry is wrong?');
        WRITE('   Enter the subscript number:  ');
        REPEAT
          x := WhereX ; y := WhereY ; i := ReadInteger ;
          IF  NOT (i IN [ bottom..top ])  THEN
            ErrorHandler(x,y,7,'Invalid subscript entered')
        UNTIL  i IN [ bottom..top ];
        WRITE('Enter the correct value of  a[',i,']:  ');
        p[degree-i] := ReadReal;
        DisplayOneScreen
      END
  UNTIL  answer IN ['Y','y']
END; (* VerifyOneScreen *)


BEGIN (* Verify *)
  IF  ((degree + 1) MOD 15) = 0  THEN
    n := (degree + 1) DIV 15
  ELSE
    n := ((degree + 1) DIV 15) + 1;
  changed := FALSE;
  FOR  k := 1  TO  n  DO
    BEGIN
      bottom := degree - 15*(k - 1) - 14;
      IF  bottom < 0  THEN
        bottom := 0;
```

```
      top := degree - 15*(k - 1);
      DisplayOneScreen;
      VerifyOneScreen
    END
END; (* Verify *)

(*****************************************************************************)
(*      Module 2   Input and Output                                           *)
(*****************************************************************************)

PROCEDURE PrintCoefficients;
VAR
  i : INTEGER;
BEGIN
  WRITELN(OutFile,'The coefficients of the original polynomial are:');
  WRITELN(OutFile);
  FOR  i := degree  DOWNTO  0  DO
    WRITELN(OutFile,'    a[',i,'] = ',p[degree-i]:F:D);
  WRITELN(OutFile) ; WRITELN(OutFile) ;
  CLOSE(OutFile)
END; (* PrintCoefficients *)

PROCEDURE PutCoefficients;
VAR
  answer : CHAR;
BEGIN
  WRITE('Do you want the coefficients printed?  (Y or N) :  ');
  Echo(answer);
  IF  answer IN ['Y','y']  THEN
    BEGIN
      RequestToPrint;
      PrintCoefficients
    END
END; (* PutCoefficients *)

PROCEDURE GetCoefficients;
VAR
  i, x, y : INTEGER;
BEGIN
  WRITE('Enter the degree of the polynomial:  ');
  REPEAT
    x := WhereX ; y := WhereY ; degree := ReadInteger ;
    IF  (degree < 1) OR (degree > order)  THEN
      ErrorHandler(x,y,7,'Invalid degree entered')
  UNTIL  (1 <= degree) AND (degree <= order);
  FOR  i := degree  DOWNTO  0  DO
    BEGIN
      WRITE('Enter   a[',i,'] :  ');
      IF  i = degree  THEN
        REPEAT
          x := WhereX ; y := WhereY ; p[0] := ReadReal ;
          IF  p[0] = 0.0  THEN
            ErrorHandler(x,y,20,'Leading coefficient is zero')
        UNTIL  p[0] <> 0.0
      ELSE
        p[degree-i] := ReadReal
    END;
  Verify;
  changed := TRUE;
  PutCoefficients
END; (* GetCoefficients *)
```

```
PROCEDURE DataIn; (* the degree and coefficients of the polynomial *)
VAR
  answer : CHAR;
  i : INTEGER;
  FileName : String14;
BEGIN
  ClrScr;
  WRITELN('Do you want to enter the coefficients from the keyboard');
  WRITE('or read them from a file?  Enter  K or F   ');
  Echo(answer);
  IF  answer IN ['F','f']  THEN
    IF  GetFile(OutFile)  THEN
      BEGIN
        degree := -1;
        WHILE  NOT EOF(OutFile)  DO
          BEGIN
            degree := degree + 1;
            READLN(OutFile,p[degree])
          END;
        CLOSE(OutFile);
        Verify;
        PutCoefficients
      END
    ELSE
      answer := 'K';
  IF  NOT (answer IN ['F','f'])  THEN
    GetCoefficients;
  IF  changed  THEN
    BEGIN
      WRITE('Do you want to write the coefficients to a file?  (Y or N)  ');
      Echo(answer);
      IF  answer IN ['Y','y']  THEN
        BEGIN
          CreateFile(OutFile,FileName);
          FOR  i := degree  DOWNTO  0  DO
            WRITELN(OutFile,p[degree-i],'  = a[',i,']');
          CLOSE(OutFile)
        END
    END
END; (* DataIn *)

PROCEDURE DataOut( VAR answer : CHAR );
VAR
  FileName : String14;
BEGIN
  WRITELN;
  WRITELN('Do you want to write the roots  (1) to the Console');
  WRITELN('                                (2) to the Printer');
  WRITELN('                                (3) to a File');
  WRITE('Enter 1, 2, 3 or any other number, if none of these  ');
  Echo(answer);
  IF  answer IN ['1','2','3']  THEN
    BEGIN
      CASE  answer  OF
            '1' : BEGIN
                    ASSIGN(OutFile,'Con') ; REWRITE(OutFile)
                  END;
            '2' : BEGIN
                    ASSIGN(OutFile,'Prn') ; REWRITE(OutFile)
                  END;
```

```
            '3' : CreateFile(OutFile,FileName)
      END;
      ClrScr ; PrintFlag := TRUE
    END
  ELSE
    PrintFlag := FALSE
END; (* DataOut *)

PROCEDURE RootsOut;
VAR
  FileName : String14;
  answer : CHAR;
  i : INTEGER;
BEGIN
  WRITELN;
  WRITE('Do you want to write the roots to a file?  (Y or N)  ');
  Echo(answer);
  IF  answer IN ['Y','y']  THEN
    BEGIN
      CreateFile(OutFile,FileName);
      FOR  i := 1  TO  count  DO
        BEGIN
          WRITE(OutFile,'Root[',i,'] = ',Re[i]);
          IF  Im[i] > 0.0  THEN
            WRITELN(OutFile,' + j ',Im[i])
          ELSE
            IF  Im[i] < 0.0  THEN
              WRITELN(OutFile,' - j ',-Im[i])
            ELSE
              WRITELN(OutFile)
        END;
      IF  count < degree  THEN
        WRITELN(OutFile,'Only ',count,' root(s) were found for the ',
                        'polynomial of degree ',degree);
      CLOSE(OutFile)
    END
END; (* RootsOut *)

(****************************************************************************)
(*      Module 3   Main program and root-finder                              *)
(****************************************************************************)

BEGIN
  Greeting;
  REPEAT
    DataIn ; DataOut(answer) ;
    JenkinsTraub(degree,p,PrintFlag,Re,Im,count,OutFile);
    IF  PrintFlag  THEN
      Close(OutFile);
    IF  NOT (answer IN ['F','f'])  THEN
      RootsOut;
    WRITELN;
    WRITE('Do another polynomial?  (Y or N)  ');
    Echo(answer)
  UNTIL  NOT (answer IN ['Y','y'])
END.
```

Chapter

4

Solution of Nonlinear Equations

In Chapter 3 we considered special methods for solving polynomial equations. In this chapter we develop methods for more general nonlinear algebraic or transcendental equations. In addition, methods for systems of nonlinear equations are discussed. These include Newton's method and quasi-Newton methods.

4.1 INTRODUCTION

Finding the zeros of a function is a common problem in science, engineering, and mathematics. For example, in elementary calculus when we want to find the maximum or minimum of a differentiable function of one variable, $f(x)$, we compute its first derivative, $f'(x)$, and then solve the equation $f'(x) = 0$. If $f'(x)$ is a linear or quadratic polynomial in x then the equation is easily solved. However, if $f'(x)$ is a transcendental function, then we may not have a formula that we can use to find the zeros of $f'(x)$. In these cases we can try to find an approximation to the desired solution by using a numerical method. The particular numerical method we choose to solve our equation should be able to approximate any zero to the accuracy we want with a reasonable amount of computational effort.

In Chapter 3 we discussed methods to solve polynomial equations of the form

$$\sum_{k=0}^{n} a_k x^k = 0$$

One of the methods briefly mentioned was Newton's method, which can also be used for nonpolynomial equations. Other methods from Chapter 3, such as Laguerre's method and the Jenkins-Traub method, are especially designed for polynomials and should be used when the roots of a polynomial are required. In this chapter we concentrate on other types of nonlinear equations.

When we try to solve an equation numerically, we are faced with several problems. These include (1) the need to know a starting estimate for a zero of the function, (2) the need to decide when an approximation to a zero is acceptable, (3) an upper bound on the amount of computational effort to be expended in obtaining a good approximation to a zero, and finally, (4) the choice of a numerical method. The choice of a numerical method often depends on such factors as the properties of the function (such as differentiability), and the user's familiarity with the particular numerical method.

We shall examine several numerical methods for solving equations in one independent variable. In each of these methods we shall use a common notation. $F(x)$ denotes a real-valued function of the real variable x; a zero of $F(x)$ is sought on the interval of real numbers $[a, b]$. It is assumed that $F(x)$ is a continuous function. For those methods that require the first derivative of $F(x)$, we assume that $F(x)$ is differentiable and that its derivative with respect to x is denoted by $F'(x)$. For a method that requires a zero to be within the interval $[a, b]$ we assume that the numbers a and b have been found so that $F(a)F(b) < 0$; that is, $F(a)$ and $F(b)$ have opposite signs. In order to control the computational effort, we associate with each numerical method a stopping criterion, $S(x)$; if $S(x)$ is true then the numerical method is terminated, otherwise it continues. We shall denote the accepted solution to $F(x) = 0$ as z, whereas the true solution is denoted as x_*.

In general, iterative methods for finding a zero of a function, $F(x)$, can be expressed in terms of an iteration formula of the form

$$x_{n+1} = G(x_n, \ldots, x_{n-k}), \qquad n = 0, 1, 2, \ldots \tag{4.1–1}$$

where $x_n, \ldots x_{n-k}$ are previously determined estimates for the zero of $F(x)$. Usually, one restricts the iteration formula so that $k = 0$ or 1. The iteration formulas that we shall examine will be of this type. To use (4.1–1) we know, or are given, $x_n, \ldots, x_{n-k}$; these values are then substituted into (4.1–1) to give the next iterate, x_{n+1}. The indexing parameter, n, is then increased by 1 and (4.1–1) is used again. Although we have not demonstrated it explicitly, G will depend on the function $F(x)$ and on the particular numerical method chosen to find a zero of $F(x)$.

The iteration method given in (4.1–1) will generate an infinite sequence of estimates for the sought-after zero of $F(x)$, x_*. If

$$\lim_{n\to\infty} x_n = x_*$$

we say that the numerical method *converges*; otherwise, the numerical method is said to *diverge*. The convergence or divergence of a particular numerical method depends upon G and F, in addition to the initial guesses $x_0, x_{-1}, \ldots, x_{-k}$. For practical computations, we would not let (4.1–1) generate an infinite sequence; instead, we would terminate the iteration process when the stopping condition, S, was satisfied. This stopping condition might include

1. a bound on the number of evaluations of the function $F(x)$ to restrict the amount of computational effort to be expended
2. a test on the magnitude of the difference between successive iterates; if $| x_{n+1} - x_n | < r$, for some non-negative number r, we accept x_{n+1} as the zero
3. if $| F(x_n) | < s$, for some non-negative number s, we accept x_n as the zero
4. a test to see if the sequence of iterates being generated by equation (4.1–1) is diverging (Is the value of $| x_n - x_0 |$ becoming very large?)

An important property of an iteration method like (4.1–1) is its rate of convergence. The convergence rate of an iterative method is measured by

$$t = \lim_{n\to\infty} \left[\frac{\ell og(E_{n+1})}{\ell og(E_n)} \right] \tag{4.1–2}$$

where $E_n = | x_n - x_* |$, provided the limit exists. The limit number, t, tells us the approximate number of "digits" of accuracy we gain in each iteration of a converging iterative method. Iterative methods with $t = 1$ are called *linear methods*; those with $t = 2$ are called *quadratic methods*. Methods with $1 < t < 2$ are called *superlinear methods*. Equation (4.1–2) is frequently given in the form

$$\lim_{n\to\infty} \left[\frac{E_{n+1}}{E_n^{\,t}} \right] = L \tag{4.1–3}$$

where L is a constant, sometimes called the *asymptotic error constant*. Note that the larger t is, the faster the method converges.

4.2 THE BISECTION METHOD

The method of bisection is the simplest conceptually. We want to determine a real zero of the real-valued continuous function $F(x)$ on the interval $[a, b]$, where it is known that $F(a)F(b) < 0$. Consequently, $F(x)$ has at least one zero on the interval $[a, b]$. At each stage in the bisection method we decrease the length of the interval containing the zero by a factor of two. The method requires knowledge only of $F(x)$ at each stage. The Bisection method can be summarized as follows.

Step 0. Set $a_0 = a$, $b_0 = b$, $n = 0$.
Step 1. Given a_n, b_n such that $F(a_n)F(b_n) \le 0$, choose $x_n = s_1a_n + s_2b_n$, where $s_1 = s_2 = 0.5$. Thus, x_n is the midpoint of the interval $[a_n, b_n]$.
Step 2. Examine the stopping criterion, S. If $S(x_n)$ is true, terminate the iterations and accept x_n as the solution; otherwise, go to Step 3.

Step 3. *If* $F(a_n)F(x_n) \le 0$, set $b_{n+1} = x_n$, $a_{n+1} = a_n$; otherwise set $a_{n+1} = x_n$, $b_{n+1} = b_n$. Replace n by $n + 1$, and go to Step 1.

It is easy to see that $F(x)$ has a zero in the interval $[a_n, b_n]$ for all n. Furthermore, the length of the interval is $(b - a)/2^n$ for $n = 0, 1, 2, \ldots$.

A convenient stopping criterion is the following. Select non-negative numbers r, s, and a positive integer N. Then $S(x_n)$ is true if $p > N$ (where p is the number of evaluations of the function $F(x)$), or $|F(x_n)| < s$, or $|b_n - a_n| \le r$; otherwise $S(x_n)$ is false. The bound, N, is used so that the number of function evaluations does not become too large if $r = 0$.

See Appendix 4.1 for a listing of a Pascal program that uses the Bisection method, as well as other methods, for Example 4.1.

EXAMPLE 4.1

Find a point of intersection of the functions

$$f(x) = 2.4 - x \quad \text{and} \quad g(x) = x^{1.25}$$

for $x > 0$. The graphs of the functions are shown below. It is clear that there is only one intersection point, which we denote as x_*. Hence, $f(x_*) = g(x_*)$. In order to use the Bisection method to find x_* we form the new function

$$F(x) = g(x) - f(x) = x^{1.25} + x - 2.4$$

and observe that $F(x_*) = g(x_*) - f(x_*) = 0$. When we applied the Bisection method with $a = 1.0$, $b = 2.0$, $r = 0$, $s = 10^{-11}$, and $N = 50$ we found that the method had converged after 36 function evaluations, with

$$z = 1.17572106848820, \quad F(z) = 5.05 \times 10^{-12}, \text{ and } |z - x_*| \le 2.91 \times 10^{-11}$$

See Table 4.1.

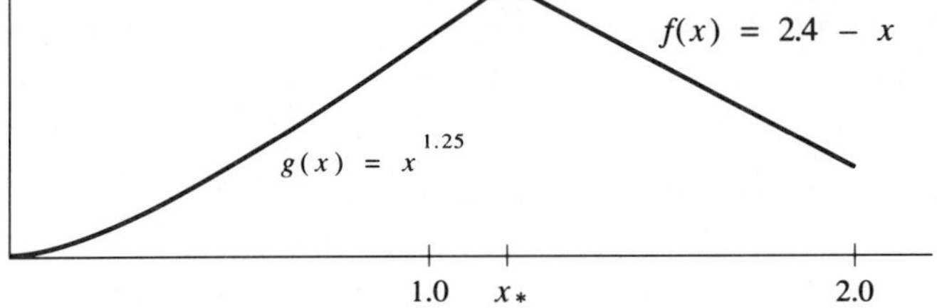

Table 4.1 EXAMPLE 4.1, BISECTION METHOD

Function Evaluation k	x_k	$F(x_k)$	$\mid b_k - a_k \mid$
1	1.0	-0.4	1.0
2	1.5	0.76	0.5
3	1.25	0.1717	0.25
4	1.125	-0.1164	0.125
5	1.1875	2.713×10^{-2}	6.25×10^{-2}
6	1.15625	-4.476×10^{-2}	3.125×10^{-2}
7	1.171875	-8.850×10^{-3}	1.563×10^{-2}
8	1.1796875	9.131×10^{-3}	7.813×10^{-3}
9	1.17578125	1.385×10^{-4}	3.906×10^{-3}
10	1.173828125	-4.356×10^{-3}	1.953×10^{-3}
11	1.1748046875	-2.109×10^{-3}	9.766×10^{-4}
12	1.17529296875	-9.853×10^{-4}	4.883×10^{-4}
13	1.175537109375	-4.234×10^{-4}	2.441×10^{-4}
14	1.1756591796875	-1.424×10^{-4}	1.221×10^{-4}
15	1.17572021484375	-1.965×10^{-6}	6.104×10^{-5}
20	1.17572212219238	2.425×10^{-6}	1.907×10^{-6}
30	1.17572106979787	3.019×10^{-9}	1.863×10^{-9}
36	1.17572106848820	5.048×10^{-12}	2.910×10^{-11}

One advantage of the Bisection method is the automatic error bound one obtains for the accepted solution, z. We are guaranteed that the error at the nth iteration is $\mid x_n - x_* \mid \leq (b - a)/2^n$. Many other methods do not have this property. However, the Bisection method can be very slow to converge, as you can see by Example 4.1.

EXERCISE 4.1

Adapt the program in Appendix 4.1 to find, by the Bisection method, the zero of the function $F(x) = \sin(x) - 1 + \sqrt{x}$ in the interval [0.25, 1]. Use the values of r, s, and N from Example 4.1. Note the number of function evaluations to obtain convergence.

Answer 0.38728606820586 in 37 function evaluations.

4.3 REGULA FALSI (THE METHOD OF FALSE POSITION)

In the Bisection method we select the new point $x_n = 0.5a_n + 0.5b_n$, the average of the two end point values. Instead of using the average we can use the more general weighted average

$$x_n = s_1 a_n + s_2 b_n$$

where $s_1 + s_2 = 1$. The method of false position, Regula Falsi, weights the contribution of either a_n or b_n to x_n according to the value of $F(b_n)$ and $F(a_n)$, respectively. Thus,

$$s_1 = \frac{F(b_n)}{F(b_n) - F(a_n)} \tag{4.3–1}$$

$$s_2 = \frac{-F(a_n)}{F(b_n) - F(a_n)} \tag{4.3–2}$$

Because $F(a_n)F(b_n) \le 0$, we have the two cases: (1) $F(a_n) < 0$ and $F(b_n) \ge 0$, (2) $F(a_n) \ge 0$ and $F(b_n) < 0$. In either case, both s_1 and s_2 are non-negative. Geometrically, we can see that s_1 and s_2 are determined by choosing x_n to be the point of intersection of the line L with the x-axis in Fig. 4.1, where the equation of line L is

$$y = m(x - a_n) + F(a_n) \tag{4.3–3}$$

$$m = \frac{F(b_n) - F(a_n)}{b_n - a_n} \tag{4.3–4}$$

Setting $y = 0$ and $x = x_n$ in (4.3–3), and solving for x_n, gives equations (4.3–1) and (4.3–2). The line L in Fig. 4.1 is a secant line.

The Regula Falsi method produces a sequence of iteration points that lie only

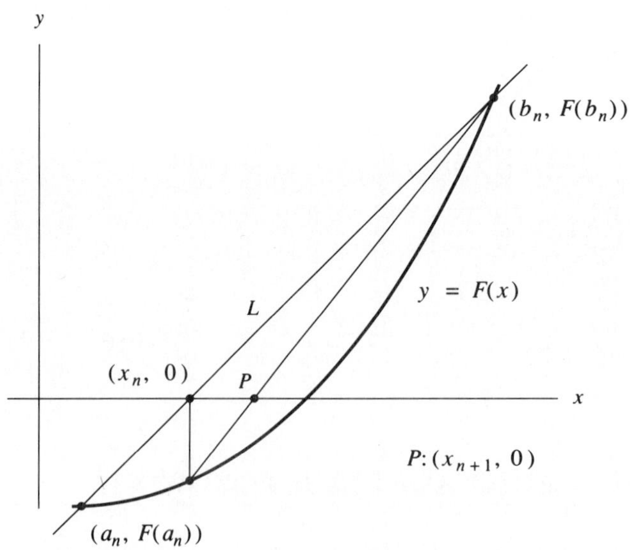

Figure 4.1 Method of Regula Falsi.

Table 4.2 EXAMPLE 4.1, REGULA FALSI METHOD

Function Evaluation k	x_k	$F(x_k)$
1	1.0	-0.4
2	2.0	1.978
3	1.16817928305074	-1.735×10^{-2}
4	1.17541084240940	-7.140×10^{-4}
5	1.17570832980275	-2.932×10^{-5}
6	1.17572054544028	-1.204×10^{-6}
7	1.17572104701000	-4.943×10^{-8}
8	1.17572106760421	-2.030×10^{-9}
9	1.17572106844980	-8.333×10^{-11}
10	1.17572106848452	-3.421×10^{-12}

on one side of the zero of $F(x)$. Although the zero, x_*, lies within the interval $[a_n, b_n]$, the length of the interval does not necessarily tend to zero as $n \to \infty$. Consequently, we do not have a convenient measure of the error, $|x_n - x_*|$, as in the Bisection method. However, the rate of convergence of the Regula Falsi method is one, whereas the Bisection method may converge very much more slowly.

The algorithm describing Regula Falsi is the same as that for the Bisection method, but with s_1 and s_2 determined by (4.3–1) and (4.3–2).

When we used the Regula Falsi method on the function in Example 4.1, with the same initial values and input parameters, we found that the method had converged after $p = 10$ function evaluations, with $z = 1.17572106848452$, $F(z) = -3.42 \times 10^{-12}$. As you can see, the convergence rate of this method is quite noticeable compared with the Bisection method. See Table 4.2.

EXERCISE 4.2

Repeat Exercise 4.1 for the method of Regula Falsi.

Answer 0.38728606820592 in 12 function evaluations.

4.4 MODIFIED REGULA FALSI

In the Regula Falsi method the zero of $F(x)$ is usually approached from one side so that when the method converges to the zero, x_*, we do not have the convenient error estimate that is available with the Bisection method. We want to modify the Regula Falsi method so that it will give error estimates like the Bisection method and will also have superlinear convergence. The modification involves replacing the secant lines used in Regula Falsi by straight lines whose slopes tend to zero, until eventually a point x_n is obtained on the opposite side of the zero of $F(x)$. See Fig. 4.2, which shows an example of the iteration. The algorithm describing the Modified Regula Falsi method is

Step 0. Set $a_0 = a$, $b_0 = b$, $x_0 = a$, $K = F(a)$, $M = F(b)$, $n = 0$.
Step 1. Given a_n, b_n such that $F(a_n)F(b_n) \leq 0$, choose $x_{n+1} = (Ma_n - Kb_n)/(M - K)$.
Step 2. If $F(a_n)F(x_{n+1}) \leq 0$ then set $a_{n+1} = a_n$, $b_{n+1} = x_{n+1}$, $M = F(x_{n+1})$ and if $F(x_n)F(x_{n+1}) > 0$ then replace K by $K/2$; otherwise, set $a_{n+1} = x_{n+1}$, $b_{n+1} = b_n$, $K = F(x_{n+1})$ and if $F(x_n)F(x_{n+1}) > 0$ then replace M by $M/2$. Now $F(x)$ has a zero in the interval $[a_{n+1}, b_{n+1}]$.
Step 3. Examine the stopping criterion, S. If $S(x_{n+1})$ is true, terminate the iterations and accept x_{n+1} as the solution; otherwise, go to Step 4.
Step 4. Replace n by $n + 1$, and go to Step 1.

The computations in Step 2 are designed to prevent the retention of a point which causes the linear convergence of the Regula Falsi method; consequently, the Modified Regula Falsi method will have a superlinear convergence rate of approximately $3^{1/3} \simeq 1.442$.

When the Modified Regula Falsi method was used to find the zero of the function in Example 4.1, with the same initial conditions and parameters, it was

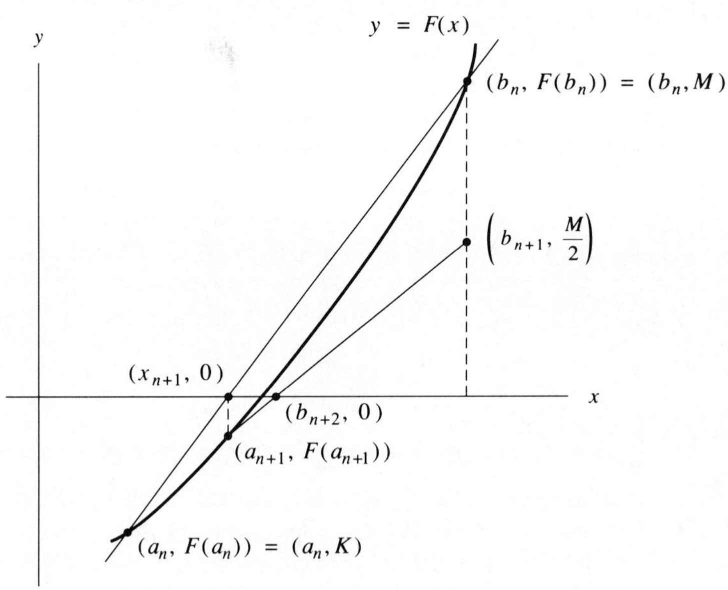

Figure 4.2 Modified Regula Falsi method.

Table 4.3 EXAMPLE 4.1, MODIFIED REGULA FALSI METHOD

Function Evaluation k	x_k	$F(x_k)$	$\lvert b_k - a_k \rvert$
1	1.0	-0.4	1.0
2	2.0	1.978	1.0
3	1.16817928305074	-1.735×10^{-2}	0.8318
4	1.18251774815055	1.565×10^{-2}	1.434×10^{-2}
5	1.17571798587716	-7.095×10^{-6}	6.800×10^{-3}
6	1.17572106722862	-2.894×10^{-9}	6.797×10^{-3}
7	1.17572106974237	2.892×10^{-9}	2.514×10^{-9}
8	1.17572106848601	2.220×10^{-16}	1.257×10^{-9}

found that convergence was achieved after eight function evaluations, with

$$z = 1.17572106848601, \quad F(z) = -2.22 \times 10^{-16} \text{ and } \lvert z - x_* \rvert \leq 1.26 \times 10^{-9}$$

See Table 4.3.

EXERCISE 4.3

Repeat Exercise 4.1 for the Modified Regula Falsi method.

Answer 0.38728606820556 in 9 function evaluations.

4.5 SECANT METHOD

The Secant method is a form of Regula Falsi. When it converges, it can locate rapidly a point where $\lvert F(x) \rvert$ is small. The method does not require the values of $F(x)$ at successive iterates to be of opposite signs, nor does it give an interval in which the zero is guaranteed to lie. In addition, the method requires two starting points, x_{-1} and x_0. The successive iterates are found from

$$x_{n+1} = x_n - m^{-1}F(x_n) \tag{4.5–1}$$

$$m = \frac{F(x_n) - F(x_{n-1})}{x_n - x_{n-1}} \tag{4.5–2}$$

Equation (4.5–1) is obtained from (4.3–3) and (4.3–4) by replacing a_n by x_n and b_n by x_{n-1}. The number m in (4.5–2) is the slope of the secant line joining the point $(x_{n-1}, F(x_{n-1}))$ to the point $(x_n, F(x_n))$; see Fig. 4.3. If $F(x)$ is differentiable, m is the value of the derivative of $F(x)$ at some point between x_{n-1} and x_n; this follows from the Mean Value Theorem.

The Secant method algorithm is

Step 0. Select x_{-1} and x_0. Set $n = 0$.
Step 1. Given x_{n-1} and x_n, compute x_{n+1} using (4.5–1).
Step 2. Examine the stopping criterion, S. If $S(x_{n+1})$ is true, terminate the iterations and accept x_{n+1} as the solution; otherwise, go to Step 3.
Step 3. Replace n by $n + 1$, and go to Step 1.

Figure 4.3 shows an example of the Secant method iteration.

If $F(x)$ has a continuous second derivative, then it can be shown that

$$E_{n+1} = LE_nE_{n-1} \qquad \text{as} \quad n \to \infty \tag{4.5–3}$$

where $L = [F'(x_*)]^{-1}F''(x_*)/2$ and $E_n = |x_n - x_*|$. To determine the rate of convergence, we assume that $E_{n+1} = kE_n^{\,t}$. Substitute this into (4.5–3), use $E_{n-1} = k^{-1/t} E_n^{\,1/t}$, and equate the exponents of E_n, to get the following equation for t,

$$t = 1 + \frac{1}{t} \tag{4.5–4}$$

The positive zero of (4.5–4) is $(1 + \sqrt{5})/2 \simeq 1.618$; hence, the Secant method is superlinear.

The Secant method was applied to the function in Example 4.1, with $x_{-1} = 1.0$ and $x_0 = 2.0$. The method converged after $p = 6$ function evaluations with $z = 1.17572106848338$ and $F(z) = -6.05 \times 10^{-12}$. This example shows how

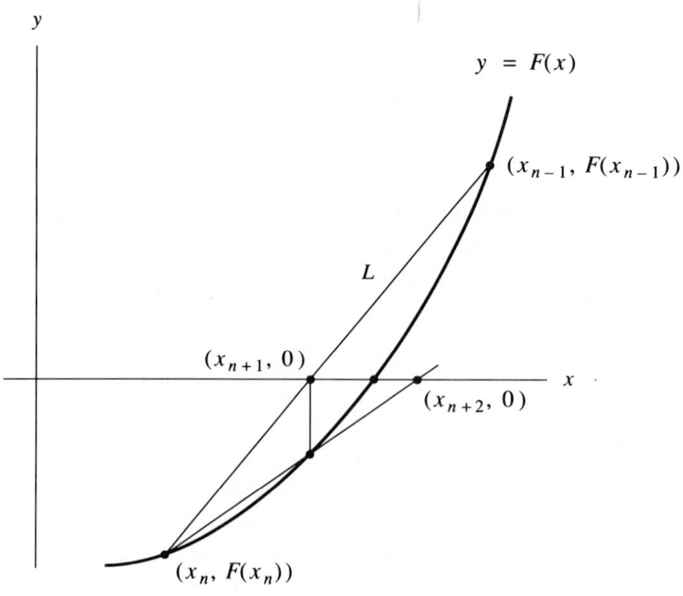

Figure 4.3 Secant method.

Table 4.4 EXAMPLE 4.1, SECANT METHOD

Function Evaluation k	x_k	$F(x_k)$
1	1.0	-0.4
2	2.0	1.978
3	1.16817928305074	-1.735×10^{-2}
4	1.17541084240940	-7.140×10^{-4}
5	1.17572120946083	3.245×10^{-7}
6	1.17572106848338	-6.053×10^{-12}

rapidly an iterative method can converge if the initial guesses, x_{-1} and x_0, are sufficiently close to the solution. See Table 4.4.

As we have shown above, the factor m in (4.5–1) can be thought of as the derivative of $F(x)$ at some point. Newton's method, discussed in the next section, uses the derivative of $F(x)$ in the iteration formula.

EXERCISE 4.4

Repeat Exercise 4.1 for the Secant method.

Answer 0.38728606820556 in 8 function evaluations.

4.6 NEWTON'S METHOD (NEWTON-RAPHSON METHOD)

When $F(x)$ is a differentiable function we can use Newton's method

$$x_{n+1} = x_n - [F'(x_n)]^{-1}F(x_n) \tag{4.6–1}$$

Comparing (4.6–1) to (4.5–1) we see that the slope of the secant line, m, in (4.5–1) has been replaced by the slope of the curve $y = F(x)$ at $x = x_n$ (see Fig. 4.4). One advantage of Newton's method is its quadratic convergence property when the iterates converge to a simple zero. A *simple zero* of $F(x)$ is one where $F(x_*) = 0$, but $|F'(x_*)| > 0$. A disadvantage of Newton's method is the need to compute the derivative at each iteration. For certain problems, $F'(x)$ may not be easily obtained; in these cases the Secant method or the Modified Regula Falsi method may be more practical. Another disadvantage of Newton's method is the requirement that x_0 be sufficiently close to x_* in order for the method to converge, provided that certain conditions are satisfied (see Theorem 4.1, below).

If $F(x)$ has a continuous second derivative, we can obtain the rate of convergence of Newton's method as follows. Use Taylor's formula to expand $F(x_*)$

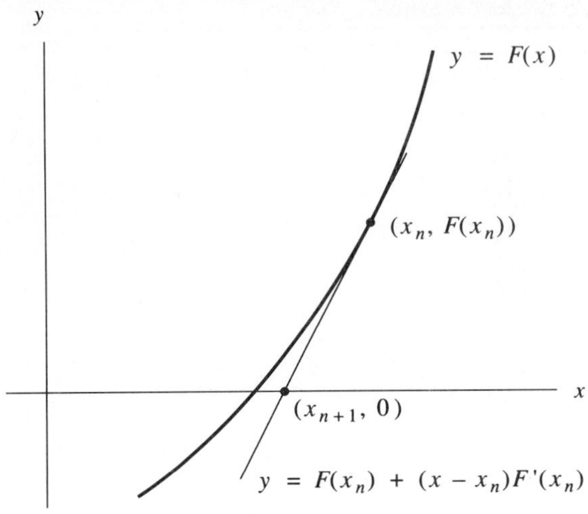

Figure 4.4 Newton's method.

about x_n; this gives

$$0 = F(x_n) + F'(x_n)(x_* - x_n) + 0.5F''(\xi)(x_* - x_n)^2$$

where ξ is between x_n and x_*. Solve for $[F'(x_n)]^{-1}F(x_n)$ and substitute into (4.6–1) after subtracting x_* from both sides of (4.6–1); this gives

$$E_{n+1} = L(E_n^2) \qquad \text{as} \quad n \to \infty$$

where $E_n = |x_n - x_*|$ and $L = [F'(x_*)]^{-1}F''(x_*)/2$. Therefore, Newton's method is quadratic provided x_* is a simple zero of $F(x)$ and $F''(x_*)$ is bounded. These conditions are in terms of the zero x_* which is not usually known when we want to apply Newton's method. However, convergence of Newton's method is guaranteed if the conditions of the following theorem are satisfied.

Theorem 4.1

Let $F(x)$ be twice continuously differentiable. Suppose there exist non-negative constants A, B, C, and h defined by

1. $|[F'(x_0)]^{-1}| \leq A$
2. $|[F'(x_0)]^{-1}F(x_0)| \leq B$
3. $|F''(x)| \leq C$ for all x satisfying the inequality in Condition 5
4. $h = ABC \leq 0.5$
5. $|x - x_0| \leq QB/h$, where $Q = 1 - (1 - 2h)^{1/2}$.

Then, Newton's method, (4.6–1), converges and

$$| x_k - x_* | \leq \frac{(2h)^m B}{2^{k-1}}, \qquad \text{where } m = 2^k - 1.$$

Condition 1 of the theorem excludes multiple zeros of $F(x)$. Although the theorem only gives sufficient conditions for convergence, Newton's method often converges when some of the conditions of the theorem are not satisfied.

Because $F'(x)$ is used in the iteration formula (4.6–1), Newton's method does not retain its quadratic convergence for multiple zeros. (x_* is a *multiple zero of order*, or *multiplicity*, m if $F(x_*)$ and all its derivatives up to the $(m - 1)$th vanish at x_*, but the mth derivative is nonzero.) If it is known that $F(x)$ has a multiple zero at x_*, then we can define $H(x)$ as

$$H(x) = [F'(x)]^{-1}F(x) \tag{4.6–2}$$

$H(x)$ has a simple zero at x_*. Consequently, (4.6–1) can be used with the function $H(x)$.

Frequently, one makes a poor initial guess for x_0, and Newton's method diverges. The method can be modified to try to improve its performance in this case. We change (4.6–1) so that it looks like

$$x_{n+1} = x_n - c[F'(x_n)]^{-1}F(x_n) \tag{4.6–3}$$

where c is chosen so that $| F(x_{n+1}) | < | F(x_n) |$. A convenient choice for c is $c = 2^{-k}$, where $k = 0, 1, 2, \ldots$. The integer k is chosen to be as small as possible so that the magnitude of the function being zeroed is decreased between successive iterations. This technique, (4.6–3), may fail due to numerical inaccuracies in computing $F(x)$ and $[F'(x)]^{-1}$; we could find a valuc of c so small that $| x_{n+1} - x_n | < r$ and $| F(x_n) |$ is not small. If this occurs, we could start Newton's method with a different initial x_0, or we could try a different numerical method.

Newton's method can fail due to a poor choice of x_0. Fig. 4.5 shows two kinds of failures; an oscillation failure that cycles on the two-term sequence $\{x_0, x_1\}$ and a diverging sequence due to very small values of $| F'(x_n) |$. This latter type of failure might be avoided by using (4.6–3).

There is another form of Newton's method that is often used,

$$x_{n+1} = x_n - qF(x_n) \tag{4.6–4}$$

where $q = [F'(x_0)]^{-1}$. This modified Newton's method has a convergence rate of one; however, it does have the advantage of not requiring the value of $F'(x)$ at each iteration.

In each of the other methods, we used the number of function evaluations as a measure of the computational effort. Newton's method, (4.6–1), requires the evaluation of both $F(x)$ and $F'(x)$ at each iteration. We must also count the derivative evaluations in Newton's method in order to get a fairer comparison with the other methods. Generally, the evaluation of the derivative requires approximately the same amount of computational effort as the evaluation of the function.

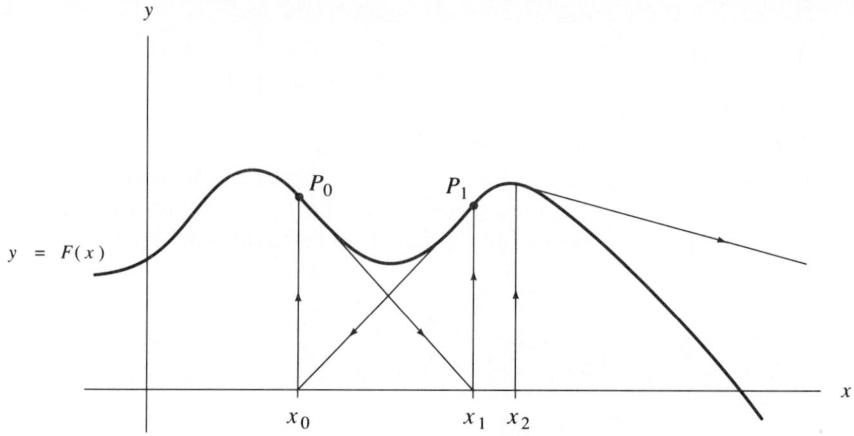

Figure 4.5 Some ways that Newton's method might fail. **(a)** Oscillation between points P_0 and P_1. **(b)** Divergence because $|F'(x_2)|$ is too small.

Consequently, a more accurate measure of the amount of computational effort for Newton's method would be obtained by doubling the number of function evaluations. We shall refer to this as the "effective number of function evaluations."

When Newton's method, (4.6–1), was applied to the function in Example 4.1, convergence was obtained for $p = 4$ function evaluations, with $z = 1.17572106848601$ and $F(z) = 8.88 \times 10^{-15}$ for $x_0 = 1.0$. Hence, the effective number of function evaluations is 8. See Table 4.5.

EXERCISE 4.5

Repeat Exercise 4.1 for Newton's method.

Answer 0.38728606820556 in 10 effective function evaluations.

If $F(x)$ has complex zeros, then Newton's method can be modified to find them. This requires replacing x by the complex variable z, which makes $F(z)$ and

Table 4.5 EXAMPLE 4.1, NEWTON'S METHOD

Iteration k	x_k	$F(x_k)$	$F'(x_k)$
1	1.5	0.76	2.383
2	1.18111183236271	1.241×10^{-2}	2.303
3	1.17572281062578	4.010×10^{-6}	2.302
4	1.17572106848619	4.203×10^{-13}	2.302

$F'(z)$ complex numbers. A computer program implementing Newton's method would have to be coded to deal with complex numbers.

4.7 THE FIXED-POINT METHOD

Suppose we have a function $G(x)$ such that for x in the interval $[a, b]$, $G(x)$ is in the same interval. Clearly, the graph of $G(x)$ lies inside a square of side length $|b - a|$ in the (x, y)-plane. Because of the restrictions on $G(x)$, it is clear that the function $H(x) = x - G(x)$ has the property that $H(a) \leq 0$, because $a \leq G(x)$, and $H(b) \geq 0$, because $G(x) \leq b$. If $G(x)$ is a continuous function, $H(x)$ must have a zero in the interval $[a, b]$ because $H(a)H(b) \leq 0$. If we denote this zero by x_*, then we see that

$$x_* = G(x_*) \tag{4.7–1}$$

The point x_* is called a *fixed point* of $G(x)$, and $G(x)$ is said to have the fixed point property. Equation (4.7–1) suggests an iterative method similar to (4.1–1),

$$x_{n+1} = G(x_n) \tag{4.7–2}$$

If we assume that $G(x)$ is also differentiable, then an error analysis of (4.7–2) shows that if there is a non-negative number v such that $v < 1$, and

$$\max\{|G'(x)|\} \leq v < 1, \qquad \text{where } x \text{ is in } [a, b] \tag{4.7–3}$$

then the iteration scheme (4.7–2) converges to the fixed point of $G(x)$, provided x_0 is in the interval $[a, b]$. For this reason, $G(x)$ is sometimes called a *contraction mapping*. Fig. 4.6 shows several iterations of the Fixed-Point method.

We observe that Newton's method can be considered as a special case of the Fixed-Point method with $G(x) = x - [F(x)]^{-1}F(x)$.

A weaker condition than (4.7–3) is the following, which does not require $G(x)$ to be differentiable,

$$|G(x) - G(y)| \leq v|x - y| \quad \text{for all } x, y \text{ in } [a, b] \tag{4.7–4}$$

and $0 \leq v < 1$.

The problem with using the Fixed-Point method to find a zero of a function $F(x)$ is to pick a suitable function $G(x)$ with the fixed point property. One method for obtaining an iteration function, when $F(x)$ is differentiable, is to define $G(x)$ as

$$G(x) = x - cF(x) \tag{4.7–5}$$

where c is chosen so that $|G'(x)| \leq v < 1$ for all x in $[a, b]$.

If $G(x)$ is differentiable, we can obtain the convergence rate of the Fixed-Point method by subtracting (4.7–2) and (4.7–1), then using the Mean Value Theorem to get

$$x_{n+1} - x_* = G'(\xi)(x_n - x_*)$$

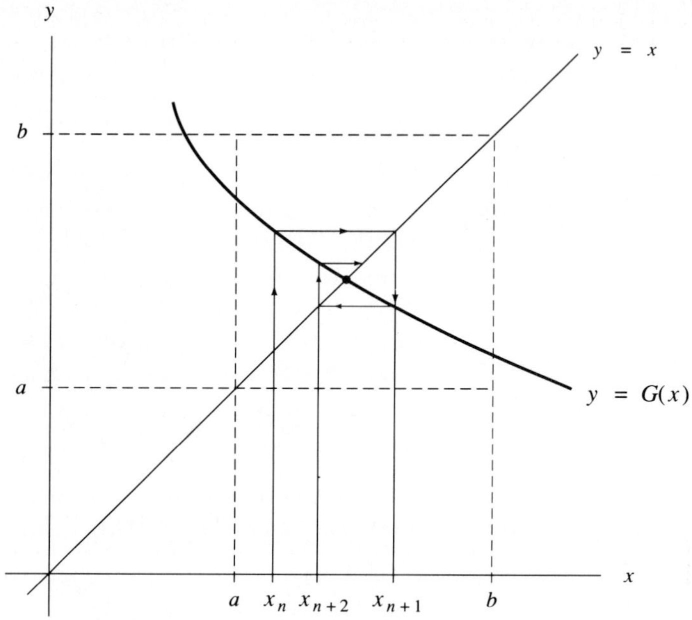

Figure 4.6 Fixed-Point method.

where ξ is between x_n and x_*. Using (4.7–3) we have

$$E_{n+1} \leq vE_n, \quad \text{where } E_n = |x_n - x_*|$$

Consequently, the convergence rate for the Fixed-Point method is one. Furthermore,

$$E_n \leq v^n E_0 \quad \text{and} \quad E_n \leq \frac{v^n |x_1 - x_0|}{1 - v}$$

This shows how slowly the Fixed-Point method converges if v is close to 1.

To construct $G(x)$ for the function in Example 4.1 according to Eq. (4.7–5), we see that for $c = 0.5$,

$$G'(x) = 0.5 - 0.625x^{0.25}, \quad \text{so that} \quad |G'(x)| < 0.25 \quad \text{for} \quad x \text{ in } [1, 2].$$

Equation (4.7–5), with $c = 0.5$, has the fixed point property. With $c = 0.5$ and $x_0 = 1.0$, the method converged after $p = 14$ function evaluations with $z = 1.1757210684855$ and $F(z) = 3.9 \times 10^{-12}$. See Table 4.6.

EXERCISE 4.6

Repeat Exercise 4.1 for the Fixed-Point method, using (4.7–5) with $c = 0.5$.

Answer 0.38728606820585 in 14 function evaluations.

Table 4.6 EXAMPLE 4.1, FIXED-POINT METHOD

Function Evaluation k	x_k	$x_k - G(x_k)$
1	1.5	0.38
2	1.11998856022476	-6.392×10^{-2}
3	1.18390872238613	9.427×10^{-3}
4	1.17448163491212	-1.426×10^{-3}
5	1.17590788462400	2.150×10^{-4}
6	1.17569289180455	-3.243×10^{-5}
7	1.17572531783558	4.890×10^{-6}
8	1.17572042762841	-7.375×10^{-7}
9	1.17572116513551	1.112×10^{-7}
10	1.17572105391002	-1.677×10^{-8}
11	1.17572107068425	2.530×10^{-9}
12	1.17572106815448	-3.815×10^{-10}
13	1.17572106853601	5.754×10^{-11}
14	1.17572106847847	-8.678×10^{-12}

4.8 AN EFFICIENT METHOD FOR SOLVING NONLINEAR EQUATIONS

The computational methods given in Sections 4.2–4.7 show that the most computationally efficient methods seem to be ones that do not require the derivative of thc function. The Pascal procedure, ZeroIn, given in Appendix 4.2 is a very efficient method for finding a zero of a nonlinear function.

ZeroIn is based on an algorithm of Brent. ZeroIn carries out an iteration in which three values of x are given, $\{a, b, c\}$ where b is the latest iterate and the closest approximation to the zero x_*, a is the previous iterate, and c is an older iterate such that $F(b)F(c) \leq 0$. At every step the zero x_* is within the interval $[\min(b, c), \max(b, c)]$, with $|F(b)| \leq |F(c)|$. The user supplies an error tolerance, $r \geq 0$, such that when

$$|b - c| \leq r + 4u|b|$$

where u is the machine epsilon (see Chapter 1), then b is returned as the estimated zero.

Given $\{a, b, c\}$, ZeroIn selects the next iterate from two possible candidates: one is obtained from the Bisection method and the other from an interpolation method. The main interpolation scheme used is Inverse Quadratic Interpolation, which is a method whereby the data points $\{a, F(a)\}$, $\{b, F(b)\}$, $\{c, F(c)\}$ are fitted to a quadratic polynomial. If the quadratic approximates $F(x)$ fairly closely on the interval containing $\{a, b, c\}$, then a root of the quadratic should be a good approximation to x_*, the zero of $F(x)$. However, using inverse quadratic interpolation requires that the points $\{a, b, c\}$ be distinct; this may not always be true. In the cases where $\{a, b, c\}$ are not distinct, ZeroIn uses linear interpolation in

Table 4.7 EXAMPLE 4.1, ZEROIN

Function Evaluation k	x_k	$F(x_k)$
1	1.0	−0.4
2	2.0	1.978
3	1.16817928305074	-1.735×10^{-2}
4	1.17573874311257	4.068×10^{-5}
5	1.17572106045492	-1.848×10^{-8}
6	1.17572106848600	-1.976×10^{-14}
7	1.17572106848601	2.220×10^{-16}

the form of the Secant method to estimate the next iterate. Because the Bisection method may converge very slowly, we would want to select the next iterate from the interpolation methods. However, the interpolation methods could give an estimate that is outside the interval [min(b, c), max(b, c)], or they could give a point that is too close to the end points of the interval. In these cases, ZeroIn would use the estimate obtained from the Bisection method.

For a function $F(x)$ with a continuous second derivative near a simple zero, ZeroIn converges to the zero at a rate at least $(1 + \sqrt{5})/2$.

When ZeroIn was used on the function in Example 4.1, with $a = 1.0$, $b = 2.0$, $r = 0$, the method converged to $z = 1.17572106848601$ with $F(z) = 2.22 \times 10^{-16}$ and $|z - x_*| < 1.1 \times 10^{-15}$. This required eight function evaluations. See Table 4.7.

In terms of computational efficiency and accuracy of approximation, ZeroIn is clearly the method of choice (see Example 4.3).

Appendix 4.1 contains a listing of the Pascal program used to obtain the numerical results cited in Sections 4.2–4.8.

EXERCISE 4.7

Repeat Exercise 4.1 using the ZeroIn procedure.

Answer 0.38728606820556 in 7 function evaluations.

4.9 SOME EXAMPLE PROBLEMS

EXAMPLE 4.2

The equation

$$x - e{\cdot}\sin(x) = m(t - t_0) \tag{4.9–1}$$

where $0 < e < 1$, is known as *Kepler's equation*. It gives the position of a point on an ellipse in terms of an angle x, measured between two fixed lines that pass

EXAMPLE 4.2 (cont'd)

through the center of the ellipse, given m, t, and t_0; m, t_0, and e are constants (e is the eccentricity of the ellipse). The variable t measures time. If the ellipse describes a planetary orbit under Newtonian gravitation, then x is called the *eccentric anomaly*, and $m(t - t_0)$ is called the *mean anomaly*. With e, m, and t_0 known, we can find x for any given value of t.

Using the values $e = 0.1$ and $m(t - t_0) = 5.0$, it is easy to see that the function

$$F(x) = x - 5.0 - 0.1 \sin(x) \tag{4.9–2}$$

has a zero in the interval [4, 5]. The function

$$G(x) = 5.0 + 0.1 \sin(x) \tag{4.9–3}$$

has the fixed point property, because $|G'(x)| = 0.1 |\cos(x)| \leq 0.1$. Taking $a = 4.0$, $b = 5.0$, $N = 50$, $r = 0.0$, $s = 10^{-11}$ we solved the problem using each of the methods described in Sections 4.2–4.8. For Newton's method and the Fixed-Point method we chose x_0 as $(a + b)/2$. The results are summarized in Table 4.8.

Table 4.8 COMPUTATIONAL RESULTS FOR EXAMPLE 4.2; p = NUMBER OF FUNCTION EVALUATIONS

Method	z	$\lvert F(z)\rvert$	$\lvert b_p - a_p\rvert$	p
Bisection	4.90178824885515	3.72×10^{-12}	5.82×10^{-11}	36
Regula Falsi	4.90178824886005	1.08×10^{-12}		10
Modified RF	4.90178824885896	1.19×10^{-14}	1.65×10^{-4}	7
Secant	4.90178824885895	4.03×10^{-15}		6
Newton	4.90178824885895	4.67×10^{-13}		4
Fixed-Point	4.90178824885897	1.07×10^{-12}		7
ZeroIn	4.90178824885895	3.33×10^{-16}		7

EXAMPLE 4.3

The equation $F(x) = 0$, where $F(x) = (\sqrt{x})\,[1.8\,\ell og(R\sqrt{x}) - 0.5] - 1$, is obtained from a problem in turbulent flow of a fluid in a smooth pipe. The number R is the Reynolds number and x is the friction factor. For $R = 10^4$, $F(x)$ has a zero in the interval $[10^{-5}, 0.5] = [a, b]$. Using $r = 0.0$, $s = 10^{-11}$, $N = 50$ we found that Newton's method required 18 effective function evaluations to converge to $z = 0.00735805675670000$ with $F(z) = 2.86 \times 10^{-13}$; ZeroIn required eleven function evaluations to converge to essentially the same answer. However, the Secant method diverged; a negative value of x was found after iteration 1 was completed.

EXAMPLE 4.4

The infinite series

$$\left(\frac{x}{2}\right)^q \left[\frac{1}{q!} - \frac{(x/2)^2}{(q+1)!} + \frac{(x/2)^4}{2!(q+2)!} - \frac{(x/2)^6}{3!(q+3)!} + \cdots\right]$$

defines a function, $J_q(x)$, called the *Bessel function* of order q, where q is a nonnegative integer. Bessel functions commonly occur in the solutions of certain differential equations. The Bessel function $J_q(x)$ is known to have an infinite number of zeros. The first three positive zeros of $J_2(x)$ lie within the intervals [5, 6], [8, 9], and [11, 12]. Using ZeroIn, with $r = 0$, the results shown in Table 4.9 were obtained.

Table 4.9 COMPUTATIONAL RESULTS FOR EXAMPLE 4.4

z	$\lvert J_2(z) \rvert$	p
5.13562230184068	3.77×10^{-16}	8
8.41724414039989	6.31×10^{-16}	8
11.6198411721492	4.42×10^{-14}	12

See Appendix 4.3 for a listing of the Pascal function used to evaluate the Bessel function $J_q(x)$.

4.10 SYSTEMS OF NONLINEAR EQUATIONS

A system of nonlinear equations can be written in the form $\mathbf{f}(\mathbf{x}) = \mathbf{0}$, where $\mathbf{f}^T = (f_1, f_2, \ldots, f_n)$, $\mathbf{x}^T = (x_1, x_2, \ldots, x_n)$, and $f_i = f_i(x_1, \ldots, x_n)$; the superscript T denotes transposition. Thus, $\mathbf{f}$ and $\mathbf{x}$ are column vectors, and we now have to solve n equations in n unknowns. Methods such as ZeroIn cannot be used because they are based on the particular properties of one function, which may not hold for the n functions $f_1, \ldots, f_n$ considered simultaneously. The Secant method, Newton's method, and the Fixed-Point method can be generalized for systems of equations. We shall give a brief discussion of Newton's method for systems of equations.

Newton's method for systems of equations is

$$\mathbf{x}_{k+1} = \mathbf{x}_k - [\mathbf{J}(\mathbf{x}_k)]^{-1}\mathbf{f}(\mathbf{x}_k), \qquad k = 0, 1, \ldots \tag{4.10–1}$$

where $\mathbf{x}_k$ is the kth iterate of the vector $\mathbf{x}$, and $\mathbf{J}(\mathbf{x})$ is the Jacobian matrix of $\mathbf{f}(\mathbf{x})$. $\mathbf{J}(\mathbf{x})$ is an $n \times n$ matrix whose (i, j)th entry is the partial derivative of f_i with respect to x_j. If the initial guess, $\mathbf{x}_0$, is sufficiently close to the zero, then (4.10–1) will give a sequence of vectors that converges quadratically to the zero. The theorem in Section 4.6 applies to systems of equations provided the absolute values are replaced by appropriate vector and matrix norms (see Section 2.5). Because the theorem gives sufficient conditions for the quadratic convergence of Newton's method, the method may still converge if the theorem conditions are not satisfied if $\mathbf{x}_0$ is a "good" approximation to a zero of $\mathbf{f}(\mathbf{x})$. Of course, it may be very difficult to find a good starting value of $\mathbf{x}_0$ so that Newton's method converges. Another associated difficulty is the need to find the Jacobian matrix $\mathbf{J}(\mathbf{x})$ and to solve the linear system of equations

$$\mathbf{J}(\mathbf{x}_k)\mathbf{d}_k = -\mathbf{f}(\mathbf{x}_k) \tag{4.10–2}$$

for the correction $\mathbf{d}_k$ to be added to $\mathbf{x}_k$ to obtain $\mathbf{x}_{k+1}$. Methods of solving (4.10–2) were studied in Chapter 2. However, finding $\mathbf{J}(\mathbf{x})$ may be a formidable task, especially if $f_1, f_2, \ldots, f_n$ are complicated functions.

Newer methods have been proposed whereby the Jacobian matrix $\mathbf{J}(\mathbf{x})$ is not computed at each iteration; one uses an approximation for the Jacobian matrix that is updated at each iteration. These are known as *quasi-Newton* methods. In a quasi-Newton method we denote an approximation to $\mathbf{J}(\mathbf{x}_k)$ as $\mathbf{A}_k$. Usually we take $\mathbf{A}_0 = \mathbf{J}(\mathbf{x}_0)$. The matrix $\mathbf{A}_k$ is chosen so that

$$\mathbf{A}_k\mathbf{s}_k = \mathbf{y}_k \tag{4.10–3}$$

where

$$\mathbf{s}_k = \mathbf{x}_k - \mathbf{x}_{k-1}$$

$$\mathbf{y}_k = \mathbf{f}(\mathbf{x}_k) - \mathbf{f}(\mathbf{x}_{k-1})$$

Eq. (4.10–3) does not define $\mathbf{A}_k$ uniquely because we do not know what $\mathbf{A}_k$ does to vectors that are orthogonal (perpendicular) to $\mathbf{s}_k$. Therefore, we impose the additional condition,

$$\mathbf{A}_k\mathbf{z} = \mathbf{A}_{k-1}\mathbf{z} \quad \text{for all vectors } \mathbf{z} \text{ such that } \mathbf{z}^T\mathbf{s}_k = 0. \tag{4.10–4}$$

$\mathbf{A}_k$ is now uniquely determined by (4.10–3) and (4.10–4). The quasi-Newton method is then

$$\mathbf{A}_k = \mathbf{A}_{k-1} + \frac{(\mathbf{y}_k - \mathbf{A}_{k-1}\mathbf{s}_k)[\mathbf{s}_k]^T}{[\mathbf{s}_k]^T\mathbf{s}_k} \tag{4.10–5}$$

$$\mathbf{x}_{k+1} = \mathbf{x}_k - [\mathbf{A}_k]^{-1}\mathbf{f}(\mathbf{x}_k) \tag{4.10–6}$$

All vectors are column vectors of dimension n, so that the term

$$(\mathbf{y}_k - \mathbf{A}_{k-1}\mathbf{s}_k)[\mathbf{s}_k]^T$$

is an $n \times n$ matrix. Equation (4.10–5) is the updating formula for the approximation to the Jacobian, while (4.10–6) is the Newton method. It is easily seen that (4.10–3) and (4.10–4) are verified by multiplying (4.10–5) on the right by $\mathbf{s}_k$ and $\mathbf{z}$, respectively. Equation (4.10–5) is called a *rank-one updating formula* because the matrix added to $\mathbf{A}_{k-1}$ has rank equal to 1.

The computational efficiency of the quasi-Newton method can be further improved by using the *Sherman-Morrison* matrix inversion formula. If $\mathbf{B}$ is a nonsingular matrix, then the matrix $(\mathbf{B} + \mathbf{v}\mathbf{w}^T)$ is nonsingular provided $\mathbf{w}^T\mathbf{B}^{-1}\mathbf{v} \neq -1$, and

$$(\mathbf{B} + \mathbf{v}\mathbf{w}^T)^{-1} = \mathbf{B}^{-1} - \frac{\mathbf{B}^{-1}\mathbf{v}\mathbf{w}^T\mathbf{B}^{-1}}{1 + \mathbf{w}^T\mathbf{B}^{-1}\mathbf{v}}$$

This formula can be verified by multiplying the right-hand side by the matrix $(\mathbf{B} + \mathbf{v}\mathbf{w}^T)$ and showing that the result is the identity matrix. We want to apply the Sherman-Morrison matrix inversion formula with $\mathbf{B} = \mathbf{A}_{k-1}$, $\mathbf{v} = (\mathbf{y}_k - \mathbf{A}_{k-1}\mathbf{s}_k)/([\mathbf{s}_k]^T\mathbf{s}_k)$, and $\mathbf{w} = \mathbf{s}_k$. This gives

$$1 + \mathbf{w}^T\mathbf{B}^{-1}\mathbf{v} = 1 + \frac{[\mathbf{s}_k]^T[\mathbf{A}_{k-1}]^{-1}(\mathbf{y}_k - \mathbf{A}_{k-1}\mathbf{s}_k)}{[\mathbf{s}_k]^T\mathbf{s}_k} = \frac{[\mathbf{s}_k]^T[\mathbf{A}_{k-1}]^{-1}\mathbf{y}_k}{[\mathbf{s}_k]^T\mathbf{s}_k}$$

so that

$$\frac{\mathbf{B}^{-1}\mathbf{v}\mathbf{w}^T\mathbf{B}^{-1}}{1 + \mathbf{w}^T\mathbf{B}^{-1}\mathbf{v}} = \frac{[\mathbf{A}_{k-1}]^{-1}(\mathbf{y}_k - \mathbf{A}_{k-1}\mathbf{s}_k)[\mathbf{s}_k]^T[\mathbf{A}_{k-1}]^{-1}}{[\mathbf{s}_k]^T[\mathbf{A}_{k-1}]^{-1}\mathbf{y}_k}$$

Consequently, the updating formula for $[\mathbf{A}_k]^{-1}$ is

$$[\mathbf{A}_k]^{-1} = [\mathbf{A}_{k-1}]^{-1} + \frac{(\mathbf{s}_k - [\mathbf{A}_{k-1}]^{-1}\mathbf{y}_k)[\mathbf{s}_k]^T[\mathbf{A}_{k-1}]^{-1}}{[\mathbf{s}_k]^T[\mathbf{A}_{k-1}]^{-1}\mathbf{y}_k} \qquad \textbf{(4.10–7)}$$

Eq. (4.10–7) can be used with (4.10–6). It avoids the problem of having to solve the linear system $\mathbf{A}_k\mathbf{s}_{k+1} = -\mathbf{f}(\mathbf{x}_k)$ for $\mathbf{s}_{k+1}$. If $[\mathbf{s}_k]^T[\mathbf{A}_{k-1}]^{-1}\mathbf{y}_k = 0$ then (4.10–7) cannot be used; in that case, we would solve the equation $\mathbf{A}_k\mathbf{s}_{k+1} = -\mathbf{f}(\mathbf{x}_k)$ or invert $\mathbf{A}_k$ and hope that on the next iteration (4.10–7) can be used.

A particular quasi-Newton method is Powell's hybrid method, in which the Jacobian matrix is computed by finite differences and updated as required. The method tries to obtain $\mathbf{x}_{k+1}$ such that

$$\| \mathbf{f}(\mathbf{x}_{k+1}) \| < \| \mathbf{f}(\mathbf{x}_k) \|$$

In addition, there are imposed constraints so that large step-sizes are avoided. A user's manual for a Pascal version of Powell's method is given in Appendix 4.4 along with a sample main program in Appendix 4.5 for the following problem.

EXAMPLE 4.5

$$f_1(x_1, x_2) = 0.1\, x_1x_2 - 1$$

$$f_2(x_1, x_2) = \exp(-x_1 10^{-5}) + \exp(-x_2) - 1.0001$$

Starting with the initial guess $(x_1, x_2) = (0.0, 1.0)$, Powell's method required 168 evaluations of the function $\mathbf{f}^T = (f_1, f_2)$ to obtain

$$x_1 = 1.098159329700 \quad \text{and} \quad x_2 = 9.106146739868$$

EXAMPLE 4.5 (cont'd)

with $\| \mathbf{f}(x_1, x_2) \| = 1.066 \times 10^{-14}$. Choosing different initial values may accelerate the convergence. For example, starting from the initial guess $(x_1, x_2) = (0.0, 10.0)$ required only 14 function evaluations to converge to the same solution (see Table 4.10). Other, nearby, starting points may not have so dramatic an effect on the convergence. For example, starting from $(x_1, x_2) = (1, 10)$ required 108 function evaluations to converge.

Table 4.10 EXAMPLE 4.5, USING THE PASCAL PROCEDURE POWELL

k	nfe*	x_1	x_2	$\| \mathbf{f}(x_1, x_2) \|$
0	1	0.0	10.0	1.0
1	4	1.00000000000000	8.57710024355574	0.1423
2	5	1.08129455900745	9.54545236830636	0.03214
3	6	1.08117082211255	9.22487207310721	2.634×10^{-3}
4	7	1.10004766608175	9.08236522299714	8.965×10^{-4}
5	8	1.09793007807353	9.10790024285397	1.624×10^{-5}
6	9	1.09815109789321	9.10618623321534	3.159×10^{-6}
7	10	1.09815910278376	9.10614740902434	1.331×10^{-7}
8	11	1.09815932316117	9.10614675985766	3.759×10^{-9}
9	12	1.09815932971772	9.10614673981233	1.035×10^{-11}
10	13	1.09815932969991	9.10614673986590	1.199×10^{-14}
11	14	1.09815932969989	9.10614673986589	6.661×10^{-16}

* nfe = number of function evaluations

EXERCISE 4.8

Adapt the program in Appendix 4.5 to solve the equations

$$f_1(x_1, x_2) = x_1^4 + x_2^4 - 6$$

$$f_2(x_1, x_2) = \exp(-x_1) + \sin(x_2) - 2$$

given the initial guess $(x_1, x_2) = (0.0, 0.0)$.

Answer $(x_1, x_2) = (-1.6311847919841 \times 10^{-5}, 1.56508458007329)$.

EXAMPLE 4.6

A biologist, who is studying a new strain of bacteria, wants to determine the growth rate of a bacteria culture. At every h minutes, starting at $t = t_0$, she counts the number of bacteria within a given field of vision of her microscope. She makes a total of 51 such determinations. For each determination, bacteria move into and

EXAMPLE 4.6 (cont'd)

out of the field of vision so that the growth rate, which depends on the bacteria count, is subject to errors. Let $N(t_0 + hk)$ denote the number of bacteria present at time $t_0 + hk$, for $k = 0, 1, 2, \ldots, 50$. The biologist has reason to believe that the number of bacteria at time t is given by a logistic law

$$N(t) = \frac{a}{1 + b\exp(-t/c)}$$

for positive numbers a, b, and c with $a \simeq 100$. Because h and t_0 are fixed, $N(t)$ can be rewritten as

$$N(t_0 + hk) = \frac{a}{1 + b\exp[-(t_0 + hk)/c]} = \frac{a}{1 + b\exp(-t_0/c)\exp(-kh/c)}$$

In order to simplify the above equation, the biologist defines the new quantities x_1, x_2, x_3, and y_k by

$$y_k = \frac{N(t_0 + hk)}{100}$$

$$x_1 = \frac{a}{100}$$

$$x_2 = b\exp(-t_0/c)$$

$$x_3 = \frac{c}{h}$$

The problem is now to find x_1, x_2, x_3 so that the data, given below, "best" fits the transformed logistic law

$$y_k = \frac{x_1}{1 + x_2\exp(-k/x_3)}$$

The biologist decides that she will choose x_1, x_2, x_3 so that the error function

$$E(x_1, x_2, x_3) = \sum_{k=0}^{50}\left[y_k - \frac{x_1}{1 + x_2\exp(-k/x_3)}\right]^2$$

is minimized. She knows that $E(x_1, x_2, x_3)$ has a local extremum when the three simultaneous nonlinear equations

$$f_i = \frac{\partial E}{\partial x_i} = 0, \qquad i = 1, 2, 3 \tag{4.10–8}$$

are satisfied. The biologist thinks that $x_2 \simeq 2$, $x_3 \simeq 10$, while $x_1 \simeq 1$ because $a \simeq 100$. She uses Powell's method, with the initial conditions $(x_1, x_2, x_3) = (1, 2, 10)$, to solve the three simultaneous nonlinear equations (4.10–8). Procedure Powell converged to the solution

EXAMPLE 4.6 (cont'd)

$$x_1 = 0.973920713954408, \; x_2 = 2.44173653810710, \; x_3 = 7.93005485205731$$

in twenty-two evaluations of the partial derivative functions (4.10–8) with a final value of $\| (f_1, f_2, f_3) \| = 5.6 \times 10^{-16}$. Because the original data are accurate to at most three significant digits, the biologist accepted the values $x_1 = 0.974$, $x_2 = 2.44$, and $x_3 = 7.93$.

k	$100y_k$	k	$100y_k$	k	$100y_k$
0	35	17	78	34	88
1	26	18	72	35	103
2	28	19	83	36	102
3	37	20	81	37	95
4	34	21	80	38	99
5	45	22	84	39	102
6	52	23	89	40	90
7	59	24	80	41	95
8	44	25	88	42	88
9	55	26	97	43	92
10	57	27	93	44	97
11	56	28	98	45	101
12	57	29	98	46	97
13	73	30	87	47	93
14	76	31	93	48	98
15	73	32	88	49	90
16	63	33	86	50	106

4.11 ADDITIONAL TOPICS IN SOLVING NONLINEAR EQUATIONS

If you have a system of nonlinear equations to be solved, it is important to examine the problem thoroughly before rushing off to use your favorite numerical method. Often, a little thought may simplify the computations. For example, many numerical methods for solving the single equation $F(x) = 0$ require that we know two values of x, a, and b, with the property that $F(a)F(b) < 0$. This ensures that there is at least one zero of $F(x)$ in the interval $[a, b]$, provided $F(x)$ is a continuous function. However, this is not enough to guarantee that there is one, and only one, zero in the interval $[a, b]$. It may happen that your favorite numerical method works very well if there is only one zero in $[a, b]$, but it fails if there is more than one zero; the method may require much more computational effort than you want to spend. In some problems the function $F(x)$ may not be given in a closed form, but it is obtained from the numerical solution of another problem, such as a linear or nonlinear differential equation. In such a case, a method requiring the first derivative of $F(x)$ may be inconvenient, if not impractical, to use. If $F(x)$ is differentiable, examination of $F'(x)$ may help to determine if a zero of $F(x)$ exists, or does not exist, in a given interval; if you know that $F(a) > 0$ and $F'(x) > 0$ for all x in $[a, b]$ then $F(x)$ cannot have a real zero inside the interval.

If you are required to solve a system of equations, examine them carefully. It may be possible to eliminate one or more of the variables, thereby reducing the number of equations and unknowns. For example, it would be silly to solve the system $f_1(x_1, x_2) = 0$, $f_2(x_1, x_2) = 0$, where

$$f_1(x_1, x_2) = \{x_2 - \arcsin(x_1)\}\{x_2 + 3\sin(x_1)\}$$

$$f_2(x_1, x_2) = \{x_2 - \tanh(x_1 + 12)\}\cos(x_1 + x_2 + 5)$$

by Powell's method because each function, f_i, is the product of two factors. By setting each factor to zero, we can obtain all possible solutions in terms of x_1. For example, setting

$$x_2 = \arcsin(x_1)$$

makes $f_1(x_1, x_2) = 0$ and setting

$$\cos(x_1 + \arcsin(x_1) + 5) = 0$$

makes $f_2(x_1, x_2) = 0$. Solving

$$\cos(x_1 + \arcsin(x_1) + 5) = 0$$

for x_1 gives

$$x_1 + \arcsin(x_1) + 5 = (m + 0.5)\pi$$

for some integer m.

In some cases, scaling the variables may help accelerate the convergence of a numerical method to a solution of $\mathbf{f}(\mathbf{x}) = \mathbf{0}$. The scaling should be done so that the components of the new variable, $\mathbf{y}$, have nearly the same magnitude at the solution. In Example 4.5, we made the change-of-variables $y_1 = 10^5 x_1$ and $y_2 = 0.1\ x_2$. Powell's method required 113 function evaluations to converge to the solution from the starting point $(y_1, y_2) = (0.0, 0.1)$; the unscaled case required 185 function evaluations. When the starting point was $(y_1, y_2) = (0.0, 1.0)$, 19 function evaluations were needed; the corresponding unscaled case was slightly faster, requiring 13 function evaluations.

Remember, the time spent in careful consideration of a problem and its method of solution can result in savings in both human and computer time.

4.12 PROBLEMS

1. $F(x) = x - \tan(x)$; $a = 4.0$, $b = 4.5$.
   Use the Pascal program and procedures given in Appendices 4.1 and 4.2 to find the zero in the interval $[a, b]$. Use the function $G(x) = x + F(x)/50$ for

the Fixed-Point method. For both Newton's method and the Fixed-Point method use the starting value 4.5.

2. $F(x) = x\cos(x) - \sin(x)$; $a = 4.0$, $b = 4.5$.
Repeat Problem 1, but use $G(x) = x - F(x)/5$ in the Fixed-Point method. Compare your results with Problem 1.

3. $F(x) = x - \tan^{-1}(x)$; $a = 4.0$, $b = 4.5$.
Repeat Problem 1, but use $G(x) = \tan^{-1}(x)$ in the Fixed-Point method. Compare your results with Problems 1 and 2.
Warning: Most computer implementations of the function $\tan^{-1}(x)$ return values in the range $[-\pi/2, \pi/2]$; you will have to use ArcTan$(x) + \pi$, where ArcTan(x) represents the computer implementation of the inverse tangent function.

4. The starting value $x_0 = 4.2$ in Problem 1 causes Newton's method to diverge. Write a Pascal procedure to implement the Modified Newton's method, (4.6–3). Try your procedure on the function in Problem 1 with $x_0 = 4.2$.

5. $F(x) = (\sqrt{x})[1.8\ \ell og(R\sqrt{x}) - 0.5] - 1$; $a = 10^{-5}$, $b = 0.5$.
Find the zeros of $F(x)$ for a range of R values in the interval $[10^4, 10^6]$ and thus obtain a graph of x vs. $\ell og_{10}(R)$. (See Example 4.3)

6. $P(x) = 0.55x^6 + 57.18x^5 + 82.15x^4 + 306.23x^3 + 796.76x^2 + 1135.91x + 418.68$.
This polynomial is the denominator term of the Laplace Transform of a certain function. $P(x)$ has real roots in the intervals $[-0.6, -0.5]$ and $[-103, -102]$. Find these roots using ZeroIn.

7. In a certain controlled mechanism, the controlling function, $u(t)$, is full-on $(u(t) = 1)$ when a certain function (the switching function) is positive; $u(t)$ is full-off $(u(t) = 0)$ when the switching function is negative. Hence, $u(t)$ switches from full-on to full-off, and vice versa, at the zeros of the switching function. Suppose that the switching function is $F(t) = \cos(t) - \tanh(t)$. Let the switching times be t_k, $k = 1, 2, \ldots$; they are the zeros of $F(t)$. If the controlled mechanism begins operating at $t = 0$, find the first three positive switching times, t_1, t_2, t_3, using ZeroIn.

8. When the Earth is assumed to be a spheroid, the gravitational attraction at a point on its surface at the latitude angle ϕ is

$$g(\phi) = 9.78049\ (1 + 0.0052884\ \sin^2(\phi) - 0.0000059\ \sin^2(2\phi))\ \text{m/sec}^2$$

At what latitude(s), in degrees, is $g(\phi) = 9.8$?

9. Find the first three positive zeros of the Bessel function $J_0(x)$. Use the function given in Appendix 4.3 to compute $J_0(x)$.

10. A sequence $\{x_n\}$ that is obtained from an iterative method, with a convergence rate equal to one, can be transformed into a more rapidly converging sequence

$\{y_n\}$ by the method of Aitken Extrapolation,

$$y_n = x_n - \frac{(x_n - x_{n-1})^2}{x_n - 2x_{n-1} + x_{n-2}} \tag{4.12–1}$$

The extrapolation formula can be directly incorporated into an iterative method such as the Fixed-Point method. Given x_0, x_1, and x_2, three successive iterates found by the Fixed-Point method, we compute y by (4.12–1). Now apply the Fixed-Point method to y. This means that we use the Fixed-Point method to obtain three estimates of the fixed point; we then apply the Aitken Extrapolation to accelerate the convergence of the sequence, and finally we restart the Fixed-Point method with the Aitken value. By using the Aitken Extrapolation method in conjunction with the Fixed-Point method we found that Example 4.1 required only $p = 7$ function evaluations to converge; this is half the number needed when the Fixed-Point method was used alone.

(a) Modify the Pascal procedure for the Fixed-Point method in Appendix 4.1 so that it uses the Aitken Extrapolation method.

(b) Use your procedure from part (a) to find the zero of the function in Problem 1.

11. When a body of mass M falls through a resisting medium at a speed above approximately 30 m/sec, the drag force is approximately described by taking it as proportional to the square of the speed. If the body starts at rest at time $t = 0$, it falls a distance of $q \cdot \ell og(\cosh(gt/q))$ meters after t sec. The constant q is proportional to M. If $q = 30$ meters, $g = 9.8$ m/sec^2, find the time t required for the body to fall from 50 meters to 200 meters.

12. The following equation arises in the theory of nuclear reactions:

$$\tan(A) = \frac{\tan(x)}{1 + (B \cdot \sec(x))^{1/2}}$$

If $A = \pi/4$ and $B = 1$, find x in [0, 1.5].

13. The function $F(x) = A\cosh(x) - xB\sinh(x) + C$ arises in the theory of bending of rectangular steel plates. If $A = 6$, $B = 5$, and $C = 23$ find $x_* > 0$ such that $F(x_*) = 0$.

14. $F(x) = x - \sinh(\sin(x))$ has a zero at $x = 0$ of multiplicity 5.

(a) Use Newton's method, Equation (4.6–1), and ZeroIn to compute this zero, given the initial interval $[-1, 2]$ and $x_0 = 2$.

(b) Repeat part (a) using the function $H(x)$, defined in Equation (4.6–2), for this $F(x)$.

(c) Equation (4.6–3), the modified Newton's method can be used for multiple zeros if the multiplicity is known; set c equal to the multiplicity of the zero. Try this technique with $F(x)$. Of course, in most problems we would not know the multiplicity of a zero before we found the zero.

15. Write a Pascal procedure to solve a system of n equations in n unknowns using Newton's method, (4.10–1). Use your procedure to solve each of the following systems of equations. Solve the systems using Powell's method (Appendix 4.4).

(a) $x = 1 + \beta(e^{y\sqrt{x}} + 5x^3y^2)$

$y = -2 + \beta \tan(e^y + x^2)$

$\beta = 0.005,\ x_0 = 1 \quad \text{and} \quad y_0 = -2.$

(b) $75 = 10x^{1.5} + 20(x - y)$
$25 = 30y - 20x + 20e^{0.1y}$
$x_0 = 5 \quad \text{and} \quad y_0 = 10.$

16. Certain iteration methods with a convergence rate of three can be written as

$$x_{n+1} = x_n - [F'(x_n)]^{-1}F(x_n)H(t_n)$$
$$t_n = [F'(x_n)]^{-2}F(x_n)F''(x_n)$$

where $H(t)$ is a differentiable function satisfying the conditions $H(0) = 1$ and $H'(0) = 0.5$. Notice that the effective number of function evaluations is three per iteration. If $H(t) = 1/(1 - 0.5t)$ we obtain Halley's method, named after the astronomer Edmund Halley who used it in 1694 to find the roots of polynomials. If $H(t) = 1 + 0.5t$ we obtain the quadratic inverse interpolation method. Write Pascal procedures for Halley's method and the quadratic inverse interpolation method. Try your procedures on

(a) Problems 1, 2 and 3 for the zero near 4.5.

(b) Problem 5.

(c) Problem 14.

17. Find the zero of $F(x)$ in the interval $[-1, 0]$, where

$$F(x) = x + (x^2 + 1)\frac{[(2x - 3)e^{2(x-1)} + 1]}{g(x)},$$
$$g(x) = 2(x - 1)[e^{2(x-1)} - 1].$$

18. Find the zero of

$$F(x) = \ell og(2 - e^{-x}) - x + \alpha$$

in the interval $[\alpha, \alpha + \ell og(2)]$ for various values of α, say $\alpha = 0.25$ to 6.0 in steps of 0.25.

19. (a) Find the zero of

$$F(x) = 2\{\sqrt{7} + \tan(\sqrt{7}x/4)\}\{3e^{2.5(x-1)} + 2\} - \{\sqrt{7} - \tan(\sqrt{7}x/4)\}\{8 - 3e^{2.5(x-1)}\}$$

that lies in the interval [0, 1].

(b) Find the zero of

$$F(x) = \{\sqrt{7} + \tan(\sqrt{7}(x - 1)/4)\}\{6 - e^{2.5x}\} - \{\sqrt{7} + 5\tan(\sqrt{7}(x - 1)/4)\}\{1.5 + e^{2.5x}\}$$

that lies in the interval [0, 1].

20. $x^2 + y^2 = 5$

$e^x + y = 3$

(a) Find both real solutions to this system of equations.

(b) In order to find complex solutions, set

$$x = x_1 + jx_2 \quad \text{and} \quad y = x_3 + jx_4,$$

where $x_1, \ldots, x_4$ are real and $j = \sqrt{(-1)}$. This gives the four equations

$$5 = x_1^2 - x_2^2 + x_3^2 - x_4^2$$

$$0 = x_1x_2 + x_3x_4$$

$$3 = \cos(x_2)\exp(x_1) + x_3$$

$$0 = \sin(x_2)\exp(x_1) + x_4.$$

It is easy to see that if $x = a + jb$ and $y = c + jd$ are solutions, then so are $x = a - jb$ and $y = c - jd$. Find at least three solutions that are not complex conjugates, different from those found in part (a).

21. A nonlinear resistor is described by the following relationship between the current, i, through it and the voltage, v, across it,

$$v = 5e^{0.2i}.$$

The resistor is connected in series with a 10-ohm resistor and a constant voltage of 120 volts is applied to the combination. Determine the current through the circuit.

22. A large gasoline storage tank is in the shape of a right-circular cylinder of length 6 meters and radius 1 meter. Each end of the cylinder is closed by an identical spherical cap, of radius 2 meters; the center of each of these spheres lies on the symmetry axis of the cylinder. (See the diagram below.) The capacity of the tank is approximately 20,000 liters. A long rod is used to determine the volume of gasoline in the tank by inserting the rod at an access valve. The part of the rod that is stained by the gasoline gives the depth of gasoline in the tank. The owner of the tank wants to calibrate the rod so that it measures the volume of gasoline in units of 1000 liters. The volume of gasoline, in units of 1000 liters, whose depth is x meters, is given by

$$V(x) = 4\left[\frac{f_1}{\sqrt{3}} + \frac{(1 - x)f_2}{3}\right] + 3\pi - 6f_3, \qquad 0 \le x \le 2,$$

where

$$f_1 = \int_{1-x}^{1} [3 + t^2]^{-1}[1 - t^2]^{3/2}\, dt$$

$$f_2 = \int_{\sqrt{3}}^{\sqrt{a}} [t^2 + (1 - x)^2]^{-1}[a - t^2]^{3/2}\, dt$$

$$f_3 = \tan^{-1}\left[\frac{(1 - x)}{b}\right] + (1 - x)b$$

$$a = (1 + x)(3 - x), \qquad b = [x(2 - x)]^{1/2}.$$

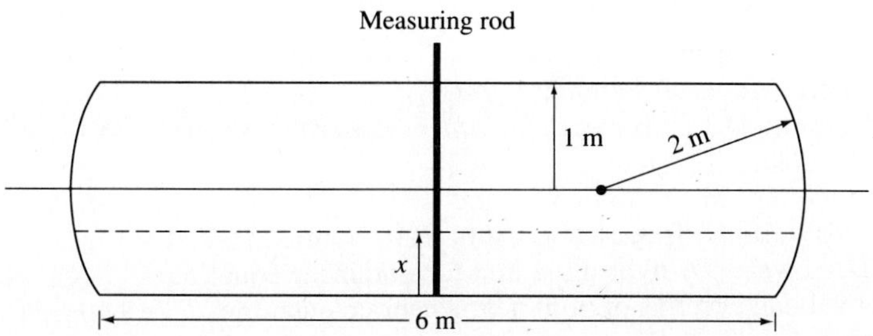

Find x_k so that $V(x_k) = k$, for $k = 1, 2, \ldots, 19$.
Hint: Use the Pascal procedure ZeroIn to find the zero of the function

$$F(x) = \frac{V(x)}{k} - 1$$

for each k. The integrals can be evaluated using the Pascal procedure QUANC8 discussed in Chapter 6.

23. The maximum propulsion efficiency of a rocket is given by

$$E(x) = (1 - x)\left[1 + \frac{[\ell\text{og}(1 - x)]^2}{x}\right] \qquad \text{for } 0 < x < 1.$$

In $E(x)$, x is the ratio of the final mass to the initial mass.

(a) Show that $\lim_{x \to 0} E(x) = 1$.

(b) Show that $\lim_{x \to 1} E(x) = 0$.

(c) Find x so that $E(x) = 0.7$.

24. Find all real x such that $\sin(x) = 1 - x^2$.

25. Solve the simultaneous equations

$$3.1y - \cos(xz) - 0.6 = 0$$

$$1.03e^{-xy} + 19.5z + 11 = 0$$

$$y^2 - 83(x + 0.11)^2 + \sin(z) + 0.97 = 0$$

by using the procedure Powell, with the starting values $x = y = z = 0$.

4.13 REFERENCES

F. S. Acton. *Numerical Methods that Work.* New York: Harper & Row, 1970.

R. L. Burden, J. D. Faires, and A. C. Reynolds. *Numerical Analysis.* 2d Ed. Boston: Prindle, Weber & Schmidt, 1981.

S. D. Conte and C. de Boor. *Elementary Numerical Analysis.* 3d Ed. New York: McGraw-Hill, 1980.

G. E. Forsythe, M. A. Malcolm, and C. B. Moler. *Computer Methods for Mathematical Computations.* Englewood Cliffs, NJ: Prentice-Hall, 1977.

M. J. D. Powell. "A hybrid method for nonlinear equations." "A FORTRAN subroutine for solving systems of nonlinear algebraic equations." In *Numerical Method for Nonlinear Algebraic Equations*, edited by P. Rabinowitz. New York: Gordon and Breach, 1970.

J. R. Rice. *Numerical Methods, Software, and Analysis.* New York: McGraw-Hill, 1983.

L. F. Shampine and R. C. Allen, Jr. *Numerical Computing: an Introduction.* Philadelphia: W.B. Saunders, 1973.

APPENDIX 4.1

Listing of the Pascal Program for Example 4.1

```
PROGRAM NonLinearEquationSolver;

VAR
  a, b, x, fx, err, r, s : REAL;
  Evaluations, N : INTEGER;
  terminate : BOOLEAN;

FUNCTION F( x : REAL ) : REAL;
BEGIN
  IF  x < 0.0  THEN
    BEGIN
      terminate := TRUE ; F := 1.0E+10
    END
  ELSE
    F := x - 2.4 + x*Sqrt(Sqrt(x))
END;

FUNCTION Fp( x : REAL ) : REAL; (*  derivative of F(x)  *)
BEGIN
  Fp := 1.0 + 1.25*Sqrt(Sqrt(x))
END;
```

```
FUNCTION G( x : REAL ) : REAL; (* function for fixed-point method *)
BEGIN
  G := x - F(x)/2.0
END;

(*$I ZeroIn.Pas *)

PROCEDURE Bisection(a, b, r, s : REAL; N : INTEGER; VAR x, fx, err : REAL;
                             VAR Evaluations : INTEGER);
VAR
  fa, fb : REAL;
BEGIN
  fa := F(a) ; Evaluations := 1 ; terminate := FALSE ;
  REPEAT
    x := (a + b)/2.0 ; fx := F(x) ; Evaluations := Evaluations + 1 ;
    IF  fx*fa <= 0.0  THEN
      BEGIN
        b := x ; fb := fx
      END
    ELSE
      BEGIN
        a := x ; fa := fx
      END;
    err := Abs(b - a)
  UNTIL  (err <= r) OR (Abs(fx) < s) OR (Evaluations > N) OR terminate
END; (* Bisection *)

PROCEDURE RegulaFalsi(a, b, s : REAL ; N :INTEGER ; VAR x, fx : REAL ;
                                      VAR Evaluations : INTEGER);
VAR
  fa, fb : REAL;
BEGIN
  fa := F(a) ; fb := F(b) ; Evaluations := 2 ; terminate := FALSE ;
  REPEAT
    x := (fb*a - fa*b)/(fb - fa);
    fx := F(x) ; Evaluations := Evaluations + 1 ;
    IF  fx*fa <= 0.0  THEN
      BEGIN
        b := x ; fb := fx
      END
    ELSE
      BEGIN
        a := x ; fa := fx
      END
  UNTIL  (Abs(fx) < s) OR (Evaluations > N) OR terminate
END; (* RegulaFalsi *)

PROCEDURE ModifiedRegulaFalsi(a, b, r, s : REAL ; N : INTEGER ;
                      VAR x, fx, err : REAL ; VAR Evaluations : INTEGER);
VAR
  fa, fb, foldx : REAL;
BEGIN
  fa := F(a) ; fb := F(b) ; foldx := fa ;
  Evaluations := 2 ; terminate := FALSE ;
  REPEAT
    x := (fb*a - fa*b)/(fb - fa);
    fx := F(x) ; Evaluations := Evaluations + 1 ;
    IF  fx*fa <= 0.0  THEN
      BEGIN
        b := x ; fb := fx ;
```

```
          IF  fx*foldx > 0.0  THEN
            fa := fa/2.0
        END
      ELSE
        BEGIN
          a := x ; fa := fx ;
          IF  fx*foldx > 0.0  THEN
            fb := fb/2.0
        END;
      foldx := fx ; err := Abs(b - a)
    UNTIL  (err <= r) OR (Abs(fx) < s) OR (Evaluations > N) OR terminate
  END; (* ModifiedRegulaFalsi *)


PROCEDURE Secant(a, b, r, s : REAL ; N : INTEGER ; VAR x, fx : REAL ;
                                    VAR Evaluations : INTEGER);
VAR
  fa, fb : REAL;
BEGIN
  fa := F(a) ; fb := F(b) ; Evaluations := 2 ; terminate := FALSE ;
  REPEAT
    x := b - fb*((b - a)/(fb - fa)) ; fx := F(x) ;
    Evaluations := Evaluations + 1 ; a := b ; fa := fb ; b := x ; fb := fx
  UNTIL  (Abs(b - a) <= r) OR (Abs(fx) < s) OR (Evaluations > N) OR terminate
END; (* Secant *)

PROCEDURE Newton(a, r, s : REAL ; N : INTEGER ; VAR x, fx : REAL ;
                                          VAR Evaluations : INTEGER);
VAR
  fpx, oldx : REAL;
BEGIN
  Evaluations := 0 ; x := a ; terminate := FALSE ;
  REPEAT
    oldx := x ; fx := F(x) ; Evaluations := Evaluations + 1 ;
    fpx := Fp(x) ; x := x - fx/fpx
  UNTIL  (Abs(x - oldx) <= r) OR (Abs(fx) < s) OR
           (Evaluations > N) OR terminate
END; (* Newton *)

PROCEDURE FixedPoint(a, r, s : REAL ; N : INTEGER ; VAR x, err : REAL ;
                                VAR Evaluations : INTEGER);
VAR
  Gx, oldx : REAL;
BEGIN
  Evaluations := 0 ; x := a ; terminate := FALSE ;
  REPEAT
    oldx := x ; Gx := G(x) ; Evaluations := Evaluations + 1 ;
    err := Abs(x - Gx) ; x := Gx
  UNTIL (Abs(x - oldx) <= r) OR (err < s) OR (Evaluations > N) OR terminate
END; (* FixedPoint *)


BEGIN (* main program *)
  a := 1.0 ; b := 2.0 ; N := 50 ; r := 0.0 ; s := 1.0E-11 ;
  WRITELN('  Method',' ':18,'z',' ':15,'F(z)',' ':8,'err', ' ':8,'p');
  WRITELN;
  Bisection(a,b,r,s,N,x,fx,err,Evaluations);
    IF  NOT terminate  THEN
      WRITELN('Bisection:    ',x,'  ',fx:10,'  ',err:10,'  ',Evaluations:3);
```

```
  RegulaFalsi(a,b,s,N,x,fx,Evaluations);
    IF  NOT terminate  THEN
      WRITELN('Regula Falsi: ',x,'  ',fx:10,' ':14,Evaluations:3);
  ModifiedRegulaFalsi(a,b,r,s,N,x,fx,err,Evaluations);
    IF  NOT terminate  THEN
      WRITELN('Modified RF:  ',x,'  ',fx:10,'  ',err:10,'  ',Evaluations:3);
  Secant(a,b,r,s,N,x,fx,Evaluations);
    IF  NOT terminate  THEN
      WRITELN('Secant:       ',x,'  ',fx:10,' ':14,Evaluations:3);
  Newton(a,r,s,N,x,fx,Evaluations);
    IF  NOT terminate  THEN
      WRITELN('Newton:       ',x,'  ',fx:10,' ':14,Evaluations:3);
  FixedPoint(a,r,s,N,x,err,Evaluations);
    IF  NOT terminate  THEN
      WRITELN('Fixed Point:  ',x,'  ',err:10,' ':14,Evaluations:3);
  ZeroIn(a,b,0.0,x,Evaluations);
    IF  NOT terminate  THEN
      WRITELN('ZeroIn:       ',x,'  ',F(x):10,' ':14,Evaluations:3)
END.
```

APPENDIX 4.2

Listing of the Pascal Procedure ZeroIn, which is in the file ZeroIn.Pas

```
PROCEDURE ZeroIn( Left,Right,tol : REAL ; VAR root : REAL ;
                                   VAR Evaluations : INTEGER );

(*   A zero of the function  F(x)  is computed in the interval [Left,Right].
     It is assumed that  F(Left)*F(Right) < 0.0 without a check.  ZeroIn
     returns a zero  x  in the given interval [Left,Right] to within a
     tolerance of 4*(machine epsilon)*abs(x) + tol.

   Input:
      Left  =  left endpoint of the initial interval
      Right =  right endpoint of the initial interval
      tol   =  desired length of the interval of uncertainty
                 of the final result  ( tol >= 0.0 )

   The user supplies the following to evaluate the function whose zero
     is sought on the interval  [Left,Right]
             FUNCTION F( x : REAL ) : REAL;

   Output:
      root  =  abscissa approximating a zero of  F  in the interval
                  [Left,Right]
      Evaluations = number of function evaluations

Procedure ZeroIn is based on the FORTRAN subroutine ZEROIN that appeared in
     G.E. Forsythe, M.A. Malcolm and C.B. Moler
      "Computer Methods for Mathematical Computations"
      Prentice-Hall. 1977    *)

CONST
  base = 2.0; (* Machine number base.  This is for Turbo Pascal with 8087 *)

VAR
  Eps, toll, a, b, fa, fb, c, fc, d, e, s, p, q, r, xm, factor : REAL;
  flag : BOOLEAN;
```

```
FUNCTION MachineEpsilon : REAL;
VAR
  a, b : REAL;
BEGIN
  a := 1.0;
  REPEAT
    a := a/base ; b := 1.0 + a
  UNTIL  b = 1.0;
  MachineEpsilon := a*base
END; (* MachineEpsilon *)


PROCEDURE Initialization;
BEGIN
  a := Left ; b := Right ;
  fa := F(a) ; fb := F(b) ;
  flag := TRUE ; Evaluations := 2 ;
  Eps := MachineEpsilon
END; (* Initialization *)


PROCEDURE LinearInterpolation;
BEGIN
  s := fb/fa ; p := 2.0*xm*s ; q := 1.0 - s
END; (* LinearInterpolation *)


PROCEDURE InverseQuadraticInterpolation;
BEGIN
  q := fa/fc ; r := fb/fc ; s := fb/fa ;
  p := s*(2.0*xm*q*(q-r) - (b-a)*(r-1.0));
  q := (q-1.0)*(r-1.0)*(s-1.0)
END; (* InverseQuadraticInterpolation *)


PROCEDURE Bisection;
BEGIN
  d := xm ; e := d
END; (* Bisection *)


PROCEDURE StepCompletion;
BEGIN
  a := b ; fa := fb ;
  IF  Abs(d) > toll  THEN
    b := b + d
  ELSE
    BEGIN
      IF  xm > 0.0  THEN
        b := b + toll
      ELSE
        b := b - toll
    END;
  fb := F(b) ; Evaluations := Evaluations + 1 ;
  IF  fb*(fc/Abs(fc)) > 0.0  THEN
    flag := TRUE
  ELSE
    flag := FALSE
END; (* StepCompletion *)
```

```
PROCEDURE Finalize;
BEGIN
  IF  flag  THEN
    BEGIN
      c := a ; fc := fa ; d := b - a ; e := d
    END;
  IF  Abs(fc) < Abs(fb)  THEN
    BEGIN
      a := b ; b := c ; c := a ; fa := fb ; fb := fc ; fc := fa
    END;
  toll := 2.0*Eps*Abs(b) + 0.5*tol;
  xm := 0.5*(c - b)
END; (* Finalize *)


PROCEDURE MainLoop;
VAR
  bisect : BOOLEAN;
BEGIN
  bisect := TRUE;
  IF  (Abs(e) >= toll) AND (Abs(fa) > Abs(fb))  THEN
    BEGIN
      IF  a = c  THEN
        LinearInterpolation
      ELSE
        InverseQuadraticInterpolation;
      IF  p > 0  THEN
        q := -q
      ELSE
        p := Abs(p);
      factor := 3.0*xm*q - Abs(toll*q);
  (* Is interpolation acceptable? *)
      IF  (2.0*p < factor) AND (p < Abs(0.5*e*q))  THEN
        BEGIN
          e := d ; d := p/q ; bisect := FALSE
        END
    END;
  IF  bisect  THEN
    Bisection;
  StepCompletion
END; (* MainLoop *)


BEGIN (* ZeroIn *)
  Initialization;
  Finalize;
  WHILE  (Abs(xm) > toll) AND (fb <> 0.0)  DO
    BEGIN
      MainLoop;
      Finalize
    END;
  root := b
END; (* ZeroIn *)
```

APPENDIX 4.3

Listing of the Pascal Function to evaluate the Bessel Function of Order *q*

```
FUNCTION Bessel( q : INTEGER ; x : REAL ) : REAL;
(* Bessel function of order  q  ,  q = 0,1,2,...
    Due to roundoff error in computing the series, this method of obtaining
    the value of the Bessel function at  x  should only be used for moderate
    values of abs(x).   *)
VAR
  k, kp1 : INTEGER;
  OldSum, sum, t, xBy2, y : REAL;
BEGIN
  t := 1.0;
  FOR  k := 1  TO  q  DO
    t := t/k;     (* At the termination of the loop, t = 1/q! *)
  xBy2 := x/2.0 ; y := xBy2 * xBy2 ; sum := t ; k := 0 ;
  REPEAT
    kp1 := k + 1;
    OldSum := sum;
    t := -y*t/(kp1*(kp1 + q));
    sum := sum + t;
    k := kp1
  UNTIL  OldSum = sum;
  Bessel := sum*Exp(q*Ln(xBy2))
END; (* Bessel *)
```

APPENDIX 4.4

User's Manual for the Pascal Procedure Powell, which is in the file Powell.Pas

The procedure Powell is a variant of M. J. D. Powell's hybrid algorithm for the solution of n equations in n unknowns. The algorithm is a variation of Newton's method, which uses a finite difference approximation to the Jacobian. Precautions are taken to avoid large step sizes or increasing residuals.

```
PROCEDURE Powell( n : INTEGER ; VAR x, f : VectorN ;
                   digits, MaxIterations, print : INTEGER ;
                    VAR flag : INTEGER ; VAR OutFile : TEXT );
```

The following must be declared in the calling program

```
        TYPE
          VectorN = ARRAY[ 1..MaxN ] OF REAL;
          MatrixNN = ARRAY[ 1..MaxN , 1..MaxN ] OF REAL;
          LongVector = ARRAY[ 1..n2 ] OF REAL;

    where  n2 = MaxN*(MaxN + 1)/2 .
```

The user supplies the following procedure to evaluate the functions

```
        PROCEDURE Fpowell( n : INTEGER ; VAR x, f : VectorN )

    where   f[j]  is the value of the  j-th function at  x[1],...,x[n].
```

```
Input:
     n           The number of equations and the number of unknowns
     x           The initial guess at the unknowns
   digits        The number of digits of accuracy desired in the computed zero
 MaxIterations   The maximum allowable number of iterations
   print         Output indicator
                    = 0  No output except for the "converged" solution, and
                           the initial guess.
                    = m  Output each  m  iterations, including the last one.
  OutFile        Output is written to the file OutFile.  Before calling
                  Powell, the user should ensure that OutFile has been
                  properly opened.

Output:
     x           The "converged" solution
     f           The values of the functions at the "converged" solution
    flag         A status indicator
                      0   Convergence achieved.
                      1   The maximum number of function evaluations,
                           MaxIterations*(n + 1), has been reached.
                      2   The input variable  digits  is too large.  No
                           further improvement in the approximate solution
                           is possible.  The user should decrease  digits.
                      3   The method has not made good progress.  Try
                           a new initial guess.
```

Procedure Powell contains the following machine dependent constant which is set for Turbo Pascal with 8087 support.

```
                giant = 1.0E+307
```

Procedure Powell is based on the IMSL FORTRAN subroutine ZSPOW.

APPENDIX 4.5

Listing of a Program using the Procedure Powell for Example 4.5

```
PROGRAM SolutionOfSimultaneousEquations;

CONST
  d = 'Con'; (* output to the console *)
  MaxN = 2 ; n2 = 3 ; (*  n2 = MaxN*( MaxN + 1 )/2  *)
  MaxIterations = 200 ; digits = 12 ;

TYPE
  VectorN = ARRAY[ 1..MaxN ] OF REAL;
  MatrixNN = ARRAY[ 1..MaxN , 1..MaxN ] OF REAL;
  LongVector = ARRAY[ 1..n2 ] OF REAL;
```

```
VAR
  n, print, flag : INTEGER;
  x, f : VectorN;
  OutFile : TEXT;

PROCEDURE Fpowell( n : INTEGER ; VAR x, f : VectorN );
BEGIN
  f[1] := x[1]*x[2]/10.0 - 1.0;
  f[2] := Exp(-x[1]*1.0E-05) + Exp(-x[2]) - 1.0001
END;

(*$I Powell.Pas *)

BEGIN
  ASSIGN(OutFile,d) ; REWRITE(OutFile) ;
  n := 2 ; print := 0 ;
  x[1] := 0.0 ; x[2] := 1.0 ;
  Powell(n,x,f,digits,MaxIterations,print,flag,OutFile);
  CLOSE(OutFile)
END.
```

Chapter 5

Approximation of Functions: Interpolation and Curve Fitting

We examine methods for approximating a given function by other functions, which could be polynomials or cubic splines. Interpolation as well as fitting to the trend of the data is discussed. The least squares method is introduced and applied by using orthogonal polynomials or the singular value decomposition.

5.1 INTRODUCTION

In mathematics, science, or engineering we often must approximate a function $f(x)$ by another function $g(x)$. Usually $g(x)$ is chosen so that it retains some of the properties of $f(x)$ that we think are important. In addition, $g(x)$ is usually a simpler function than $f(x)$, such as a polynomial or a rational function (the ratio of polynomials). Suppose we know the values of $f(x)$ for each x in the set $I = \{x_0, x_1, \ldots, x_n\}$, where $x_k < x_{k+1}$ for $k = 0, 1, \ldots, n - 1$. Set $g(x_i) = f(x_i)$ for $i = 0, 1, \ldots, n$; then $g(x)$ is an *exact approximation* to $f(x)$. In this case, we say that $g(x)$ *interpolates* $f(x)$. When we want to determine an approximate value for $f(z)$, where $x_0 < z < x_n$, we compute $g(z)$, the *interpolated* value of $f(z)$. If z is not in the interval $[x_0, x_n]$ then $g(z)$ gives an *extrapolated* value for $f(z)$. If $g(x_i)$ is not equal to $f(x_i)$ for one or more of the data points in I, then we have *inexact approximation* (curve fitting). Inexact approximation would be used in problems where the data $\{x_i, f(x_i)\}$ are corrupted by noise so that it is more important to approximate the trend than to exactly mimic the data points. We shall first discuss various methods for exact approximation and then methods for inexact approximation.

Interpolation was more commonly used before the widespread use of digital computers, when various functions such as sines, cosines, and logarithms appeared in books containing tabulated values of functions. When the value of the function was desired, it would be sought in the appropriate table of values. Often the value desired was not at one of the tabulated points; the value would have to be estimated from adjacent points, usually by using a degree one (linear) interpolating polynomial, $p(x)$, that passed through the two adjacent points. Thus, the value of the function at the intermediate data point, z, was estimated by calculating $p(z)$. For example, linear interpolation between the data points $f(0.2) = 1.22140$ and $f(0.3) = 1.34986$ gives $f(0.26) \simeq 1.29848$.

The method of table-lookup with interpolation is much too slow for digital computers. Standard mathematical functions such as $\sin(x)$, $\cos(x)$, $\ell og(x)$, and e^x are recomputed each time they are needed. In fact, each of these functions is computed by an approximation which is usually accurate to within several units of the machine epsilon (the least positive machine floating point number u such that $1 + u > 1$). Although table-lookup with interpolation is no longer used extensively, there are many numerical methods that are based on interpolation. Some of these are discussed in other chapters.

We shall mention one interesting use of interpolation in the display of graphical data on a microcomputer terminal. Suppose that we are given the discrete points $\{x_i, y_i\}$, $i = 0, \ldots, n$ which we know are obtained from a "smooth" curve. If we plot the points on the screen and draw straight lines between the points, we will obtain a curve with corners. The piecewise straight line curve interpolates the data but fails to give a reasonable shape to the curve. There are other interpolating functions (that we will study later) which will interpolate the given data and give more reasonable shapes to the curve that is being approximated.

A polynomial, $p(x)$, can be simply represented in a digital computer because only the coefficients need to be known. The value of the polynomial function at $x = t$ can be computed by Horner's rule using only additions and multiplications (see Section 3.2.1). Addition, subtraction, multiplication, integration, and differentiation of polynomials also can be performed by simple machine arithmetic operations. Because of this simplicity, polynomials are often used to interpolate continuous functions. The following theorem shows that a continuous function can be approximated arbitrarily closely by a polynomial of degree n.

Theorem 5.1 (Weierstrass Approximation Theorem)

If $f(x)$ is any continuous function on the finite interval $[a, b]$, then for every $\epsilon > 0$ there exists a polynomial $p(x)$, of degree n depending on ϵ, such that

$$\underset{a \le x \le b}{\text{maximum}}\{\, | f(x) - p(x) | \,\} < \epsilon$$

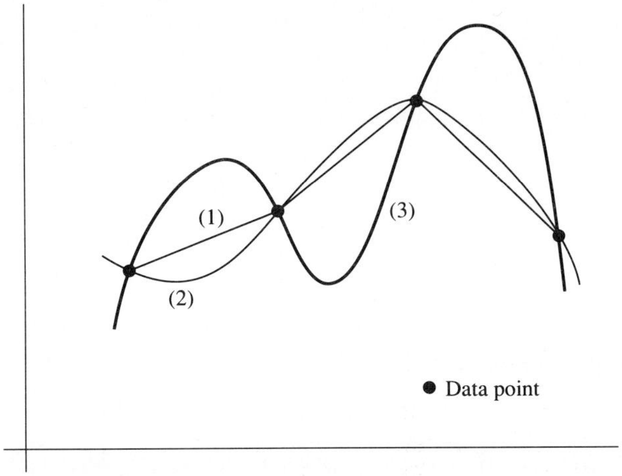

Figure 5.1 Examples of interpolation polynomials. (1) Piecewise linear; (2), (3) Polynomials of degree at least 3.

In the Weierstrass Theorem the degree of the approximating polynomial, n, may be so large as to be unsatisfactory for practical usage. Also, the theorem gives no information as to a *satisfactory* interpolating polynomial for a given set of data. Figure 5.1 shows a function defined at four data points; the function has been interpolated by a piecewise linear polynomial, and two other polynomials of degrees at least three. Although each of the approximating functions interpolates the given data, the question of which is "best" depends on additional information. One of the high degree polynomials has large oscillations between the data points, and, therefore, may not be suitable. This shows the difficulty of selecting a satisfactory interpolating function. In the next few sections we shall discuss various methods of constructing satisfactory interpolating polynomials.

Before we begin looking at various forms of interpolating polynomials, we need the following important theorem.

Theorem 5.2

Let $\{x_0, x_1, \ldots, x_n\}$ be a set of $n + 1$ distinct points, and let $\{y_0, y_1, \ldots, y_n\}$ be another set of $n + 1$ points, not necessarily distinct. If $p(x)$ is a degree n polynomial with the property that

$$p(x_i) = y_i \quad \text{for} \quad i = 0, 1, 2, \ldots, n$$

then $p(x)$ is unique.

Proof

Let $q(x)$ be another polynomial of degree n with the property that

$$q(x_i) = y_i \quad \text{for} \quad i = 0, 1, 2, \ldots, n$$

Consider the polynomial $h(x) = p(x) - q(x)$, which has degree at most n. We know that $h(x_i) = p(x_i) - q(x_i) = y_i - y_i = 0$, for $i = 0, 1, \ldots, n$. Thus, $h(x)$, a polynomial of degree at most n, has $n + 1$ distinct roots. But any degree n polynomial has at most n roots, unless that polynomial is the zero polynomial. Consequently, $h(x) = 0$ for all x, and thus $p(x) = q(x)$ for all x.

To see why Theorem 5.2 is important, consider a polynomial

$$p(x) = \sum_{k=0}^{n} a_k x^k$$

where $p(x_i) = y_i$ for $i = 0, 1, \ldots, n$ and the points $\{x_i\}$ are distinct. Suppose we are given another polynomial

$$q(x) = b_0(x - b_1)(x - b_2) \ldots (x - b_n)$$

where $b_1, b_2, \ldots, b_n$ are the roots of $q(x)$. Clearly, the forms of $p(x)$ and $q(x)$ are very different. However, if we know that $q(x)$ also has the property that $q(x_i) = y_i$ for $i = 0, 1, \ldots, n$, then according to Theorem 5.2, $q(x)$ and $p(x)$ are the same polynomial.

5.2 LAGRANGE INTERPOLATION

Suppose we are given the data $\{x_i, y_i\}$, $i = 0, 1, \ldots, n$. The points x_i must be distinct, but we can order them so that $x_k < x_{k+1}$ for $k = 0, 1, \ldots, n - 1$. We want to obtain the degree n polynomial, $L(x)$, that passes through all $n + 1$ data points; thus,

$$L(x_i) = y_i \quad \text{for} \quad 0 \le i \le n.$$

A simple form for $L(x)$, the Lagrange interpolation polynomial, is

$$L(x) = \sum_{i=0}^{n} y_i L_i(x) \tag{5.2–1}$$

where each $L_i(x)$ is a polynomial of degree at most n and

$$L_i(x_k) = \begin{cases} 1 & \text{if } k = i \\ 0 & \text{if } k \ne i \end{cases}$$

Consequently, $L_i(x)$ has the factors

$$(x - x_0) \ldots (x - x_{i-1})(x - x_{i+1}) \ldots (x - x_n)$$

Notice that the term $(x - x_i)$ is missing from this degree n polynomial. Because $L_i(x_i) = 1$ we must divide the product by

$$(x_i - x_0) \ldots (x_i - x_{i-1})(x_i - x_{i+1}) \ldots (x_i - x_n)$$

Therefore, we can write

$$L_i(x) = \frac{Q_i(x)}{Q_i(x_i)} \tag{5.2–2}$$

$$Q_i(x) = (x - x_0)(x - x_1) \ldots (x - x_{i-1})(x - x_{i+1}) \ldots (x - x_n) \tag{5.2–3}$$

If the data points were obtained from a function $f(x)$ we can determine the error made when (5.2–1) is used instead of $y = f(x)$. Suppose that $f(x)$ is $n + 1$ times continuously differentiable, and $y_i = f(x_i)$ for $i = 0, 1, \ldots, n$ then the error made when the Lagrange interpolation polynomial $L(x)$ is used instead of $f(x)$ is

$$f(x) - L(x) = \frac{f^{(n+1)}(\xi)\cdot Z(x)}{(n + 1)!} \tag{5.2–4}$$

where $f^{(k)}(x)$ denotes the kth derivative of $f(x)$, $x_0 \le \xi \le x_n$, and

$$Z(x) = (x - x_0)(x - x_1) \ldots (x - x_n)$$

Observe that if $x = x_k$ in (5.2–4), $0 \le k \le n$, then (5.2–4) holds for any number ξ in the interval $[x_0, x_n]$. If $x \ne x_k$, $0 \le k \le n$, define the function $g(t)$ as

$$g(t) = f(t) - L(t) - \frac{[f(x) - L(x)]Z(t)}{Z(x)}$$

where x is fixed, $x_0 < x < x_n$. The function $g(t)$ is $n + 1$ times continuously differentiable because $f(t)$ is, and $L(t)$ and $Z(t)$ are polynomials. Furthermore, $g(t) = 0$ for $t = x$ and $t = x_k$, $0 \le k \le n$. Thus, $g(t)$ has at least $n + 2$ distinct zeros in the interval $[x_0, x_n]$. If we apply Rolle's Theorem to $g(t)$ we find that $g^{(1)}(t)$ has at least $n + 1$ distinct zeros in the interval $x_0 < t < x_n$. Apply Rolle's Theorem to $g^{(1)}(t)$ to see that $g^{(2)}(t)$ has at least n distinct zeros in the interval $x_0 < t < x_n$. Continue in this manner until we find that $g^{(n+1)}(t)$ has at least one zero, ξ, in the interval $x_0 < \xi < x_n$. Thus, $g^{(n+1)}(\xi) = 0$, and

$$g^{(n+1)}(t) = f^{(n+1)}(t) - L^{(n+1)}(t) - \frac{[f(x) - L(x)]Z^{(n+1)}(t)}{Z(x)}$$

But, $L^{(n+1)}(t) = 0$ for all t because $L(t)$ is a degree n polynomial. $Z(t)$ is a degree $n + 1$ polynomial, so that $Z^{(n+1)}(t) = (n + 1)!$ for all t. Hence, (5.2–4) also holds for $x \ne x_k$, $0 \le k \le n$.

EXERCISE 5.1

Find the degree 2 polynomial that interpolates the data {(0.0, 0.0), (0.2, 0.3894), (0.5, 0.8415)}.

Answer $x(2.123 - 0.880x)$.

In elementary calculus we study a polynomial that can be used to approximate a function that is $n + 1$ times continuously differentiable. That polynomial is the Taylor polynomial, obtained by retaining the first $n + 1$ terms in the Taylor series expansion of a function about some point. The Taylor polynomial of $f(x)$, expanded about $x = x_0$, is

$$P(x) = \sum_{k=0}^{n} \frac{f^{(k)}(x_0)\cdot(x - x_0)^k}{k!} \tag{5.2–5}$$

The error made when (5.2–5) is used instead of $f(x)$ is

$$f(x) - P(x) = \frac{f^{(n+1)}(\tau)\cdot(x - x_0)^{n+1}}{(n + 1)!} \tag{5.2–6}$$

where τ is between x_0 and x. Notice that (5.2–6) is very similar to (5.2–4); they agree if we take $x_0 = x_i$ for $i = 1, 2, \ldots, n$ in (5.2–4).

The Lagrange interpolation polynomial (5.2–1) uses information about $f(x)$ at $n + 1$ points in the interval $[x_0, x_n]$, whereas the Taylor polynomial (5.2–5) uses information about $f(x)$ and its derivatives only at x_0. Figure 5.2 shows the errors made by the degree three interpolation polynomial ($n = 3$ in (5.2–1)) and the degree three Taylor polynomial when the function in Example 5.1, below, is used. In Fig. 5.2 the Lagrange interpolation polynomial error curve has been drawn with the error values multiplied by 100 because these errors are very much smaller, in magnitude, than the errors of the Taylor polynomial. Notice that the

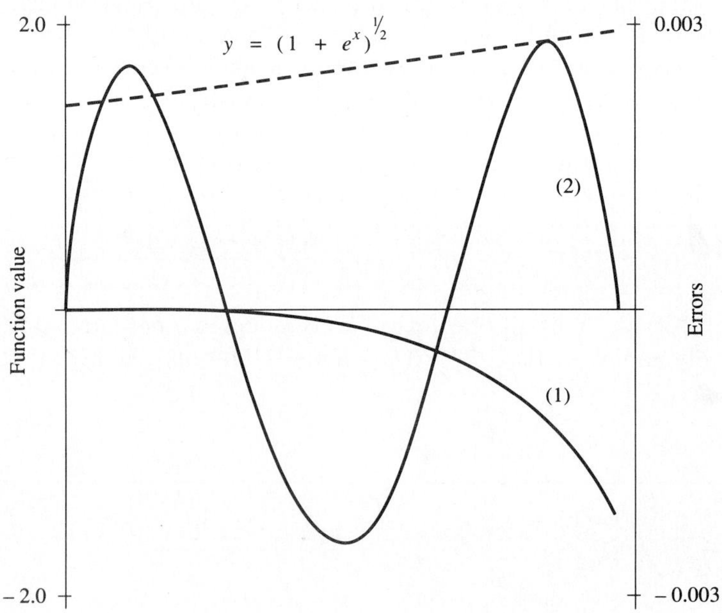

Figure 5.2 Errors in degree-3 polynomials approximating $f(x) = (1 + e^x)^{1/2}$ on the interval [0, 1]. (1) Taylor polynomial, (2) Lagrange interpolation polynomial ($\times 100$).

error in the Taylor polynomial is very small for x near zero, and decreases to -0.0022 as x increases to 1. The error due to the Lagrange interpolation polynomial attains maxima and minima between the data point values; its maximum absolute error is 2.8×10^{-5}. This example shows that an interpolation polynomial can be considerably better than a Taylor polynomial in approximating a function over an interval. However, the penalty one pays for increased accuracy in approximating the function is decreased accuracy in the derivative of the function.

EXAMPLE 5.1

$$f(x) = (1 + e^x)^{1/2}$$

with the interpolation points $x_0 = 0.0$, $x_1 = 0.3$, $x_2 = 0.7$, $x_3 = 1.0$.

One can create a table of values of a function in which interpolation is to be done to within a specified accuracy. We use the error formula (5.2–4) to estimate the error. For example, suppose we want to construct a table of values of the twice continuously differentiable function $f(x)$ at equally spaced x values, and do linear interpolation within the table to within an accuracy of 10^{-4}. Because linear interpolation is to be used, set $n = 1$ in (5.2–4) and take absolute values, giving $|E| \le |f^{(2)}(\xi)| \cdot |(x - x_0)(x - x_1)|/2$, where x_0 and x_1 denote adjacent points. Suppose that $\max\{|f^{(2)}(x)|\} \le M$ for all x to be included in the table of function values. Then,

$$|E| \le \frac{M}{2}\,[\underset{x_0 \le x \le x_1}{\text{maximum}}(|(x - x_0)(x - x_1)|)] = \frac{M}{8}(x_1 - x_0)^2 \le 10^{-4}$$

Consequently, the spacing between adjacent data points in the table should be $|x_1 - x_0| \le 10^{-2}(8/M)^{1/2}$.

5.3 NEWTON DIVIDED DIFFERENCES INTERPOLATING POLYNOMIAL

There is another form for a degree n polynomial that interpolates the data points $\{x_i, y_i\}$ for $0 \le i \le n$. It is the Newton divided differences polynomial, which has the form

$$P(x) = a_0 + a_1(x - x_0) + a_2(x - x_0)(x - x_1) + \ldots + a_n(x - x_0)(x - x_1)\ldots(x - x_{n-1}) \quad \textbf{(5.3–1)}$$

Setting $x = x_0$ shows that $a_0 = y_0$; setting $x = x_1$ gives

$$a_1 = \frac{y_1 - y_0}{x_1 - x_0}$$

We now give a systematic method for finding the coefficients a_i. In order to obtain these coefficients we introduce the notation

$$f[i, i+1, \ldots, i+k] = \frac{f[i+1, i+2, \ldots, i+k] - f[i, i+1, \ldots, i+k-1]}{x_{i+k} - x_i} \tag{5.3–2}$$

where $f[i] = f(x_i)$, $i = 0, 1, \ldots, n$. The right-hand side of (5.3–2) is a *divided difference*; hence the name of the Newton interpolating polynomial. The coefficients in (5.3–1) are given by

$$a_k = f[0, 1, \ldots, k] \quad \text{for} \quad 0 \le k \le n.$$

If $f(x)$ is k times continuously differentiable, then

$$f[i, i+1, \ldots, i+k] = \frac{f^{(k)}(\xi)}{k!}$$

for some ξ in the smallest interval containing the points $x_i, x_{i+1}, \ldots, x_{i+k}$. Therefore, the coefficient a_k is approximately the kth derivative of $f(x)$, divided by $k!$.

To see how one would compute divided differences, consider the function in Example 5.1. We have

$$f[0] = f(x_0) = f(0.0) \simeq 1.41421$$

$$f[1] = f(x_1) = f(0.3) \simeq 1.53292$$

$$f[2] = f(x_2) = f(0.7) \simeq 1.73606$$

$$f[3] = f(x_3) = f(1.0) \simeq 1.92828$$

Forming the first divided differences gives

$$f[0, 1] = \frac{f[1] - f[0]}{x_1 - x_0} = \frac{1.53292 - 1.41421}{0.3 - 0.0} \simeq 0.39570$$

$$f[1, 2] = \frac{f[2] - f[1]}{x_2 - x_1} = \frac{1.73602 - 1.53292}{0.7 - 0.3} \simeq 0.50773$$

$$f[2, 3] = \frac{f[3] - f[2]}{x_3 - x_2} = \frac{1.92828 - 1.73602}{1.0 - 0.7} \simeq 0.64089$$

The second divided differences are computed as

$$f[0, 1, 2] = \frac{f[1, 2] - f[0, 1]}{x_2 - x_0} = \frac{0.50773 - 0.39570}{0.7 - 0.0} \simeq 0.16003$$

$$f[1, 2, 3] = \frac{f[2, 3] - f[1, 2]}{x_3 - x_1} = \frac{0.64089 - 0.50773}{1.0 - 0.3} \simeq 0.19029$$

and the third divided difference, for this example, is

$$f[0, 1, 2, 3] = \frac{f[1, 2, 3] - f[0, 1, 2]}{x_3 - x_0} = \frac{0.19029 - 0.16003}{1.0 - 0.0} \simeq 0.03020$$

The divided differences (5.3–2) can be arranged in an array as shown in Table 5.1 for the function in Example 5.1. The coefficients of the Newton interpolating polynomial are the first entries in the columns labelled $f[i]$, $f[i, i + 1]$, . . . , $f[i, i + 1, i + 2, i + 3]$. The entries in column $f[i, i + 1]$, etc., are obtained by taking divided differences according to Eq. (5.3–2). For example, the table entry 0.16003 is $f[0, 1, 2]$; note that $f^{(2)}(0.35)/2! = 0.16119$.

EXERCISE 5.2

Form the Newton divided-differences array, similar to Table 5.1, for the cubic $f(x) = 2 + x - x^2 + x^3$, using the interpolation points 0, 1, 2, 3, 4.

Answer

x_i	$f[i]$	$f[i, i+1]$	$f[i, i+1, i+2]$	$f[i, i+1, i+2, i+3]$
0	2			
		1		
1	3		2	
		5		1
2	8		5	
		15		1
3	23		8	
		31		
4	54			

The Newton interpolation polynomial (5.3–1) has a simple form when the interpolation points $\{x_i\}$ are equally spaced. If $x_k = x_0 + kh$ for $k = 0, 1, \ldots, n$, and we want to do interpolation at $x = x_0 + zh$, for some number z, (5.3–1) becomes the Newton forward-difference interpolation formula

$$P(x_0 + zh) = \sum_{k=0}^{n} z(z - 1)(z - 2) \ldots (z - k + 1)h^k f[0, 1, \ldots, k]$$

where $f[0, 1, 2, \ldots, k]$ is given by (5.3–2). If the interpolation points are reordered as $x_n, x_{n-1}, \ldots, x_0$ and we do interpolation at $x = x_n - zh$, the polynomial is

Table 5.1 THE NEWTON DIVIDED-DIFFERENCES ARRAY FOR $f(x) = (1 + e^x)^{1/2}$

i	x_i	$f[i]$	$f[i, i+1]$	$f[i, i+1, i+2]$	$f[i, i+1, i+2, i+3]$
0	0.0	1.41421			
			0.39570		
1	0.3	1.53292		0.16003	
			0.50773		0.03020
2	0.7	1.73602		0.19024	
			0.64089		
3	1.0	1.92828			

the Newton backward-difference interpolation polynomial

$$P(x_n - zh) = \sum_{k=0}^{n} z(z-1)(z-2)\ldots(z-k+1)h^k f[n-k,\ldots,n-1,n]$$

where $f[n - k, \ldots, n - 1, n]$ is also given by (5.3–2).

The degree n Lagrange interpolation polynomial and the degree n Newton divided differences interpolation polynomial, when applied to the same set of $n + 1$ points, $\{x_i, f(x_i)\}$, will give the same polynomial, although the form is different. This follows from Theorem 5.2. Therefore, the degree n polynomial interpolating the data $\{x_i, f(x_i)\}$ is unique, and the right-hand side of the error expression (5.2–4) gives the error when the Newton form is used instead of the Lagrange form of the interpolating polynomial.

The Newton polynomial (5.3–1) can be efficiently evaluated at any x by Horner's rule, whereas the Lagrange polynomial cannot. Notice that the Newton interpolation polynomial only contains $x_0, x_1, \ldots, x_{n-1}$ although the coefficients a_i are determined by using all the data $\{x_i, f(x_i)\}$, where $i = 0, 1, \ldots, n$. Again, we assume the x_k are all distinct. If we set a_n to zero in (5.3–1), we obtain the degree $n - 1$ polynomial interpolating the data at the points $x_0, \ldots, x_{n-1}$.

Appendix 5.1 contains a Pascal procedure and function which obtains the Newton divided difference interpolating polynomial for a given data set. The Pascal procedure NewtonInterpolation is used to generate the coefficients a_i. The procedure uses a one-dimensional array to form the entries $f[i, i + 1, \ldots, i + k]$ because the entries $f[i, i + 1, \ldots, i + k - 1]$ are not needed once the entries $f[i, i + 1, \ldots, i + k]$ are formed. The value of the Newton interpolation polynomial can be computed by using the Pascal function EvaluateNewton. Appendix 5.2 contains a Pascal program that uses the Newton interpolation polynomial.

EXERCISE 5.3

Adapt the program in Appendix 5.2 to determine the coefficients of the Newton interpolation polynomial that interpolates $f(x) = x - \tan(x)$ at the points 0.0, 0.3, 0.5, 0.7, and 0.9.

Answer $a_0 = 0.0$, $a_1 = -0.03112$, $a_2 = -0.3074$, $a_3 = -0.6147$, $a_4 = -0.7721$, to four significant digits.

EXAMPLE 5.2

The function $f(x) = 1/(1 + 25x^2)$, $-1 \le x \le 1$, is called *Runge's function.* We generated the degree n interpolating polynomial for Runge's function by using the equally spaced data points $x_i = -1 + 2i/n$, for $0 \le i \le n$ using the program in Appendix 5.2. The error of maximum absolute value was obtained for each interpolating polynomial. Some of these errors are shown in Table 5.2.

EXAMPLE 5.2 (cont'd)

Table 5.2 ERRORS IN VARIOUS POLYNOMIALS INTERPOLATING $f(x) = 1/(1 + 25x^2)$ ON THE INTERVAL $[-1, 1]$

Degree	The Error of Maximum Absolute Value
5	0.433
10	1.92
15	2.10
20	58.3
25	75.8

As you can see in Table 5.2, contrary to intuition, the error in approximation does not go down as the number of data points increases. This phenomenon, not unique to Runge's function, requires some discussion.

Consider the problem of finding interpolating polynomials for a function that is continuous on the interval $[a, b]$. Define the triangular array x_{ij}, $0 \le i$, $0 \le j \le i$, called an *interpolation array*, with the property that for each i, all x_{ij} are distinct. Let $P_{fn}(x)$ denote the degree n polynomial that interpolates $f(x)$ at the points x_{nj}, $j = 0, 1, \ldots, n$. The result we observed for Runge's function is a special case of the following theorem.

Theorem 5.3

Given any interpolation array (x_{ij}) there exists a continuous function $g(x)$ and a point z in $[a, b]$ such that

$$\lim_{n\to\infty} |\, P_{gn}(z) - g(z) \,| \neq 0$$

The problem with Runge's function occurred because we took the interpolation points x_i with equal spacing; we may be able to avoid the effects of Theorem 5.3 by choosing an unequal spacing.

One method of selecting an unequal spacing for the interpolation points $\{x_i\}$ is to use the roots of the *Chebyshev* polynomials. The degree n Chebyshev polynomial is

$$T_n(x) = \cos[n\cdot\arccos(x)], \quad \text{for} \quad -1 \le x \le 1 \tag{5.3–3}$$

For a degree $n - 1$ interpolating polynomial, we select the x_i to be the roots of the degree n Chebyshev polynomial, which are located at

$$x_i = -\cos\left[\frac{(2i + 1)\pi}{2n}\right] \qquad i = 0, 1, \ldots, n - 1. \tag{5.3–4}$$

The minus sign in (5.3–4) orders the roots as $-1 < x_0 < x_1 < \cdots < x_{n-1} < 1$. One reason for selecting the x_i at these points is related to a property of Chebyshev polynomials. If $p(x)$ is any degree n polynomial, whose coefficient of x^n is one, then

$$2^{1-n} = \underset{-1\le x\le 1}{\text{maximum}}[2^{1-n} \mid T_n(x) \mid] \le \underset{-1\le x\le 1}{\text{maximum}}[\mid p(x) \mid] \tag{5.3–5}$$

The error term (5.2–4) for the degree n interpolation polynomial, either the Lagrange or the Newton form, contains the polynomial

$$Z(x) = (x - x_0)(x - x_1) \ldots (x - x_n)$$

$Z(x)$ is a degree $n + 1$ polynomial whose coefficient of x^{n+1} is one. According to (5.3–5) the least value of $\max\{ \mid Z(x) \mid \}$, $-1 \le x \le 1$, occurs if $Z(x) = 2^{-n}T_{n+1}(x)$; hence, $Z(x)$ and $T_{n+1}(x)$ have the same roots. Choosing the interpolation points according to (5.3–4), with n replaced by $n + 1$, will minimize the maximum absolute error, (5.2–4). If we want to interpolate the function $f(t)$ on the interval $[a, b]$ with a degree n polynomial, using the Chebyshev spacing, we locate the interpolation points at

$$t_i = 0.5(b - a)x_i + 0.5(b + a)$$

where the x_i are the roots of the degree $n + 1$ Chebyshev polynomial. We shall discuss some other important properties of the Chebyshev polynomials in Section 5.9.

When we used the Chebyshev spacing for the interpolation points with Runge's function, we obtained the results shown in Table 5.3. For comparison we also give the results using the equally spaced interpolation points. These results show the great improvement attained when the Chebyshev spacing is used for Runge's function. Remember, according to Theorem 5.3 there will be a continuous function for which the Chebyshev spacing can fail.

Table 5.3 ERRORS IN VARIOUS POLYNOMIALS INTERPOLATING $f(x) = 1/(1 + 25x^2)$ ON THE INTERVAL $[-1, 1]$

	The Error of Maximum Absolute Value	
Degree	Equal Spacing	Chebyshev Spacing
5	0.433	0.556
10	1.92	0.109
15	2.10	0.0831
20	58.3	0.0153
25	75.8	0.0114

EXERCISE 5.4

Repeat Exercise 5.3 for the interpolation points $x_i = 0.5s_i + 0.5$, where s_i are the roots of the degree 5 Chebyshev polynomial.

Answer $a_0 = -4.886 \times 10^{-6}$, $a_1 = -0.01632$, $a_2 = -0.2758$, $a_3 = -0.6468$, $a_4 = -0.9292$, to four significant digits.

5.4 HERMITE INTERPOLATION

The degree n Taylor polynomial, (5.2–5), agrees with the function $f(x)$ and its first n derivatives at x_0. Such a polynomial is called an *osculating* polynomial. We can generalize the Taylor polynomial by defining an osculating polynomial, $P(x)$, approximating a function $f(x)$ that has m continuous derivatives, as that polynomial of least degree satisfying the conditions

$$P^{(k)}(x_i) = f^{(k)}(x_i), \quad 0 \le i \le n, \quad 0 \le k \le m_i \le m \tag{5.4–1}$$

The degree of $P(x)$ is at most $n + \sum_{i=0}^{n} m_i$. We obtain the *Hermite osculating polynomials* if $m_i = 1$ for $i = 0, 1, \ldots, n$. The Hermite polynomial interpolating $f(x)$ at $x_0, x_1, \ldots, x_n$ agrees with $f(x)$ and its first derivative at the interpolation points $\{x_i\}$. The Hermite polynomial has degree at most $2n + 1$; it can be written in the form

$$H_{2n+1}(x) = \sum_{j=0}^{n} f(x_j)R_j(x) + \sum_{j=0}^{n} f'(x_j)W_j(x) \tag{5.4–2}$$

where $f'(x)$ denotes the derivative of $f(x)$,

$$R_j(x) = [1 - 2(x - x_j)L_j'(x_j)]\cdot[L_j(x)]^2,$$

$$W_j(x) = (x - x_j)\cdot[L_j(x)]^2,$$

$L_j(x)$ is defined by (5.2–2), and $L_j'(x)$ is the derivative of $L_j(x)$. Eq. (5.4–2) generalizes the Lagrangian interpolation polynomials, (5.2–1). Appendix 5.3 contains a listing of a Pascal function for Hermite interpolation of the data $\{x_i, f(x_i), f'(x_i)\}$, $i = 0, 1, \ldots, n$.

When we applied Hermite interpolation to the function in Example 5.1, we found the maximum absolute error was 4.8×10^{-10} with four interpolation points. The error for the degree three Newton interpolation polynomial was 2.8×10^{-5}. The Hermite polynomial used here was of degree nine, $(2 \times 4 + 1)$. The results obtained for Runge's function, Example 5.2, are summarized in Table 5.4. Notice how the errors grow as n increases when equally spaced data points are used; the Chebyshev spacing also helps here as it did with the Newton interpolation polynomial.

Table 5.4 ERRORS IN HERMITE INTERPOLATION OF THE FUNCTION $f(x) = 1/(1 + 25x^2)$ ON THE INTERVAL $[-1, 1]$

	The Error of Maximum Absolute Value	
n	Equal Spacing	Chebyshev Spacing
5	0.222	0.309
6	0.400	0.142
7	0.625	0.153
8	1.14	0.0722
9	1.99	0.0725
10	3.84	0.0358
11	7.10	0.0334
12	13.6	0.0172

5.5 INVERSE INTERPOLATION

Suppose we have the data $\{x_i, y_i\}$, where $y_i = f(x_i)$ and $x_k < x_{k+1}$ for $0 \le k \le n - 1$. We want to find $f(b)$ when b is not one of the interpolation points. We know that an interpolation polynomial will allow us to find an approximation to $f(b)$. There are cases where we know the value of $f(c)$ but we do not know c. For example, if we are trying to find a zero of the function $f(x)$ we know $f(c) = 0$ but c is unknown. When c is determined by interpolation, we call that process *inverse interpolation*. Recall that the procedure ZeroIn discussed in Chapter 4 uses inverse interpolation to estimate the sought-for zero of a function.

Because some functions have inverses that are multivalued, we would have to restrict the range of the inverse so that a single-valued inverse was used. For example, the inverse of the cosine function is multivalued, but we can restrict its inverse so that $0 \le \cos^{-1}(x) \le \pi$.

Table 5.5 INVERSE INTERPOLATION USING THE DEGREE THREE NEWTON POLYNOMIAL FOR THE FUNCTION $f(x) = (1 + e^x)^{1/2}$

$f(x)$	x	Absolute Value of the Error
1.5	0.222592759639	5.5×10^{-4}
1.6	0.445615890896	9.3×10^{-4}
1.7	0.637156506126	5.8×10^{-4}
1.8	0.805356443976	1.1×10^{-3}
1.9	0.958357543097	9.9×10^{-4}

If the function $f(x)$ has an inverse, $f^{-1}(x)$, on the interval $[x_0, x_n]$ we want to interpolate $f^{-1}(x)$ at the interpolation points $y_0, y_1, \ldots, y_n$. We use the mathematical identity $f^{-1}(f(x)) = x$ and organize the data as $\{y_i, x_i\}$. Hence, the roles of x and y are interchanged and the previously developed interpolation procedures can be used for inverse interpolation.

When inverse interpolation was used on the function in Example 5.1 to estimate x values for some given function values, the results shown in Table 5.5 were obtained.

EXERCISE 5.5

Use inverse interpolation with a degree 4 Newton interpolation polynomial to estimate the value of x for which $f(x) = 0.5$, where $f(x) = \sin(x)$ and the interpolation points are $\{0.0, 0.3, 0.5, 0.7, 0.9\}$.

Answer $x = 0.52352688$.

5.6 PIECEWISE POLYNOMIAL INTERPOLATION

So far we have only considered interpolating a function $f(x)$ by the same polynomial over the whole domain of interpolation points $\{x_i\}$. There is no reason why we could not use different interpolating polynomials over different subsets of the points $\{x_i\}$. The different polynomials would agree at the interpolation points, $\{x_i\}$, or *knots*. In this case, the interpolation function is a piecewise polynomial. The simplest case of piecewise polynomial interpolation occurs when successive points are joined by straight lines; this is illustrated in Fig. 5.1 which shows a piecewise linear interpolation polynomial joining the points. Can there be an advantage in using several low-degree piecewise polynomials interpolating $f(x)$ at the knots $\{x_i\}$ rather than a single high-degree polynomial? The answer is yes, because a high-degree polynomial may have large oscillations. When a high-degree polynomial is used for interpolation, these oscillations may occur between the knots, giving large errors. A low-degree polynomial, when interpolating fewer data points, will not have these oscillations. For example, consider the following function in Example 5.3.

EXAMPLE 5.3

$$f(x) = \sin(10x), \quad -1 \le x \le 1.$$

Using equally spaced knots we obtained the Newton interpolation polynomials of degrees 5, 10, 15, and 20. For comparison, we obtained the piecewise cubic Hermite interpolation polynomial for the corresponding knots; that is, we used the cubic Hermite polynomial, $p_i(x)$, that passes through the points $(x_i, f(x_i))$ and $(x_{i+1}, f(x_{i+1}))$. The polynomial $p_k(x)$ was used for interpolation on the interval

EXAMPLE 5.3 (cont'd)

$[x_k, x_{k+1}]$ for each $k = 0, 1, \ldots, n - 1$. The results are shown in Table 5.6. The piecewise cubic Hermite polynomial is very much better than the Newton polynomials for degrees 5, 10, and 15, and quite comparable for the degree 20 polynomial.

Table 5.6

Newton Interpolation Polynomial		Piecewise Cubic Hermite Polynomial	
Degree	Error	Knots	Error
5	2.71	6	0.501
10	2.47	11	0.0374
15	0.111	16	0.00790
20	0.00170	21	0.00255

The error of maximum absolute value in interpolating $f(x) = \sin(10x)$ on the interval $[-1, 1]$, using equally spaced knots and high-degree polynomials or piecewise cubic Hermite polynomials.

Piecewise polynomial interpolation can also be used to "cure" the problem associated with Theorem 5.3. We know that equally spaced knots caused both the Newton interpolation polynomials and the Hermite interpolation polynomials to diverge from Runge's function (Example 5.2). When we applied piecewise cubic Hermite polynomials to Runge's function we obtained the errors shown in Table 5.7; for comparison we also give the errors for various high-degree Hermite in-

Table 5.7

Knots	Hermite Interpolation Polynomial Error	Piecewise Cubic Hermite Polynomial Error
6	0.222	0.250
7	0.400	0.0908
8	0.625	0.114
9	1.14	0.0355
10	1.99	0.0556
11	3.84	0.0129
12	7.10	0.0293
13	13.6	0.00408

The error of maximum absolute value in interpolating $f(x) = 1/(1 + 25x^2)$ on the interval $[-1, 1]$, using equally spaced knots and high-degree Hermite polynomials or piecewise cubic Hermite polynomials.

terpolation polynomials. Note that if there are $n + 1$ knots, the Hermite interpolation polynomial has degree $2n + 1$. Table 5.7 shows that a piecewise low-degree interpolation polynomial approximation can be very much better than a high-degree interpolation polynomial approximation. There is an interesting effect occurring for the piecewise cubic Hermite interpolation polynomials: the error for $2k - 1$ knots is smaller than the error for $2k$ knots. When the number of knots is odd, the knots are symmetrically placed relative to the y-axis; this helps the piecewise polynomial approximation because Runge's function is an even function. The effect also occurs when Chebyshev spacing is used with high-degree Newton interpolation polynomials. The error of maximum absolute value for the interpolation polynomial of degree $2k$ is less than that for the degree $2k + 1$ polynomial.

To use the cubic Hermite polynomial as the basic unit for piecewise polynomial interpolation requires knowledge of the function $f(x)$ and its first derivative, $f'(x)$, at each of the knots. There are many problems in which the derivative of $f(x)$ is unknown; we may only know the values $\{x_i, f(x_i)\}$. We want to use piecewise cubic polynomials to interpolate the data. Following the technique used for piecewise cubic Hermite interpolation polynomials, we use a different polynomial on each interval $[x_i, x_{i+1}]$, $x_i < x_{i+1}$, $i = 0, 1, \ldots, n - 1$. All of these polynomials have the same degree; in addition, they are required to agree at the knots $x_1, \ldots, x_{n-1}$. This type of piecewise polynomial interpolation function is called a *spline*. In the next section we discuss the most commonly used spline, the cubic spline.

5.7 CUBIC SPLINES

Suppose we are given the data $\{x_i, y_i\}$, where $y_i = f(x_i)$ and $i = 1, 2, \ldots, n$. We want all x_i to be distinct; they can be ordered so that $x_i < x_{i+1}$, $1 \le i \le n - 1$. Let $S(x)$ denote the cubic spline that interpolates the data; that is, $S(x_i) = y_i$ for $i = 1, \ldots, n$. $S(x)$ will have the form

$$S(x) = y_i + a_i(x - x_i) + b_i(x - x_i)^2 + c_i(x - x_i)^3, \qquad x_i \le x \le x_{i+1}, \quad i = 1, 2, \ldots, n - 1. \tag{5.7-1}$$

In addition,

$$S(x_n) = y_n \tag{5.7-2}$$

There are $3(n - 1)$ unknowns in (5.7–1); $\{a_i, b_i, c_i\}$, $i = 1, 2, \ldots, n - 1$. To determine these unknowns we must use other information about $S(x)$. At each of the interior knots $\{x_2, x_3, \ldots, x_{n-1}\}$ we want the cubics on either side of the interior knot to agree; in addition, we require their first and second derivatives to agree also. These conditions enforce continuity and smoothness of the spline at the interior knots. We use the notation $g(v^+)$ to denote the limit of the function

$g(x)$ as $x \to v$ from the right; $g(v^-)$ denotes the limit of $g(x)$ as $x \to v$ from the left. Using this notation, the conditions at the interior knots can be written as

$$S(x_i^-) = S(x_i^+),\ S'(x_i^-) = S'(x_i^+),$$
$$S''(x_i^-) = S''(x_i^+), \qquad i = 2, 3, \ldots, n-1. \tag{5.7–3}$$

We have $3n - 5$ equations in (5.7–2) and (5.7–3); there are $3n - 3$ unknowns to be determined. This means that we must find two more equations; they can be chosen in any way that allows us to uniquely determine the unknowns.

If we set

$$S''(x_1) = S''(x_n) = 0 \tag{5.7–4}$$

we obtain the *natural* cubic spline.

If the function $f(x)$ is differentiable, then the conditions

$$S'(x_1) = f'(x_1), \qquad S'(x_n) = f'(x_n) \tag{5.7–5}$$

give a cubic spline with a *clamped boundary*.

We can also obtain the two extra equations by using the first four and the last four data points. We pass cubics $p(x)$ and $q(x)$ through each of these sets of four points; $p(x)$ and $q(x)$ are uniquely determined. Now set

$$S^{(3)}(x_1) = p^{(3)}(x_1), \qquad S^{(3)}(x_n) = q^{(3)}(x_n) \tag{5.7–6}$$

EXERCISE 5.6

Data are given at the points $x = -1, 0, 1, 2$. A natural cubic spline (5.7–4) is constructed and has the form

$$s(x) = \begin{cases} 2 - x + x^2 + x^3/3 & -1 \le x \le 0 \\ 2 - x + x^2 + x^3 & 0 \le x \le 1 \\ a + bx + cx^2 + dx^3 & 1 \le x \le 2 \end{cases}$$

Determine the values of a, b, c, and d.

Answer $a = 13/3$, $b = -8$, $c = 8$, $d = -4/3$.

Whatever extra equations are chosen, (5.7–2) and (5.7–3) are needed to find the coefficients $\{a_i, b_i, c_i\}$. We shall rewrite these equations in a simpler form. When (5.7–1) is substituted into (5.7–2) and (5.7–3) the following equations are obtained:

$$a_i + b_ih_i + c_ih_i^2 = \Delta_i, \qquad i = 1, 2, \ldots, n-1 \tag{5.7–8}$$

$$a_i + 2b_ih_i + 3c_ih_i^2 = a_{i+1}, \qquad i = 1, 2, \ldots, n-2 \tag{5.7–9}$$

$$2b_i + 6c_ih_i = 2b_{i+1}, \qquad i = 1, 2, \ldots, n-2 \tag{5.7–10}$$

where $h_i = x_{i+1} - x_i$, $\Delta_i = (y_{i+1} - y_i)/h_i$, $1 \le i \le n - 1$. Equations (5.7–8) to (5.7–10) and the two extra equations, (5.7–4 or 5.7–5 or 5.7–6), are to be solved for the unknowns a_i, b_i, and c_i. The solution of these equations can be simplified by introducing the new variables, σ_i, $i = 1, 2, \ldots, n$. In terms of the new variables we find that

$$a_i = \Delta_i - h_i(2\sigma_i + \sigma_{i+1}) \tag{5.7–11}$$

$$b_i = 3\sigma_i \tag{5.7–12}$$

$$c_i = (\sigma_{i+1} - \sigma_i)/h_i \tag{5.7–13}$$

$$i = 1, 2, \ldots, n - 1.$$

Equations (5.7–8) to (5.7–10) now reduce to

$$h_i\sigma_i + 2(h_i + h_{i+1})\sigma_{i+1} + h_{i+1}\sigma_{i+2} = \Delta_{i+1} - \Delta_i$$

$$i = 1, 2, \ldots, n - 2 \tag{5.7–14}$$

The remaining two equations come from the extra equations, (5.7–4 or 5.7–5 or 5.7–6), which can also be written in terms of the σ_i.

Appendix 5.4 contains a user's manual for the Pascal procedure, Spline, that solves (5.7–14) and (5.7–6) for the σ_i, and then determines $\{a_i, b_i, c_i\}$ using (5.7–11 to 5.7–13). A Pascal function, EvaluateSpline, is also described in the user's manual; as its name implies, EvaluateSpline evaluates the spline (5.7–1) at any desired x. Appendix 5.5 contains a listing of a Pascal program using Spline and EvaluateSpline to do interpolation with the function in Example 5.2. Table 5.8 compares the cubic spline with the piecewise cubic Hermite interpolation polynomial. The results obtained for the cubic spline are not as good as those of the piecewise cubic Hermite polynomial because the Hermite polynomial uses information about the derivative of $f(x)$ at each of the knots. However, the cubic spline can be used for functions whose derivative is unknown, as in Example 5.4.

Table 5.8

Knots	Cubic Spline Interpolation Polynomial Error	Piecewise Cubic Hermite Polynomial Error
6	0.431	0.250
7	0.130	0.0908
8	0.245	0.114
9	0.0560	0.0355
10	0.143	0.0556
11	0.0220	0.0129
12	0.0841	0.0293
13	0.00686	0.00408

The error of maximum absolute value in interpolating $f(x) = 1/(1 + 25x^2)$ on the interval $[-1, 1]$, using equally spaced knots and a cubic spline or piecewise cubic Hermite polynomials

EXAMPLE 5.4

We are given the following data:

i	1	2	3	4	5	6	7
x_i	−1.00	−0.95	−0.85	−0.80	0.20	0.50	0.90
y_i	−1.00	−0.05	0.90	1.00	0.90	0.50	−0.30

For purposes of plotting, we want to obtain a function that interpolates the data. We first obtain the degree 6 Newton interpolation polynomial, and then the cubic spline interpolation polynomial. The results are shown in Fig. 5.3. The cubic spline would be close to the curve most people would draw freehand to pass through the data points, whereas the Newton polynomial exhibits the large oscillations that can occur for high-degree polynomials.

EXERCISE 5.7

Adapt the program in Appendix 5.5 to obtain the coefficients of the cubic spline interpolating the function in Exercise 5.3.

Answer (To four significant digits)

i	a_i	b_i	c_i
1	−0.02583	0.1668	−0.6147
2	−0.09174	−0.3865	−0.3948
3	−0.2937	−0.6234	−1.538
4	−0.7277	−1.546	−1.310
5	−1.503	−2.332	−1.310

Thus far we have only discussed exact approximations; that is, interpolating functions, polynomials, or piecewise polynomials, have been found that exactly agree with the given data. This assumes that the data was precisely known. Often this is not the case; data may be corrupted by noise so that it does not make sense to obtain an exact approximation. Rather, we should try to obtain approximations that fit the trend of the data; this is inexact approximation. One of the techniques we shall examine uses orthogonal polynomials, which are discussed in the next section.

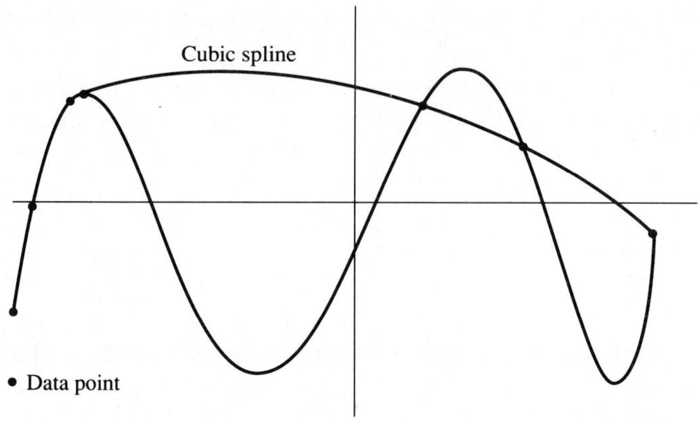

Figure 5.3 Example 5.4, comparing the cubic spline and the degree-6 interpolating polynomial.

5.8 ORTHOGONAL POLYNOMIALS

Orthogonal polynomials are special cases of orthogonal functions. In order to define orthogonal functions we need a method of determining if two functions are orthogonal, or perpendicular. We generalize the standard concept of the dot, or inner, product of two vectors. If the functions are defined on an interval $[a, b]$, and the functions are integrable, then the inner product of $f(x)$ and $g(x)$, denoted by the symbol $\langle f, g\rangle$, is defined as

$$\langle f, g\rangle = \int_a^b w(x)f(x)g(x)\,dx \tag{5.8–1}$$

where $w(x) > 0$ is a weighting function. If the functions $f(x)$ and $g(x)$ are only defined on a finite set, $\{x_i\}$, $i = 1, 2, \ldots, m$, the integral in (5.8–1) is replaced by a sum, to give

$$\langle f, g\rangle = \sum_{i=1}^{m} w_i f(x_i)g(x_i) \tag{5.8–2}$$

where $w_i > 0$, for $i = 1, 2, \ldots, m$, are the weights.

Definition 5.1

Two functions, $f(x)$ and $g(x)$, are called *orthogonal* with respect to a given inner product if $\langle f, g\rangle = 0$ when $f \neq g$.

It is often convenient to select $f(x)$ and $g(x)$ to be polynomials because orthogonal polynomials have a very nice property; they can be generated by a three-term recurrence relation of the form

$$P_{n+1}(x) = a_{n+1}(x - b_{n+1})P_n(x) - c_{n+1}P_{n-1}(x), \qquad n = 1, 2, \ldots \tag{5.8–3}$$

if $P_0(x)$ and $P_1(x)$ are known. If $P_0(x)$ has degree zero and $P_1(x)$ has degree one, then $P_n(x)$ is degree n.

We have already used one set of orthogonal polynomials, the Chebyshev polynomials, which were defined by (5.3–3). Table 5.9 lists some sets of orthogonal polynomials. There are, of course, many more orthogonal polynomials. The choice of a set of polynomials to be used depends on the nature of the problem. For many approximation problems, the Chebyshev polynomials are important. The Legendre polynomials are used in numerical quadrature methods, discussed in Chapter 6. For inexact approximation (curve fitting) it may not be convenient to use a given set of orthogonal polynomials; rather, we would use (5.8–2) and (5.8–3) to generate the set of polynomials based on the data. The Hermite orthogonal polynomials should not be confused with the Hermite interpolation polynomials used in Section 5.4.

We can generate a set of orthogonal polynomials with the inner product (5.8–2) and the recurrence relation (5.8–3) by first defining $P_0(x) = 1$ and taking $a_{n+1} = 1$ for all n. Set $s_0 = \langle P_0, P_0 \rangle = \sum_{i=1}^{m} w_i > 0$, and define $P_1(x) = (x - b_1)P_0(x)$, where b_1 is chosen so that $P_1(x)$ is orthogonal to $P_0(x)$. Substituting this expression for $P_1(x)$ into (5.8–2) and setting the inner product to zero gives $b_1 = \langle xP_0, P_0 \rangle / s_0$. Suppose we have computed the mutually orthogonal polynomials $P_0(x), P_1(x), \ldots, P_k(x)$. Set $s_k = \langle P_k, P_k \rangle$; $P_{k+1}(x)$ is determined from (5.8–3) with $n = k$ and $a_{k+1} = 1$; the constants b_{k+1} and c_{k+1} are determined by making $P_{k+1}(x)$ orthogonal to $P_k(x)$ and $P_{k-1}(x)$. This gives $b_{k+1} = \langle xP_k, P_k \rangle / s_k$ and $c_{k+1} = \langle xP_k, P_{k-1} \rangle / s_{k-1}$. A short calculation shows that $\langle xP_k, P_{k-1} \rangle = s_k$, so that the

Table 5.9 SOME SETS OF ORTHOGONAL POLYNOMIALS DEFINED BY (5.8–1)

Name	a	b	w(x)	Three-Term Recurrence
Chebyshev	-1	1	$(1 - x^2)^{-1/2}$	$T_{n+1}(x) = 2xT_n(x) - T_{n-1}(x)$ $T_0(x) = 1,\ T_1(x) = x$
Legendre	-1	1	1	$P_{n+1}(x) = (2n + 1)xP_n(x)/(n + 1) - nP_{n-1}(x)/(n + 1)$ $P_0(x) = 1,\ P_1(x) = x$
Hermite	$-\infty$	∞	e^{-x^2}	$H_{n+1}(x) = 2xH_n(x) - 2nH_{n-1}(x)$ $H_0(x) = 1,\ H_1(x) = 2x$
Laguerre	0	∞	e^{-x}	$L_{n+1}(x) = (2n + 1 - x)L_n(x) - n^2L_{n-1}(x)$ $L_0(x) = 1,\ L_1(x) = 1 - x$

orthogonal polynomials can be generated from (5.8–3), with $a_{n+1} = 1$, by

$$P_0(x) = 1 \tag{5.8–4}$$

$$P_1(x) = (x - b_1)P_0(x) \tag{5.8–5}$$

$$s_n = \langle P_n, P_n \rangle \qquad n = 0, 1, 2, \ldots \tag{5.8–6}$$

with

$$s_0 = \sum_{i=1}^{m} w_i$$

$$b_{n+1} = \langle xP_n, P_n \rangle / s_n \qquad n = 0, 1, 2, \ldots \tag{5.8–7}$$

$$c_{n+1} = s_n / s_{n-1} \qquad n = 1, 2, 3, \ldots \tag{5.8–8}$$

We shall use (5.8–4) to (5.8–8) in Section 5.10 for inexact approximation problems. Before doing that we discuss some properties of Chebyshev polynomials because of their importance in approximating functions.

EXERCISE 5.8

Find $P_2(x)$, given that $P_0(x) = 1$ and $P_1(x) = x$, using (5.8–4) to (5.8–8) with the inner product (5.8–2), where $m = 5$, $w_i = 1$ for $1 \leq i \leq 5$, and $x_1 = -0.9$, $x_2 = -0.4$, $x_3 = 0.0$, $x_4 = 0.4$, $x_5 = 0.9$.

Answer $P_2(x) = x^2 - 0.388$.

5.9 CHEBYSHEV POLYNOMIALS

Chebyshev polynomials were introduced first in Section 5.3, where we gave the closed form of $T_n(x)$ in (5.3–3). Also, (5.3–5) shows that of all degree n polynomials whose coefficient of x^n is one, $2^{1-n}T_n(x)$ has the least absolute value on the interval $[-1, 1]$. This property showed that selecting the interpolation points so that they are distributed according to the roots of the degree n Chebyshev polynomial minimizes the maximum absolute error in the degree $n - 1$ interpolating polynomial. We can also exploit this *minimax* property of Chebyshev polynomials to "economize" other polynomials; that is, replace a degree n polynomial by a degree $n - 1$ polynomial while minimizing the maximum error made.

5.9.1 Economization of Polynomials

Suppose we have the polynomial

$$p(x) = \sum_{i=0}^{n} a_i x^i, \; a_n \neq 0$$

and we want to replace $p(x)$ by a polynomial of degree $n - 1$ on the interval $[-1, 1]$. If we were given the function $F(t)$ on the interval $[a, b]$, we could use the transformation

$$t = 0.5(b - a)x + 0.5(b + a)$$

and define $p(x) = F(0.5(b - a)x + 0.5(b + a))$. Of course, we could merely delete the term in x^n from $p(x)$; instead, we use Chebyshev polynomials to write

$$x^n = \sum_{j=0}^{n} b_j T_j(x)$$

Hence,

$$p(x) = \sum_{i=0}^{n-1} a_i x^i + \sum_{j=0}^{n-1} a_n b_j T_j(x) + a_n b_n T_n(x)$$

If we now drop the term in $T_n(x)$ we will have a degree $n - 1$ polynomial, $q(x)$, that approximates $p(x)$ with an error $|p(x) - q(x)| \le |a_n b_n|$ because $|T_n(x)| \le 1$ on $[-1, 1]$. The following example illustrates the economization of a degree 3 polynomial.

EXAMPLE 5.5

If $p(x) = 1 + x/4 + 3x^2/32 + 7x^3/384$, find a quadratic, $q(x)$, that economizes $p(x)$ on the interval $[-1, 1]$. We need to write x^3 in terms of Chebyshev polynomials of at most degree three. The first four Chebyshev polynomials are $T_0(x) = 1$, $T_1(x) = x$, $T_2(x) = 2x^2 - 1$, and $T_3(x) = 4x^3 - 3x$. Clearly, $x^3 = 0.75T_1(x) + 0.25T_3(x)$. Substituting for x^3 into $p(x)$, and collecting terms gives

$$p(x) = q(x) + \left(\frac{7}{1536}\right)T_3(x)$$

where

$$q(x) = 1 + \left(\frac{135}{512}\right)x + \left(\frac{3}{32}\right)x^2$$

The error is

$$|p(x) - q(x)| \le \frac{7}{1536}, \quad \text{for} \quad -1 \le x \le 1.$$

If we just drop the cubic term in $p(x)$ the error is 7/384 when $-1 \le x \le 1$. The error made using Chebyshev economization is 1/4 the error made by directly dropping the cubic term.

5.9.2 Chebyshev Interpolation

We can obtain a degree n interpolation polynomial in terms of Chebyshev polynomials if we are allowed to choose the interpolation points at the roots of a

Chebyshev polynomial; see (5.3–4). To obtain the degree n polynomial interpolating the data $\{x_i, f(x_i)\}$, where the x_i are the roots of the degree $n + 1$ Chebyshev polynomial, we use the following identity for Chebyshev polynomials:

$$\sum_{i=0}^{n} T_j(x_i)T_k(x_i) = \begin{cases} 0 & \text{if } j \neq k \\ (n + 1)/2 & \text{if } j = k \neq 0 \\ n + 1 & \text{if } j = k = 0 \end{cases} \tag{5.9–1}$$

If the interpolating polynomial is

$$p(x) = \sum_{j=0}^{n} a_j T_j(x), \qquad p(x_k) = f(x_k), \quad k = 0, 1, \ldots, n \tag{5.9–2}$$

then multiply $p(x_i)$ by $T_k(x_i)$, sum over i, and use Eq. (5.9–1) to obtain the coefficients

$$a_0 = \frac{\sum_{i=0}^{n} f(x_i)}{n + 1} \tag{5.9–3}$$

$$a_j = \frac{2\left[\sum_{i=0}^{n} f(x_i)T_j(x_i)\right]}{n + 1}, \qquad j = 1, 2, \ldots, n \tag{5.9–4}$$

One advantage in writing an interpolating polynomial in the form (5.9–2) is that the coefficients a_j generally decrease in magnitude. When a polynomial is expressed in powers of x, or in the form of the Newton interpolating polynomial, the coefficients may not have this nice property. For example, when Runge's function, $f(x) = 1/(1 + 25x^2)$, is interpolated at the roots of the degree 11 Chebyshev polynomial, the degree 10 Newton interpolation polynomial and the degree 10 Chebyshev interpolation polynomial have the coefficients listed in Table 5.10. The Newton polynomial coefficients generally increase in magnitude, while

Table 5.10 COEFFICIENTS OF DEGREE 10 POLYNOMIALS INTERPOLATING $f(x) = 1/(1 + 25x^2)$ ON THE INTERVAL $[-1, 1]$

i	Newton Polynomial	Chebyshev Polynomial
0	0.03923	0.2011
1	0.08589	0.0
2	0.16988	−0.2745
3	0.40425	0.0
4	1.36189	0.1905
5	−0.51416	0.0
6	−10.20	−0.1371
7	26.78	0.0
8	−40.12	0.1057
9	46.16	0.0
10	−46.63	−0.09108

the Chebyshev coefficients decrease. Because Runge's function is an even function, the coefficients of the odd degree Chebyshev polynomials, $T_{2j+1}(x)$, are all zero. Remember, both of these interpolation polynomials represent the same degree 10 polynomial.

Chebyshev polynomials and other orthogonal polynomials are used in the next section for inexact approximation.

5.10 LEAST SQUARES CURVE FITTING USING ORTHOGONAL POLYNOMIALS

Suppose we have the data $\{x_i, y_i\}$, $i = 1, 2, \ldots, m$, and we know that they have been corrupted by noise. Because of the noise it would not be appropriate to find a function that interpolates the data. Instead we want to fit the trend of the data by finding a function $F(x)$ so that some measure of the errors at the points $\{x_i, y_i\}$ is minimized. We could select the form of $F(x)$ and introduce some undetermined parameters which would be chosen so that the error measure is minimized. For example, we could select a linear function $F(x) = ax + b$, an exponential function $F(x) = ae^{bx}$, or some other convenient function. Once the form of $F(x)$ is fixed, we must select an error criterion. There are many error functions we could use. We might want to minimize the maximum error (the minimax error),

$$\text{minimum}[\underset{1 \le i \le m}{\text{maximum}}(\,|\,y_i - F(x_i)\,|\,)]$$

but the resulting mathematics problem is extremely difficult. A somewhat easier mathematical problem occurs if we try to minimize the absolute deviation,

$$\text{minimize} \sum_{i=1}^{m} |\,y_i - F(x_i)\,|$$

but this would cause problems because the absolute value function is not differentiable when its argument is zero. We can avoid these differentiation problems if we try to minimize the sum of the squares of the errors,

$$\text{minimize} \sum_{i=1}^{m} [y_i - F(x_i)]^2 \tag{5.10–1}$$

Eq. (5.10–1) is the basis of the method of least squares. If $F(x) = ax + b$ we have the simplest case of linear least squares: to select a and b so that

$$S = \sum_{i=1}^{m} (y_i - ax_i - b)^2 \tag{5.10–2}$$

is minimized. By taking the partial derivative of S with respect to a and to b, and setting each derivative to zero, we obtain the *normal equations* for the parameters

a and b:

$$a \sum_{i=1}^{m} x_i^2 + b \sum_{i=1}^{m} x_i = \sum_{i=1}^{m} x_i y_i \tag{5.10–3}$$

$$a \sum_{i=1}^{m} x_i + bm = \sum_{i=1}^{m} y_i \tag{5.10–4}$$

EXAMPLE 5.6

A linear least squares curve fit to the data

i	1	2	3	4
x_i	0.00	0.35	0.70	1.00
y_i	1.41	1.56	1.74	1.93

using the approximating function $F(x) = ax + b$ gives the values $a = 0.5179$ and $b = 1.3946$.

We can generalize the linear least squares (5.10–2) by choosing the approximating function in the form

$$g(x) = \sum_{i=1}^{n} a_i \phi_i(x) \tag{5.10–5}$$

where the functions $\{\phi_i(x)\}$ form a *linearly independent set* and $\{a_i\}$ are coefficients to be determined. The functions $\{\phi_i(x)\}$ form a linearly independent set if the only solution to the equation

$$\sum_{i=1}^{n} b_i \phi_i(x) = 0, \quad \text{for all } x \text{ in some interval,} \tag{5.10–6}$$

is $b_i = 0$, $i = 1, 2, \ldots, n$. If (5.10–6) has a solution with at least one of the coefficients $b_i \neq 0$, then the functions associated with the nonzero b_i are linear combinations of the remaining functions, and they can be omitted from (5.10–5), thereby decreasing n.

If $n = m$ and $\phi_k(x)$ is a degree $k - 1$ polynomial, then (5.10–5) is the degree $n - 1$ interpolating polynomial. The number of functions is usually chosen to be much less than the number of data points, $n \ll m$, because we want to fit the trend of the data and not to interpolate the data.

Eq. (5.10–2) can be further generalized by introducing the weights, $w_i > 0$, so that some data can contribute more to the sum of the squares of the errors. We would do this if we knew that some data were more accurately determined than others; the more accurate data would be assigned larger weights than the less accurate data. The new weighted sum of squares of the errors is

$$S = \sum_{j=1}^{m} w_j \left[y_j - \sum_{i=1}^{n} a_i \phi_i(x_j) \right]^2 \tag{5.10–7}$$

Minimizing S in (5.10–7) gives the weighted least squares approximation to the data. We compute $\partial S/\partial a_k$, for $k = 1, 2, \ldots, n$, and set each partial derivative to zero to get the following linear system of equations, the normal equations,

$$\mathbf{Ba} = \mathbf{c} \tag{5.10–8}$$

where **B** is an $n \times n$ matrix whose (i, j)th entry is b_{ij}, **a** is a column vector whose jth entry is a_j, and **c** is a column vector whose ith entry is c_i, with

$$b_{ij} = \sum_{k=1}^{m} w_k \phi_i(x_k) \phi_j(x_k) \tag{5.10–9}$$

$$c_i = \sum_{k=1}^{m} w_k y_k \phi_i(x_k) \tag{5.10–10}$$

In Example 5.6 we had $m = 4$, $n = 2$, $w_1 = w_2 = 1$, $\phi_1(x) = 1$, and $\phi_2(x) = x$.

It is not advisable to directly solve (5.10–8) because the matrix **B** is usually ill-conditioned, especially if the data involves polynomials using powers of x, $\{1, x, x^2, \ldots, x^n\}$. We need a more efficient method of solving the linear system, (5.10–8). Eq. (5.10–9) looks like (5.8–2), the inner product over a finite set. If we choose the functions $\{\phi_i(x)\}$ to be orthogonal with respect to the inner product, which is the right-hand side of (5.10–9), then **B** is a diagonal matrix and $a_j = c_j/b_{jj}$ for $j = 1, 2, \ldots, n$. In order to solve (5.10–8) efficiently we can generate the functions $\{\phi_i(x)\}$ as orthogonal polynomials using (5.8–3) to (5.8–8), with the inner product defined by (5.10–9).

Choosing the functions $\{\phi_i(x)\}$ as orthogonal polynomials has some advantages. The orthogonal polynomials can be easily generated by (5.8–3) to (5.8–8). If this is done, then $\phi_{k+1}(x)$ is the degree k orthogonal polynomial; the function $g(x)$ in (5.10–5) is a degree $n - 1$ polynomial given in terms of a sum of orthogonal polynomials. The form of (5.10–5) and the choice of $\{\phi_i(x)\}$ as orthogonal polynomials easily allows us to obtain lower degree least squares approximations. For example, to obtain the least squares degree $n - 2$ polynomial, we just omit the term $a_n\phi_n(x)$ from (5.10–5).

Appendix 5.6 contains a Pascal procedure and function for obtaining orthogonal polynomials. Appendix 5.7 contains a Pascal program for least squares curve fitting with orthogonal polynomials for the function in Example 5.7. The orthogonal polynomials are generated by computing the coefficients b_i and c_i in (5.8–3) and the coefficients a_i in (5.10–5) in the Pascal procedure OrthogonalPolynomial. The approximating function (5.10–5) can be evaluated using the associated Pascal function EvaluateOrthogonalPolynomial.

EXAMPLE 5.7

Data are generated by evaluating the function $f(x) = e^x$ at 11 equally spaced points on the interval $[-1, 1]$. The data are $\{x_i, f_i\}$, where f_i retains only the first two significant digits in $f(x_i)$. We allowed the number of terms in (5.10–5) to vary from 2 to 11. For each integer n, $2 \le n \le 11$, we computed

$$\underset{-1\le x\le 1}{\text{maximum}}\left[\left|e^x - \sum_{i=1}^{n} a_i\phi_i(x)\right|\right].$$

The following results were obtained.

n	2	3	4	5	6	7	8	9	10	11
Max. error	0.3856	0.0635	0.0142	0.0093	0.0089	0.0101	0.0108	0.0101	0.0235	0.0666

The maximum error is least for $n = 6$, a degree 5 polynomial.

EXAMPLE 5.8

It is believed that the following set of data comes from a process that behaves like $f(x) = e^{x/2}$. The data were obtained from an experiment that measured $y_i = f(x_i)$ with an experimental error of about five parts in one thousand.

x_i	−1.0	−0.8	−0.6	−0.4	−0.2	0.0	0.2	0.4	0.6	0.8	1.0
y_i	0.61	0.67	0.74	0.82	0.90	1.00	1.11	1.22	1.35	1.49	1.65

We adapted the program in Appendix 5.7 to fit the data to orthogonal polynomials of degrees 1 to 10 ($2 \le n \le 11$). When the degree k orthogonal polynomial, $p_k(x)$, was compared with the function $f(x) = e^{x/2}$, it was found that

$$\underset{1\le k\le 10}{\text{minimum}}[\underset{-1\le x\le 1}{\text{maximum}}(|p_k(x) - e^{x/2}|)] = 0.0024 \quad \text{for } k = 3 (n = 4)$$

when the weights were $w_i = 1$ for $i = 1, 2, \ldots, 11$. When the weights were $w_i = i$ the minimax error was 0.0004 for $k = 3 (n = 4)$. This shows that changing the weights can change the error and, in some cases, may change the "best" polynomial fit to the data.

EXERCISE 5.9

Use the orthogonal polynomials and inner product from Exercise 5.8 to determine the least squares fit $p(x) = c_0P_0(x) + c_1P_1(x) + c_2P_2(x)$ to the data

x_i	-0.9	-0.4	0.0	0.4	0.9
y_i	1.0	1.3	1.1	1.4	2.1

Answer $c_0 = 1.38$, $c_1 = 0.53093$, $c_2 = 0.43525$.

5.11 LEAST SQUARES CURVE FITTING USING THE SINGULAR VALUE DECOMPOSITION

The functions $\{\phi_i(x)\}$ in (5.10–5) could have been given *a priori*. These functions may not be orthogonal with respect to the inner product (5.8–2) that forms the entries of the matrix **B** in the normal equations (5.10–8). We could use the Gram-Schmidt process to generate orthogonal functions $\{\psi_i(x)\}$ from the given functions $\{\phi_i(x)\}$ using the given inner product. When this is done the old functions can be written in terms of the new functions, at the points $\{x_i\}$, as $\mathbf{\Phi} = \mathbf{\Psi S}$, where $\mathbf{\Phi}$ is an $m \times n$ matrix whose (i, j)th entry is $\phi_j(x_i)$, $\mathbf{\Psi}$ is an $m \times n$ matrix whose (i, j)th entry is $\psi_j(x_i)$, and **S** is an upper triangular $n \times n$ matrix with ones on the main diagonal and the coefficients from the Gram-Schmidt process in the upper part of the matrix. If **g** is the column vector whose ith entry is $g(x_i)$, where $g(x)$ is given by (5.10–5), then $\mathbf{g} = \mathbf{\Phi a} = \mathbf{\Psi a}_*$, where **a** is the unknown vector to be found in (5.10–8), $\mathbf{a}_*$ is the solution of (5.10–8) when the orthogonal functions $\{\psi_i(x)\}$ are used in (5.10–9) and (5.10–10). Hence, $\mathbf{Sa} = \mathbf{a}_*$. But **S** is upper triangular so that the solution of $\mathbf{Sa} = \mathbf{a}_*$ is obtained easily.

The Gram-Schmidt process can be influenced by roundoff errors, so that the method may not give as accurate answers as one would wish. A better computational method involves the use of the *singular value decomposition* of a matrix. Before discussing the singular value decomposition, we need to reexamine the problem of curve fitting when the functions $\{\phi_i(x)\}$ in (5.10–5) are specified *a priori*.

If we are given the functions $\{\phi_i(x)\}$ we can evaluate (5.10–5) at each point x_k to give the over-determined system of equations

$$v_k y_k = v_k \sum_{i=1}^{n} a_i \phi_i(x_k), \qquad 1 \le k \le m \tag{5.11–1}$$

where the positive weights are $w_k = v_k^2$, $v_k \neq 0$. We have left the nonzero factor v_k on both sides of (5.11–1) in order to show how these equations relate to the normal equations (5.10–8). The m equations in (5.11–1) can be written as

$$\mathbf{y} = \mathbf{Da} \tag{5.11–2}$$

where the kth entry of the column vector **y** is $v_k y_k$, the (k, i)th entry of the $m \times n$ matrix **D** is $v_k \phi_i(x_k)$, and the ith entry of the column vector **a** is a_i. The set of functions $\{\phi_i(x)\}$ is assumed to be linearly independent, so that the matrix $\mathbf{D}^T\mathbf{D}$ is invertible. When (5.11–2) is multiplied on the left by $\mathbf{D}^T$ the resulting system is (5.10–8) with $\mathbf{B} = \mathbf{D}^T\mathbf{D}$ and $\mathbf{c} = \mathbf{D}^T\mathbf{y}$. We do not want to do this matrix multiplication, because it leads to the normal equations which can be ill-conditioned. Instead, we use the singular value decomposition to express the matrix **D** in a form in which a "solution" of the over-determined system is obtained.

The singular value decomposition of a real $m \times n$ matrix **H**, $m \geq n$, is a matrix factorization of the form

$$\mathbf{H} = \mathbf{U}\mathbf{S}\mathbf{V}^T \tag{5.11–3}$$

where **U** is an $m \times m$ orthogonal matrix, **V** is an $n \times n$ orthogonal matrix, and **S** is an $m \times n$ matrix with $s_{ij} = 0$ if $i \neq j$ and $s_{ii} = \sigma_i \geq 0$ for $i = 1, 2, \ldots, n$. (An $n \times n$ real matrix **P** is an orthogonal matrix if $\mathbf{P}^T\mathbf{P}$ is the identity matrix.) The numbers σ_i are called the *singular values* of **H**: the columns of **U** are the *left singular vectors* of **H**; the columns of **V** are the *right singular vectors* of **H**. The matrices $\mathbf{H}\mathbf{H}^T$ and $\mathbf{H}^T\mathbf{H}$ have the same nonzero eigenvalues. Recall that the number λ, real or complex, is an *eigenvalue* of the $n \times n$ matrix **A** if there is a nonzero vector **x**, called an *eigenvector*, such that $\mathbf{A}\mathbf{x} = \lambda\mathbf{x}$ (see Appendix A.4). The singular values of **H** are the positive square roots of the eigenvalues of $\mathbf{H}^T\mathbf{H}$. See Appendix 5.10 for a derivation of the singular value decomposition (5.11–3).

The singular values of **H** can be ordered so that

$$\sigma_1 \geq \sigma_2 \geq \sigma_3 \geq \cdots \geq \sigma_n \geq 0. \tag{5.11–4}$$

If we have an $n \times m$ matrix, **K**, with $n \leq m$, set $\mathbf{K}^T = \mathbf{H}$, find the singular value decomposition of **H**, and then transpose (5.11–3) to obtain that of **K**. Because we have assumed that $m \geq n$, only thc first n columns of **U** need be computed. If the singular values are ordered according to (5.11–4) and $\sigma_r > 0$ but $\sigma_{r+1} = 0$, then only the first r columns of **U** need be found.

EXAMPLE 5.9

The matrix

$$\mathbf{H} = \begin{bmatrix} 1.0 & 2.0 & 3.0 \\ 6.0 & 7.0 & 8.0 \\ 11.0 & 12.0 & 13.0 \\ 16.0 & 17.0 & 18.0 \\ 21.0 & 22.0 & 23.0 \end{bmatrix}$$

has the singular values $\sigma_1 = 54.0132$, $\sigma_2 = 1.6034$, and $\sigma_3 = 0.0$. The first three columns of the matrix **U** are (to four significant digits)

−0.06537	−0.7718	−0.04435
−0.22551	−0.4991	0.3584
−0.38564	−0.2265	−0.06324
−0.54577	0.04623	−0.7712
−0.70590	0.3189	0.5204

EXAMPLE 5.9 (cont'd)

and the matrix **V** is

$$\mathbf{V} = \begin{bmatrix} -0.5409 & 0.7354 & 0.4082 \\ -0.5766 & 0.02915 & -0.8165 \\ -0.6123 & -0.6771 & 0.4082 \end{bmatrix}$$

To "solve" (5.11–2) we obtain the singular value decomposition of **D** as $\mathbf{D} = \mathbf{USV}^T$; substitute this into (5.11–2) and multiply by $\mathbf{U}^T$. The resulting equation is $\mathbf{Sr} = \mathbf{U}^T\mathbf{y}$, where $\mathbf{r} = \mathbf{V}^T\mathbf{a}$. The last $m - n$ rows of **S** are zero; consequently, only the first n rows of $\mathbf{U}^T\mathbf{y}$ contribute to the determination of the entries in **r**. The last $m - n$ rows of $\mathbf{U}^T\mathbf{y}$ involve the last $m - n$ columns of **U** which are not actually required. Recall that the first n rows and n columns of **S** form a diagonal matrix with the $\{\sigma_i\}$ on the main diagonal and ordered according to (5.11–4). If $\sigma_j = 0$, $k + 1 \leq j \leq n$, then the ith entry of **r** is given by $r_i = (\mathbf{U}^T\mathbf{y})_i/\sigma_i$ for $i = 1, 2, \ldots, k$ and r_i is arbitrary for $i = k + 1, k + 2, \ldots, n$; we can set $r_i = 0$ for $i > k$. Then $\mathbf{a} = \mathbf{Vr}$. This constitutes a linear least squares solution of the problem because the sum of the squares of the errors can be written as

$$(\mathbf{Da} - \mathbf{y})^T(\mathbf{Da} - \mathbf{y}) = (\mathbf{USV}^T\mathbf{a} - \mathbf{y})^T(\mathbf{USV}^T\mathbf{a} - \mathbf{y}) = (\mathbf{Sr} - \mathbf{U}^T\mathbf{y})^T(\mathbf{Sr} - \mathbf{U}^T\mathbf{y})$$

This is minimized if **r** is computed by the method described above.

If $m = n$ and the matrix **D** is invertible, the least squares singular value decomposition solution gives the unique solution to (5.11–2).

Often some of the singular values are nearly zero; if we use them to compute $r_i = (\mathbf{U}^T\mathbf{y})_i/\sigma_i$ then we might be introducing large errors into the solution. To avoid this we select a tolerance, $\epsilon > 0$. If $(\sigma_j/\sigma_1) \leq \epsilon$ for some $j > 1$, we treat the σ_i, $j \leq i \leq n$, as if they were zero in the calculation of **r**.

Appendix 5.8 contains a user's manual for a Pascal procedure to find the singular value decomposition of a real matrix. Appendix 5.9 contains a listing of a Pascal program using singular value decomposition with Example 5.10. The numerical methods used to compute the singular values and the matrices **U** and **V** in (5.11–3) are also used to compute the eigenvalues and eigenvectors of real symmetric matrices. These techniques are discussed in Chapter 8.

EXAMPLE 5.10

We use the data from Example 5.8. The functions $\{\phi_i(x)\}$ were chosen as $\phi_{k+1}(x) = x^k$. With the weights $w_i = 1$ and $g(x) = \sum_{k=1}^{n} a_k x^{k-1}$, it was found that $\underset{2 \leq n \leq 11}{\text{minimum}} \left[\underset{-1 \leq x \leq 1}{\text{maximum}} \left(| g(x) - e^{x/2} | \right)\right] = 0.0024$ for $n = 4$. This is the same as found in Example 5.8.

EXAMPLE 5.11

In many engineering and science problems we may have a good idea that a given type of mathematical model describes a physical system that we want to study. The model usually contains some parameters which must be determined from measurements of the actual behavior of the system. One way to do this is to use the method of least squares to find the "best" values of the parameters. For example, the following data is believed to come from a system that evolves according to the differential equation

$$\frac{dy}{dt} = ay(1 - by)$$

where a and b are the parameters to be determined. The measurements were taken at $t = kh$, for some constant h.

k	$y(hk)$	k	$y(hk)$	k	$y(hk)$
0	0.333	17	0.712	34	0.931
1	0.355	18	0.732	35	0.937
2	0.379	19	0.751	36	0.943
3	0.402	20	0.769	37	0.948
4	0.427	21	0.786	38	0.952
5	0.451	22	0.803	39	0.957
6	0.476	23	0.818	40	0.961
7	0.501	24	0.832	41	0.964
8	0.526	25	0.846	42	0.967
9	0.551	26	0.858	43	0.970
10	0.576	27	0.870	44	0.973
11	0.600	28	0.881	45	0.976
12	0.624	29	0.891	46	0.978
13	0.647	30	0.900	47	0.980
14	0.669	31	0.909	48	0.982
15	0.691	32	0.917	49	0.983
16	0.712	33	0.924	50	0.985

We want to find the values of a and b so that

$$S(a, b) = \sum_{k=0}^{50} \left[\frac{dy}{dt}(hk) - ay(hk)\,[1 - by(hk)]\right]^2$$

is a minimum. Although we know $y(hk)$ we do not know its derivatives which are needed in $S(a, b)$. We shall fit the data to a cubic spline and estimate the derivative of $y(t)$ by using the derivative of the spline function. The Pascal function EvaluateSpline in Appendix 5.4 can be adapted to return a value of the derivative. Let $y_k = y(hk)$ and let z_k denote the estimated value of the derivative of y at t

EXAMPLE 5.11 (cont'd)

$= hk$. Instead of minimizing $S(a, b)$ we shall minimize

$$f(x_1, x_2) = \sum_{k=0}^{50} [z_k - x_1 y_k + x_2 (y_k)^2]^2$$

where $x_1 = a$ and $x_2 = ab$. The partial derivatives of f are

$$\frac{\partial f}{\partial x_1} = -2 \sum_{k=0}^{50} [z_k - x_1 y_k + x_2 (y_k)^2] y_k$$

$$\frac{\partial f}{\partial x_2} = 2 \sum_{k=0}^{50} [z_k - x_1 y_k + x_2 (y_k)^2] (y_k)^2$$

When the simultaneous equations

$$\frac{\partial f}{\partial x_1} = 0 \qquad \frac{\partial f}{\partial x_2} = 0$$

are solved for x_1 and x_2 we find that $x_1 = 0.0997676$ and $x_2 = 0.0998410$. Because the original data were correct to only three significant digits, we accept the values $a = 0.0998$ and $b = 1.00$.

Example 5.11 can also be solved using the singular value decomposition. In this case, we want to solve the system of linear equations $\mathbf{Ax} = \mathbf{b}$, where $\mathbf{x} = (x_1, x_2)^T$, and

$$\mathbf{A} = \begin{bmatrix} y_0 & -(y_0)^2 \\ y_1 & -(y_1)^2 \\ \vdots & \vdots \\ y_{50} & -(y_{50})^2 \end{bmatrix} \qquad \mathbf{b} = \begin{bmatrix} z_0 \\ z_1 \\ \vdots \\ z_{50} \end{bmatrix}$$

The program in Appendix 5.9 was adapted to solve this problem; the computed solution for x_1 and x_2 agreed with the solution given above.

5.12 RICHARDSON EXTRAPOLATION

If we approximate a function $f(x)$ over an interval $[a, b]$ by another function $g(x)$ and we want to compute $g(z)$ for some z outside the interval $[a, b]$, then we are trying to use $g(x)$ to *extrapolate* to the value $f(z)$. Extrapolation does not work very well for cases where $g(x)$ is constructed as an interpolating function. However, there is a case where extrapolation can be used to obtain more accurate estimates of an extrapolation value.

Suppose we know the value of a function of x, $f(x)$ as a series in powers of x, and we want to estimate $f(0)$; that is, we want to extrapolate the known value $f(x)$ to obtain the value $f(0)$. This is known as *Richardson extrapolation to the limit* or *deferred approach to the limit*. It is often used in the numerical evaluation of integrals or in the numerical solution of differential equations.

The function $f(x)$ is assumed to have the form

$$f(x) = a_0 + a_p x^p + a_{p+1} x^{p+1} + \cdots \tag{5.12–1}$$

where $|x|$ is small enough so that the coefficients a_k, $k \geq 1$, can be considered as constants. The constant a_0 is to be estimated. If we drop the terms in powers of x on the right-hand side of (5.12–1), a_0 will be given by $f(x)$, accurate to terms containing x^p. We want a better estimate of a_0. To do this, we compute $f(qx)$, where $0 < q < 1$,

$$f(qx) = a_0 + a_p (qx)^p + a_{p+1} (qx)^{p+1} + \cdots \tag{5.12–2}$$

Subtracting (5.12–2) from (5.12–1) and dividing by $(1 - q^p)$ gives

$$a_p x^p = \frac{f(x) - f(qx)}{1 - q^p} + O(x^{p+1}) \tag{5.12–3}$$

In (5.12–3) the expression $O(x^{p+1})$ denotes terms bounded in magnitude by Cx^{p+1}, where C is independent of x. When (5.12–3) is substituted into (5.12–1) we obtain

$$a_0 = f(x) - \frac{f(x) - f(qx)}{1 - q^p} + O(x^{p+1}) \tag{5.12–4}$$

Equation (5.12–4) gives $f(0)$ accurate to $O(x^{p+1})$.

EXAMPLE 5.12

The following illustrates the Richardson extrapolation method.

$$f(x) = \begin{cases} \dfrac{e^x - 1}{x} & \text{if } x \neq 0 \\ 1 & \text{if } x = 0 \end{cases}$$

We want to compute $f(x)$ for various values of $x \neq 0$ so that $f(0)$ can be estimated. It is easy to see that $p = 1$ in (5.12–1) for this function. When we calculate $f(0.25)$ and $f(0.125)$ we find their values as 1.1361017 and 1.0651876, respectively. Applying (5.12–4), with $p = 1$ and $q = 0.5$, gives $f(0) \simeq 0.9942736$, with an error bounded by 0.006. The error in $f(0.125)$ is approximately 0.065; Richardson extrapolation has significantly improved the result.

5.13 ADDITIONAL TOPICS IN APPROXIMATING FUNCTIONS

There are several things to keep in mind if you want to use polynomials or other functions for approximation. We shall discuss these in turn.

The first point to consider is the suitability of the given function for approximation by the chosen functions. We have seen cases where a single polynomial is not appropriate for the entire interval of interest; a piecewise polynomial may be better. There may be properties of the function to be approximated that would restrict the choice of the approximating functions. For example, the function and/or its lower order derivatives may have a singularity or discontinuity in the approximation interval, including the end points. An approximation should be broken at such points, and separate approximations obtained on either side of the discontinuity or singularity, if possible. It may be possible to transform the function so as to remove a discontinuity or singularity. Also, a function $f(t)$ may not be suitable for polynomial approximation, but a change of variables $t = g(x)$ might give a function $F(x) = f(g(x))$ that is suitable for polynomial approximation; for example, changing variables so that the independent variable is a logarithm, or changing both the independent variable and the dependent variable to be logarithms.

If we want to use polynomials as approximating functions, does the input data define a polynomial satisfactorily? We have seen that high-degree polynomials can have large oscillations, so that the trend of the data may be lost. Perhaps curve fitting by orthogonal polynomials or fitting the data with other functions and using the singular value decomposition is more suitable. There is a good rule-of-thumb to bear in mind when considering the use of a degree n polynomial to approximate data at $m + 1$ equally spaced points: choose n so that $n < 2\sqrt{m}$. This rule comes from the construction of a certain set of polynomials orthogonal with respect to the inner product

$$\langle f, g \rangle = \sum_{i=0}^{m} f(x_i)g(x_i), \qquad \text{where } x_i = -1 + \frac{2i}{m}.$$

These polynomials have large oscillations between the data points if $n \gg \sqrt{m}$.

Once the type of approximating functions is chosen, we must ensure that the parameters to be found are suitable. For example, suppose the function to be approximated is an even function and the interval on which the approximation is sought is symmetric with respect to the y-axis. Then we should select the approximating functions to be even functions. Using odd functions increases the amount of computation without improving the accuracy of approximation.

Although we have concentrated on polynomials for approximation, other functions can be used. If the approximating functions are linear combinations of $\sin(kx)$ and/or $\cos(kx)$, where k is an integer, we have Fourier approximations. The Fourier approximation can be written as

$$p(x) = \sum_{k=0}^{n-1} b_k e^{jkx}, \qquad \text{where} \quad j = \sqrt{(-1)}$$

If we want $p(x)$ to interpolate $f(x)$ at the interpolation points $\{x_i\}$, and we can select $x_i = 2\pi i/n$, then the problem is one of finding the Fourier Transform of

$f(x)$. It is greatly simplified if $n = 2^m$, for some positive integer m; the Fast Fourier Transform (FFT) can be used.

Rational functions are often used for approximation because they give greater accuracy, with fewer parameters, than high-degree polynomials. The Remes' algorithm can be used to find the rational approximation that minimizes the error of maximum absolute value.

5.14 PROBLEMS

1. The function $f(x) = \tanh(10x)$, $-1 \le x \le 1$ is smooth for real values of x because $\tanh(u) = (e^u - e^{-u})/(e^u + e^{-u})$.

(a) Find $p_n(x)$, the polynomial of degree n interpolating $f(x)$ at the equally spaced interpolation points $x_i = -1 + 2i/n$, $i = 0, 1, \ldots, n$, for $n = 5$, 10, 15. For each n, obtain the error of maximum absolute value by computing $|p_n(x) - f(x)|$ at approximately 100 points.

(b) Repeat part (a) using the Chebyshev interpolation points

$$x_i = -\cos\left(\frac{(2i + 1)\pi}{2n + 2}\right), \quad i = 0, 1, \ldots, n.$$

Compare your results with those of part (a).

(c) Piecewise cubic Hermite polynomials are to be used with equally spaced interpolation points, $n = 5, 10, 15$. Compare your results to those obtained previously.

(d) Repeat part (c) using the cubic spline.

(e) Repeat part (a) using orthogonal polynomials.

2. The following set of data was obtained from an experiment that measured the collapse and rebound of a spherical bubble in a compressible liquid: x_i and y_i measure time and bubble radius, respectively, in nondimensional units. A polynomial fit to the data points is required for purposes of plotting.

(a) Use the Newton divided differences interpolating polynomial to fit the data. Check the suitability of the interpolating polynomial by plotting the values predicted by the polynomial at approximately 100 points.

(b) If the results of part (a) are not satisfactory, write x and y as functions of the variable t. Consider $x = x(t)$ and $y = y(t)$ and obtain interpolation polynomials for x and y separately. Choose the t data points at the roots of the appropriate Chebyshev polynomial. Plot your results on an (x, y) graph.

(c) Try dividing the full x-interval, $[a, c]$, into two subintervals, $[a, b]$ and $[b, c]$, where $a < b < c$. Obtain separate interpolating polynomials on each subinterval using the (x, y) data directly. Plot your results on the same graph.

(d) Repeat part (c) with x and y parameterized as in part (b). Plot your results on the same graph.

(e) Comment on the results obtained in parts (a)–(d),

i	x_i	y_i
1	0.0000	5.0000
2	1.2943	4.8239
3	2.1572	4.4877
4	3.1527	3.7884
5	3.5386	3.3719
6	4.0362	2.5880
7	4.4744	1.0491
8	4.5008	0.7846
9	4.5243	0.3166
10	4.5282	0.1032
11	4.5363	0.3011
12	4.5833	0.7042
13	4.6600	1.0161
14	4.9966	1.6798
15	5.3004	2.0083
16	6.0803	2.4522
17	6.9163	2.5770

3. The viscosity of a certain liquid at a pressure of one atmosphere changes with temperature. The following table gives the viscosity at various temperatures.

Temperature degrees Celsius	Viscosity gm/(cm sec)
0	1.868
10	1.743
20	1.635
30	1.328
40	0.963
50	0.730
60	0.570
70	0.457
80	0.372
90	0.311
100	0.263

The following empirical formulas give the viscosity, η, in terms of the temperature T:

$$\ell og_{10}(\eta(T)) = \frac{3000}{3000 + (T - 25)(8 + 0.001(T - 25))} - 0.8 \quad \text{for } 0 \le T \le 25$$

$$\ell og_{10}(\eta(T)) = 0.2 - \frac{(T - 25)(2 + 0.001(T - 25))}{T + 100} \quad \text{for } 25 \le T \le 100.$$

(a) Find the degree 10 polynomial that interpolates the data. Choose approximately 100 data points in the interval [0, 100] to evaluate the polynomial, and thus determine the error of maximum absolute value between the polynomial value and the empirical formula value.

(b) Use orthogonal polynomials of degrees 1 to 10 to fit the data. For each polynomial, determine the error of maximum absolute value between the polynomial value and the empirical formula value. Which polynomial is best according to this measure of "best fit"?

4. Suppose you are given a table of 11 data points $[x_i, y_i]$, where $x_i = -1 + i/5$, $y_i = f(x_i)$, $f(x) = \ell og(\cosh(10x))$, $i = 0, 1, 2, \ldots, 10$. It is desired to obtain the degree 10 interpolating polynomial, $p_{10}(x)$, and the cubic spline interpolating polynomial, $s(x)$, based on the given 11 knots. Use the Newton divided differences form of the degree 10 interpolating polynomial.

(a) On the same graph, plot the error functions

$$p_{10}(x) - f(x) \quad \text{and} \quad s(x) - f(x).$$

Which is the better interpolating function?

(b) It is desired to estimate the derivative of the function represented by the table of values at various points in the interval $[-1, 1]$.

(i) Use the degree 10 interpolating polynomial, $p_{10}(x)$, to estimate the derivative of the function.

(ii) Use the cubic spline to estimate the derivative of the function.

For the two approximations given by (i) and (ii), plot (on the same graph) the error functions

$$p'_{10}(x) - f'(x) \quad \text{and} \quad s'(x) - f'(x).$$

Which form of the interpolation is better for derivative estimation?

Note 1. The Pascal function EvaluateSpline in Appendix 5.4 can be adapted to compute the derivative of the spline.

Note 2. The Pascal function EvaluateNewton in Appendix 5.1 can be adapted to compute the derivative of the Newton polynomial as follows. The

Newton interpolation polynomial of degree n can be written as

$$p(x) = a_0 + \sum_{j=1}^{n} a_j q_j(x)$$

where $q_{k+1}(x) = (x - x_k)q_k(x)$, $k = 0, 1, \ldots, n - 1$, $q_0(x) = 1$. The derivatives of $\{q_i(x)\}$ satisfy the recurrence relation

$$q'_{k+1}(x) = q_k(x) + (x - x_k)q_k'(x), \; k = 0, 1, \ldots, n - 1.$$

Therefore, the derivative of $p(x)$ is

$$p'(x) = \sum_{j=1}^{n} a_j q_j'(x).$$

5. This problem shows the sensitivity of the coefficients of interpolating polynomials to small changes in the data. Generate the 11 data points $\{x_i, y_i\}$ by $x_i = -1 + i/5$, $y_i = x_i + 0.5$, for $i = 0, 1, \ldots, 10$.

(a) Obtain the degree 10 Newton interpolating polynomial and the cubic spline interpolating polynomial. Print out the coefficients $\{a_i\}$ for the Newton polynomial and the coefficients $\{a_i, b_i, c_i\}$ for the cubic spline.

(b) Repeat part (b) but change (x_5, y_5) from (0.0, 0.5) to (0.0, 0.55).

6. The following data values were obtained from the Bessel function of order zero, $J_0(x)$. Use inverse interpolation to estimate x_*, where $J_0(x_*) = 0$.

i	x_i	$J_0(x_i)$
0	5.3	−0.07580
1	5.4	−0.04121
2	5.5	−0.00684
3	5.6	0.02697
4	5.7	0.05992
5	5.8	0.09170

7. The following table gives the resistance of a certain metal wire for various wire diameters, at 20°.

(a) Find the degree 13 interpolating polynomial and use it to obtain a graph of resistance as a function of wire diameter.

(b) Repeat part (a) with the cubic spline.

(c) Use inverse interpolation to estimate wire diameters for the resistances, $R_i = i/10$, $7 \le i \le 12$, for both the degree 13 interpolating polynomial and the cubic spline.

Wire diameter mm	Resistance ohms per Km
14.0	0.653
13.0	0.671
12.0	0.692
11.0	0.719
10.0	0.753
9.0	0.796
8.0	0.854
7.0	0.932
6.0	1.04
5.0	1.22
4.0	1.50
3.0	2.04
2.0	3.33
1.5	4.85

8. The following set of points lie on a smooth curve. The curve is to be plotted in the (x, y)-plane by finding functions interpolating the data. The data points are such that neither y is a function of x nor x is a function of y. To obtain the plotting functions we assume that $x = x(t)$ and $y = y(t)$ for the new independent variable t. We select the knots as $t_i = i$, for $i = 1, 2, \ldots, 20$. Use the cubic spline to obtain an interpolating polynomial for the (t_i, x_i) data points and another cubic spline for the (t_i, y_i) data points. Use approximately 100 points to obtain $(x(t), y(t))$ from their respective spline polynomials; plot each of these points to obtain the original curve.

i	x_i	y_i	i	x_i	y_i
1	6.4	19.5	11	5.8	5.7
2	8.7	18.8	12	3.8	4.0
3	10.9	18.7	13	2.0	3.9
4	13.9	19.6	14	1.3	6.1
5	14.1	20.9	15	3.5	7.9
6	12.8	21.4	16	6.0	7.5
7	10.5	20.1	17	8.6	6.0
8	9.3	18.2	18	10.9	3.8
9	8.0	14.3	19	13.0	2.4
10	6.7	7.8	20	15.6	2.5

There are many applications where data are obtained in the form $\{x(t), y(t)\}$ and we want to plot y in terms of x, when y is not a function of x. The variable t is used as the independent variable to obtain separate interpolation polynomials, which are then used to generate a sufficient number of points so that a satisfactory curve can be plotted in the (x, y)-plane.

9. The following table gives the surface tension, σ, of a liquid as a function of the temperature, T. Find a and b so that $\sigma = aT + b$ best fits the data in the least squares sense. Plot the resulting graph of $\sigma = aT + b$.

Temperature deg. Celsius	Surface Tension dynes/cm	Temperature deg. Celsius	Surface Tension dynes/cm
−10.0	82.8	30.0	75.6
−5.0	82.0	40.0	74.3
0.0	81.5	50.0	73.6
5.0	80.9	60.0	72.5
10.0	79.0	70.0	71.9
15.0	78.4	80.0	70.4
20.0	77.8	90.0	69.8
25.0	76.3	100.0	68.9

10. We believe that the set of data, given below, comes from a process which behaves like

$$f(x) = 1 + x + x^2 + x^3 + x^4 + x^5.$$

The data have been obtained by an experiment which measured $y_i = f(x_i)$, $1 \le i \le 21$, with an experimental error of about one part in one thousand. Fit the polynomial

$$p(x) = d_0 + d_1x + d_2x^2 + d_3x^3 + d_4x^4 + d_5x^5$$

to the data in the least squares sense, using orthogonal polynomials with the weights $w_i = 1$ for all i. Then express the result in powers of x and print out your values of d_i, $0 \le i \le 5$. Comment on how well your values of d_i compare with the theoretical values of $d_i = 1$.
Hint. The Pascal procedure OrthogonalPolynomial in Appendix 5.6 returns the coefficients $\{a_i, b_i, c_i\}$, where b_i and c_i are used in (5.8–3) to generate the orthogonal polynomials $p_i(x)$ such that

$$q(x) = a_1p_0(x) + \ldots + a_{n+1}p_n(x),$$

for a degree n polynomial $q(x)$. To write

$$q(x) = d_0 + d_1x + \ldots + d_nx^n,$$

you must express each $p_k(x)$ in terms of powers of x. If the coefficients of $p_k(x)$ are $v_{k,i}$ when $p_k(x)$ is written in powers of x, then $v_{k,j}$ can be expressed in terms of $v_{k-1,i}$ $(i = j - 1 \text{ or } j)$, $v_{k-2,j}$, and the appropriate coefficients b_m,

c_m. The coefficients d_k can then be written in terms of the coefficients $v_{k,i}$ and a_j.

x_i	y_i	x_i	y_i	x_i	y_i
0.0	0.99975	0.7	2.94329	1.4	16.31257
0.1	1.11045	0.8	3.68905	1.5	20.77943
0.2	1.24870	0.9	4.68114	1.6	26.30708
0.3	1.42632	1.0	6.00311	1.7	33.07555
0.4	1.65906	1.1	7.71481	1.8	41.28834
0.5	1.96858	1.2	9.92388	1.9	51.12392
0.6	2.38450	1.3	12.75130	2.0	62.94270

11. The function $f(x) = \sin(x)/x$ is an even function on the interval $[-a, a]$. We define $f(0)$ to be 1 so that $f(x)$ is continuous on $[-a, a]$. We want to obtain a least squares fit to $f(x)$ on the interval $[-2, 2]$ in the form $p(x) = a_1 + a_2x^2 + a_3x^4$. Use the singular value decomposition to implement the least squares method; you can adapt the program in Appendix 5.9 so that it will find a_i. Pick about 100 equally spaced points in $[-2, 2]$ to evaluate the error $|p(x) - f(x)|$. Determine the maximum error.

12. The following data are believed to come from a process of the form

$$y = axe^{-x/b},$$

for $a > 0$. Take logarithms to obtain the function

$$\ell og(y) = \ell og(a) + \ell og(x) - \frac{x}{b}.$$

Set $z = \ell og(y/x)$; the data are to be fitted to the curve $z = a_1 + a_2x$ using the method of least squares. Find a and b.

x_i	y_i
1.0	5.636
2.0	8.470
3.0	9.548
4.0	9.567
5.0	8.986
6.0	8.104
7.0	7.105
8.0	6.102
9.0	5.158
10.0	4.307
11.0	3.560

13. An astronomy research team has obtained the following data. They believe the data come from a model of the form $y = ax^b$. The x and y data points have been given in terms of nondimensional variables. Use the singular value decomposition to find least squares estimates for a and b, given that experimental errors in the data lead the astronomers to assign different weights, w_i, to the data points in proportion to the amount of certainty in each experimental measurement. Recall that an equation is multiplied by $\sqrt{w_i}$, in the singular value decomposition method, so that the square of the error for that equation is weighted by w_i.

i	x_i	y_i	w_i
1	0.2409	0.3871	1.00
2	0.6152	0.7233	1.00
3	1.0000	1.0000	1.00
4	1.8808	1.5237	1.00
5	11.8613	5.204	0.04
6	29.4569	9.581	0.04
7	84.0081	19.1	0.01
8	164.7837	30.1	0.01
9	248.3463	39.3	0.01

14. The splitting-tensile strength of concrete cylinders that have been cured for the same length of time by two different methods has been measured. It is believed that the strengths of the cylinders can be modeled by an equation of the form $y = ax + b$, where x and y are strengths corresponding to each curing method. Use the method of least squares to determine a and b.

x_i	y_i	x_i	y_i
450.0	403.0	243.0	229.0
461.0	345.0	272.0	199.0
335.0	350.0	212.0	111.0
332.0	345.0	222.0	162.0
321.0	316.0	225.0	147.0
318.0	269.0	189.0	138.0
329.0	212.0	181.0	64.0
294.0	238.0	179.0	122.0
243.0	117.0	167.0	124.0
269.0	235.0	181.0	76.0
296.0	215.0	155.0	64.0

15. Many engineering processes are determined by measuring the output of the system for a given set of inputs. Given a mathematical model of the process containing a number of unknown parameters, the data are fitted to the model, using linear least squares, provided the parameters appear linearly in the model.

(a) Assume a model of the form $x_{k+2} = a_1x_{k+1} + a_2x_k + a_3$, where $k \geq 0$. Obtain a_1, a_2, and a_3 so that the following data $\{x_j\}$ will fit the model in the least squares sense.

(b) Suppose that the model has the form

$$\begin{aligned} x_{k+1} &= a_1x_k + a_2y_k + a_3 \\ y_{k+1} &= b_1x_k + b_2y_k + b_3 \end{aligned}$$

Find a_i and b_i for $i = 1, 2, 3$ so that the data $\{x_j, y_j\}$ will fit the model in the least squares sense.

k	x_k	y_k
0	0.000000	0.000000
1	0.004528	0.086106
2	0.016429	0.148410
3	0.033587	0.192010
4	0.054344	0.220990
5	0.077409	0.238650
6	0.101790	0.247620
7	0.126710	0.249990
8	0.151620	0.247430
9	0.176280	0.241270
10	0.199790	0.232540

16. A steel company uses a small computer to control the temperature of a blast furnace by adjusting the flow of gas into the burner. The steady-state flow vs. temperature curve of the furnace is highly nonlinear and very expensive to calibrate. The calibration process would require that the furnace be unavailable for production. Therefore, it was decided to measure the gas flow for only a few strategic temperatures and hold the resulting data in the computer's memory. Gas flows corresponding to any arbitrary temperature can then be determined by interpolation, based on these stored values. The following table gives the measured data.

Temperature (deg. C)	Flow Rate (arb. unit)
2012	10
2547	15
3289	25
4125	40
5346	124

Determine the gas flow for a temperature of 3012°.

17. Neville's polynomial interpolation algorithm is quite useful if one only wants to obtain interpolated values rather than the interpolation polynomial. Suppose we have the data $\{x_i, y_i\}$ for $0 \le i \le n$, where the points x_i are all distinct. We want the interpolated value at $x = t$. Define $r(i, j)$ as

$$r(i, 0) = y_i \quad \text{for} \quad i = 0, 1, \ldots, n$$

and, for $j = 1, 2, \ldots, i$ with $i = 0, 1, \ldots, n$

$$r(i, j) = r(i, j-1) + \frac{[r(i, j-1) - r(i-1, j-1)](t - x_i)}{x_i - x_{i-j}}$$

The interpolated value at $x = t$ is $r(n, n)$.

(a) Write a Pascal function to compute the Neville interpolation value at t given the data $\{x_i, y_i\}$ for $i = 0, 1, \ldots, n$, where the points x_i are all distinct.

(b) Repeat Problem 16 using your version of Neville's algorithm.

18. In many linear least squares problems we determine the n-dimensional vector $\mathbf{x}$ from the linear system $\mathbf{Ax} = \mathbf{B}$, where $\mathbf{A}$ is $m \times n$, $\mathbf{B}$ is $m \times 1$ and $m \ge n$. The least squares solution is $\mathbf{x} = [\mathbf{A}^T\mathbf{A}]^{-1}\mathbf{A}^T\mathbf{B}$, where we assume that $\mathbf{A}$ has rank n so that the $n \times n$ matrix $\mathbf{A}^T\mathbf{A}$ is nonsingular. Because $\mathbf{x}$ is determined from m measurements ($\mathbf{A}$ is $m \times n$) we denote it as $\mathbf{x}_m$. Similarly, we use $\mathbf{A}_m$ and $\mathbf{B}_m$ to emphasize that m measurements were required to obtain the $\mathbf{A}$ and $\mathbf{B}$ matrices. Now suppose that one more measurement is taken so that the linear system to be solved is $\mathbf{A}_{m+1}\mathbf{x}_{m+1} = \mathbf{B}_{m+1}$, where

$$\mathbf{A}_{m+1} = \begin{bmatrix} \mathbf{A}_m \\ \mathbf{a}^T \end{bmatrix}, \qquad \mathbf{B}_{m+1} = \begin{bmatrix} \mathbf{B}_m \\ b \end{bmatrix},$$

$\mathbf{a}$ is $n \times 1$ and b is a scalar. The least squares solution of this new system is

$$\mathbf{x}_{m+1} = [\mathbf{A}_{m+1}^T\mathbf{A}_{m+1}]^{-1}\mathbf{A}_{m+1}^T\mathbf{B}_{m+1}.$$

Finding $\mathbf{x}_{m+1}$ this way requires repetition of much of the work done to obtain $\mathbf{x}_m$. There are situations in which computer storage and time are limited, so that repeating the calculations is too expensive. The method of recursive least squares is designed to avoid repeating much of the calculations. For the method of recursive least squares we define the variables

$$\mathbf{P}_m = [\mathbf{A}_m{}^T\mathbf{A}_m]^{-1} \tag{5.14–1}$$

$$\mathbf{K}_{m+1} = \frac{\mathbf{P}_m\mathbf{a}}{1 + \mathbf{a}^T\mathbf{P}_m\mathbf{a}} \tag{5.14–2}$$

$$\mathbf{P}_{m+1} = \mathbf{P}_m - \frac{[\mathbf{P}_m\mathbf{a}]\,[\mathbf{P}_m\mathbf{a}]^T}{1 + \mathbf{a}^T\mathbf{P}_m\mathbf{a}}. \tag{5.14–3}$$

The updated value of $\mathbf{x}$ is given by

$$\mathbf{x}_{m+1} = \mathbf{x}_m + \mathbf{K}_{m+1}[b - \mathbf{a}^T\mathbf{x}_m]. \tag{5.14–4}$$

$\mathbf{K}_{m+1}$ is called a *gains matrix*; $[b - \mathbf{a}^T\mathbf{x}_m]$ is the error between the actual measured output b and the predicted output $\mathbf{a}^T\mathbf{x}_m$. To see that these relationships are correct, consider

$$\mathbf{A}_{m+1}^T\mathbf{A}_{m+1} = \mathbf{A}_m{}^T\mathbf{A}_m + \mathbf{a}\mathbf{a}^T$$

and

$$\mathbf{A}_{m+1}^T\mathbf{B}_{m+1} = \mathbf{A}_m{}^T\mathbf{B}_m + \mathbf{a}b.$$

Using the definition of $\mathbf{P}_m$, (5.14–1), we have

$$[\mathbf{P}_{m+1}]^{-1} = [\mathbf{P}_m]^{-1} + \mathbf{a}\mathbf{a}^T$$

so that

$$[\mathbf{P}_{m+1}]^{-1}\mathbf{x}_m = [\mathbf{P}_m]^{-1}\mathbf{x}_m + (\mathbf{a}\mathbf{a}^T)\mathbf{x}_m.$$

Now,

$$\begin{aligned}
\mathbf{x}_{m+1} &= \mathbf{P}_{m+1}\mathbf{A}_{m+1}^T\mathbf{B}_{m+1} \\
&= \mathbf{P}_{m+1}(\mathbf{A}_m{}^T\mathbf{B}_m + \mathbf{a}b) \\
&= \mathbf{P}_{m+1}([\mathbf{P}_m]^{-1}\mathbf{x}_m + \mathbf{a}b) \\
&= \mathbf{P}_{m+1}([\mathbf{P}_{m+1}]^{-1}\mathbf{x}_m - [\mathbf{a}\mathbf{a}^T]\mathbf{x}_m + \mathbf{a}b) \\
&= \mathbf{x}_m + \mathbf{P}_{m+1}\mathbf{a}(b - \mathbf{a}^T\mathbf{x}_m).
\end{aligned}$$

A short calculation shows that $\mathbf{P}_{m+1}\mathbf{a} = \mathbf{K}_{m+1}$, which verifies (5.14–4). The recursive least squares method is started by choosing $m = k$ so that $\mathbf{P}_k$ can be found from (5.14–1) and $\mathbf{x}_k = \mathbf{P}_k\mathbf{A}_k{}^T\mathbf{B}_k$. As each new measurement $\{\mathbf{a}, b\}$ is determined, the new $\mathbf{x}$ is computed according to the recursive least squares algorithm (5.14–2) to (5.14–4). One possible choice for k is $k = n$.

(a) Write a Pascal program that implements the recursive least squares method (5.14–1) to (5.14–4).

(b) Test your program on the following data, which is believed to satisfy the relationship $z_i = (x_1 - x_2 y_i) y_i$ for $k = 1, 2, \ldots, 20$.

i	y_i	$100z_i$	i	y_i	$100z_i$
1	0.333	2.22	11	0.576	2.44
2	0.355	2.29	12	0.600	2.40
3	0.379	2.35	13	0.624	2.35
4	0.402	2.40	14	0.647	2.28
5	0.427	2.45	15	0.669	2.21
6	0.451	2.48	16	0.691	2.14
7	0.476	2.49	17	0.712	2.05
8	0.501	2.50	18	0.732	1.96
9	0.526	2.49	19	0.751	1.87
10	0.551	2.47	20	0.769	1.78

(c) Repeat Problem 15(a) using recursive least squares.

19. A certain highway code gives the values of the distance s, in meters, traveled by a typical automobile before coming to rest from a speed v, in meters/sec.

v	30	50	60	80	100
s	10	20	35	50	75

(a) Find a degree 4 polynomial of the form

$$s = \sum_{k=0}^{4} a_k \phi_k(v),$$

where $\{\phi_k(v)\}$ are orthogonal polynomials, that interpolates the data.

(b) Obtain the interpolating polynomial in the form $s = \sum_{k=0}^{4} c_k v^k$.

(c) For what values of v can the automobile be stopped within a distance of 30 meters?

5.15 REFERENCES

F. S. Acton. *Numerical Methods that Work*. New York: Harper & Row, 1970.

R. L. Burden, J. D. Faires, and A. C. Reynolds. *Numerical Analysis*. 2d Ed. Boston: Prindle, Weber & Schmidt, 1981.

S. D. Conte and C. de Boor. *Elementary Numerical Analysis*. 3d Ed. New York: McGraw-Hill, 1980.

G. Dahlquist, A. Bjorck, and N. Anderson. *Numerical Methods*. Englewood Cliffs, NJ: Prentice-Hall, 1974.

G. E. Forsythe, M. A. Malcolm, and C. B. Moler. *Computer Methods for Mathematical Computations*. Englewood Cliffs, NJ: Prentice-Hall, 1977.

J. Ll. Morris. *Computational Methods in Elementary Numerical Analysis*. New York: John A. Wiley, 1983.

J. R. Rice. *Numerical Methods, Software, and Analysis*. New York: McGraw-Hill, 1983.

L. F. Shampine and R. C. Allen, Jr. *Numerical Computing: An Introduction*. Philadelphia: W.B. Saunders, 1973.

G. Strang. *Linear Algebra and Its Applications*. New York: Academic Press, 1976.

APPENDIX 5.1

Listing of a Pascal Procedure and function for Newton interpolation, which are in the file NewIntrp.Pas

```
(* The following must be declared in the calling program

              TYPE
                VectorNp1 = ARRAY[ 0..MaxN ] OF REAL;

   where the degree of the interpolating polynomial is less than,
   or equal to  MaxN.  *)

PROCEDURE NewtonInterpolation( n : INTEGER ; VAR x, f, a : VectorNp1 );
(*  To compute the coefficients  a[0],a[1],...,a[n]  for Newton's
    divided differences interpolating polynomial

  p(t) = a[0] + a[1]*(t - x[0]) + a[2]*(t - x[0])*(t - x[1]) + ...
                    + a[n]*(t - x[0])*(t - x[1])*...*(t - x[n-1])

  where  a[0] = f[0]  &  all  x[i]  are distinct.   To evaluate  p(t)
  use the function  EvaluateNewton. *)
VAR
  i, j : INTEGER;
BEGIN
  FOR  j := 0  TO  n  DO
    a[j] := f[j];
  FOR  j := 1  TO  n  DO
    FOR  i := n  DOWNTO  j  DO
      a[i] := (a[i] - a[i-1])/(x[i] - x[i-j])
END; (* NewtonInterpolation *)

FUNCTION EvaluateNewton( t : REAL ; VAR x,a : VectorNp1 ; n : INTEGER ) : REAL;
(* Evaluates the Newton divided differences interpolating polynomial, given
   that the coefficients  a[0], a[1], ..., a[n]  have been determined by the
   procedure NewtonInterpolation *)
VAR
  p : REAL;
  j : INTEGER;
BEGIN
  p := a[n];
  FOR  j := n - 1  DOWNTO  0  DO
    p := p*(t - x[j]) + a[j];
  EvaluateNewton := p
END; (* EvaluateNewton *)
```

APPENDIX 5.2

Listing of a Pascal program for Example 5.2, using the Newton Interpolation Polynomial

```
PROGRAM PolynomialInterpolation;

CONST
  MaxN = 25 ; m = 100 ;
  start = -1.0 ; finish = 1.0 ;

TYPE
  VectorNp1 = ARRAY[ 0..MaxN ] OF REAL;

VAR
  x, y, a : VectorNp1;
  n : INTEGER;

(*$I NewIntrp.Pas *)

FUNCTION F( x : REAL ) : REAL;
BEGIN
  F := 1.0/(1.0 + Sqr(5.0*x))
END;

PROCEDURE GenerateDataPoints( n : INTEGER );
VAR
  dx : REAL;
  i : INTEGER;
BEGIN
  dx := (finish - start)/n;
  FOR i := 0 TO n DO
    BEGIN
      x[i] := start + i*dx ; y[i] := F(x[i])
    END
END; (* GenerateDataPoints *)


FUNCTION MaxError( n : INTEGER ) : REAL;
VAR
  t, dt, error, Large : REAL;
  i : INTEGER;
BEGIN
  NewtonInterpolation(n,x,y,a);
  dt := (finish - start)/m ; Large := 0.0 ;
  FOR i := 0 TO m DO
    BEGIN
      t := start + i*dt;
      error := Abs(EvaluateNewton(t,x,a,n) - F(t));
      IF error > Large THEN
        Large := error
    END;
  MaxError := Large
END; (* MaxError *)

BEGIN (* Main program *)
  WRITELN('Error of maximum absolute value in various polynomials');
  WRITELN('interpolating the function f(x) = 1/( 1 + 25*x*x )');
```

```
  WRITELN;
  WRITELN('degree    Maximum error');
  WRITELN;
  FOR  n := 5  TO  MaxN  DO
    BEGIN
      GenerateDataPoints(n);
      WRITELN(n:4,' ':4,MaxError(n):15:10)
    END
END.
```

APPENDIX 5.3

Listing of a Pascal Function for Hermite Interpolation

```
FUNCTION Hermite( t : REAL ; n : INTEGER ; VAR x, y, yp : VectorNp1 ) : REAL;
VAR
  j, k : INTEGER;
  herm, Lj, Ljp, s, xj : REAL;
BEGIN
  herm := 0.0;
  FOR  j := 0  TO  n  DO
    BEGIN
      xj := x[j] ; Lj := 1.0 ; Ljp := 0.0 ;
      FOR  k := 0  TO  n  DO
        IF  k <> j  THEN
          BEGIN
            s := 1.0/(xj - x[k]);
            Lj := Lj*s*(t - x[k]);
            Ljp := Ljp + s
          END;
      s := t - x[j];
      herm := herm + Sqr(Lj)*(s*yp[j] + (1.0 - 2.0*s*Ljp)*y[j])
    END;
  Hermite := herm
END; (* Hermite *)
```

APPENDIX 5.4

User's Manual for the Pascal Procedure Spline and the Pascal Function EvaluateSpline, which are in the file Spline.Pas

```
The following must be declared in the calling program

      TYPE
        VectorN = ARRAY[ 1..MaxN ] OF REAL;

    where  n, the number of data points, is less than or equal to  MaxN.

PROCEDURE Spline( n : INTEGER ; VAR x, y, a, b, c : VectorN );
```

```
 The procedure Spline is used to compute the coefficients a[i], b[i], and
 c[i], i=1,2,...,n  for a cubic interpolating spline
     s(t) = y[i] + (t - x[i])*(a[i] + (t - x[i])*(b[i] + (t - x[i])*c[i]))
 with  x[i] ≤ t ≤ x[i+1] .
  Input:
    n = The number of data points or knots (n ≥ 2)
    x = The abscissas of the knots in strictly increasing order
           x[i+1] > x[i]   for all  i.
    y = The ordinates of the knots
  Output:
    a, b, c = arrays of spline cofficients as defined above, with
            y[i] = s(x[i])
            a[i] = s'(x[i])
            b[i] = s''(x[i])/2.0
            c[i] = s'''(x[i])/6.0  (Derivative from the right)


FUNCTION EvaluateSpline( n : INTEGER ; t : REAL ;
                               VAR x, y, a, b, c : VectorN ) : REAL;

 Evaluates the cubic spline function at  t.  The coefficients a, b, c have
 been determined by the procedure Spline.
  Input:
    n = The number of data points
    t = The abscissa at which the spline is to be evaluated
    x, y = The arrays of data abscissas and ordinates
    a, b, c = Arrays of spline coefficients computed by  Spline
  Output:
    The value of the spline at  t  is returned.


Spline and EvaluateSpline are based on the FORTRAN subroutine SPLINE and
function SEVAL that appeared in
   G.E. Forsythe, M.A. Malcolm and C.B. Moler,
     Computer Methods for Mathematical Computations
     Prentice-Hall, 1977
```

APPENDIX 5.5

Listing of a Pascal Program for Example 5.2, using Cubic Spline Interpolation

```
PROGRAM PolynomialInterpolation;

CONST
  MaxN = 21 ; m = 100 ;
  start = -1.0 ; finish = 1.0 ;

TYPE
  VectorN = ARRAY[ 1..MaxN ] OF REAL;

VAR
  x, y : VectorN;
  k : INTEGER;

(*$I Spline.Pas *)
```

```
FUNCTION F( x : REAL ) : REAL;
BEGIN
  F := 1.0/(1.0 + Sqr(5.0*x))
END;

PROCEDURE GenerateDataPoints( n : INTEGER );
VAR
  dx : REAL;
  i : INTEGER;
BEGIN
  dx := 2.0/(n - 1);
  FOR  i := 1  TO  n  DO
    BEGIN
      x[i] := start + (i-1)*dx ; y[i] := F(x[i])
    END
END; (* GenerateDataPoints *)

FUNCTION MaxError( n : INTEGER ) : REAL;
VAR
  a, b, c : VectorN;
  t, dt, error, Large : REAL;
  i : INTEGER;
BEGIN
  Spline(n,x,y,a,b,c);
  dt := (finish - start)/m ; Large := 0.0 ;
  FOR  i := 0  TO  m  DO
    BEGIN
      t := start + i*dt;
      error := Abs(EvaluateSpline(n,t,x,y,a,b,c) - F(t));
      IF  error > Large  THEN
        Large := error
    END;
  MaxError := Large
END; (* MaxError *)

BEGIN (* Main program *)
  WRITELN('Error of maximum absolute value in various cubic');
  WRITELN('splines interpolating f(x) = 1/(1 + 25*x*x)');
  WRITELN ; WRITELN('   knots      error') ; WRITELN ;
  FOR  k := 2  TO  21  DO
    BEGIN
      GenerateDataPoints(k);
      WRITELN(' ':4,k:2,' ':6,MaxError(k):8:5)
    END
END.
```

APPENDIX 5.6

Listing of a Pascal Procedure and Function for Orthogonal Polynomials, which are in the file OrthPoly.Pas

```
(* The following must be declared in the calling program

                TYPE
                  VectorN = ARRAY[ 1..MaxN ] OF REAL;
                  VectorM = ARRAY[ 1..MaxM ] OF REAL;
```

```
   where
         m, the number of data points, is less than or equal to  MaxM.
         The degree of the polynomial is 1 less than  n, which is less
          than or equal to  MaxN.  *)

PROCEDURE OrthogonalPolynomial( m, n : INTEGER ;
                  VAR x, f, w, error : VectorM ; VAR a, b, c : VectorN );
(* Evaluates the coefficients  a[*], b[*], c[*]  for the polynomial

     p(t) = a[1]*p (t) + a[2]*p (t) + ... + a[n]*p (t)
                  1            2                 n

  where  p (t)  is of degree  k-1.  The polynomial, p(t), minimizes
          k


        Q =  sum( Sqr( f[i] - p(x[i]) )*w[i] ) .
              i

  The polynomials  p (t)  are computed by
                    k

     p   (t) = (t - b[j+1])*p   (t)   -   c[j+1]*p (t)     j = 0,1,...,n-2
      j+2                    j+1                  j

     p (t) = 0  ,  p (t) = 1  ,  c[1] = 0   .
      0             1


  The data points are  (x[i],f[i]) , i = 1,2,...,m  .
  The weights are  w[i] , i =1,2,...,m  .
  error[i] = f[i] - p(x[i])

  Reference: S.D. Conte and C. de Boor,
             "Elementary Numerical Analysis", Second Edition
             McGraw-Hill, 1972,  page 260   *)

  VAR
    pjm1, pj : VectorM;
    s : VectorN;
    i, j : INTEGER;
    p : REAL;
  BEGIN
    FOR j := 1 TO n DO
      BEGIN
        a[j] := 0.0 ; b[j] := 0.0 ; c[j] := 0.0 ; s[j] := 0.0
      END;
    FOR i := 1 TO m DO
      BEGIN
        a[1] := a[1] + f[i]*w[i] ; b[1] := b[1] + x[i]*w[i] ;
        s[1] := s[1] + w[i]
      END;
    a[1] := a[1]/s[1];
    FOR i := 1 TO m DO
      error[i] := f[i] - a[1];
    IF n > 1 THEN
      BEGIN
        b[1] := b[1]/s[1];
        FOR i := 1 TO m DO
```

```
            BEGIN
              pjml[i] := 1.0 ; pj[i] := x[i] - b[1]
            END;
          FOR  j := 2  TO  n  DO
            BEGIN
              FOR  i := 1  TO  m  DO
                BEGIN
                  p := pj[i]*w[i] ; a[j] := a[j] + error[i]*p ;
                  p := p*pj[i] ; b[j] := b[j] + x[i]*p ;
                  s[j] := s[j] + p
                END;
              a[j] := a[j]/s[j];
              FOR  i := 1  TO  m  DO
                error[i] := error[i] - a[j]*pj[i];
              IF  j < n  THEN
                BEGIN
                  b[j] := b[j]/s[j] ; c[j] := s[j]/s[j-1] ;
                  FOR  i := 1  TO  m  DO
                    BEGIN
                      p := pj[i];
                      pj[i] := (x[i] - b[j])*pj[i] - c[j]*pjml[i];
                      pjml[i] := p
                    END
                END
            END
        END
    END; (* OrthogonalPolynomial *)

FUNCTION EvaluateOrthogonalPolynomial( n : INTEGER ; VAR t : REAL ;
                      VAR a, b, c : VectorN ) : REAL;
(* Evaluates the orthogonal polynomial computed by the procedure
   OrthogonalPolynomial. *)
VAR
  k : INTEGER;
  p, prev, prev2 : REAL;
BEGIN
  p := a[n] ; prev := 0.0 ;
  FOR  k := n-1  DOWNTO  1  DO
    BEGIN
      prev2 := prev ; prev := p ;
      p := a[k] + (t - b[k])*prev - c[k+1]*prev2
    END;
  EvaluateOrthogonalPolynomial := p
END; (* EvaluateOrthogonalPolynomial *)
```

APPENDIX 5.7

Listing of a Pascal Program for Example 5.7, using Least Squares with Orthogonal Polynomials

```
PROGRAM FittingData;

CONST
  MaxN = 11 ; MaxM = 11 ;
  md = 100 ; start = -1.0 ; finish = 1.0 ;
```

```
TYPE
  VectorN = ARRAY[ 1..MaxN ] OF REAL;
  VectorM = ARRAY[ 1..MaxM ] OF REAL;

VAR
  x, y, w, error : VectorM;
  a, b, c : VectorN;
  i, n, m : INTEGER;
  MaxError : REAL;

(*$I OrthPoly.Pas *)

PROCEDURE GenerateDataPoints( n : INTEGER );
VAR
  dx : REAL;
  i : INTEGER;
BEGIN
  dx := (finish - start)/(n - 1);
  FOR i := 1 TO n DO
    BEGIN
      x[i] := start + (i - 1)*dx;
      y[i] := Trunc(100.0*Exp(x[i]))/100.0
    END
END; (* GenerateDataPoints *)

PROCEDURE Data( n : INTEGER );
VAR
  t, dt, err : REAL;
  i : INTEGER;
BEGIN
  OrthogonalPolynomial(m,n,x,y,w,error,a,b,c);
  dt := (finish - start)/md;
  MaxError := 0.0;
  FOR i := 0 TO md DO
    BEGIN
      t := start + i*dt;
      err := Abs(EvaluateOrthogonalPolynomial(n,t,a,b,c) - Exp(t));
      IF err > MaxError THEN
        MaxError := err
    END
END; (* Data *)

BEGIN (* Main program *)
  m := 11;
  FOR i := 1 TO m DO
    w[i] := 1.0; (* set the weights *)
  GenerateDataPoints(m);
  WRITELN('Data values truncated after two significant digits');
  WRITELN;
  WRITELN('    n        error' );
  WRITELN;
  FOR n := 2 TO MaxN DO
    BEGIN
      Data(n);
      WRITELN('   ',n:2,' ',MaxError:11:4)
    END
END.
```

APPENDIX 5.8

User's Manual for the Pascal Procedure SingularValueDecomposition, which is in the file SVD.Pas

```
PROCEDURE SingularValueDecomposition( m, n : INTEGER ; VAR a : MatrixMN ;
            VAR sigma : VectorN ; VAR u : MatrixMN ; MatrixU : BOOLEAN ;
            VAR v : MatrixNN ; MatrixV : BOOLEAN ; VAR error : INTEGER );
```

This procedure determines the singular value decomposition of an m×n real matrix A = U*S*(V–transpose), with m ≤ n.

The following must be declared in the calling program

```
TYPE
  MatrixMN = ARRAY[ 1..MaxM , 1..MaxN ] OF REAL;
  VectorN  = ARRAY[ 1..MaxN ] OF REAL;
  MatrixNN = ARRAY[ 1..MaxN , 1..MaxN ] OF REAL;
```

where m ≤ MaxM, n ≤ MaxN, and MaxN ≤ MaxM

Input:

```
        m = number of rows of the matrix
        n = number of columns of the matrix
        a = the matrix to be decomposed
  MatrixU = TRUE if the  u  matrix is to be found; FALSE, otherwise
  MatrixV = TRUE if the  v  matrix is to be found; FALSE, otherwise
```

Output:

```
      a  is unaltered, unless overwritten by  u
  sigma  contains the  n  non-negative singular values of  a, the
          diagonal elements of  S.  If an error exit is made, the
          singular values should be correct for indices error+1,
          error+2, ..., n.  The singular values are ordered so that
          sigma[error+1] ≥ sigma[error+2] ≥ ... ≥ sigma[n].
      u  contains the first  n  orthogonal column vectors for the  m×m
          orthogonal matrix  U, provided  MatrixU is  TRUE.  If an error
          exit is made, the columns of  u corresponding to indices of
          correct singular values should be correct.
      v  contains the  n×n  orthogonal matrix  V, provided MatrixV is
          TRUE.  If an error exit is made, the columns of v
          corresponding to indices of correct singular values should be
          correct.
  error  is set to 0 for normal return, otherwise it is set to  k  if
          the k-th singular value has not been found after
          MaxIterations iterations.
```

Procedure SingularValueDecomposition is based on the FORTRAN subroutine SVD that appeared in

G.E. Forsythe, M.A. Malcolm, and C.B. Moler
Computer Methods for Mathematical Computations
Prentice-Hall, 1977

APPENDIX 5.9

Listing of a Pascal program for Example 5.10, using Singular Value Decomposition

```
PROGRAM LeastSquaresUsingSingularValueDecomposition;

CONST
  MaxM = 11 ; MaxN = 11 ; eps = 1.0E-08 ;
  mt = 100 ; start = -1.0 ; finish = 1.0 ;


TYPE
   VectorM = ARRAY[ 1..MaxM ] OF REAL;
   VectorN = ARRAY[ 1..MaxN ] OF REAL;
  MatrixMN = ARRAY[ 1..MaxM , 1..MaxN ] OF REAL;
  MatrixNN = ARRAY[ 1..MaxN , 1..MaxN ] OF REAL;


VAR
  x, y : VectorM;
  a, sigma : VectorN;
  phi, u : MatrixMN;
  v : MatrixNN;
  n, m, error : INTEGER;
  tolerance : REAL;

(*$I SVD.Pas *)


FUNCTION F( x : REAL ) : REAL;
BEGIN
  F := Exp(x/2.0)
END;


PROCEDURE GenerateDataPoints;
VAR
  i, k : INTEGER;
BEGIN
  x[1] := -1.0 ; y[1] := 0.61 ; x[2] := -0.8 ; y[2] := 0.67 ;
  x[3] := -0.6 ; y[3] := 0.74 ; x[4] := -0.4 ; y[4] := 0.82;
  x[5] := -0.2 ; y[5] := 0.90 ; x[6] := 0.0 ; y[6] := 1.0 ;
  x[7] := 0.2 ; y[7] := 1.11 ; x[8] := 0.4 ; y[8] := 1.22 ;
  x[9] := 0.6 ; y[9] := 1.35 ; x[10] := 0.8 ; y[10] := 1.49 ;
  x[11] := 1.0 ; y[11] := 1.65 ;
  FOR i := 1 TO m DO
    BEGIN
      phi[i,1] := 1.0 ;
      FOR k := 2 TO n DO
        phi[i,k] := phi[i,k-1]*x[i]
    END
END; (* GenerateDataPoints *)


FUNCTION Evaluate( t : REAL ) : REAL;
VAR
  p : REAL;
  i : INTEGER;
```

```
BEGIN
  p := a[n];
  FOR  i := n-1  DOWNTO  1  DO
    p := p*t + a[i];
  Evaluate := p
END; (* Evaluate *)

FUNCTION MaxError : REAL;
VAR
  t, dt, err, sum, Large : REAL;
  i, j, k : INTEGER;
BEGIN
  GenerateDataPoints;
  SingularValueDecomposition(m,n,phi,sigma,u,TRUE,v,TRUE,error);
  IF  error <> 0  THEN
    WRITELN('error = ', error);
  tolerance := sigma[1]*eps;
  FOR  i := 1  TO  n  DO
    BEGIN
      t := 0.0;
      FOR  j := 1  TO  n  DO
        IF  sigma[j] > tolerance  THEN
          BEGIN
            sum := 0.0;
            FOR  k := 1  TO  m  DO
              sum := sum + u[k,j]*y[k];
            t := t + v[i,j]*sum/sigma[j]
          END;
      a[i] := t
    END;
  dt := (finish - start)/mt;
  Large := 0.0;
  FOR  i := 0  TO  mt  DO
    BEGIN
      t := start + i*dt;
      err := Abs(Evaluate(t) - F(t));
      IF  err > Large  THEN
        Large := err
    END;
  MaxError := Large
END; (* MaxError *)

BEGIN (* Main program *)
  WRITELN(' n       error');
  WRITELN;
  m := MaxM;
  FOR  n := 2  TO  MaxN  DO
    WRITELN(n:2,'  ',MaxError:11:4)
END.
```

APPENDIX 5.10

The Singular Value Decomposition

The singular value decomposition, as used in Section 5.11, is given by the following theorem.

Theorem 5.4

Any $m \times n$ $(m \geq n)$ real matrix $\mathbf{A}$ can be factored into the form

$$\mathbf{A} = \mathbf{MDN}^T,$$

where $\mathbf{M}$ is an $m \times m$ orthogonal matrix, $\mathbf{N}$ is an $n \times n$ orthogonal matrix, $\mathbf{D}$ is an $m \times n$ real matrix of the form

$$\mathbf{D} = \begin{bmatrix} \mathbf{S} & \mathbf{0} \\ \mathbf{0} & \mathbf{0} \end{bmatrix}$$

and $\mathbf{S} = \text{diag}(\sigma_1, \sigma_2, \ldots, \sigma_r)$. The n singular values of $\mathbf{A}$ are

$$\{\sigma_1, \sigma_2, \ldots, \sigma_r, 0, 0, \ldots, 0\}$$

Proof

The $n \times n$ real symmetric matrix $\mathbf{A}^T\mathbf{A}$ has an orthonormal set of eigenvectors (mutually orthogonal unit vectors)

$$\{\mathbf{v}_1, \mathbf{v}_2, \ldots, \mathbf{v}_n\}$$

corresponding to the eigenvalues $\lambda_1, \lambda_2, \ldots, \lambda_n$. These n vectors form the columns of the orthogonal matrix $\mathbf{N}$. Thus,

$$\mathbf{A}^T\mathbf{A}\mathbf{v}_i = \lambda_i\mathbf{v}_i \qquad 1 \leq i \leq n$$

with

$$\mathbf{v}_i{}^T\mathbf{v}_k = \begin{cases} 0 & \text{if } k \neq i \\ 1 & \text{if } k = i \end{cases}$$

Clearly $\lambda_i = \mathbf{v}_i{}^T\mathbf{A}^T\mathbf{A}\mathbf{v}_i = (\mathbf{A}\mathbf{v}_i)^T(\mathbf{A}\mathbf{v}_i) \geq 0$. Suppose that $\lambda_1, \lambda_2, \ldots, \lambda_r$ are positive, while the remaining $(n - r)$ λ's are zero; that is, $\lambda_k = 0$ for $k > r$. Thus, $\mathbf{A}\mathbf{v}_k = \mathbf{0}$ for $k > r$. Now set

$$\sigma_i = \sqrt{\lambda_i} \quad \text{and} \quad \mathbf{y}_i = \frac{\mathbf{A}\mathbf{v}_i}{\sigma_i} \quad \text{for} \quad 1 \leq i \leq r.$$

If $1 \leq i \leq r$ and $1 \leq k \leq r$, then

$$\mathbf{y}_k{}^T\mathbf{y}_i = \frac{(\mathbf{A}\mathbf{v}_k)^T(\mathbf{A}\mathbf{v}_i)}{\sigma_k\sigma_i} = \frac{\mathbf{v}_k{}^T(\mathbf{A}^T\mathbf{A}\mathbf{v}_i)}{\sigma_k\sigma_i} = \frac{\mathbf{v}_k{}^T\lambda_i\mathbf{v}_i}{\sigma_k\sigma_i} = \frac{\lambda_i\mathbf{v}_k{}^T\mathbf{v}_i}{\sigma_k\sigma_i} = \begin{cases} 0 & \text{if } k \neq i \\ 1 & \text{if } k = i \end{cases}$$

so that $\{y_1, y_2, \ldots, y_r\}$ is an orthonormal set of vectors in a vector space of dimension m. Therefore, we can extend the set

$$\{\mathbf{y}_1, \mathbf{y}_2, \ldots, \mathbf{y}_r\}$$

to an orthonormal basis

$$\{\mathbf{y}_1, \mathbf{y}_2, \ldots, \mathbf{y}_r, \mathbf{y}_{r+1}, \ldots, \mathbf{y}_m\}$$

for the vector space by using the Gram-Schmidt process (see Appendix A.1). The vectors

$$\mathbf{y}_1, \mathbf{y}_2, \ldots, \mathbf{y}_r, \mathbf{y}_{r+1}, \ldots, \mathbf{y}_m$$

form the columns of the orthogonal matrix $\mathbf{M}$. The (k, i)th entry of the matrix $\mathbf{M}^T\mathbf{A}\mathbf{N}$ is

$$(\mathbf{M}^T\mathbf{A}\mathbf{N})_{ki} = \mathbf{y}_k{}^T\mathbf{A}\mathbf{v}_i = \begin{cases} 0 & \text{if } i > r \quad \text{because} \quad \mathbf{A}\mathbf{v}_j = \mathbf{0} \quad \text{for} \quad j > r \\ \sigma_i \mathbf{y}_k{}^T\mathbf{y}_i & \text{if } i \leq r \end{cases}$$

where

$$\mathbf{y}_k{}^T\mathbf{y}_i = \begin{cases} 1 & \text{if } k = i \\ 0 & \text{if } k \neq i \end{cases}$$

Hence,

$$(\mathbf{M}^T\mathbf{A}\mathbf{N})_{ki} = \begin{cases} 0 & \text{if } i > r \quad \text{or} \quad k \neq i \\ \sigma_i & \text{if } 1 \leq i \leq r \quad \text{and} \quad k = i \end{cases}$$

Therefore, $\mathbf{M}^T\mathbf{A}\mathbf{N} = \mathbf{D}$, so that $\mathbf{A} = \mathbf{M}\mathbf{D}\mathbf{N}^T$.

Chapter 6

Numerical Evaluation of Integrals and Derivatives

Methods for the numerical evaluation of definite integrals are given. These include the Mid-Point, Trapezoidal, and Simpson Rules, in addition to Romberg quadrature and Gaussian quadrature. The method of adaptive quadrature is discussed. Also, methods for numerical differentiation of functions are given.

6.1 INTRODUCTION

In elementary calculus courses we learn how to evaluate the definite integral of certain functions; this is done by finding an antiderivative in terms of elementary functions and then using the fundamental theorem of calculus. Students soon learn that there are functions, such as e^{-x^2}, that do not have an antiderivative in terms of elementary functions. Integrals involving such functions often arise in engineering and science problems. In addition to these integrals, we often encounter functions that are only defined on a set of discrete points; integrals of such discrete functions may also be required. Therefore, when we encounter a definite integral whose integrand is not recognizable as the derivative of some function or is given at discrete points, we may be able to use some numerical approximation method to estimate the definite integral. In this chapter, we look at various numerical methods for the estimation of definite integrals for functions of one independent variable. The evaluation of a definite integral is often called *quadrature*. We are interested in the numerical estimation of the value of a definite integral, or *nu-*

The complementary problem of finding derivatives of a function is also encountered in practice. Many methods for solving numerical problems require the approximation of derivatives. For example, Powell's method in Section 4.10 requires the numerical estimation of a Jacobian matrix. If the closed form of the function is available, then the differentiation rules learned in elementary calculus can be used. However, a closed form for the function may not be available, or the function may be so complicated that the time spent in doing the differentiation becomes prohibitive. In the latter case, the increasing availability of symbolic mathematics computer programs will remove the drudgery associated with such computations. However, in the former case, numerical estimation of the derivative can be done. In this chapter we shall consider some methods of numerical differentiation.

6.2 THE RIEMANN INTEGRAL

Suppose we are given a function $f(x)$ that is defined on the interval $[a, b]$; we assume that the definite integral

$$I = \int_a^b f(x)\,dx \tag{6.2–1}$$

exists in the sense of Riemann; that is, (6.2–1) is the *Riemann integral* of $f(x)$. The Riemann integral is obtained in terms of the Riemann sum, in which the interval $[a, b]$ is divided into n subintervals by the $n + 1$ distinct points $\{x_i\}$, where $0 \le i \le n$, and $x_0 = a$, $x_n = b$. The points $\{x_i\}$ must form a partition of $[a, b]$.

Definition 6.1

A set of distinct points $\{x_i\}$ in the interval $[a, b]$ is called a *partition* of the interval $[a, b]$ if $\lim_{n\to\infty} (\text{maximum}_{1\le i\le n} \mid x_i - x_{i-1} \mid) = 0$. Once a partition of $[a, b]$ is chosen, a set of intermediate points $\{z_i\}$ is selected so that z_i is in the subinterval $[x_{i-1}, x_i]$, for $i = 1, 2, \ldots, n$. The Riemann sum is defined as

$$I_n = \sum_{i=1}^{n} f(z_i)\cdot(x_i - x_{i-1}) \tag{6.2–2}$$

If $\lim_{n\to\infty} I_n$ exists for all possible partitions of $[a, b]$ and all possible choices of the intermediate points $\{z_i\}$, then that limit is defined to be the Riemann integral of $f(x)$, given by (6.2–1).

In elementary calculus, we learn that continuous functions or functions with a finite number of finite-value jump discontinuities are Riemann integrable. Certain improper integrals exist in the Riemann sense; they are defined in terms of the limit of a sequence of Riemann integrals. These could include integrals with $f(x)$ unbounded within $[a, b]$, such as $f(x) = 1/\sqrt{x}$ on $[0, 1]$, or integrals with one or both of $\mid a \mid$ and $\mid b \mid$ taken as infinite.

In order to estimate a definite integral numerically, we try to approximate (6.2–1) by a finite sum similar to (6.2–2); this is a numerical quadrature method based on a partition $\{x_i\}$ of the interval $[a, b]$. The region between the graph of $f(x)$ and the subinterval $[x_{i-1}, x_i]$ on the x-axis is called a *panel*. Thus, the Riemann sum (6.2–2) approximates (6.2–1) on n panels. Figure 6.1 shows a typical function on $[a, b]$, with the interval divided into n panels. Let $I(n)$ denote the approximation to (6.2–1) obtained by using only n panels. We want the numerical approximation, $I(n)$, to satisfy an absolute error criterion

$$|I - I(n)| < \epsilon \tag{6.2–3}$$

or a relative error criterion

$$|I - I(n)| < \epsilon|I|, \quad \text{provided} \quad I \neq 0 \tag{6.2–4}$$

where ϵ is a preassigned, positive error tolerance. We can estimate the *global discretization error* in $I(n)$ by estimating the *local discretization error* made in the approximation on each panel. These estimates can then be used to ensure that either (6.2–3) or (6.2–4) is satisfied.

In the following sections we consider some methods for quadrature on a finite interval $[a, b]$; that is, both end points are finite. Later, we discuss methods on an infinite interval as well as other improper integrals.

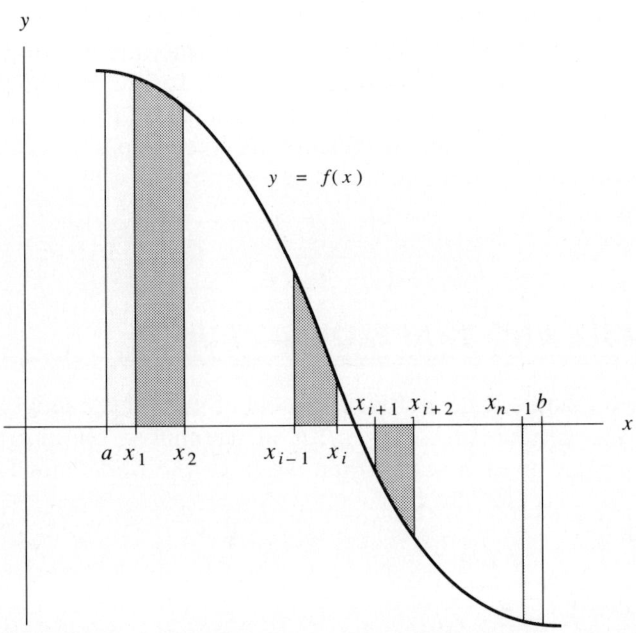

Figure 6.1 Finding the integral of $f(x)$ on $[a, b]$ by using n panels.

6.3 THE RECTANGLE RULE

The simplest form of quadrature is the Rectangle Rule, in which we use n panels of equal base length $h = (b - a)/n$, then choose the partition points as $x_i = a + ih$, $i = 0, 1, \ldots, n$, and take $z_i = x_{i-1}$ for $i = 1, 2, \ldots, n$ in (6.2–2). This gives the rule

$$I_R = h \sum_{k=0}^{n-1} f(x_k) \tag{6.3–1}$$

To obtain the error made by using (6.3–1) to estimate (6.2–1), we consider the interval $[x_k, x_{k+1}]$. For $x_k \le t \le x_{k+1}$ we have

$$f(t) = f(x_k) + f^{(1)}(x_k)\cdot(t - x_k) + \tfrac{1}{2} f^{(2)}(x_k)\cdot(t - x_k)^2 + \cdots \tag{6.3–2}$$

from Taylor's formula, where $f^{(j)}(x)$ denotes the jth derivative of $f(x)$ at x. If we integrate both sides of (6.3–2) over the interval $[x_k, x_{k+1}]$ and sum over $k = 0, 1, \ldots, n - 1$ we obtain

$$\int_a^b f(x)\,dx = h \sum_{k=0}^{n-1} f(x_k) + \left(\frac{h}{2}\right) \sum_{k=0}^{n-1} hf^{(1)}(x_k) + 0(h^3) \tag{6.3–3}$$

The notation $0(h^p)$ denotes terms that are bounded in magnitude by Ch^p, where C is independent of h. Equation (6.3–3) can be written as

$$I = I_R + \left(\frac{h}{2}\right) \sum_{k=0}^{n-1} hf^{(1)}(x_k) + 0(h^3) \tag{6.3–4}$$

The summation term in (6.3–4) is approximately the integral of $f^{(1)}(x)$ over the interval $[a, b]$. Thus, the global discretization error for the Rectangle Rule is $0(h)$. We refer to a method by its global error; the Rectangle Rule is an $0(h)$ method.

Although the Rectangle Rule is conceptually very simple, it has a poor global error. We can do considerably better than a method of order h, as will be seen in the following sections.

6.4 THE MID-POINT RULE AND TRAPEZOIDAL RULE

Similar to the Rectangle Rule we use n panels of equal base length $h = (b - a)/n$; thus, $x_i = a + ih$, $i = 0, 1, 2, \ldots, n$. If we choose the intermediate points at the base-midpoint of each panel, then we have the Mid-Point Rule.

Mid-Point Rule

$$I_{MP} = h \sum_{i=1}^{n} f\left(\frac{x_i + x_{i-1}}{2}\right) \tag{6.4–1}$$

Fig. 6.2 shows a typical panel for the Mid-Point Rule. The panel's area is ap-

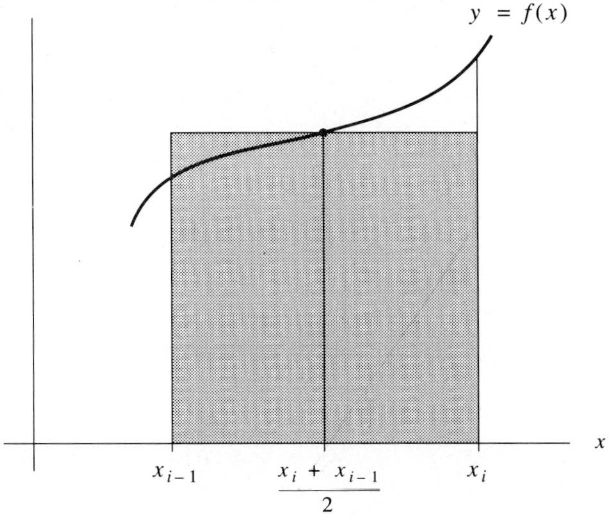

Figure 6.2 Typical panel for the Mid-Point Rule.

proximated by the area of the rectangle whose height is $f(t_i)$, where $t_i = (x_i + x_{i-1})/2$, and whose width is the panel width.

We can estimate the error made by this approximation if $f(x)$ is sufficiently differentiable so that Taylor's formula gives

$$f(x) = f(t_i) + f^{(1)}(t_i)\cdot(x - t_i) + \tfrac{1}{2}f^{(2)}(t_i)\cdot(x - t_i)^2 + \tfrac{1}{6}f^{(3)}(t_i)\cdot(x - t_i)^3 + \cdots$$

where $f^{(k)}(x)$ denotes the kth derivative of $f(x)$. Then

$$\int_{x_{i-1}}^{x_i} f(x)\,dx = hf(t_i) + \tfrac{1}{24}h^3 f^{(2)}(t_i) + 0(h^5) \qquad \textbf{(6.4–2)}$$

so that the local discretization error for the Mid-Point Rule is $0(h^3)$. The global discretization error is obtained by summing (6.4–2) over $i = 1, 2, \ldots, n$:

$$I = I_{MP} + \frac{h^2}{24}\left[\sum_{i=1}^{n} hf^{(2)}(t_i)\right] + 0(h^5) \qquad \textbf{(6.4–3)}$$

The term inside the brackets in (6.4–3) is approximately the integral of $f^{(2)}(x)$ over the interval $[a, b]$. Thus the global discretization error for the Mid-Point Rule is $0(h^2)$.

Another quadrature rule can be obtained by using trapezoids, rather than rectangles, to determine the area of a panel. Fig. 6.3 shows a typical panel whose area is approximated by a trapezoid. The area of the trapezoid is

$$h[f(x_i) + f(x_{i-1})]/2.$$

Adding the areas of all the trapezoids gives the Trapezoidal rule.

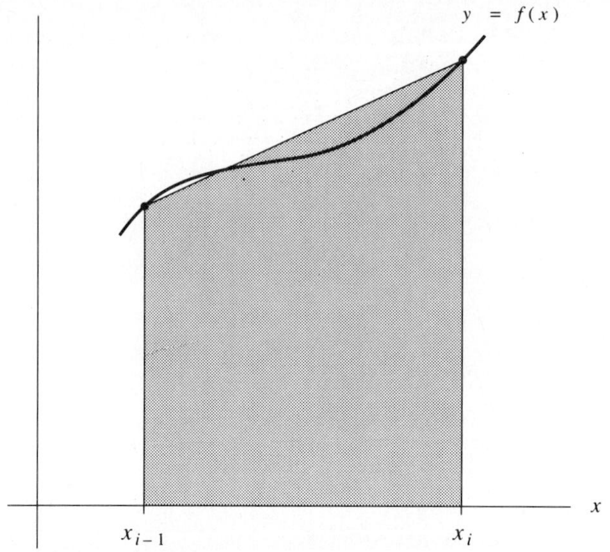

Figure 6.3 Typical panel for the Trapezoidal Rule.

Trapezoidal Rule

$$I_T = \frac{h}{2} \sum_{i=1}^{n} [f(x_i) + f(x_{i-1})]$$

or

$$I_T = \frac{h}{2} \left[f(a) + 2 \sum_{i=1}^{n-1} f(x_i) + f(b) \right] \tag{6.4–4}$$

An error analysis, similar to that for the Mid-Point Rule, gives

$$I = I_T - \frac{h^2}{12} \left[\sum_{i=1}^{n} hf^{(2)}(t_i) \right] + 0(h^5). \tag{6.4–5}$$

If $| h |$ is sufficiently small, then the magnitude of the error made by using the Trapezoidal Rule is approximately twice that of the Mid-Point Rule.

The Mid-Point and Trapezoidal rules are improvements over the Rectangle Rule.

EXAMPLE 6.1

Use $n = 2^k$ panels, $k = 3, 4, 5, 6$, and 7, to estimate the value of the integral

$$I = \int_0^{\pi/2} \sin(\cos(x)) \cdot \sin(2x)\, dx$$

EXAMPLE 6.1 (cont'd)

using the Mid-Point and the Trapezoidal rules. The exact value of I is $2\{\sin(1) - \cos(1)\}$ so that the error of each method can be computed. The following results were obtained.

Table 6.1

Panels	Mid-Point Rule	Trapezoidal Rule
8	2.72×10^{-3}	-5.43×10^{-3}
16	6.77×10^{-4}	-1.35×10^{-3}
32	1.69×10^{-4}	-3.38×10^{-4}
64	4.23×10^{-5}	-8.45×10^{-5}
128	1.06×10^{-5}	-2.11×10^{-5}

Errors in the numerical quadrature of $\sin(\cos(x))\cdot\sin(2x)$ on the interval $[0, \pi/2]$ using the Mid-Point and the Trapezoidal rules.

As expected, the magnitude of the Trapezoidal Rule error is approximately twice that of the Mid-Point Rule.

Appendix 6.1 contains a listing of a Pascal program using the Mid-Point and Trapezoidal rules, with another quadrature method, for this example problem.

EXERCISE 6.1

Adapt the program in Appendix 6.1 to estimate the value of the integral

$$I = \int_0^{0.5} f(x)\, dx$$

where $f(x) = 1/[(1 - x)(4 - x)]$, using the Mid-Point and the Trapezoidal rules with 64 panels.

Answer $I = \ell\text{og}(7/4)/3$.

6.5 SIMPSON'S RULE

We have seen that the single panel version of the Mid-Point Rule can be written as

$$\int_{x_{i-1}}^{x_i} f(x)\, dx = hf(t_i) + Kh^3 + 0(h^5)$$

with $t_i = (x_i + x_{i-1})/2$ while the corresponding Trapezoidal Rule is

$$\int_{x_{i-1}}^{x_i} f(x)\,dx = \tfrac{1}{2}h[f(x_i) + f(x_{i-1})] - 2Kh^3 + 0(h^5)$$

Multiply the first equation by 2, add the result to the second equation, and divide by 3 to obtain the quadrature rule

$$\int_{x_{i-1}}^{x_i} f(x)\,dx = \tfrac{1}{6}h[f(x_i) + 4f(t_i) + f(x_{i-1})] + 0(h^5) \tag{6.5–1}$$

whose local discretization error is $0(h^5)$. Summing (6.5–1) over i, and dropping the $0(h^5)$ term, gives the famous *Simpson's Rule* for numerical quadrature

$$I_S = \left(\frac{h}{6}\right)\sum_{i=1}^{n}\left[f(x_i) + 4f\left(\frac{x_i + x_{i-1}}{2}\right) + f(x_{i-1})\right] \tag{6.5–2}$$

whose global discretization error is $0(h^4)$. We would expect Simpson's Rule to be more accurate than either the Mid-Point Rule or the Trapezoidal Rule for functions that have continuous fourth derivatives. Figure 6.4 shows a typical panel for Simpson's rule. The function $f(x)$ is approximated by a parabola.

Simpson's Rule can also be considered to be applied on $2n$ panels; in this case we take

$$h_* = \frac{b - a}{2n}$$

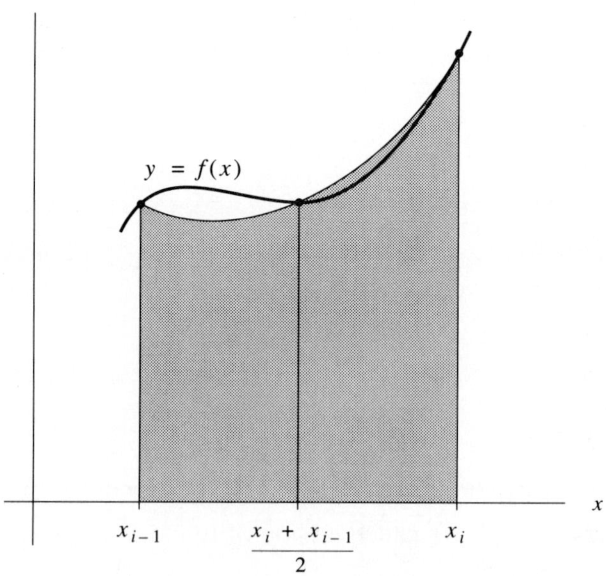

Figure 6.4 Typical panel for Simpson's Rule.

Table 6.2

Panels	Mid-Point Rule	Simpson's Rule
8	2.72×10^{-3}	8.42×10^{-5}
16	6.77×10^{-4}	5.18×10^{-6}
32	1.69×10^{-4}	3.22×10^{-7}
64	4.23×10^{-5}	2.01×10^{-8}
128	1.06×10^{-5}	1.26×10^{-9}

Errors in the numerical quadrature of $\sin(\cos(x))\cdot\sin(2x)$ on the interval $[0, \pi/2]$ using the Mid-Point and Simpson rules.

so that (6.5–2) becomes

$$I_S = \left(\frac{h_*}{3}\right)\left[f(a) + f(b) + 2\sum_{k=0}^{n-1} f(a + (2k + 1)h_*) + 4\sum_{i=1}^{n-1} f(a + 2ih_*)\right] \quad \textbf{(6.5–3)}$$

Equation (6.5–3) is the form of Simpson's Rule that is used in the Pascal program in Appendix 6.1. When we used Simpson's Rule on the integral in Example 6.1 we obtained the results shown in Table 6.2. For comparison, the Mid-Point Rule results from Table 6.1 are given again.

EXERCISE 6.2

Repeat Exercise 6.1 using Simpson's rule, (6.5–3).

The quadrature methods discussed so far have been based on panels of equal width. These formulas can be written for panels of different widths by moving h inside the summations in (6.4–1), (6.4–4), and (6.5–2); then replace h by h_i, where $h_i = x_i - x_{i-1}$. We call these *variable-width*, or *variable-step*, methods to distinguish them from the equal width versions. Appendix 6.2 contains a listing of a Pascal program using the variable-step Mid-Point, Trapezoidal, and Simpson's rules for the integral in Example 6.2.

EXAMPLE 6.2

Use $n = 2^k$ panels, $k = 3, 4, 5, 6$, and 7, to estimate the value of the integral

$$I = \int_0^1 \sin(\sqrt{x})\,dx$$

using the equal panel-width versions of the Mid-Point, Trapezoidal, and Simpson's rules. The problem is to be repeated using the variable panel-width versions of these quadrature methods, when $x_i = (i/n)^2$ for n panels. This particular choice for x_i is motivated by the fact that the argument of the sine function, $\sqrt{x}$, in the integrand, is changing most rapidly near $x = 0$. Using $x_i = (i/n)^2$ concentrates more points near $x = 0$ than $x = 1$ because the spacing of the points is $x_{i+1} - x_i = (2i + 1)n^{-2}$. The value of I is $2\{\sin(1) - \cos(1)\}$ so that the error of each method can be computed. The results, (see Table 6.3), show that the equal-width versions do not perform very well with this integral because the graph of the integrand, $\sin(\sqrt{x})$, has a vertical tangent at $x = 0$. The function $\sin(\sqrt{x})$ changes rapidly near $x = 0$ and less rapidly as $x = 1$ is approached. The equal-width methods treat all parts of the interval as though the behavior of the integrand was uniform. The variable-width methods concentrate more panels near $x = 0$ than $x = 1$.

Table 6.3

	Mid-Point Rule		Trapezoidal Rule		Simpson's Rule	
Panels	Equal Width	Variable Width	Equal Width	Variable Width	Equal Width	Variable Width
8	2.5×10^{-3}	1.4×10^{-3}	-8.8×10^{-3}	-3.0×10^{-3}	-3.6×10^{-3}	-7.2×10^{-5}
16	9.1×10^{-4}	3.6×10^{-4}	-3.2×10^{-3}	-7.4×10^{-4}	-1.3×10^{-3}	-9.2×10^{-6}
32	3.3×10^{-4}	9.1×10^{-5}	-1.1×10^{-3}	-1.9×10^{-4}	-4.5×10^{-4}	-1.2×10^{-6}
64	1.2×10^{-4}	2.3×10^{-5}	-4.0×10^{-4}	-4.7×10^{-5}	-1.6×10^{-4}	-1.5×10^{-7}
128	4.1×10^{-5}	5.8×10^{-6}	-1.4×10^{-4}	-1.2×10^{-5}	-5.6×10^{-5}	-1.8×10^{-8}

Errors in the numerical quadrature of $\sin(\sqrt{x})$ on the interval [0, 1] using the equal panel-width and variable panel-width versions of the Mid-Point, Trapezoidal, and Simpson's rules.

Example 6.2 shows that using a constant panel width over the whole interval may not be as computationally efficient as allowing the panel widths to change. We will want to study quadrature methods that allow variable-width panels but do not require specialized choices for the points $\{x_i\}$.

EXERCISE 6.3

Adapt the variable-step Trapezoidal rule procedure given in Appendix 6.2 to estimate the integral

$$I = \int_{x0}^{x5} f(x)\,dx$$

where $f(x)$ is given by the tabular data

i	0	1	2	3	4	5
x_i	−1.589	−1.052	−0.866	−0.787	0.201	1.147
$f(x_i)$	−1.862	−3.705	−2.688	0.486	2.210	0.086

Answer $I \simeq 0.24156$.

6.6 QUADRATURE METHODS WITH ERROR ESTIMATION

The quadrature methods discussed so far have been used to estimate certain integrals numerically. Examples 6.1 and 6.2 show that increasing the number of panels increases the accuracy of the result. Suppose we have computed the estimate $I_r(n)$ using method "r" on n panels. To see if this estimate for the integral I is sufficiently accurate we could compute $I_r(2n)$. If $|\,I_r(2n) - I_r(n)\,|$ is sufficiently small, then we could accept $I_r(2n)$ as the value of the integral. We show that the error can be estimated by using the difference $I_r(2n) - I_r(n)$.

Suppose that the global error for method "r" is $0(h^p)$. Then

$$I = I_r(n) + Ch^p + 0(h^{p+q}), \quad \text{where} \quad q \geq 1, \quad \text{and}$$

$$I = I_r(2n) + C\left(\frac{h}{2}\right)^p + 0(h^{p+q})$$

The factor C is independent of h. Subtract these two expressions for I to get

$$I_r(n) - I_r(2n) = -Ch^p(1 - 2^{-p}) + 0(h^{p+q})$$

The magnitude of the approximate estimated error in $I_r(2n)$ is

$$\left|\, C\left(\frac{h}{2}\right)^p \,\right| = \frac{|\,I_r(n) - I_r(2n)\,|}{2^p - 1} \tag{6.6–1}$$

The right-hand side of (6.6 – 1) can be computed after each doubling of the number of panels; it can be used in conjunction with either of the error bounds (6.2–3) or (6.2–4).

Appendix 6.3 contains a listing of a Pascal procedure, called Simpson, that uses Eq. (6.6–1) with Simpson's Rule. The Pascal procedure Simpson uses a relative error bound of the form of (6.2–4). Recall that for Simpson's Rule, $p = 4$ in (6.6–1). Appendix 6.4 contains a listing of a Pascal program that uses the procedure Simpson. With RelErr $= 5 \times 10^{-5}$, the method required 65 evaluations of the integrand function to obtain an estimate of the integral in Example 6.2 to within an error of -1.6×10^{-4}. This is the same as the error obtained with the equal-width version of Simpson's Rule (6.5–3) on 64 panels (129 integrand function evaluations). This shows that estimation of the error can lead to a more efficient quadrature method.

EXERCISE 6.4

Adapt the program in Appendix 6.4 to obtain an estimate of the integral in Exercise 6.1. Use RelErr $= 10^{-5}$.

6.7 NEWTON-COTES QUADRATURE

The Mid-Point, Trapezoidal, and Simpson's rules can be generalized by considering quadrature formulas, approximating (6.2–1), of the form

$$I(n) = \sum_{i=0}^{n} c_i f(x_i) \tag{6.7–1}$$

where c_i are constants and $\{x_i\}$ is a set of distinct points in the interval $[a, b]$. If $f(x)$ is interpolated at $\{x_i\}$ by the degree n polynomial $P(x)$, then

$$P(x) = \sum_{k=0}^{n} f(x_k)L_k(x) \tag{6.7–2}$$

where $L_k(x)$ is the kth degree-n Lagrange interpolation polynomial (see Section 5.2). When (6.7–2) is integrated over the interval $[a, b]$ we obtain the coefficients c_i as

$$c_i = \int_a^b L_i(x)\, dx \tag{6.7–3}$$

The error in approximating I by (6.7–1), with c_i given by (6.7–3), is $0(h^p)$, where $p = n + 3$ if n is even, $p = n + 2$ if n is odd.

If the points x_i are chosen with equal spacing in the interval $[a, b]$ then we have the *Newton-Cotes* family of quadrature methods. There are two subfamilies of Newton-Cotes quadrature formulas, the *Open* and the *Closed*. Quadrature

Table 6.4

Some Closed Newton-Cotes Formulas: $h = (b - a)/n$

n	s_0	s_1	s_2	s_3	s_4
0	1				
1	1/2	1/2			
2	1/3	4/3	1/3		
3	3/8	9/8	9/8	3/8	
4	14/45	64/45	8/15	64/45	14/45

methods with unequal spacing of the points x_i are called *Gaussian*; they are considered in the next section.

The Closed Newton-Cotes formulas have panel widths $h = (b - a)/n$, with $x_i = a + ih$, $i = 0, 1, 2, \ldots, n$. The Trapezoidal Rule and Simpson's Rule are Closed Newton-Cotes formulas, with $n = 1$ for the Trapezoidal Rule and $n = 2$ for Simpson's Rule. In the Trapezoidal Rule the interpolation polynomial has degree one; in Simpson's Rule, the polynomial has degree two.

The Open Newton-Cotes formulas have panel widths $h = (b - a)/(n + 2)$, with $x_i = a + (i + 1)h$, $i = 0, 1, 2, \ldots, n$; here, $x_0 = a + h$ and $x_n = b - h$. The Mid-Point Rule is an Open Newton-Cotes formula with $n = 0$; the interpolating polynomial has degree zero.

Because the Newton-Cotes formulas are based on equal spacing of the points x_i, the quadrature method can be written as

$$I(n) = h \sum_{i=0}^{n} s_i f(x_i) \qquad \textbf{(6.7–4)}$$

The coefficients s_i for some Closed Newton-Cotes formulas are given in Table 6.4; Table 6.5 shows s_i for some Open Newton-Cotes formulas. As you can see,

Table 6.5

Some Open Newton-Cotes Formulas: $h = (b - a)/(n + 2)$

n	s_0	s_1	s_2	s_3	s_4
0	1				
1	3/2	3/2			
2	8/3	−4/3	8/3		
3	55/24	5/24	5/24	55/24	
4	33/10	−21/5	39/5	−21/5	33/10

some of the Open Newton-Cotes formulas have negative s_i. The Closed Newton-Cotes formulas also can have negative s_i. Appendix 6.5 contains a listing of a Pascal program using an 8-panel Closed Newton-Cotes quadrature method for the integral in Example 6.1. This Newton-Cotes method determined the integral in Example 6.1 with an error of -7.3×10^{-6}.

EXERCISE 6.5

Adapt the program in Appendix 6.5 to estimate the value of the integral in Exercise 6.1.

One problem associated with high-order Newton-Cotes quadrature formulas (large n) is that they are based on interpolating polynomials at equally spaced interpolation points. Such polynomials can be highly oscillatory between the interpolation points (see Chapter 5). Consequently, these quadrature formulas may behave poorly for certain functions.

EXAMPLE 6.3

$$I = \int_{-1}^{1} f(x)\,dx, \qquad f(x) = \frac{1}{1 + 25x^2}$$

The integrand is Runge's function from Example 5.2. If Runge's function is interpolated at equally spaced points in $[-1, 1]$ by a high-degree interpolation polynomial, $P(x)$, it is known that $|P(x) - f(x)|$ can become very large. We used the Newton-Cotes method given in Appendix 6.5 to estimate the integral; the error in the estimated value was -0.249. However, Runge's function can be interpolated at equally spaced points by a piecewise polynomial function with much greater accuracy. When the interval $[-1, 1]$ was subdivided into k subintervals and the Newton-Cotes method applied to each subinterval, the error in the estimated value of the integral decreases; for example, when $k = 16$ the error in the integral was 2.43×10^{-9}.

6.8 GAUSSIAN QUADRATURE

We shall now describe a quadrature method of the form (6.7–1) with maximal error order; that is, if the error is $0(h^p)$, then p is to be as large as possible. In

(6.7–1) there are $2(n + 1)$ constants that can be chosen, c_i and x_i. Therefore, it is possible to make (6.7–1) exact for all polynomials of degree at most $2n + 1$. This is done by using the Hermite interpolation polynomial (5.4–2). When the points x_i are chosen at the roots of a degree $n + 1$ orthogonal polynomial, we have *Gaussian quadrature*. The particular orthogonal polynomial depends on the form of the integral whose value is sought. For example, the change-of-variable

$$x = 0.5(b - a)t + 0.5(b + a)$$

transforms (6.2–1) into

$$I = 0.5(b - a) \int_{-1}^{1} F(t)\, dt \tag{6.8–1}$$

where $F(t) = f(0.5(b - a)t + 0.5(b + a))$. In (6.8–1), the integral to be approximated has the form

$$\int_{-1}^{1} g(x)\, dx \tag{6.8–2}$$

If (6.8–2) is to be approximated by (6.7–1), we choose x_i to be the roots of the degree $n + 1$ Legendre polynomial because the Legendre polynomials are orthogonal on $[-1, 1]$ with respect to the inner product

$$\langle f, g \rangle = \int_{-1}^{1} f(x)g(x)\, dx$$

(see Section 5.8). We combine the name of the set of orthogonal polynomials with the Gauss name; if the points x_i are selected at the roots of Legendre polynomials we have *Gauss-Legendre* quadrature. The Gauss-Legendre quadrature formulas for $n = 1$ and 2 are

$$n = 1 \quad x_i = -1/\sqrt{3}, \qquad 1/\sqrt{3}$$

$$c_i = 0.5,\ 0.5$$

$$n = 2 \quad x_i = -\sqrt{0.6}, \qquad 0,\ \sqrt{0.6}$$

$$c_i = 5/9,\ 8/9,\ 5/9.$$

Appendix 6.6 contains a listing of a Pascal program using Gauss-Legendre quadrature ($n = 9$) on the integral in Example 6.1. The value of the integral was found with an error of 1.20×10^{-13}.

EXERCISE 6.6

Adapt the program in Appendix 6.6 to estimate the value of the integral in Exercise 6.1.

The Gaussian quadrature methods using (6.7–1) are associated with integrals of the form

$$\int_{\alpha}^{\beta} w(x)g(x)\,dx \tag{6.8–3}$$

where $w(x)$ is a positive weighting function. The appropriate orthogonal polynomials are orthogonal with respect to the inner product

$$\langle F, G\rangle = \int_{\alpha}^{\beta} w(x)F(x)G(x)\,dx$$

If $-\alpha = \beta = 1$ and $w(x) = (1 - x^2)^{-1/2}$ we have *Gauss-Chebyshev* quadrature. *Gauss-Laguerre* quadrature is used if $\alpha = 0$, $\beta = \infty$ and $w(x) = e^{-x}$; *Gauss-Hermite* quadrature requires $-\alpha = \beta = \infty$ and $w(x) = e^{-x^2}$.

Appendix 6.7 contains a listing of a Pascal program that uses Gauss-Laguerre quadrature ($n = 9$) to estimate the integral in Example 6.4.

EXAMPLE 6.4

$$\int_0^{\infty} e^{-x}\sin(x)\,dx$$

The Gauss-Laguerre program returned a value of 0.500000204964850 for the integral. The exact value is 0.5.

To determine the coefficients c_i, after the points x_i are selected as the roots of the appropriate degree $n + 1$ orthogonal polynomial, recall that the quadrature formula is exact for all polynomials of degree at most $2n + 1$. Therefore, the coefficients c_i satisfy the equations

$$\int_{\alpha}^{\beta} w(x)P_k(x)\,dx = \sum_{i=0}^{n} c_i P_k(x_i), \qquad 0 \le k \le n$$

where $P_k(x)$ is a degree k polynomial. Gaussian quadrature has all $c_i > 0$, whereas in Newton-Cotes quadrature some of the coefficients, s_k, may be negative.

Gaussian quadrature, (6.7–1), is ideal for polynomials of degree at most $2n + 1$ because it only requires $n + 1$ values of the polynomial to give the exact value of the integral. However, a disadvantage of Gaussian quadrature is that the points x_i and the constants c_i are generally not rational numbers, and therefore may not be exactly representable as machine floating point numbers. Also, if the integrand function is given at discrete points, the abscissas may not coincide with the x_i required for Gaussian quadrature. Gaussian quadrature would then be awkward to use.

Gaussian quadrature allows us to tailor a quadrature formula to a particular integral of the form of (6.8–3). We would want to approximate (6.8–3) by (6.7–1) so that the quadrature formula is exact if $g(x)$ is a polynomial of degree at most $2n + 1$. Given $w(x)$ and the integer n we would need to find the weights $\{c_i\}$ and the abscissas $\{x_i\}$. Such a specialized quadrature formula is useful if the numerical value of the integral, or similar integrals, were required very many times during the solution of a larger problem. Although the computational effort required to obtain $\{c_i\}$ and $\{x_i\}$ may be significant, the expense may be justified if the larger problem can be solved more efficiently.

6.9 ROMBERG QUADRATURE

In Section 6.5 we found that by combining two quadrature methods of the same error order, the Mid-Point and Trapezoidal rules, we could obtain a quadrature rule with a higher-order error, Simpson's rule. This can be generalized as follows.

Suppose we have a quadrature method "r" with global error $0(h^p)$, I_r. When (6.2–1) is approximated on n panels by quadrature method "r," we have $I_r(n) = I + C_p h^p + C_{p+q} h^{p+q} + C_{p+2q} h^{p+2q} + \dots$, where $h = (b - a)/n$ and $q \geq 1$. Replacing n by $2n$, h by $h/2$, and then subtracting the two expressions gives

$$C_p\left(\frac{h}{2}\right)^p = \frac{I_r(n) - I_r(2n)}{2^p - 1} + 0(h^{p+q})$$

Consequently,

$$I = \frac{2^p I_r(2n) - I_r(n)}{2^p - 1} + 0(h^{p+q})$$

The approximation $(2^p I_r(2n) - I_r(n))/(2^p - 1)$ to I is more accurate than either $I_r(n)$ or $I_r(2n)$. If the approximation $I_r(4n)$ were available, we could combine it with $I_r(2n)$ to obtain another $0(h^{p+q})$ approximation. The two $0(h^{p+q})$ approximations could then be used to obtain a more accurate approximation. This technique uses Richardson extrapolation (see Section 5.12) to obtain more accurate estimates to I. Usually n is chosen as a power of two. The general procedure is called *Romberg quadrature*.

In Romberg quadrature we form the Romberg array $(R_{k,j})$, where

$$R_{k,0} = I_r(2^k), \qquad k = 0, 1, 2, \dots$$

$$R_{k,j+1} = \frac{2^p R_{k,j} - R_{k-1,j}}{2^p - 1}, \qquad 0 \leq j \leq k$$

The Romberg array is lower triangular, with the entries in column $j + 1$ determined by the entries in column j. The second subscript index in the array $R_{k,j}$ denotes the extrapolation level. The triangular Romberg array is

$$\begin{array}{llllll} R_{0,0} \\ R_{1,0} & R_{1,1} \\ R_{2,0} & R_{2,1} & R_{2,2} \\ R_{3,0} & R_{3,1} & R_{3,2} & R_{3,3} \\ \vdots & \vdots & \vdots & \vdots & \vdots & \vdots \; . \end{array}$$

If $\lim_{k\to\infty} R_{k,0} = I$, then $\lim_{k\to\infty} R_{k,k} = I$ also. The standard implementation of Romberg quadrature uses the Trapezoidal Rule; therefore, $R_{k,0} = I_T(2^k)$, (6.4–4), so that $R_{k,0}$ can be expressed as

$$R_{k,0} = \tfrac{1}{2}\left(R_{k-1,0} + \tfrac{1}{2} I_{MP}(2^{k-1})\right)$$

where $I_{MP}(m)$ is the Mid-Point Rule on m panels, (6.4–1).

Appendix 6.8 contains a listing of the Pascal procedure called RombergQuadrature. The procedure tries to compute an approximation to within user-specified error tolerances. Appendix 6.9 contains a program using the procedure RombergQuadrature. When the program was used to estimate the integral in Example 6.3, with a relative error tolerance of 10^{-6}, the procedure required seven extrapolations (257 evaluations of the integrand function) to obtain the estimate with an actual error of 9.10×10^{-11}. The integral in Example 6.2 required 11 extrapolations (4097 evaluations of the integrand) to determine an estimate with an actual error of -2.62×10^{-7}. Recall that the integrand in Example 6.2 has a vertical tangent at $x = 0$ so that the Romberg method requires many extrapolation levels to meet the imposed error tolerance.

EXERCISE 6.7

Adapt the program in Appendix 6.9 to estimate the value of integral in Exercise 6.1, using RelErr $= 10^{-5}$ and AbsErr $= 0.0$.

6.10 ADAPTIVE QUADRATURE

In the Simpson's Rule Pascal procedure given in Appendix 6.3, we estimate the accuracy of an integral approximation with respect to a user-supplied error tolerance. If the tolerance condition is met, then the approximation is accepted as the value of the integral. If the tolerance is not satisfied, the number of panels is doubled and a new error estimate obtained. The process continues by doubling the number of panels until an acceptable approximation is obtained or the maximum allowable number of doublings is reached. The Romberg procedure in Appendix 6.8 works in a similar way. Each of these methods treats the integrand, $f(x)$, in (6.2–1) as if its behavior were uniform in the whole interval $[a, b]$. The amount of effort required to approximate the integral over one part of $[a, b]$ is exactly the same as any other part. If $f(x)$ is changing slowly over a particular panel we might be able to approximate the integral on the panel with much less effort than that required for another panel where $f(x)$ changes rapidly. This is the basic idea of *adaptive quadrature*; expend less effort on regions where $f(x)$ changes slowly, and more effort on regions where $f(x)$ changes more quickly. We want a quadrature procedure to be able to do this.

Suppose that we want to approximate (6.2–1) to within a specified error tolerance, ϵ. We first obtain an estimate of the error in evaluating the integral on $[a, b]$. If the error estimate exceeds ϵ, obtain estimates of the integral on $[a, (a + b)/2]$ and $[(a + b)/2, b]$, I_1 and I_2, and see if the error estimates on each subinterval are each bounded by $\epsilon/2$. If so, accept $I_1 + I_2$ as the integral on $[a, b]$. If either I_1 or I_2 passes its error test retain it; for the other integral, halve the subinterval and check the approximations on both halves to see if their error estimates are bounded by $\epsilon/4$. Acceptable approximations are added to the sum for the original integral; unacceptable approximations are subjected to further bisections of the panels, with a consequent halving of the error tolerance to be satisfied. Continue in this manner until all subintervals give acceptable approximations, or the maximum level of interval bisection is reached. Adaptive quadrature therefore operates at various levels. At level 0 the error tolerance is ϵ and the panel width is $| b - a |$; at level m ($m \geq 0$) the error tolerance is $\epsilon/2^m$ and the panel width is $| b - a |/2^m$.

An advantage of adaptive quadrature is its efficiency in approximating integrals; adaptive methods frequently require much fewer evaluations of the integrand functions than a Romberg quadrature or the Simpson's Rule method given in Appendix 6.3.

Appendix 6.10 contains a user's manual for the adaptive quadrature Pascal procedure QUANC8; the method implements the eight panel Closed Newton-Cotes formula that was used in the program given in Appendix 6.5. The error tolerance used in QUANC8 is Max{AbsErr, $| J | \times$ RelErr}, where AbsErr and RelErr are user-supplied error tolerances, and J is an estimate of the integral. QUANC8 returns a status indicator, *flag*. On a normal return, *flag* is set to 0.0. If there is an abnormal return, *flag* is a number of the form *xxx.yyy*. The integer part of *flag* is the number of panels on which the error criterion could not be attained; if this is small, then the computed value of the integral is probably reasonably accurate provided the fractional part is zero. If the fractional part is nonzero, then QUANC8 could not do the integral on the whole interval. The place at which trouble occurred is located at approximately $x = b - s(b - a)$, where s is the fractional part of the indicator *flag*.

We used QUANC8 to find the integral in Example 6.5. The program listing is in Appendix 6.11.

EXAMPLE 6.5

$$\int_0^1 g(x)\,dx, \qquad g(x) = \frac{1}{x^2 - x + 0.2501}$$

With RelErr $= 10^{-8}$ and AbsErr $= 0$, QUANC8 required 225 integrand function evaluations to find the value of the integral. The estimated error was 2.46×10^{-8}, while the actual error was 2.19×10^{-8}. The Simpson's Rule procedure with error estimation, Appendix 6.3, required 1025 function evaluations to satisfy the error criterion. The Romberg quadrature procedure, Appendix 6.8, required 4097 function evaluations to satisfy the error tolerance. It is interesting to examine the distribution of the points in [0, 1] at which QUANC8 evaluated the integrand function; these points are shown in Table 6.6. The adaptive quadrature procedure concentrates its efforts in the interval [0.4, 0.6] where $g(x)$ changes most rapidly.

Table 6.6 DISTRIBUTION OF THE POINTS AT WHICH QUANC8 EVALUATED THE INTEGRAND IN EXAMPLE 6.5

Interval	Number of Points
[0.0, 0.1]	7
[0.1, 0.2]	6
[0.2, 0.3]	10
[0.3, 0.4]	16
[0.4, 0.5]	73
[0.5, 0.6]	74
[0.6, 0.7]	16
[0.7, 0.8]	10
[0.8, 0.9]	6
[0.9, 1.0]	7

EXERCISE 6.8

Adapt the program in Appendix 6.11 to estimate the value of the integral in Exercise 6.1, using RelErr $= 10^{-5}$ and AbsErr $= 0.0$.

QUANC8 is very useful in the cases where it has trouble computing a value for an integral, as the following example shows.

EXAMPLE 6.6

$$\int_0^2 \tan(x)\, dx$$

When QUANC8 was used to estimate this integral, using RelErr $= 10^{-11}$ and AbsErr $= 0$, it returned with flag $= 43.213$ after 3889 function evaluations. Thus, on 43 panels the error criterion was not satisfied. However, the fractional part of flag indicates that the trouble occurred near $x = 1.574$. The integrand is, of course, unbounded at $x = \pi/2 \simeq 1.5707$.

Numerical estimation of a definite integral is often part of a larger engineering or science problem. If such integrals must be done very many times, then it may be computationally more efficient to use a particular quadrature formula for the whole interval rather than an automatic method such as the Pascal procedure QUANC8. The reason for this is the much larger overhead cost in using QUANC8. The advantage of QUANC8 is that it relieves the casual user from much work.

6.11 IMPROPER INTEGRALS

Most numerical quadrature methods have difficulty with improper integrals; that is, (a) integrals where one or both limits of integration are infinite, or (b) integrands with discontinuities or singularities within the interval of integration. For example,

$$\int_0^1 \left(\frac{\cos(x)}{\sqrt{x}}\right) dx \tag{6.11–1}$$

is an improper integral because the integrand is unbounded at $x = 0$. We first examine integrals with discontinuities or singularities of the integrand.

Often we can use a change-of-variables to obtain a more tractable integral; the substitution $x = t^2$ changes (6.11–1) to the integral

$$2 \int_0^1 \cos(t^2)\, dt \tag{6.11–2}$$

Another technique that often helps is integration-by-parts. When it is applied to (6.11–1) we obtain

$$2 \cos(1) + 2 \int_0^1 (\sqrt{x}) \sin(x)\, dx \tag{6.11–3}$$

Sometimes the singularity can be subtracted out; (6.11–1) can be written as

$$\int_0^1 \left\{\frac{(\cos(x) - 1)}{\sqrt{x}}\right\} dx + \int_0^1 \left(\frac{1}{\sqrt{x}}\right) dx \tag{6.11–4}$$

The integrand in the first integral of (6.11–4) approaches zero as $x \to 0$, so it is well-behaved at $x = 0$. The second integral has the value 2. The integrals in (6.11–2), (6.11–3), and the first integral in (6.11–4) were approximated using QUANC8 with RelErr $= 10^{-9}$ and AbsErr $= 0$. QUANC8 required 33 function evaluations to obtain (6.11–2) and 97 function evaluations for each of the other integrals.

Integrals containing integrands with finite-value jump discontinuities should be written as the sum of integrals where the "break" points occur at the discontinuities.

Integrals with infinite limits of integration can be handled by a change-of-variables. A change-of-variables would convert the infinite interval of integration into a finite interval. For example, the transformation $x = 1/t$ converts the integral

$$\int_1^\infty g(x)\,dx \quad \text{where} \quad g(x) = \frac{1}{x^5 + x^4 + 1} \tag{6.11–5}$$

into

$$\int_0^1 f(t)\,dt \quad \text{where} \quad f(t) = \frac{t^3}{t^5 + t + 1} \tag{6.11–6}$$

With RelErr $= 10^{-10}$ and AbsErr $= 0$, QUANC8 required 49 function evaluations to obtain the value 0.116490349675400 for (6.11–6).

Another way to approximate an integral with infinite limits of integration is to approximate the "tail." We need to write the integral as a finite-interval part and a "tail." The finite-interval part is chosen so that the error made by neglecting the "tail" is small, usually less than the machine epsilon (the smallest positive machine number u such that $1.0 + u > 1.0$). The following example illustrates this technique.

EXAMPLE 6.7

$$\int_0^\infty \cos(x)e^{-x^3}\,dx \tag{6.11–7}$$

can be written as

$$\int_0^c \cos(x)e^{-x^3}\,dx + \int_c^\infty \cos(x)e^{-x^3}\,dx \tag{6.11–8}$$

If $1 \le c \le x$, then $c^2x \le x^3$, so that $e^{-x^3} \le e^{-c^2x}$. The "tail" is bounded by

$$\left|\int_c^\infty \cos(x)e^{-x^3}\,dx\right| \le \int_c^\infty e^{-x^3}\,dx \le \int_c^\infty e^{-c^2x}\,dx = \frac{e^{-c^3}}{c^2}$$

When $c \ge 3.3$, $e^{-c^3}/c^2 < 2.3 \times 10^{-17}$. Therefore, (6.11–7) can be approximated very closely by the first integral in (6.11–8) when $c = 3.3$.

EXERCISE 6.9

Estimate (6.11–7) by integrating the first integral in (6.11–8) with $c = 3.3$ using QUANC8, with RelErr $= 10^{-5}$ and AbsErr $= 0.0$.

Answer $I \simeq 0.738315$.

6.12 NUMERICAL DIFFERENTIATION

Many computational problems require an estimate of the derivative of a function. If the function, $f(x)$, is known in closed form, then we could do the differentiation by hand and insert the derivative function into our computer program. If the function is more complicated, then the work required to do the differentiation may be long and tedious, thereby increasing the chances of error. If we have access to a symbolic mathematics program, we could have the differentiation done by a computer. If such a program is unavailable, or if the function is only known at tabulated points, then we must estimate the derivative numerically.

There are several ways to estimate the derivative of $f(x)$ by using Taylor's formula:

$$f(x + h) = f(x) + \sum_{i=1}^{n} Z_i(x) + R_{n+1} \qquad (6.12\text{–}1)$$

where

$$Z_k(x) = \frac{h^k}{k!}\frac{d^k f(x)}{dx^k}, \qquad R_{n+1} = Z_{n+1}(x + \theta h), \qquad 0 < \theta < 1$$

The first derivative of $f(x)$ can then be estimated by one of the following formulas. In each case we assume that $h > 0$.

$$\text{Forward difference} \quad f'(x) = \frac{f(x+h) - f(x)}{h} + 0(h) \qquad (6.12\text{–}2)$$

$$\text{Backward difference} \quad f'(x) = \frac{f(x) - f(x-h)}{h} + 0(h) \qquad (6.12\text{–}3)$$

$$\text{Central difference} \quad f'(x) = \frac{f(x+h) - f(x-h)}{2h} + 0(h^2) \qquad (6.12\text{–}4)$$

Eq. (6.12–2) can be obtained directly from (6.12–1); the other two formulas are obtained by replacing h by $-h$ in (6.12–1). The estimate of $f'(x)$ is obtained by calculating the difference-quotient in (6.12–2) to (6.12–4). The Jacobian matrix used in Powell's method for nonlinear equations (Section 4.10) is computed by forming difference-quotients.

The forward and backward differences require one additional function evaluation (assuming that $f(x)$ is known), whereas the central difference requires two extra evaluations. The central difference is more accurate if $f(x)$ is twice continuously differentiable. Once a particular formula is chosen, we need to pick a value of h. If $|h|$ is too large, the difference-quotient will be inaccurate. If $|h|$ is too small, catastrophic cancellation may occur in the numerators of the difference-quotients. A good choice for h, depending on the machine epsilon u, is

$$h = \begin{cases} \sqrt{u} & \text{if } x = 0 \\ x\sqrt{u} & \text{if } x \neq 0 \end{cases} \tag{6.12–5}$$

The choice of $\sqrt{u}$ in (6.12–5) is justified as follows. Consider the function

$$g(x) = f(x) - f'(x_0)\cdot(x - x_0)$$

We have $g(x_0) = f(x_0)$ and $g'(x_0) = 0$. Setting $h = x - x_0$, we have

$$\frac{g(x_0 + h) - g(x_0)}{h} = \frac{f(x_0 + h) - f(x_0)}{h} - f'(x_0)$$

Taylor's formula applied to $g(x)$, about the point x_0, gives

$$g(x_0 + h) - g(x_0) = \frac{h^2}{2} g''(x_0 + \theta h), \qquad 0 < \theta < 1 \tag{6.12–6}$$

because $g'(x_0) = 0$. In order to distinguish $g(x_0 + h)$ and $g(x_0)$, their difference in magnitude must not be smaller than $u\,|g(x_0)|$, assuming that $g(x_0) \neq 0$. Therefore, we have

$$|g(x_0 + h) - g(x_0)| \simeq u\,|g(x_0)| \tag{6.12–7}$$

Equations (6.12–6) and (6.12–7) give

$$u\,|g(x_0)| \simeq \frac{h^2}{2}\,|g''(x_0 + \theta h)| \simeq \frac{h^2}{2}\,|g''(x_0)|$$

Consequently, $|h|$ is proportional to $\sqrt{u}$.

We used the forward and central differences to estimate the derivative of x^3 at $x = 1$. On an IBM PC, with Turbo Pascal and an 8087 coprocessor, we computed the derivative with $h = 10^{-k}$, $k = 0, 1, 2, \ldots, 16$. It was found that the error in the derivative was least for $h = 10^{-8}$ when the forward difference formula was used; for the central difference formula, the least error occurred at $h = 10^{-6}$. (The machine epsilon is approximately 2×10^{-16}).

EXERCISE 6.10

If $f(x) = \cos(x)$, estimate $f'(2)$ by the central difference method using h given by (6.12–5). Compare your answer with the exact result.

EXERCISE 6.11

Newton's interpolation polynomial (5.3–1), with equally spaced interpolation points $x_k = x_0 + kh$, $0 \le k \le n$, is to be used to interpolate the function $f(x)$. Show that the coefficient a_2 in the interpolation polynomial is given by

$$a_2 = \frac{f(x_2) - 2f(x_1) + f(x_0)}{2h^2}$$

Consequently, a_2 is half the central differences approximation for the second derivative of $f(x)$ at $x = x_1$.

The derivative can also be estimated by using the Richardson extrapolation method (see Section 5.12); however, this may involve many evaluations of $f(x)$. If $f(x)$ is somewhat costly to evaluate in terms of computer resources, then an extrapolation method probably should not be used. Use the forward or backward differences instead.

The preceding methods assume that $f(x \pm h)$ can be computed for sufficiently many values of h near x. If $f(x)$ is only available at tabulated points, then we can try to estimate $f'(x)$ by an interpolation polynomial. The results obtained by this method can be quite poor for high-degree interpolation polynomials; cubic splines usually give better estimates of the derivative (see Chapter 5, Problem 4).

6.13 PROBLEMS

1. The integral $\int_a^b f(x)\,dx$ is to be computed using the Mid-Point Rule, the Trapezoidal Rule, and Simpson's Rule from the program in Appendix 6.3. Obtain, for each quadrature method, the difference between the computed value and the exact value, I, using 2^k, $2 \le k \le 7$, panels.

(a) $f(x) = \dfrac{(x(1-x))^4}{1+x^2}$, $a = 0$, $b = 1$, $I = \dfrac{22}{7} - \pi$.

(b) $f(x) = \dfrac{\tan(2x)}{1 + (\sec(2x))^{1/2}}$,
$a = 0$, $b = 0.5$, $I = \ell og(2) - \ell og(1 + (\cos(1))^{1/2})$.

(c) $f(x) = \dfrac{1}{1.25 + \sin(x)}$, $a = 0$, $b = 2\pi$, $I = \dfrac{8\pi}{3}$.

(d) $f(x) = 1/[1 + \sin^2(x)]$, $a = -\pi$, $b = \pi$, $I = \pi\sqrt{2}$.

(e) $f(x) = \left[\dfrac{x}{\sin(x)}\right]^2$, $a = 0$, $b = \pi/2$, $I = \pi \ell og(2)$.

2. Use the Simpson's Rule procedure in Appendix 6.3 (Simpson's Rule with error estimation) to compute the integrals in Problem 1. Take RelErr $= 10^{-8}$. For each integral, compare the result obtained with this Simpson's Rule with that of the Simpson rule in Problem 1.

3. Use the Romberg quadrature procedure in Appendix 6.8 to compute the integrals in Problem 1. Take RelErr $= 10^{-8}$ and AbsErr $= 0$. For each integral, compare the results with those of Problem 2.

4. Repeat Problem 3 using the adaptive quadrature procedure, QUANC8, in Appendix 6.10.

5. There is a version of the Trapezoidal Rule with an end correction; this quadrature rule is

$$I_{TEC} = h\left[\frac{f(b) + f(a)}{2} + \sum_{i=1}^{n-1} f(x_i)\right] - \frac{h^2}{12}[f'(b) - f'(a)].$$

The global error is $0(h^4)$, so this quadrature method is comparable to Simpson's Rule. Repeat Problem 1 for the Trapezoidal Rule with End Correction and compare the results to those obtained for the Trapezoidal Rule in Problem 1.

6. In Section 5.7 it was shown that the cubic spline was often a very satisfactory interpolation function. If we are given a function that is only known at the discrete points $\{x_i\}$ and we want its integral, we could use the Trapezoidal Rule to estimate the integral; however, by fitting the data to a cubic interpolatory, spline we could estimate the integral by integrating the spline.

Suppose the data are $\{x_i, y_i\}$, where $a = x_1 < x_2 < \ldots < x_n = b$. Let $s(x)$ denote the cubic spline that interpolates the data. When the cubic spline procedure in Appendix 5.4 is used, the spline quadrature formula is

$$I_{Sp} = \frac{1}{2}\sum_{i=1}^{n-1} h_i(y_i + y_{i+1}) - \frac{1}{12}\sum_{i=1}^{n-1} h_i^3(b_i + b_{i+1}),$$

where the coefficients $\{b_i\}$ are determined in the Pascal procedure Spline in Appendix 5.4. This Spline Quadrature formula is the Trapezoidal Rule (the first sum) minus a correction term (the second sum).

Use the Trapezoidal Rule and the Spline Quadrature formula to estimate the integral of the function given by the data

x_i	0	10	20	30	40	50	60
$f(x_i)$	187	1301	2298	2603	1799	845	464

7. The period of a simple pendulum is given, in nondimensional units, as the integral

$$4\int_0^{\pi/2} [1 - k^2 \sin^2(x)]^{-1/2}\, dx,$$

where $k = \sin(\theta_0/2)$, and $0 \le \theta_0 < \pi$. The angle θ_0 is the initial position of the pendulum when it is released from rest at time 0. Use the adaptive quadrature procedure QUANC8 with RelErr $= 10^{-8}$ and AbsErr $= 0$ to determine the nondimensional pendulum period for $\theta_0 = (\pi/180)D$, $D = 1, 2, \ldots, 179$. The period is infinite if $\theta_0 = \pi$. (This integral is known as an *elliptic integral.*)

8. The integral $\int_a^b f(x)\,dx$ is to be computed using the adaptive quadrature procedure QUANC8 for each of the following cases.

(a) $f(x) = e^{-x^2}$, $a = 0$, $b = 10$.

(b) $f(x) = \dfrac{e^{-x^2}}{1 + x^2}$, $a = 0$, $b = 10$.

(c) $f(x) = [1 + \sin^2(x)]^{1/2}$, $a = 0$, $b = \pi/2$.

(d) $f(x) = \ell og[\cos(x) + \sin(x)]$, $a = 0$, $b = \pi/4$.

(e) $f(x) = \cos(x^2)$, $a = 0$, $b = (2\pi)^{1/2}$.

(f) $f(x) = e^{-0.5\sec(x)}$, $a = 0$, $b = 0.5$.

(g) $f(x) = \dfrac{\cos[20(\sin(x) - x)]}{\pi}$, $a = 0$, $b = \pi$.

(h) $f(x) = \arcsin(x^2)$, $a = 0$, $b = 1$.

9. The response of a certain transducer to an explosion shock wave is given by the function $F(t) = 8e^{-t}I(a, b)/\pi$ for $t \ge a$, where

$$I(a, b) = \int_0^1 f(x)\,dx$$

$$f(x) = \frac{(x - x^2)^{-1/2}e^{ax}}{[1 + b(x - 0.5)]^{1.13}}.$$

Estimate $I(0.001, 0.001)$.

10. A numerical estimate of the integral $\int_a^b f(x)\,dx$ is required.

(a) $f(x) = x^4e^x(e^x - 1)^{-2}$, $a = 0$, $b = 1$.

(b) $f(x) = \dfrac{\cosh(2x) - \cos(2x) + x}{\sinh(x)}$, $a = -1$, $b = 2$.

(c) $f(x) = e^x/x$, $a = 1$, $b = e$.

(d) $f(x) = [(1 - x^2)(2 + x^2)]^{-1/2}$, $a = 0$, $b = 0.5$.

(e) $f(x) = \cos(\sin(x))$, $a = 0$, $b = 3$.

11. Repeat Problem 10 for each of the following.

(a) $f(x) = [\sin(x)]^{-1/2}$, $a = 0$, $b = 1$.

(b) $f(x) = e^{-x}\sqrt{x}$, $a = 0$, $b = \infty$.

(c) $f(x) = e^{-x^3}$, $a = 0$, $b = \infty$.

(d) $f(x) = x[1 - e^{-x}]^{-1}$, $a = 0$, $b = 2$.

(e) $f(x) = \dfrac{x^3}{e^x - 1}$, $a = 0$, $b = 25$.

(f) $f(x) = \dfrac{e^{0.5x}}{1 + e^x}$, $a = -\infty$, $b = \infty$.

(g) $f(x) = (1 + x)^{-1}e^{-x}\sqrt{x}$, $a = 0$, $b = \infty$.

12. Gauss-Chebyshev quadrature is useful for integrals of the form

$$I = \int_{-1}^{1} f(x)\,[1 - x^2]^{-1/2}\,dx.$$

The quadrature formula is

$$I_C = \left(\frac{\pi}{n}\right) \sum_{i=0}^{n-1} f(x_i),$$

where x_i are the roots of the degree n Chebyshev polynomial. Consequently, $I = I_C + E_n$, where

$$E_n = \frac{2\pi f^{(2n)}(\xi)}{2^{2n}(2n)!}, \quad |\,\xi\,| < 1.$$

(a) Write a Pascal function for Gauss-Chebyshev quadrature.

(b) Use your Pascal function from part (a) to estimate the integral I for

(i) $f(x) = \cos(x)$

(ii) $f(x) = \left[\dfrac{\arcsin(x)}{x}\right]^2.$

13. We want to estimate the value of the integral $I = \int_{-1}^{\infty} (e^{-x}/x)dx$ to within a relative error tolerance of 10^{-8}. To do this, we write $I = I_1 + I_2$, where

$$I_1 = \int_{-1}^{1} \left(\frac{e^{-x}}{x}\right) dx \quad \text{and} \quad I_2 = \int_{1}^{\infty} \left(\frac{e^{-x}}{x}\right) dx.$$

(a) Why can we not use QUANC8 directly to compute I, or I_1, or I_2? Sketch a graph of the integrand function to help with your explanation.

(b) The integral I_1 is, in fact, not well-defined because of the singularity of the integrand at $x = 0$. The value that we want to compute is the *Cauchy Principal Value* of the integral, which is defined by

$$\int_{-a}^{a} f(x)\,dx = \lim_{\epsilon \to 0} \left[\int_{-a}^{-\epsilon} f(x)\,dx + \int_{\epsilon}^{a} f(x)\,dx \right]$$

for any function $f(x)$ whose only singularity in the interval $[-a, a]$ is at $x = 0$, provided that the limit exists.
Determine the Cauchy principal value of the integral

$$J = \int_{-1}^{1} \left(\frac{1}{x}\right) dx.$$

(c) Use the result of part (b) to show that

$$I_1 = \int_{-1}^{1} \left(\frac{e^{-x} - 1}{x}\right) dx.$$

Why is this expression for I_1 more suitable for numerical quadrature than the original definition for I_1? Use QUANC8 to estimate the value of I_1 to within a relative accuracy of 10^{-8}.

(d) To compute the integral I_2, use the change-of-variable $x = 1/t$ to transform the problem into one with finite limits of integration, and hence find I_2. Finally, determine the value of the complete integral I.

14. We are given the function $f(x) = \tanh(10x)$.

(a) Estimate $f'(x)$, at $x = 0.0, 0.1, 0.2, 0.3, 0.4, 0.5$, by using forward, backward, and central differences, (6.12–2) to (6.12–4). Select h according to (6.12–5). For each x, compare each computed value with the exact value.

(b) Repeat Problem 4(b) in Section 5.14 for $f(x)$. Compare the derivatives computed from the interpolating polynomials with those found in part (a).

15. Double integrals of the form $\int_a^b \int_{g(x)}^{h(x)} f(x, y)\, dydx$ can be estimated by numerical iterated integrals; that is, we compute an estimate for

$$I = \int_a^b F(x)\, dx,$$

where

$$F(x) = \int_{g(x)}^{h(x)} f(x, y)\, dy.$$

If, for example, we use a Gauss-Legendre method to estimate the "outside" integral I, then the "inside" integral must be estimated in order to compute $F(x_i)$ for each x_i required by the Gauss-Legendre method. Appendix 6.12 contains a listing of a Pascal program for double integrals. The integrals are estimated by the 10-point Gauss-Legendre quadrature method for Appendix 6.6. Use the program in Appendix 6.12 to obtain estimates for each integral.

(a) $f(x, y) = \sin(x + y)$, $a = 0$, $b = 1$, $g(x) = 0$, $h(x) = x$,
$I = \sin(1) - \sin(2)/2$

(b) $f(x, y) = \ell og(x + y)$, $a = 3$, $b = 3.5$, $g(x) = 1$, $h(x) = 2$,
$I = 15.125\ell og(5.5) - 10.125\ell og(4.5) - 12.5\ell og(5) + 16\ell og(2) - 3/4$

(c) $f(x, y) = [(x - y)^2 + 2(x + y) + 1]^{-1/2}$, $a = 0$, $b = 2$, $g(x) = 0$, $h(x) = x$, $I = \ell og(4) - 0.5$

16. The following function arises in the determination of the surface tension of a liquid surrounding a spherical gas-filled bubble.

$$F(t) = \int_0^t \frac{P(x)}{Q(x)}\, dx, \quad \text{for} \quad 0 \le t \le 1,$$

where $P(x) = 3 + 3x + x^2$ and $Q(x) = 3 + 6x + 6x^2 + 2x^3$.

(a) Prove that $Q(x)$ has only one real root, which is negative. Therefore, $F(t)$ is a continuous function for $0 \le t \le 1$.

(b) Obtain the graph of $F(t)$ by calculating $F(t)$ for about 20 values of t in the interval [0, 1].

17. Use QUANC8, with RelErr $= 10^{-8}$ and AbsErr $= 0.0$, to estimate each of the following integrals. For each integral obtain the number of integrand function evaluations. Which integral is more more easily computed using QUANC8? Why? *Note*: $I_1 = I_2 + 2/3$.

(a) $I_1 = \displaystyle\int_0^1 [x \cdot \cos(x)]^{1/2}\, dx$

(b) $I_2 = \displaystyle\int_0^1 \{[x \cdot \cos(x)]^{1/2} - \sqrt{x}\}\, dx$

18. Estimate the value of the integral

$$I = \int_0^\pi e^{\cos(x)} \left(\frac{p(x) + q(x)}{f(x)} \right) dx$$

where
$$\begin{aligned} p(x) &= \sinh[\cos(x)] \cdot \cos[\sin(x)] \cdot \cos[x + \sin(x)] \\ q(x) &= \cosh[\cos(x)] \cdot \sin[\sin(x)] \cdot \sin[x + \sin(x)] \\ f(x) &= \sinh^2[\cos(x)] + \sin^2[\sin(x)]. \end{aligned}$$

19. Chapter 4, Problem 22.

20. The function $f(x)$ is known to have continuous derivatives up to the fourth derivative. Develop a central difference formula to compute the second derivative of $f(x)$ using equally spaced points. Is the error $0(h)$ or $0(h^2)$?

21. The torsion of a certain tapered solid circular cylinder of base radius b and length L is given by

$$\theta(x) = \left(\frac{T}{G}\right) \int_0^x [I(s)]^{-1}\, ds, \qquad 0 \le x \le L,$$

where $\theta(x)$ is the angle of twist at x, T is the constant applied torque, G is the shear modulus of the cylinder material, and

$$I(s) = \frac{\pi b^4}{2}\left[1 - 1.5\left(\frac{s}{L}\right)^2 + \left(\frac{s}{L}\right)^3\right]^4$$

is the moment of inertia of the cross-sectional area at s. If we make the change-of-variable $s = Lt$, we obtain

$$\theta(x) = \frac{2TL}{\pi G b^4} P(x/L)$$

where

$$P(y) = \int_0^y (1 - 1.5t^2 + t^3)^{-4}\, dt, \qquad 0 \le y \le 1.$$

Determine $P(y)$ for $y = 0.1k$, $k = 1, 2, \ldots, 10$.

6.14 REFERENCES

F. S. Acton. *Numerical Methods that Work*. New York: Harper & Row, 1970.

R. L. Burden, J. D. Faires, and A. C. Reynolds. *Numerical Analysis*. 2d Ed. Boston: Prindle, Weber & Schmidt, 1981.

S. D. Conte and C. de Boor. *Elementary Numerical Analysis*. 3d Ed. New York: McGraw-Hill, 1980.

G. Dahlquist, A. Bjorck, and N. Anderson. *Numerical Methods*. Englewood Cliffs, NJ: Prentice-Hall, 1974.

G. E. Forsythe, M. A. Malcolm, and C. B. Moler. *Computer Methods for Mathematical Computations*. Englewood Cliffs, NJ: Prentice-Hall, 1977.

J. Ll. Morris. *Computational Methods in Elementary Numerical Analysis*. New York: Wiley, 1983.

J. R. Rice. *Numerical Methods, Software, and Analysis*. New York: McGraw-Hill, 1983.

L. F. Shampine and R. C. Allen, Jr. *Numerical Computing: an Introduction*. Philadelphia: W.B. Saunders, 1973.

APPENDIX 6.1

Listing of a Pascal Program for Example 6.1

The program illustrates several numerical quadrature methods.

```
PROGRAM Quadrature;

CONST
   a = 0.0 ; b = 1.57079632679489661923l322 ;
   title = 'Errors in the integral of  sin(cos(x))sin(2x)  on  [0,PiBy2]';
```

```
VAR
  k, n : INTEGER;
  error, exact : REAL;

FUNCTION F( x : REAL ) : REAL;
BEGIN
  F := sin(cos(x))*sin(x+x)
END;

FUNCTION TrapezoidalRule( n : INTEGER ; a, b : REAL ) : REAL;
(* Trapezoidal rule for quadrature.  The interval  [a,b]  is subdivided
   into  n  equal subintervals.  *)
VAR
  sum, h : REAL;
  i : INTEGER;
BEGIN
  IF  n >= 1  THEN
    BEGIN
      sum := (F(a) + F(b))/2.0 ; h := (b - a)/n ;
      FOR  i := 1  TO  n-1  DO
        sum := sum + F(a + i*h);
      TrapezoidalRule := h*sum
    END
  ELSE
    TrapezoidalRule := 0.0
END; (* TrapezoidalRule *)

FUNCTION MidPointRule( n : INTEGER ; a, b : REAL ) : REAL;
(* Midpoint rule for quadrature.  The interval  [a,b]  is divided into
   n  equal subintervals.  *)
VAR
  sum, h : REAL;
  i : INTEGER;
BEGIN
  IF  n >= 1  THEN
    BEGIN
      sum := 0.0 ; h := (b - a)/(n + n) ;
      FOR  i := 1  TO  n  DO
        sum := sum + F(a + (2*i - 1)*h);
      MidPointRule := h*(sum + sum)
    END
  ELSE
    MidPointRule := 0.0
END; (* MidPointRule *)

FUNCTION SimpsonRule( n : INTEGER ; a, b : REAL ) :REAL;
(* Simpson's rule for quadrature.  The interval  [a,b]  is divided into
   2n  equal subintervals.  *)
VAR
  s1, s2, h : REAL;
  i : INTEGER;
BEGIN
  IF  n >= 1  THEN
    BEGIN
      h := (b - a)/(n + n) ; s1 := F(a + h) ; s2 := 0.0 ;
      FOR  i := 1  TO  n-1  DO
        BEGIN
          s1 := s1 + F(a + (2*i + 1)*h);
          s2 := s2 + F(a + 2*i*h)
        END;
```

```
      SimpsonRule := h*(F(a) + F(b) + 2.0*(s2 + 2.0*s1))/3.0
    END
  ELSE
    SimpsonRule := 0.0
END; (* SimpsonRule *)

BEGIN (* main program *)
  WRITELN(title) ; WRITELN ; WRITELN ;
  WRITELN('   n   Trapezoidal     Midpoint     Simpson''s');
  WRITELN('          Rule            Rule           Rule');
  WRITELN;
  n := 4 ; exact := 2.0*(sin(1.0) - cos(1.0)) ;
  FOR  k := 3  TO  7  DO
    BEGIN
      n := 2*n ; WRITE(n:4) ;
      error := TrapezoidalRule(n,a,b) - exact;
      WRITE(' ':4,error:10);
      error := MidPointRule(n,a,b) - exact;
      WRITE(' ':4,error:10);
      error := SimpsonRule(n DIV 2,a,b) - exact;
      WRITELN(' ':4,error:10)
    END
END.
```

APPENDIX 6.2

Listing of a Pascal Program for Example 6.2

The program illustrates several variable-step numerical quadrature methods.

```
PROGRAM Quadrature;

CONST
   MaxN = 128;
   title = 'Errors in the integral of  sin(Sqrt(x))  on  [0,1]';

TYPE
  VectorNp1 = ARRAY[ 0..MaxN ] OF REAL;

VAR
  i, k, n : INTEGER;
  x : VectorNp1;
  error, exact : REAL;

FUNCTION F( x : REAL ) : REAL;
BEGIN
  F := sin(Sqrt(x))
END;

FUNCTION VariableStepTrapezoidalRule( n : INTEGER ; VAR x : VectorNp1 ) : REAL;
(*  The interval  [a,b]  is subdivided into n subintervals with
    a = x[0] < x[1] < ... < x[n-1] < x[n] = b  *)
VAR
  sum, g, w : REAL;
  i : INTEGER;
```

```
BEGIN
  sum := 0.0 ; g := F(x[0]) ;
  FOR  i := 1  TO  n  DO
    BEGIN
      w := F(x[i]);
      sum := sum + (x[i] - x[i-1])*(g + w);
      g := w
    END;
  VariableStepTrapezoidalRule := sum/2.0
END; (* VariableStepTrapezoidalRule *)

FUNCTION VariableStepMidPointRule( n : INTEGER ; VAR x : VectorNp1 ) : REAL;
(*  The interval  [a,b]  is subdivided into n subintervals with
    a = x[0] < x[1] < ... < x[n-1] < x[n] = b  *)
VAR
  sum : REAL;
  i : INTEGER;
BEGIN
  sum := 0.0;
  FOR  i := 1  TO  n  DO
    sum := sum + (x[i] - x[i-1])*F((x[i-1] + x[i])/2.0);
  VariableStepMidPointRule := sum
END; (* VariableStepMidPointRule *)

FUNCTION VariableStepSimpsonRule( n : INTEGER ; VAR x : VectorNp1 ) : REAL;
(*  The interval  [a,b]  is subdivided into n subintervals with
    a = x[0] < x[1] < ... < x[n-1] < x[n] = b  *)
VAR
  sum, g, w : REAL;
  i : INTEGER;
BEGIN
  sum := 0.0 ; g := F(x[0]) ;
  FOR  i := 1  TO  n  DO
    BEGIN
      w := F(x[i]);
      sum := sum + (x[i] - x[i-1])*(g + w + 4.0*F((x[i-1] + x[i])/2.0));
      g := w
    END;
  VariableStepSimpsonRule := sum/6.0
END; (* VariableStepSimpsonRule *)

BEGIN (* main program *)
  WRITELN(title) ; WRITELN ;
  WRITELN('   n   Trapezoidal     Midpoint     Simpson''s');
  WRITELN('          Rule            Rule          Rule');
  WRITELN;
  n := 4 ; exact := 2.0*(sin(1.0) - cos(1.0)) ;
  FOR  k := 3  TO  7  DO
    BEGIN
      n := 2*n ; WRITE(n:4) ;
      FOR  i := 0  TO  n  DO
        x[i] := Sqr(i/n);
      error := VariableStepTrapezoidalRule(n,x) - exact;
      WRITE(' ':4,error:10);
      error := VariableStepMidPointRule(n,x) - exact;
      WRITE(' ':4,error:10);
      error := VariableStepSimpsonRule(n,x) - exact;
      WRITELN(' ':4,error:10)
    END
END.
```

APPENDIX 6.3

A Pascal Procedure for Simpson's Method with Error Estimation, which is in the file Simpson.Pas

```
PROCEDURE Simpson( a, b, RelErr : REAL ; VAR Result, EstErr : REAL ;
                                     VAR Evaluations, flag : INTEGER );
(* Simpson's rule to compute the integral of a function over the interval
   [a,b]  to within a relative error of  RelErr.

   Input:
      a       left end point of the interval
      b       right end point of the interval
      RelErr  relative error tolerance

   The user supplies the following to evaluate the integrand:
      FUNCTION F( x : REAL ) : REAL;

   Output:
      Result       calculated value of the integral
      EstErr       estimate of the absolute error in  Result
      Evaluations  number of function evaluations
      flag         convergence indicator
                     0 = error tolerance attained
                     1 = error tolerance not attained  *)

CONST
  Limit = 13;

VAR
  Level, i, n : INTEGER;
  Eps, Fe, Snodes, Smid, Sn, S2n, u, h, hBy4 : REAL;

FUNCTION Max( x, y : REAL ) : REAL;
BEGIN
  IF  y <= x  THEN
    Max := x
  ELSE
    Max := y
End; (* Max *)

BEGIN (* Simpson *)
  Eps := 1.0;
  REPEAT
    Eps := Eps/2.0;
    u := 1.0 + eps
  UNTIL  u = 1.0; (* Eps is machine epsilon, approximately *)
  n := 1 ; h := b - a ; Level := 0 ;
  Smid := F(a + h/2.0);
  Fe := F(a) + F(b);
  Evaluations := 3;
  Sn := h*(Fe + 4.0*Smid)/6.0;
  Snodes := 0.0;
  REPEAT
    hBy4 := h/4.0;
    Snodes := Snodes + Smid;
    Smid := 0.0;
    FOR  i := 1  TO  2*n  DO
      Smid := Smid + F(a + (2*i - 1)*hBy4);
```

```
      Evaluations := Evaluations + 2*n;
      S2n := h*(Fe + 2.0*(Snodes + 2.0*Smid))/12.0;
      u := RelErr*Max(Eps,Abs(S2n));
      EstErr := Abs(Sn - S2n)/15.0;
      Sn := S2n ; Level := Level + 1 ; n := n + n ; h := h/2.0
    UNTIL  (Level = Limit) OR (EstErr <= u);
    Result := S2n;
    IF  EstErr > u  THEN
      flag := 1
    ELSE
      flag := 0
  END; (* Simpson *)
```

APPENDIX 6.4

Listing of a Pascal Program for Example 6.2, using Simpson's Rule with Error Estimation

```
PROGRAM Quadrature;

CONST
  a = 0.0 ; b = 1.0 ;

VAR
  RelErr, Result, EstErr : REAL;
  Evaluations, flag : INTEGER;

FUNCTION F( x : REAL ) : REAL;
BEGIN
  F := sin(Sqrt(x))
END ;

(*$I Simpson.Pas *)

BEGIN (* main program *)
  RelErr := 1.0E-05;
  Simpson(a,b,RelErr,Result,EstErr,Evaluations,flag);
  WRITELN('Integral = ',Result,'     Estimated Error = ',EstErr:10);
  WRITELN('function evaluations = ',Evaluations:5);
  WRITELN('Actual error = ',Result - 2.0*(sin(1.0) - cos(1.0)):10);
  WRITELN;
  IF  flag <> 0  THEN
    WRITELN('flag = ',flag)
END.
```

APPENDIX 6.5

Listing of a Pascal Program for Example 6.1, using Newton-Cotes Quadrature

```
PROGRAM Quadrature;

CONST
  a = 0.0 ; b = 1.5707963267948966192313 22 ;
```

```
VAR
  Integral : REAL;

FUNCTION F( x : REAL ) : REAL;
BEGIN
  F := sin(cos(x))*sin(x+x)
END;

FUNCTION NewtonCotes( a, b : REAL ) : REAL;
(* Quadrature using the nine-point (n=8) Closed Newton-Cotes
     quadrature formula *)
VAR
  i : INTEGER;
  sum, h : REAL;
  s : ARRAY[ 0..8 ] OF REAL;
BEGIN
  s[0] := 3956.0/14175.0 ; s[1] := 23552.0/14175.0 ;
  s[2] := -3712.0/14175.0 ; s[3] := 41984.0/14175.0 ;
  s[4] := -18160.0/14175.0;
  FOR  i := 5  TO  8  DO
    s[i] := s[8-i];
  sum := 0.0 ; h := (b - a)/8.0 ;
  FOR  i := 0  TO  8  DO
    sum := sum + s[i]*F(a + i*h);
  NewtonCotes := h*sum
END; (* NewtonCotes *)

BEGIN (* main program *)
  WRITELN('Newton-Cotes Quadrature') ; WRITELN ;
  Integral := NewtonCotes(a,b);
  WRITELN(Integral:20:15,'   error = ',
           Integral - 2.0*(sin(1.0) - cos(1.0)):10)
END.
```

APPENDIX 6.6

Listing of a Pascal Program for Example 6.1, using Gauss-Legendre Quadrature

```
PROGRAM Quadrature;

CONST
  a = 0.0 ; b = 1.5707963267948966192313 22 ;

VAR
  Integral : REAL;

FUNCTION F( x : REAL ) : REAL;
BEGIN
  F := sin(cos(x))*sin(x+x)
END;

FUNCTION GaussLegendre( a, b : REAL ) : REAL;
(* Quadrature using the ten-point (n=9) Gauss-Legendre
     quadrature formula *)
```

```
VAR
  i : INTEGER;
  sum, c1, c2 : REAL;
  x, s : ARRAY[ 0..9 ] OF REAL;
BEGIN
  x[0] := -9.73906528517166E-001 ; s[0] := 6.66713443086915E-002 ;
  x[1] := -8.65063366688993E-001 ; s[1] := 1.49451349150566E-001 ;
  x[2] := -6.79409568299021E-001 ; s[2] := 2.19086362516005E-001 ;
  x[3] := -4.33395394129247E-001 ; s[3] := 2.69266719309976E-001 ;
  x[4] := -1.48874338981631E-001 ; s[4] := 2.95524224714767E-001 ;
  FOR i := 5 TO 9 DO
    BEGIN
      x[i] := -x[9-i] ; s[i] := s[9-i]
    END;
  sum := 0.0 ; c1 := (b - a)/2.0 ; c2 := (b + a)/2.0 ;
  FOR i := 0 TO 9 DO
    sum := sum + s[i]*F(c1*x[i]+c2);
  GaussLegendre := c1*sum
END; (* GaussLegendre *)

BEGIN (* main program *)
  WRITELN('Gauss-Legendre Quadrature') ; WRITELN ;
  Integral := GaussLegendre(a,b);
  WRITELN(Integral:20:15,'   error = ',
            Integral - 2.0*(sin(1.0) - cos(1.0)):10)
END.
```

APPENDIX 6.7

Listing of a Pascal Program for Example 6.4, using Gauss-Laguerre Quadrature

```
PROGRAM Quadrature;

FUNCTION F( x : REAL ) : REAL;
BEGIN
  F := sin(x)
END;

FUNCTION GaussLaguerre : REAL;
(* Quadrature using the ten-point (n=9) Gauss-Laguerre
     quadrature formula *)
VAR
  i : INTEGER;
  sum : REAL;
  x, s : ARRAY[ 0..9 ] OF REAL;
BEGIN
  x[0] := 0.137793470540492 ; x[1] := 0.729454549503171 ;
  x[2] := 1.80834290174032 ; x[3] := 3.40143369785483 ;
  x[4] := 5.55249614006397 ; x[5] := 8.33015274676442 ;
  x[6] := 11.8437858378998 ; x[7] := 16.2792578313785 ;
  x[8] := 21.9965858119804 ; x[9] := 29.9206970122740 ;
  s[0] := 3.08441115765019E-001 ; s[1] := 4.01119929155277E-001 ;
  s[2] := 2.18068287611804E-001 ; s[3] := 6.20874560986810E-002 ;
  s[4] := 9.50151697518125E-003 ; s[5] := 7.53008388587250E-004 ;
```

```
  s[6] := 2.82592334960373E-005 ; s[7] := 4.24931398493113E-007 ;
  s[8] := 1.83956482414385E-009 ; s[9] := 9.91182718743507E-013 ;
  sum := 0.0;
  FOR  i := 0  TO  9  DO
    sum := sum + s[i]*F(x[i]);
  GaussLaguerre := sum
END; (* GaussLaguerre *)

BEGIN (* main program *)
  WRITELN('Gauss-Laguerre Quadrature') ; WRITELN ;
  WRITELN(GaussLaguerre:20:15)
END.
```

APPENDIX 6.8

A Pascal Procedure for Romberg Quadrature, which is in the file Romberg.Pas

```
PROCEDURE RombergQuadrature( a, b, RelErr, AbsErr : REAL ;
                                  VAR Result, EstErr : REAL ;
                                     VAR Evaluations, flag : INTEGER );
(*  The user supplies the following function to evaluate the integrand at
    each  x  in the interval  [a,b]:

       FUNCTION F( x : REAL ) : REAL;

   Input:
       a        Lower limit of integration
       b        Upper limit of integration
     RelErr     Required tolerance (relative error)   Relerr >= 0.0
     AbsErr     Required tolerance (absolute error)   AbsErr >= 0.0

   Output:
      Result       The approximation to the integral.
      EstErr       Actual error tolerance used.  If EstErr <= tol, where
                    tol <= Max(AbsErr,Abs(Result)*RelErr,eps), then a normal
                    exit has been made.  eps  is the approximate machine
                    epsilon.  IF  kmax + 1  extrapolations have been made and
                    EstErr > tol  then an abnormal exit was made.
    Evaluations    The number of function evaluations.
       flag        If  EstErr <= tol,  flag  is equal to the number of
                    extrapolations.  If  EstErr > tol  and  kmax + 1
                    extrapolations have been made, then  flag = 0.

    Note on usage:
       This procedure can be used to estimate the definite integral of f(x)
       between  a  and  b, provided  f(x)  is at least three or four times
       differentiable and  f(x)  is not periodic with period Abs(b-a).

Procedure  RombergQuadrature is based on the FORTRAN subroutine ROMINT that
appeared in
     G. Fairweather,
       "Algorithm 351, Modified Romberg Quadrature [D1]"
       Communications of the A.C.M.  vol. 12  June 1969, pp. 324-325.    *)
```

```
CONST
  kmax = 13; (* bound on the number of extrapolations *)

VAR
  eps, h, r, s, s0, s1, t, tol, w : REAL;
  rm : ARRAY[ 0..kmax ] OF REAL;
  i, j, k, n, n0, n1, p, q, v : INTEGER;
  finished : BOOLEAN;

BEGIN
  RelErr := Abs(RelErr) ; AbsErr := Abs(AbsErr) ; eps := 1.0 ;
  REPEAT
    eps := eps/2.0 ; r := 1.0 + eps
  UNTIL  r = 1.0; (* eps is the approximate machine epsilon *)
  (* initial trapezoid value *)
        t := 0.5*(b - a)*(F(a) + F(b));
  (* initial midpoint value *)
        rm[0] := (b - a)*F(0.5*(a + b));
  n := 2 ; r := 4.0 ; k := 0 ; finished := FALSE ;
  REPEAT
    w := (0.5*r - 1.0)/(r - 1.0);
    t := rm[0] + w*(t - rm[0]); (* improved trapezoid value *)
    n := n + n ; s := 0.0 ; h := (b - a)/n ; i := 1 ;
    IF  n <= 32  THEN
      n0 := n
    ELSE
      n0 := 32;
    IF  n <= 512  THEN
      n1 := n
    ELSE
      n1 := 512;
    WHILE  i <= n  DO  (* calculate midpoint value *)
      BEGIN
        s1 := 0.0 ; j := i ; v := i + n1 - 1 ;
        WHILE  j <= v  DO
          BEGIN
            s0 := 0.0 ; p := j + n0 - 1 ; q := j ;
            WHILE  q <= p  DO
              BEGIN
                s0 := s0 + F(a + q*h) ; q := q + 2
              END;
            s1 := s0 + s1 ; j := j + 32
          END;
        s := s + s1 ; i := i + 512
      END ;
    rm[k+1] := 2.0*h*s; (* end of calculation for midpoint value *)
    r := 4.0;
    FOR  j := k  DOWNTO  0  DO
      BEGIN (* Form the Romberg table from the midpoint values *)
        rm[j] := rm[j+1] + (rm[j+1] - rm[j])/(r - 1.0) ; r := 4.0*r
      END;
    tol := RelErr*Abs(t + rm[0])/2.0;
    IF  tol < AbsErr  THEN
      tol := AbsErr;
    IF  tol < eps   THEN
      tol := eps;
    EstErr := Abs(t - rm[0])/2.0;
    IF  EstErr <= tol  THEN
      finished := TRUE
```

```
    ELSE
      k := k + 1
   UNTIL  finished  OR  (k = kmax);
   Evaluations := n + 1;
   Result := (t + rm[0])/2.0;
   IF  finished  THEN
     flag := k + 1
   ELSE
     flag := 0
END; (* RombergQuadrature *)
```

APPENDIX 6.9

Listing of a Pascal Program for Example 6.3, using Romberg Quadrature

```
PROGRAM Quadrature;

CONST
  a = -1.0 ; b = 1.0 ;

VAR
  RelErr, AbsErr, Result, ErrEst : REAL;
  Evaluations, flag : INTEGER;

FUNCTION F( x : REAL ) : REAL;
BEGIN
  F := 1.0/(1.0 + Sqr(5.0*x))
END ;

(*$I Romberg.Pas *)

BEGIN (* main program *)
  RelErr := 1.0E-08 ; AbsErr := 0.0 ;
  RombergQuadrature(a,b,RelErr,AbsErr,Result,ErrEst,Evaluations,flag);
  WRITELN('Integral = ',Result,'    Estimated error = ',ErrEst:10);
  WRITELN('function evaluations = ',Evaluations:5);
  WRITELN('Actual error = ',Result - 0.4*ArcTan(5.0):10);
  IF  flag = 0  THEN
    WRITELN('did not converge')
  ELSE
    WRITELN(flag,' extrapolations made')
END.
```

APPENDIX 6.10

User's manual for the Pascal Procedure QUANC8, which is in the file QUANC8.Pas

```
PROCEDURE QUANC8( a, b, AbsErr, RelErr : REAL ;
                    VAR Result, ErrorEstimate : REAL ;
                       VAR Evaluations : INTEGER ; VAR flag : REAL );
```

```
The procedure QUANC8 estimates the integral of  F(x)  from  a  to  b  to
within a user-provided tolerance.  QUANC8  is an automatic adaptive routine
based on the 8-panel Newton-Cotes rule.


  Input:

      a      The lower limit of integration
      b      The upper limit of integration (b < a is allowed)
    AbsErr   An absolute error tolerance which should be non-negative
    RelErr   A relative error tolerance which should be non-negative


The user supplies the following function to evaluate the integrand for any
x  in the interval  [a,b]

       FUNCTION F( x : REAL ) : REAL;


  Output:

       Result       An approximation to the integral
    ErrorEstimate   An estimate of the magnitude of the actual error
     Evaluations    The number of function values used to determine  Result
        flag        A reliability indicator.  If  flag  is zero, then
                      Result  probably satisfies the error tolerance.  If
                      flag = xxx.yyy, then  xxx  is the number of panels
                      which have not converged and  0.yyy  is the fraction
                      of the interval left to do when the limit on
                      Evaluations  was approached.  For example, if
                      flag = 23.169 then 23 panels failed to converge and
                      there was trouble at approximately b - 0.169×(b - a).


Procedure QUANC8 is based on the FORTRAN subroutine that appeared in
  G.E. Forsythe, M.A. Malcolm, and C.B. Moler
    Computer Methods for Mathematical Computations
    Prentice-Hall, Englewood Cliffs, N.J., 1977
```

APPENDIX 6.11

Listing of a Pascal Program for Example 6.5, using QUANC8

```
PROGRAM Quadrature;

CONST
  a = 0.0 ; b = 1.0 ;

VAR
  AbsErr, RelErr, Result, flag, ErrEst : REAL;
  Evaluations : INTEGER;
```

```
FUNCTION F( x : REAL ) : REAL;
BEGIN
  F := 1.0/(x*(x - 1.0) + 0.2501)
END;

(*$I QUANC8.Pas *)

BEGIN (* main program *)
  RelErr := 1.0E-08 ; AbsErr := 0.0 ;
  QUANC8(a,b,AbsErr,RelErr,Result,ErrEst,Evaluations,flag);
  WRITELN('Integral = ',Result,'   Estimated error = ',ErrEst:10);
  WRITELN('function evaluations = ',Evaluations:5);
  WRITELN('Actual error = ',Result - 200.0*ArcTan(50.0):10);
  IF  flag > 0.0  THEN
    WRITELN('flag = ',flag:8:3)
END.
```

APPENDIX 6.12

Listing of a Pascal Program for Double Integrals

```
PROGRAM DoubleIntegral;

CONST
  a = 0.0; (* Lower limit of integration for the outside integral *)
  b = 1.0; (* Upper limit of integration for the outside integral *)

VAR
  Integral : REAL;

FUNCTION F( x, y : REAL ) : REAL; (* Integrand for the inside integral *)
BEGIN
  F := sin(x+y)
END;

FUNCTION G( x : REAL ) : REAL;
(* Lower limit of integration for the inside integral *)
BEGIN
  G := 0.0
END; (* G *)

FUNCTION H( x : REAL ) : REAL;
(* Upper limit of integration for the inside integral *)
BEGIN
  H := x
END; (* H *)

FUNCTION GaussLegendre( a, b : REAL ) : REAL;
(* Quadrature using the ten-point Gauss-Legendre quadrature formula *)
VAR
  i : INTEGER;
  sum, t, c1, c2 : REAL;
  x, s : ARRAY[ 0..9 ] OF REAL;
```

```
FUNCTION GaussLegendreI( a, b : REAL ) : REAL;
(* To compute the inside integral *)
VAR
  i : INTEGER;
  sum, c1, c2 : REAL;
BEGIN
  sum := 0.0 ; c1 := (b - a)/2.0 ; c2 := (b + a)/2.0 ;
  FOR  i := 0  TO  9  DO
    sum := sum + s[i]*F(t,c1*x[i]+c2);
  GaussLegendreI := c1*sum
END; (* GaussLegendreI *)

BEGIN (* GaussLegendre *)
  x[0] :=  -9.73906528517166E-001 ; s[0] :=   6.66713443086915E-002 ;
  x[1] :=  -8.65063366688993E-001 ; s[1] :=   1.49451349150566E-001 ;
  x[2] :=  -6.79409568299021E-001 ; s[2] :=   2.19086362516005E-001 ;
  x[3] :=  -4.33395394129247E-001 ; s[3] :=   2.69266719309976E-001 ;
  x[4] :=  -1.48874338981631E-001 ; s[4] :=   2.95524224714767E-001 ;
  FOR  i := 5  TO  9  DO
    BEGIN
      x[i] := -x[9-i] ; s[i] := s[9-i]
    END;
  sum := 0.0 ; c1 := (b - a)/2.0 ; c2 := (b + a)/2.0 ;
  FOR  i := 0  TO  9  DO
    BEGIN
      t := c1*x[i] + c2;
      sum := sum + s[i]*GaussLegendreI(G(t),H(t))
    END;
  GaussLegendre := c1*sum
END; (* GaussLegendre *)

BEGIN (* main program *)
  WRITELN('Gauss-Legendre Quadrature for Double Integrals');
  WRITELN;
  Integral := GaussLegendre(a,b);
  WRITELN(Integral:20:15)
END.
```

Chapter

7

Numerical Solution of Ordinary Differential Equations

Numerical integration of ordinary differential equations by various one-step methods and predictor-corrector methods is introduced. The methods include Euler, Heun, modified Euler, Runge-Kutta, Adams, and backward differentiation for initial value problems. The Runge-Kutta-Fehlberg method, with automatic step-size choice based on the user-requested error, is discussed. The shooting method is developed for two-point boundary-value problems.

7.1 INTRODUCTION

The study of many physical systems often leads to a description, or model, of the system as an ordinary differential equation. Such differential equations are derived by considering the underlying physics of the system. If the system has been correctly modeled, solutions of the differential equation should mimic the behavior of the physical system. As an example, consider the problem of determining the period of the simple pendulum. The differential equation describing the motion of the pendulum is

$$\theta'' = -\left(\frac{g}{\ell}\right)\sin(\theta) \tag{7.1–1}$$

where $''$ denotes the second derivative of θ with respect to the independent variable t. In order to obtain (7.1–1) certain assumptions had to be made:

1. Motion of the pendulum was confined to one vertical plane
2. The friction due to air resistance and the friction at the supporting joint were negligible
3. The pendulum chord was massless and inextensible.

If the real pendulum is such that these assumptions are valid, or nearly valid, then the solution of (7.1–1) should closely approximate the motion of the real pendulum.

The period of the pendulum is given in terms of nonelementary functions, called *elliptic integrals*. When (7.1–1) is "solved" in elementary physics courses, it is usually by the use of the small-angle approximation: we assume that $|\theta| \ll 1$ so that $\sin(\theta)$ can be replaced by θ, thereby deducing the well-known formula, $2\pi\sqrt{\ell/g}$, for the period. This formula is only valid for very small $|\theta|$. If we want the period of the pendulum when $|\theta|$ is not small, we will need to use elliptic integrals, or we can integrate (7.1–1) numerically. In this chapter, we shall examine various methods for integrating ordinary differential equations numerically.

Usually one has an nth order ordinary differential equation of the form

$$\frac{d^n y}{dt^n} = f\left(t, y, \frac{dy}{dt}, \frac{d^2y}{dt^2}, \ldots, \frac{d^{n-1}y}{dt^{n-1}}\right), \qquad a \le t \le b. \tag{7.1–2}$$

Equation (7.1–2) along with the extra conditions that y and its first $n - 1$ derivatives are given at $t = a$ is known as an *initial value problem* because all conditions on y and its derivatives are specified at the single point $t = a$. If some of the terms

$$y, \frac{dy}{dt}, \frac{d^2y}{dt^2}, \ldots, \frac{d^{n-1}y}{dt^{n-1}}$$

in (7.1–2) are specified at $t = a$, while the remaining terms are specified at $t = b$, we have a *two-point boundary-value problem*. Two-point boundary-value problems are discussed in Section 7.14.

In elementary courses, one learns to find solutions for certain special cases of (7.1–2), such as for $n = 1$, or where f is linear in y and its derivatives. For these cases we may be able to find a particular solution, or even the general solution. However, this is not the case for most ordinary differential equations; we may not be able to obtain a solution by standard methods. Consequently, a numerical solution may be required.

Most methods dealing with ordinary differential equations are concerned with first-order equations, or systems of first-order equations. Equations like (7.1–2) can be converted to a system of first-order ordinary differential equations by introducing new variables,

$$y_1 = y, \qquad y_{k+1} = \frac{d^k y}{dt^k} \qquad \text{for} \qquad k = 1, 2, \ldots, n-1$$

so that (7.1–2) can be written as the first-order system of ordinary differential equations

$$\left.\begin{aligned} \frac{dy_k}{dt} &= y_{k+1}; \qquad k = 1, 2, \ldots, n-1 \\ \frac{dy_n}{dt} &= f(t, y_1, y_2, \ldots, y_n) \end{aligned}\right\} \tag{7.1–3}$$

For example, if we convert (7.1–1) into a system of two first-order equations we will obtain

$$\frac{dy_1}{dt} = y_2, \qquad \frac{dy_2}{dt} = -(g/\ell)\sin(y_1) \qquad \text{where} \quad y_1 = \theta.$$

The first-order system (7.1–3) is a special case of the vector differential equation

$$\frac{d\mathbf{y}}{dt} = \mathbf{f}(t, \mathbf{y}) \tag{7.1–4}$$

where $\mathbf{y}^T = [y_1, y_2, \ldots, y_n]$, $\mathbf{f}^T = [f_1, f_2, \ldots, f_n]$, $f_i = f_i(t, y_1, y_2, \ldots, y_n)$. The superscript T denotes transposition. When (7.1–3) is put into the form of (7.1–4) we have $f_i = y_{i+1}$, $1 \le i \le n-1$, and $f_n = f$.

We shall examine several different methods for numerically integrating systems of the form (7.1–4); however, we shall use the special case $n = 1$ to study each method. Consequently, we will examine the initial value problem

$$\frac{dy}{dt} = f(t, y), \qquad a \le t \le b \tag{7.1–5}$$

$$y(a) = y_0 \tag{7.1–6}$$

in the earlier parts of this chapter. After studying various methods for the $n = 1$ case, we will show how these methods can be easily extended to systems ($n > 1$).

7.2 SOME MATHEMATICAL PRELIMINARIES

In order to understand the methods required for solving initial value problems numerically, we shall need to discuss some mathematical concepts which arise from the following questions.

1. Does a solution of the initial value problem exist?
2. If the answer to (1) is yes, is the solution unique?

3. Will small changes in the differential equation or the initial conditions lead to correspondingly small changes in the solution?

These questions are concerned with both theoretical and computational aspects of differential equations. In order to see how these questions relate to the numerical solution of ordinary differential equations we need some definitions.

Definition 7.1

A function $f(t, y)$ is said to satisfy a *Lipschitz condition* on a region R of the (t, y)-plane if there exists a constant $L > 0$ such that

$$|f(t, u) - f(t, v)| \leq L|u - v| \tag{7.2–1}$$

for all (t, u) and (t, v) in R.

Definition 7.1 can be extended to systems of differential equations like (7.1–4) by replacing the absolute values by appropriate vector norms. If $\partial f/\partial y$ exists and $|\partial f/\partial y|$ is bounded, then we could take the *Lipschitz constant L* as

$$L = \text{maximum}\left\{\left|\frac{\partial f}{\partial y}\right|\right\} \quad \text{for} \quad (t, y) \quad \text{in} \quad R.$$

As an example, consider the differential equation $y' = y\cdot\cos(t)$, where $'$ denotes differentiation with respect to t. Here, $f(t, y) = y\cdot\cos(t)$, so that $|f(t, u) - f(t, v)| = |\cos(t)|\cdot|u - v| \leq |u - v|$ for all u, v, and t. Hence, the Lipschitz constant is $L = 1$ for this example.

Definition 7.2

An initial value problem is called *well-posed* if the answers to questions (2) and (3) above are "yes."

Therefore, in a well-posed problem there is only *one* solution through a given point $(a, y(a))$, and small changes in the initial conditions or $f(t, y)$ give a solution that is "close" to the original solution.

The following theorem relates the Lipschitz condition to question (1).

Theorem 7.1

If $f(t, x)$ satisfies a Lipschitz condition on a region R of the (t, x)-plane, there is at most one solution $x(t)$ of $x' = f(t, x)$ which satisfies a given initial condition $x(a) = x_0$ in R.

There exist differential equations for which the answer to question (3) is "no." Such equations may have some *unstable* solutions, or the behavior of solutions is *chaotic*. In the former case, a solution becomes unbounded as the independent variable increases, or decreases. For example, the differential equation $y' = 1 + y^2$, with $y(0) = 0$, has the solution $y(t) = \tan(t)$; clearly, $|y(t)| \to \infty$ as $|t| \to \pi/2$. Figure 7.1 shows examples of both stable and unstable solutions (trajectories). On the other hand, a chaotic solution remains bounded, but it tends to fill up some region of the solution space.

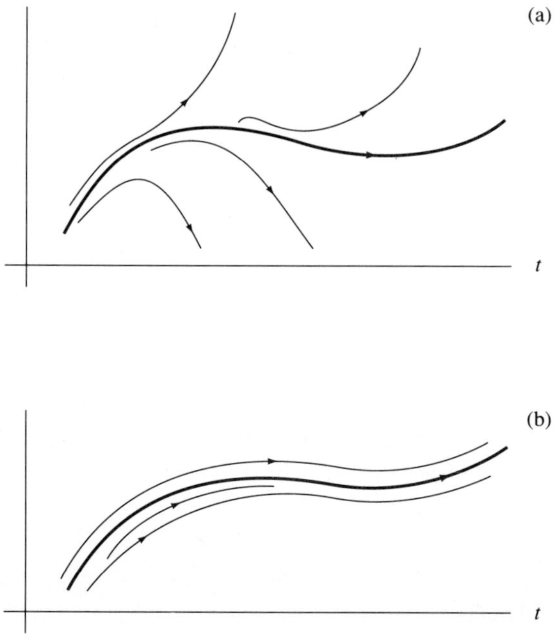

Figure 7.1 Trajectories of **(a)** unstable, and **(b)** stable differential equations.

To solve an ordinary differential equation numerically on a digital computer, we must replace a continuous process (the differential equation) by a discrete process. This occurs because derivatives must be approximated by difference quotients due to the finite precision available with digital computers. Recall that the derivative of a real-valued function with respect to a real variable is the result of a limit operation over real numbers. However, a digital computer can only handle a small, finite subset of the real numbers. We must replace the differential equation (7.1–5) by a *discrete equation* or *difference equation* of the form

$$y_{k+1} = G(t_{k+1}, y_{k+1}, t_k, y_k, \ldots, t_{k-m}, y_{k-m}; f) \tag{7.2–2}$$

where y_i is the computed approximation to $y(t_i)$, and $t_{i+1} - t_i = h_i$ is the stepsize at t_i. Equation (7.2–2) is a very general difference equation. When (t_{k-i}, y_{k-i}), $i = 0, 1, \ldots, m$, and t_{k+1} are known, y_{k+1} is determined from (7.2–2); then k is increased by one and the process is repeated to find y_{k+2}, and so on. The function G will depend on the right-hand side of the differential equation (7.1–5); consequently, we show the explicit dependence on f in (7.2–2). When G contains y_{k+1} we have an *implicit* method; otherwise, the method is *explicit*. If $m \geq 1$, (7.2–2) is called a *multistep* method; $m = 0$ gives a *single-step* method. We will discuss single-step methods in Sections 7.3–7.9 and multistep methods in Sections 7.11 and 7.12.

Because (7.2–2) is a very general difference equation, we usually restrict the function G so that it does not explicitly contain y_{k+1}. Also, we can simplify G

by using a constant step-size, h, so that $t_{i+1} - t_i = h$ for all i. If G and the step-sizes, h_i, are properly chosen, (7.2–2) should provide a "good" approximation to the solution of (7.1–5). In succeeding sections we will study different choices for G and m.

When a difference equation, such as (7.2–2), is used to approximate the differential equation (7.1–5), a type of error called a *discretization*, or *truncation*, error is introduced. Recall that this is unavoidable because we have replaced a continuous (infinite) process by a discrete (finite) process. The discretization error is due to the method used. To keep the discretization error "small" we would want to restrict the size of $\max\{ |h_i| \}$. In addition to discretization errors, each floating point calculation required to evaluate G introduces a roundoff error, due to the finite precision of the computer. Hence, to reduce the effect of roundoff errors we would want $|h_i|$ to be "large" so that the total number of computations is as small as possible. A well-designed method for the numerical integration of ordinary differential equations must maintain a balance between the cumulative effects of roundoff errors and the cumulative effects of discretization errors.

Given the approximate solution y_i at t_i, we can define the *local discretization error* at t_{i+1} as

$$D(t_{i+1}) = y_{i+1} - u(t_{i+1}) \qquad \textbf{(7.2–3)}$$

where $u(t)$ is the local solution to the differential equation

$$\frac{du}{dt} = f(t, u), \; u(t_i) = y_i \qquad \textbf{(7.2–4)}$$

It is important to keep in mind that the local discretization error is obtained by assuming we are computing with infinite precision (no roundoff error). If infinite precision arithmetic is used, then the only error present is the discretization error.

It is usually easy to estimate $|D(t_i)|$. Generally, one would like to know the *global error*, $E(t_i) = y_i - y(t_i)$, which is the cumulative effect of both the local discretization errors and the roundoff errors up to t_i. However, $E(t_i)$ is generally hard to estimate. Consequently, most implemented computer methods for differential equations only try to control the magnitude of the local discretization error. Figure 7.2 illustrates the local and global errors in terms of the local solution, (7.2–4), and the global solution, (7.1–5).

The effect of a small error in the solution of (7.1–5) can be modelled by using the corresponding *variational equation*

$$\frac{dE}{dt} = J(t)E \qquad \textbf{(7.2–5)}$$

where $J(t) = \partial f/\partial y$, evaluated at the computed solution $y(t)$ of (7.1–5). Eq. (7.2–5) is obtained by substituting $y(t) + E(t)$ into (7.1–5), expanding the right-hand side by using Taylor's formula, and dropping all terms multiplying powers of E greater than one. Suppose that $J(t)$ is not identically zero, and $E(a) \neq 0$. The solutions of (7.1–5) are stable if $\lim_{t\to\infty} E(t) = 0$, and the solutions are unstable

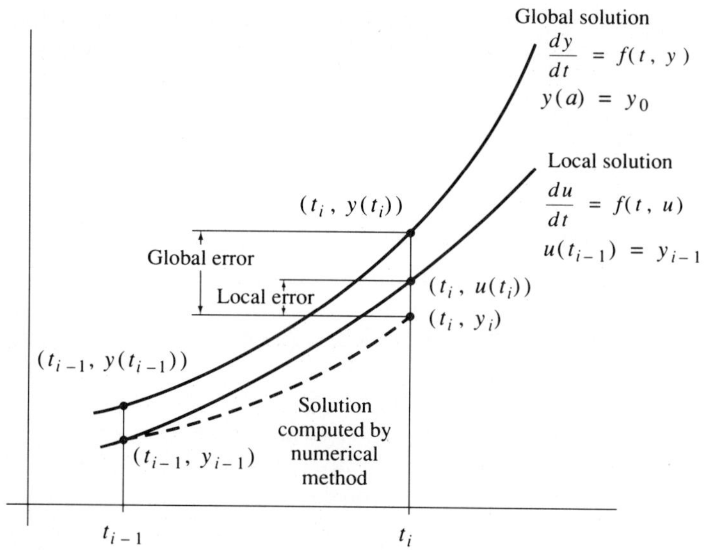

Figure 7.2 Local and global errors in the numerical solution of $dy/dt = f(t, y)$. The computed solution at t_i is y_i.

if $|E(t)| \to \infty$ as $t \to \infty$ (see Figure 7.1). If it can be shown that $|E(t)|$ remains "small" as $t \to \infty$ then we can say that small perturbations in the solution of (7.1–5) do not grow too large as t increases. For example, the differential equation

$$\frac{dy}{dt} = y \cdot \cos(t)$$

is its own variational equation; hence $|E(t)| = |E(a)| e^{\sin(t)} \leq |E(a)| e$ and therefore, $|E(t)|$ remains small if $|E(a)|$ is small.

If $J(t)$ were identically zero, (7.2–5) would give no information concerning the stability of the solutions of (7.1–5) or the first-order growth of errors.

We can use (7.2–5) to determine a bound on $|h|$, where h is the step-size, if (7.2–2) is used to approximate the solution of (7.1–5). Suppose that $J(t)$ in (7.2–5) satisfies the inequality

$$c_1 \leq J(t) \leq c_2$$

for constants c_1 and c_2. Let us assume that (7.1–5) has only stable solutions; consequently, $c_1 \leq c_2 \leq 0$. Let $\lambda = |c_1|$. Then (7.2–5) can be approximated by the differential equation

$$\frac{dx}{dt} = -\lambda x \qquad \textbf{(7.2–6)}$$

Equation (7.2–6) can also serve as an analogue for the variational equation of (7.1–4) if we interpret $-\lambda$ as largest magnitude eigenvalue of the Jacobian matrix, **J**, of **f** in (7.1–4). In this case, stability of the solutions requires that the real part of each eigenvalue of **J** be negative. The number λ, real or complex, is an *eigenvalue* of the $n \times n$ matrix **A** if there is a nonzero vector **x**, called an *eigenvector*, such that $\mathbf{Ax} = \lambda\mathbf{x}$ (see Appendix A.4).

When (7.2–2) is used to approximate the solution of (7.2–6), we obtain a difference equation whose solution is to be close to the solution of (7.2–6). This means that the difference equation solution must tend to zero as $t \to \infty$ because the solution of (7.2–6) is $x(t) = x(a)e^{-\lambda(t-a)}$. The difference equation solution will be some function of λh; the fact that the solution should tend to zero will give bounds on $|h|$, depending on λ and the particular function G in (7.2–2). We shall discuss this point further in Section 7.8, after we have studied some particular methods for numerically integrating (7.1–5). The point to remember here is that if $|h|$ is too large the computed solution, from (7.2–2), will not be close to the solution of the differential equation (7.1–5).

In the following sections we will examine various methods for the solution of initial value problems for ordinary differential equations of the form (7.1–5) and (7.1–6). Although we emphasize the case of the single equation, the extension to the vector case, (7.1–4), is straightforward. We first look at some single-step methods, $m = 0$ in (7.1–2); later we discuss some multistep methods, $m \geq 1$ in (7.2–2).

7.3 EULER'S METHOD, A SINGLE-STEP METHOD OF ORDER ONE

The simplest single-step method for the initial value problem (7.1–5) and (7.1–6) is Euler's method,

$$y_{i+1} = y_i + hf(t_i, y_i) \qquad i = 0, 1, \ldots \tag{7.3–1}$$

where $y_0 = y(a)$, $t_0 = a$, and $t_{i+1} = t_i + h$. Here, we are using a constant step-size, h. A typical step for Euler's method is shown in Fig. 7.3. In order to use (7.3–1) we set $i = 0$ and compute y_1, given the known values of t_0 and y_0. If the magnitude of the step-size is sufficiently small, then y_1 should be a good approximation to the solution $y(t_1)$. The integer i is incremented by one and (7.3–1) is used to find y_2, and so on.

We want to obtain some information on the accuracy of Euler's method; that is, to determine the local discretization error made when one step is taken with Euler's method. We shall assume that the function $f(t, y)$ in (7.1–5) is twice continuously differentiable, so that we can use Taylor's formula for the solution $y(t)$ expanded about $t = t_i$. This gives

$$y(s) = y(t_i) + y'(t_i)(s - t_i) + 0.5y''(\xi_i)(s - t_i)^2 \tag{7.3–2}$$

where ξ_i is between s and t_i. Setting $s = t_{i+1}$ in (7.3–2), and recalling that $h = t_{i+1} - t_i$, we have

$$y(t_{i+1}) = y(t_i) + hf(t_i, y(t_i)) + 0.5h^2y''(\xi_i) \tag{7.3–3}$$

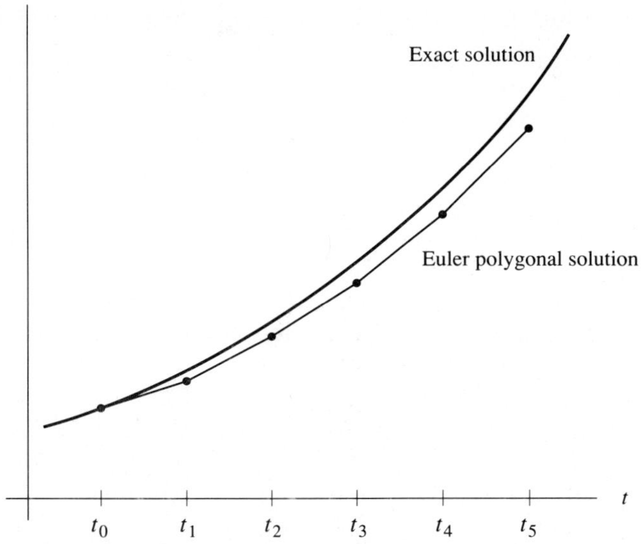

Figure 7.3 Euler's method.

The first two terms on the right-hand side of (7.3–3) are the right-hand side of (7.3–1), Euler's method. When we compare (7.3–1) and (7.3–3) we see that the local discretization error at t_{i+1} is $0.5h^2y''(\xi_i)$. If we sum (7.3–3) over $i = 0, 1, \ldots, m$ we obtain

$$y(t_{m+1}) = y(a) + h \sum_{i=0}^{m} f(t_i, y(t_i)) + E \tag{7.3–4}$$

$$E = 0.5h^2 \sum_{i=0}^{m} y''(\xi_i) = 0.5h\left[\sum_{i=0}^{m} y''(\xi_i)h\right]$$

If $|y''(x)| \leq M$, then $|E| \leq 0.5\,|h|\,M(t_{m+1} - a)$; thus, the global discretization error on any given t-interval is bounded by a constant times $|h|$. We write this as $|E| = 0(h)$; read this as "big oh of h." This means that $|E| \leq C\,|h|$, where the constant C is independent of h. A numerical method with a local discretization error $0(h^{p+1})$ is a method of order p. We refer to the method as an $0(h^p)$ method. Hence, Euler's method is $0(h)$. Generally, for an $0(h^p)$ method, if h is replaced by qh the error is changed by the factor q^p.

Fig. 7.3 shows that Euler's method approximates the solution of the initial value problem (7.1–5) by a polygonal curve made up of straight line segments joined at the points (t_i, y_i). If the step-size has been selected carefully, then the polygonal curve should closely approximate the solution curve. The following example problem shows that this may not happen unless h is properly chosen.

EXAMPLE 7.1

Consider the differential equation

$$\frac{dy}{dt} = y \cdot \cos(t), \qquad y(0) = 1$$

We want to obtain $y(t)$ at $t = 1$, 2, and 3 using Euler's method, with step-sizes $h = 1.0$, 0.5, and 0.25. The results are summarized in Table 7.1. This differential equation is integrable by the standard technique of separation-of-variables; its solution is $y(t) = e^{\sin(t)}$. See Appendix 7.1 for a listing of a Pascal program that uses Euler's method to obtain a numerical solution for this differential equation. The program also contains several integration methods yet to be discussed. As you can see in Table 7.1, Euler's method does "track" the solution, but not very well at these step-sizes. This is because the method is $0(h)$. In the next section we present methods that are $0(h^2)$.

Table 7.1 SOLUTION OF $y' = y \cdot \cos(t)$, $y(0) = 1$, BY EULER'S METHOD

	The Step-Size h			
t	1.00	0.50	0.25	Exact Solution
0.00	1.00000	1.00000	1.00000	1.00000
0.25			1.25000	1.28070
0.50		1.50000	1.55279	1.61515
0.75			1.89346	1.97712
1.00	2.00000	2.15819	2.23982	2.31978
1.25			2.54236	2.58309
1.50		2.74122	2.74278	2.71148
1.75			2.79128	2.67510
2.00	3.08060	2.83818	2.66690	2.48258
2.25			2.38944	2.17727
2.50		2.24763	2.01419	1.81934
2.75			1.61078	1.46472
3.00	1.79862	1.34729	1.23857	1.15156

EXERCISE 7.1

Adapt the program in Appendix 7.1 to integrate the differential equation

$$\frac{dy}{dt} = (1 - y)(4 - y), \qquad y(0) = 0$$

on the interval [0, 1] using Euler's method with a step-size of $h = 1/32$. Obtain the error in $y(1)$ by comparing the computed solution with the exact solution $y(t) = 4(e^{3t} - 1)/(4e^{3t} - 1)$.

Answer The magnitude of the error at $t = 1$ is 6.263×10^{-3}.

7.4 SOME SINGLE-STEP METHODS OF ORDER TWO

7.4.1 The Modified Euler Method

If we change the Euler method by first taking a step of length $h/2$ from t_i, use the predicted value of y at $t_i + h/2$ to compute $f(t, y)$, and then use this value of f in the Euler method to step from t_i to t_{i+1}, we obtain the modified Euler method

$$K_1 = f(t_i, y_i) \tag{7.4–1}$$

$$K_2 = f\left(t_i + \frac{h}{2}, y_i + \frac{h}{2}K_1\right) \tag{7.4–2}$$

$$y_{i+1} = y_i + hK_2 \tag{7.4–3}$$

Fig. 7.4 shows the modified Euler method.

EXERCISE 7.2

Repeat Exercise 7.1 for the modified Euler method.

Answer The magnitude of the error at $t = 1$ is 2.74×10^{-4}.

7.4.2 The Heun Method

The Heun method uses the average of the f values computed at t_i and t_{i+1}, where the value of y at t_{i+1} is estimated by Euler's method. Consequently, Heun's

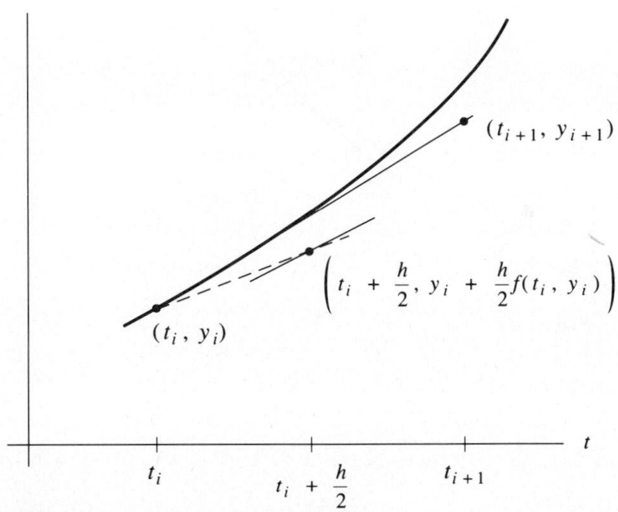

Figure 7.4 Modified Euler's method.

method differs from the modified Euler method in the manner in which K_2 and y_{i+1} are computed. Heun's method is

$$K_1 = f(t_i, y_i) \tag{7.4–4}$$

$$K_2 = f(t_i + h, y_i + hK_1) \tag{7.4–5}$$

$$y_{i+1} = y_i + \frac{h(K_1 + K_2)}{2} \tag{7.4–6}$$

Fig. 7.5 shows the Heun method.

EXERCISE 7.3

Repeat Exercise 7.1 for the Heun method.

Answer The magnitude of the error at $t = 1$ is 2.41×10^{-4}.

When error analyses of the modified Euler and the Heun methods are done by using Taylor expansions about (t_i, y_i), both methods are found to be $0(h^2)$.

When we applied the modified Euler method and the Heun method to the differential equation in Example 7.1 with the step-size $h = 0.25$, we obtained the results shown in Table 7.2, along with the Euler method results from Table 7.1 (see Appendix 7.1). Table 7.2 shows the improvement of the modified Euler and the Heun methods over the Euler method.

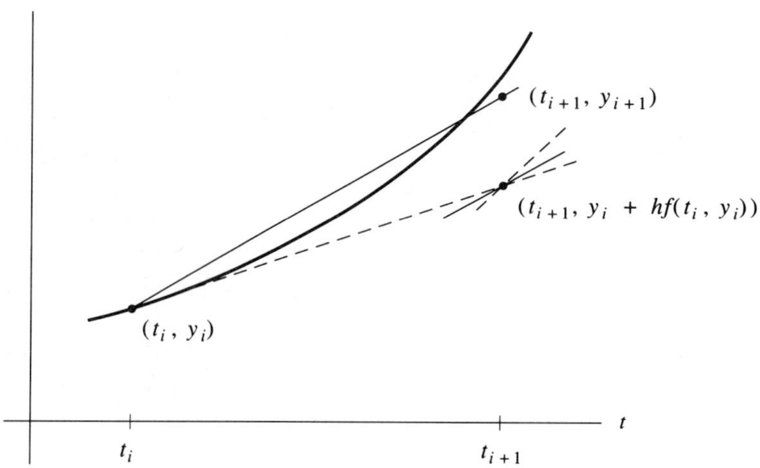

Figure 7.5 Heun method.

Table 7.2

t	Euler	Modified Euler	Heun	Exact Solution
0.00	1.00000	1.00000	1.00000	1.00000
0.25	1.25000	1.27906	1.27639	1.28070
0.50	1.55279	1.61263	1.60492	1.61515
0.75	1.89346	1.97545	1.95996	1.97712
1.00	2.23982	2.32096	2.29581	2.31978
1.25	2.54236	2.58805	2.55358	2.58309
1.50	2.74278	2.71888	2.67858	2.71148
1.75	2.79128	2.68173	2.64153	2.67510
2.00	2.66690	2.48539	2.45139	2.48258
2.25	2.38944	2.17541	2.15141	2.17727
2.50	2.01419	1.81444	1.80087	1.81934
2.75	1.61078	1.45952	1.45413	1.46472
3.00	1.23857	1.14820	1.14775	1.15156

Solution of $y' = y \cdot \cos(t)$, $y(0) = 1$, by Euler's Method, the modified Euler's method, and the Heun method, with the step-size $h = 0.25$.

Both the modified Euler method and the Heun method are special cases of a family of single-step methods with order $0(h^2)$. The members of this family (characterized by a nonzero parameter, α) are described by the equations

$$\omega = \frac{h}{2\alpha} \qquad \textbf{(7.4–7)}$$

$$K_1 = f(t_i, y_i) \qquad \textbf{(7.4–8)}$$

$$K_2 = f(t_i + \omega, y_i + \omega K_1) \qquad \textbf{(7.4–9)}$$

$$y_{i+1} = y_i + h[(1 - \alpha)K_1 + \alpha K_2] \qquad \textbf{(7.4–10)}$$

The modified Euler method has $\alpha = 1$; the Heun method has $\alpha = 0.5$. This family of methods is called the *Runge-Kutta $0(h^2)$ methods*. Runge-Kutta methods are specified by the number of K_j functions required (called *stages*) and by the order of the method. Runge-Kutta methods use a weighted average of the values of the function f at various points in the interval $[t_i, t_{i+1}]$ to estimate y_{i+1}. Each member of the family of methods given by (7.4–7) to (7.4–10) is a two-stage Runge-Kutta method of order $0(h^2)$. (Euler's method can be considered as a one-stage Runge-Kutta method with order one.) We will be examining higher order Runge-Kutta methods in other sections. Before doing that we give the local discretization error for (7.4–7) to (7.4–10).

The local discretization error is obtained by using the multivariable version of Taylor's formula to expand the various functions in (7.4–7) to (7.4–10) about the point (t_i, y_i). In addition, we have the Taylor expansion of the solution of

(7.2–4), so that

$$D(t_{i+1}) = h^3\left[Q(f_{tt} + 2ff_{ty} + f^2 f_{yy}) - \frac{f_y(f_t + ff_y)}{6}\right] \qquad \textbf{(7.4–11)}$$

$$Q = \frac{3 - 4\alpha}{24\alpha}$$

When the variables t and y are used as subscripts in (7.4–11) they denote partial differentiation with respect to the subscript variable. The functions on the right-hand side of (7.4–11) are evaluated at a point on the line joining (t_i, y_i) to (t_{i+1}, y_{i+1}).

7.5 THE CLASSICAL RUNGE-KUTTA METHOD OF ORDER FOUR

This famous method is given by the equations

$$\omega = \frac{h}{2} \qquad \textbf{(7.5–1)}$$

$$K_1 = f(t_i, y_i) \qquad \textbf{(7.5–2)}$$

$$K_2 = f(t_i + \omega, y_i + \omega K_1) \qquad \textbf{(7.5–3)}$$

$$K_3 = f(t_i + \omega, y_i + \omega K_2) \qquad \textbf{(7.5–4)}$$

$$K_4 = f(t_{i+1}, y_i + hK_3) \qquad \textbf{(7.5–5)}$$

$$y_{i+1} = y_i + \frac{h(K_1 + 2K_2 + 2K_3 + K_4)}{6} \qquad \textbf{(7.5–6)}$$

The method consists of four stages, with a local discretization error of $0(h^5)$. Notice that (7.5–6) uses a weighted average of the four computed derivatives to determine y_{i+1}. This is discussed further when we consider the general Runge-Kutta method (7.5–7) to (7.5–10), below.

The differential equation of Example 7.1 was integrated using this Runge-Kutta method. The results are shown in Table 7.3, along with the results obtained with the modified Euler and Heun methods (see Appendix 7.1). Table 7.3 shows the improved accuracy of the $0(h^4)$ Runge-Kutta method over the order two methods.

EXERCISE 7.4

Repeat Exercise 7.1 for the $0(h^4)$ classical Runge-Kutta method.

Answer The magnitude of the error at $t = 1$ is 1.33×10^{-7}.

Table 7.3

t	Modified Euler	Heun	Runge-Kutta	Exact Solution
0.00	1.00000	1.00000	1.00000	1.00000
0.25	1.27906	1.27639	1.28069	1.28070
0.50	1.61263	1.60492	1.61513	1.61515
0.75	1.97545	1.95996	1.97709	1.97712
1.00	2.32096	2.29581	2.31974	2.31978
1.25	2.58805	2.55358	2.58304	2.58309
1.50	2.71888	2.67858	2.71144	2.71148
1.75	2.68173	2.64153	2.67505	2.67510
2.00	2.48539	2.45139	2.48254	2.48258
2.25	2.17541	2.15141	2.17723	2.17727
2.50	1.81444	1.80087	1.81931	1.81934
2.75	1.45952	1.45413	1.46469	1.46472
3.00	1.14820	1.14775	1.15155	1.15156

Solution of $y' = y \cdot \cos(t)$, $y(0) = 1$, by the modified Euler's method, the Heun method, and the classical Runge-Kutta method with step-size $h = 0.25$

The general form of the Runge-Kutta methods is

$$K_1 = f(t_i, y_i) \tag{7.5–7}$$

$$W_j = \sum_{k=1}^{j-1} b_{jk}K_k \qquad 2 \le j \le m \tag{7.5–8}$$

$$K_j = f(t_i + a_jh, y_i + hW_j) \qquad 2 \le j \le m \tag{7.5–9}$$

$$y_{i+1} = y_i + h \sum_{j=1}^{m} c_jK_j \tag{7.5–10}$$

The constants a_j, b_{jk}, and c_j in (7.5–7 to 7.5–10) are determined by matching powers of h in Taylor series expansions of the functions K_j about the point (t_i, y_i). This general form of the Runge-Kutta method more clearly shows the use of weighted averages of previous stages (derivatives) in computing the derivatives at any stage.

We have seen that the Euler method is a one-stage method of order one; the Heun and modified Euler are two-stage methods of order two; the classical Runge-Kutta method is a four-stage, order four method. Is there a Runge-Kutta p-stage method of order p? The answer is "yes" for $p = 1, 2, 3, 4$; for $p = 5$ or 6 stages, the best we can do is a method of order $p - 1$. (An order five method would require at least six stages.)

7.6 TAYLOR SERIES METHOD

The preceding formulas for the modified Euler, the Heun, and the classical Runge-Kutta methods may seem unnecessarily complicated, especially in view of the fact that the local discretization error is determined by the use of Taylor's formula. Why not use Taylor's formula, truncated after m terms? After all, Euler's method is just the first two terms from the Taylor expansion of $y(s)$ about $s = t_i$. Applying Taylor's formula, with remainder, to $y(s)$ expanded about $s = t_i$ gives

$$y(t_{i+1}) = y_i + \sum_{j=1}^{m} Z_j(t_i) + R_{m+1} \tag{7.6–1}$$

where

$$Z_k(t) = \frac{h^k}{k!}\frac{d^k y}{dt^k}(t) \quad \text{and} \quad R_{m+1} = Z_{m+1}(t_i + \theta h) \qquad 0 < \theta < 1.$$

Observe that $Z_1(s) = f(s, y(s))$. The Taylor series method is obtained when we drop the remainder term, R_{m+1}, in (7.6–1) to give

$$y_{i+1} = y_i + \sum_{j=1}^{m} Z_j(t_i) \tag{7.6–2}$$

It is clear that a Taylor series method which includes terms up to the mth derivative of y is $0(h^m)$, because the local discretization error is the remainder term, R_{m+1}.

One disadvantage of the Taylor series method is the necessity of computing high derivatives of y. This may be a formidable task if the function $f(t, y)$ is complicated. However, the increasing use of computer algebraic programs which allow one to do symbolic computation makes such lengthy calculations more tractable.

To apply the Taylor series method in the numerical solution of the differential equation in Example 7.1, we need to compute the terms $Z_k(t)$ in (7.6–2). The first four $Z_k(t)$ are computed as follows. We have

$$Z_k(t) = \frac{h^k}{k!}\frac{d^k y}{dt^k}.$$

Therefore, we need only compute the first four derivatives of y. From the differential equation we have

$$\frac{dy}{dt} = y \cdot \cos(t)$$

Differentiate to get

$$\frac{d^2y}{dt^2} = \frac{dy}{dt}\cos(t) - y \cdot \sin(t)$$

and continuing for the next two derivatives,

$$\frac{d^3y}{dt^3} = \frac{d^2y}{dt^2}\cos(t) - 2\frac{dy}{dt}\sin(t) - y\cdot\cos(t)$$

$$\frac{d^4y}{dt^4} = \frac{d^3y}{dt^3}\cos(t) - 3\frac{d^2y}{dt^2}\sin(t) - 3\frac{dy}{dt}\cos(t) + y\cdot\sin(t)$$

These derivatives can be computed, at a value of t at which y is known, in the order from the first to the fourth. Each derivative is computed using all the previously computed derivatives. When the derivatives have been found at t, $Z_k(t)$, $k = 1, 2, 3, 4$, can be calculated. Thus, (7.6–2) can be used to advance the solution from t_i to t_{i+1}, when $m = 4$.

We selected $m = 3$ and $m = 4$ (four and five terms, respectively, in (7.6–2)) to integrate the differential equation in Example 7.1. The step-size was fixed at $h = 0.25$ so that the results could be compared with those previously obtained by other methods. The computational results are shown in Table 7.4. (See Appendix 7.1)

EXERCISE 7.5

Repeat Exercise 7.1 for the four-term Taylor series method ($m = 3$).

Answer The magnitude of the error at $t = 1$ is 1.31×10^{-5}.

Table 7.4

t	Taylor Series		Runge-Kutta	Exact Solution
	4 terms	5 terms		
0.00	1.00000	1.00000	1.00000	1.00000
0.25	1.28125	1.28076	1.28069	1.28070
0.50	1.61669	1.61527	1.61513	1.61515
0.75	1.97991	1.97725	1.97709	1.97712
1.00	2.32356	2.31980	2.31974	2.31978
1.25	2.58695	2.58292	2.58304	2.58309
1.50	2.71424	2.71114	2.71144	2.71148
1.75	2.67606	2.67474	2.67505	2.67510
2.00	2.48205	2.48238	2.48254	2.48258
2.25	2.17629	2.17730	2.17723	2.17727
2.50	1.81890	1.81951	1.81931	1.81934
2.75	1.46524	1.46491	1.46469	1.46472
3.00	1.15285	1.15168	1.15155	1.15156

Solution of $y' = y\cdot\cos(t)$, $y(0) = 1$, by the Taylor series method and the classical Runge-Kutta method, with step-size $h = 0.25$.

7.7 EXTRAPOLATION METHODS

We have seen that the local discretization error is $0(h^{p+1})$ for a method of order p. We can exploit this to obtain more accurate estimates of y_{i+1} in a numerical integration scheme. The method to be used is called *Richardson extrapolation to the limit*, or *deferred approach to the limit*. Richardson extrapolation is described in Section 5.12. We want to apply Richardson extrapolation to the integration of an ordinary differential equation.

Suppose we are solving a differential equation with a method of order p and we have computed several values of y at t_{i+1} with the step-sizes $h/2^k$ for $k = 0, 1, 2, \ldots$. We denote these computed values of the solution by $A_{k,0}$, $k = 0, 1, 2, \ldots$; thus, $A_{k,0} = y(t_{i+1}; h/2^k)$

$$A_{k,0} = y(t_{i+1}; 0) + c_p\left(\frac{h}{2^k}\right)^p + 0(h^{p+1}) \tag{7.7–1}$$

with $y(t_{i+1}; 0)$ denoting the solution at t_{i+1} with a step-size $h = 0$. The values, $A_{k,0}$, of the solution are then used in the Richardson extrapolation method to obtain more accurate estimates of y_{i+1} as follows. We form an array, $A_{k,j}$, $j \geq 0$, where

$$A_{k,m+1} = \frac{2^{p+m}A_{k,m} - A_{k-1,m}}{2^{p+m} - 1} \tag{7.7–2}$$

$$k = 0, 1, \ldots \quad \text{and} \quad m = 0, 1, \ldots, k - 1$$

The array is triangular and has the form

$$\begin{array}{lllll} A_{0,0} \\ A_{1,0} & A_{1,1} \\ A_{2,0} & A_{2,1} & A_{2,2} \\ \vdots & \vdots & \vdots & \vdots & \vdots \; . \end{array}$$

The second index, j, in $A_{k,j}$ gives the level of extrapolation. We are using the Richardson extrapolation over and over to successively eliminate higher and higher powers of h in the versions of (7.7–1) that arise from the particular single-step method used to solve the differential equation. It should be noted that the number of levels of extrapolation should be kept small, say less than eight, in order to avoid accumulation of roundoff errors which can occur in forming the extrapolates $A_{k,j}$.

After the extrapolation process has been carried out, it is possible that the extrapolated value is not sufficiently accurate. If this occurs, $|h|$ may have to be reduced and the extrapolation repeated.

Suppose we apply this repeated extrapolation technique to the differential equation $y' = -y$, $y(0) = 1$, integrated using Euler's method on the interval [0, 0.25]. In (7.7–2), $p = 1$. We use $h = 0.25/2^k$ so that the triangular array $A_{k,j}$, for four levels of extrapolation, is

k	$A_{k,0}$	$A_{k,1}$	$A_{k,2}$	$A_{k,3}$	$A_{k,4}$
0	0.7500000				
1	0.7656250	0.7812500			
2	0.7724762	0.7793274	0.7786865		
3	0.7756999	0.7789236	0.7787890	0.7788036	
4	0.7772652	0.7788305	0.7787994	0.7788009	0.7788007

The entries in column $A_{k,j}$ are computed from the entries in column $A_{k,j-1}$ according to (7.7–2), with $p = 1$. The final extrapolated value $A_{4,4} = 0.7788007$ is correct to within 3.6×10^{-8}, although we only show seven significant digits.

A listing of a Pascal procedure for the Extrapolation Method with Euler's method is given in Appendix 7.1. The Pascal procedure Extrapolation uses a one-dimensional array to represent $A_{k,j}$ because the values of $A_{k,j-1}$ are not required once $A_{k,j}$ is computed.

When we applied the extrapolation method, in conjunction with Euler's method, to the differential equation in Example 7.1, we obtained the results shown in Table 7.5; we also give the Euler method and the classical Runge-Kutta methods for comparison. The results in Table 7.5 show that the extrapolation method, with the level of extrapolation set to 3, compares very well to the classical Runge-Kutta method of $0(h^4)$. Although the computational results are in close agreement, we must point out that the computational cost (as measured by the number of

Table 7.5

t	Euler	Extrapolation with Euler	Runge-Kutta	Exact Solution
0.00	1.00000	1.00000	1.00000	1.00000
0.25	1.25000	1.28070	1.28069	1.28070
0.50	1.55279	1.61515	1.61513	1.61515
0.75	1.89346	1.97712	1.97709	1.97712
1.00	2.23982	2.31978	2.31974	2.31978
1.25	2.54236	2.58309	2.58304	2.58309
1.50	2.74278	2.71148	2.71144	2.71148
1.75	2.79128	2.67510	2.67505	2.67510
2.00	2.66690	2.48258	2.48254	2.48258
2.25	2.38944	2.17728	2.17723	2.17727
2.50	2.01419	1.81934	1.81931	1.81934
2.75	1.61078	1.46471	1.46469	1.46472
3.00	1.23857	1.15156	1.15155	1.15156

Solution of $y' = y{\cdot}\cos(t)$, $y(0) = 1$, by the Euler method with extrapolation (Level = 3) and the classical Runge-Kutta method, with step-size $h = 0.25$.

derivative function evaluations) for the Runge-Kutta method is much smaller than that of the extrapolation method. This is due to the extrapolation method requiring 15 evaluations of the derivative function per integration step as compared with four derivative function evaluations for the Runge-Kutta method.

EXERCISE 7.6

Repeat Exercise 7.1 for the Extrapolation method, using three levels of extrapolation.

Answer The magnitude of the error at $t = 1$ is 8.00×10^{-8}.

7.8 SELECTING THE STEP-SIZE IN A NUMERICAL INTEGRATOR

So far we have discussed basic numerical integration methods. The choice of the step-size was not discussed in any detail. Now we want to see how to select a step-size so that the numerical integration of our differential equation can be performed with as large an error as we are willing to accept. We have seen that the local discretization error is $0(h^{p+1})$ for an order p method. This suggests that $|h|$ should be small; however, the more integration steps we take, the more roundoff errors we will make. Ideally, we should select a step-size that minimizes the magnitude of the local error (the sum of the local discretization error and the local roundoff error). It will be extremely difficult to obtain the exact minimum of the local error's magnitude; the best we can hope for is a sufficiently small local error magnitude. Again, even this is unattainable because the only quantity available to us is the local discretization error. If we can estimate it and keep in mind that $|h|$ cannot be too small because of roundoff errors, we should be able to compute h so that the magnitude of the local discretization error does not exceed a predetermined constant. Consequently, if the local discretization error at t_{i+1} is given by

$$D(t_{i+1}) = C_{p+1}h^{p+1} + 0(h^{p+2}) \tag{7.8–1}$$

where C_{p+1} is independent of h, we want to choose h so that

$$|D| \simeq |C_{p+1}h^{p+1}| \leq \epsilon \tag{7.8–2}$$

The positive number ϵ is our predetermined error tolerance.

We want to use (7.8–2) to see if our step-size is acceptable. If the left-hand side of the inequality in (7.8–2) is very much smaller than ϵ, then $|h|$ is too small and should be increased. If the left-hand side is greater than ϵ, then $|h|$ should be decreased.

One method of estimating $|C_{p+1}h^{p+1}|$ is to use our numerical integration method to integrate from t_i to t_{i+1} with one step of size h. We then repeat the integration using two steps, each of size $h/2$. Denoting the value of y at t, obtained with the step-size h, as $y(t; h)$ and the true value as $u(t)$, we have

$$y(t_{i+1}; h) = u(t_{i+1}) + C_{p+1}h^{p+1} + 0(h^{p+2})$$

$$y\left(t_{i+1}; \frac{h}{2}\right) = u(t_{i+1}) + C_{p+1}\left(\frac{h}{2}\right)^{p+1} + 0(h^{p+2})$$

Subtract these to obtain

$$C_{p+1}h^{p+1} = \frac{y(t_{i+1}; h) - y(t_{i+1}; h/2)}{1 - 2^{-(p+1)}} + 0(h^{p+2})$$

Hence, the local discretization error, accurate to $0(h^{p+1})$, is given by

$$D(t_{i+1}) = \frac{y(t_{i+1}; h) - y(t_{i+1}; h/2)}{1 - 2^{-(p+1)}} \tag{7.8–3}$$

The magnitude of the right-hand side of (7.8–3) would be compared with the error tolerance ϵ to see if $|h|$ should change. Notice the similarity of this method of estimating the local discretization error with the Extrapolation method in Section 7.7.

In order to see the effectiveness of (7.8–3) we used Euler's method on the differential equation in Example 7.1. We compared the estimated local discretization error (7.8–3) to the actual local error. The results are summarized in Table 7.6, where the absolute value of each error is given. The magnitudes of the estimated errors appear to track the magnitudes of the actual errors quite well.

Table 7.6

t	Estimated Error	Actual Error
0.00	0.00000	0.00000
0.25	0.01937	0.03070
0.50	0.01548	0.02365
0.75	0.00578	0.00732
1.00	0.01012	0.01819
1.25	0.02987	0.04831
1.50	0.04793	0.07404
1.75	0.05732	0.08531
2.00	0.05332	0.07650
2.25	0.03678	0.05052
2.50	0.01416	0.01757
2.75	0.00628	0.01081
3.00	0.01908	0.02783

Comparison of the magnitudes of the estimated local discretization error and the actual local error in the numerical integration of $y' = y \cdot \cos(t)$, $y(0) = 1$, using Euler's method with $h = 0.25$.

A new step-size, H, can be selected so that

$$| H | = \left(\frac{\epsilon}{| C_{p+1} |} \right)^{\frac{1}{p+1}}$$

this expression is obtained from (7.8–2). We might want to be conservative and choose the new step-size as $0.9H$, to account for the discrepancy between the estimated error and the actual error, as illustrated in Table 7.6. In addition, we would want to have upper and lower bounds on $| h |$ such as

$$0 < h_{\min} \leq | h | \leq h_{\max} \tag{7.8–4}$$

where $h_{\min}$ and $h_{\max}$ are chosen so that neither very small steps nor very large steps are taken. Once we have a procedure that relates the step-size to the selected error tolerance, ϵ, and bounds like (7.8–4) we could write a computer program that automatically determines the step-size, rather than using a fixed step-size as was done in Example 7.1.

The error estimation method in (7.8–3) is quite expensive in machine computations. Each computation must be done twice; there is one step of size h and two steps of size $h/2$ required to estimate the local discretization error. If one step of the chosen numerical integrator requires m floating point operations, obtaining (7.8–3) requires $3m$ floating point operations. To avoid this problem we could try to use two methods of different orders, selected in such a way that an error estimate can be inexpensively obtained for the lower order method. In this way, we may be able to avoid the extra work required for (7.8–3). To see how this would work for a method $0(h^p)$ and another method $0(h^q)$, where $q \geq p + 1$, we consider the solutions obtained by each method at t_{i+1},

$$y_{i+1} = u_{i+1} + A_{p+1}h^{p+1} + 0(h^{p+2})$$

$$z_{i+1} = u_{i+1} + B_{q+1}h^{q+1} + 0(h^{q+2})$$

Subtracting these equations gives

$$y_{i+1} - z_{i+1} = A_{p+1}h^{p+1} + 0(h^{p+2})$$

because $q \geq p + 1$. Hence, an estimate of the local discretization error at t_{i+1} is obtained by subtracting the computed solutions at t_{i+1};

$$D(t_{i+1}) = y_{i+1} - z_{i+1} \tag{7.8–5}$$

In Section 7.2 we discussed the problem of determining bounds on the maximum magnitude of the step-size h in terms of numerically computing the solution of (7.2–6) by a difference equation of the form (7.2–2). When Euler's method (7.3–1) is used to solve (7.2–6), if λ is real and positive, then we find that

$$y_{i+1} = y_i - \lambda h y_i = (1 - \lambda h)y_i$$

Thus, $y_k = (1 - \lambda h)^k y_0$; we want $\lim_{k\to\infty} y_k = 0$ so that $| 1 - \lambda h | < 1$. Hence,

$$0 < \lambda h < 2$$

A similar analysis shows that the Runge-Kutta methods of order two, (7.4–7) to (7.4–10), also have the same restriction on h. Bounds on $z = \lambda h$ for the classical Runge-Kutta method of order four, (7.5–1) to (7.5–6), are obtained by using (7.5–1) to (7.5–6) on the differential equation (7.2–6), with $\lambda > 0$. This gives the difference equation

$$y_{i+1} = \frac{P(\lambda h)}{24} y_i \tag{7.8–6}$$

where

$$P(z) = 24 - 24z + 12z^2 - 4z^3 + z^4$$

The solution of the difference equation (7.8–6) is

$$y_k = \left[\frac{P(\lambda h)}{24}\right]^k y_0$$

Because we want

$$\lim_{k \to \infty} y_k = 0, \; |P(z)| < 24.$$

More generally, bounds for $|h|$ can be obtained if λ is a complex number, giving a region in the complex plane in which numerically computed solutions of (7.2–6) are bounded.

In the next section we look at a technique that is based upon Runge-Kutta methods of orders four and five. Before examining that procedure, we want to recall that each of the numerical integration methods discussed in previous sections can be extended to systems of first-order ordinary differential equations. A numerical method would be applied separately to each component of the vector differential equation (7.1–4). For example, the classical Runge-Kutta method, for (7.1–4), would be given as

$$\omega = \frac{h}{2}$$

$$\mathbf{K}_1 = \mathbf{f}(t_i, \mathbf{y}_i)$$

$$\mathbf{K}_2 = \mathbf{f}(t_i + \omega, \mathbf{y}_i + \omega \mathbf{K}_1)$$

$$\mathbf{K}_3 = \mathbf{f}(t_i + \omega, \mathbf{y}_i + \omega \mathbf{K}_2)$$

$$\mathbf{K}_4 = \mathbf{f}(t_{i+1}, \mathbf{y}_i + h \mathbf{K}_3)$$

$$\mathbf{y}_{i+1} = \mathbf{y}_i + \frac{h(\mathbf{K}_1 + 2\mathbf{K}_2 + 2\mathbf{K}_3 + \mathbf{K}_4)}{6}$$

Here, $\mathbf{y}_i$ denotes the approximation to the vector $\mathbf{y}$ at $t = t_i$. The jth component of $\mathbf{y}_i$ is the approximation to the jth component of $\mathbf{y}$ at $t = t_i$.

7.9 THE RUNGE-KUTTA-FEHLBERG METHOD

The Runge-Kutta-Fehlberg method consists of an order four and an order five Runge-Kutta method, selected so that only six $\mathbf{K}_j$ functions are required to advance the numerical solution of (7.1–4) from t_i to t_{i+1} and to estimate the local discretization error. In general, if order four and order five methods were used in this way, ten $\mathbf{K}_j$ functions would be needed. The savings in computation is quite apparent.

The order four method is

$$\mathbf{y}_{i+1} = \mathbf{y}_i + h \sum_{j=1}^{6} c_{4j}\mathbf{K}_j \tag{7.9–1}$$

and the order five method is

$$\mathbf{z}_{i+1} = \mathbf{y}_i + h \sum_{j=1}^{6} c_{5j}\mathbf{K}_j \tag{7.9–2}$$

where $c_{42} = c_{46} = c_{52} = 0$. The estimated local discretization error at t_{i+1} is obtained by subtracting (7.9–2) from (7.9–1), to give the vector form of (7.8–5),

$$\mathbf{D}(t_{i+1}) = \mathbf{y}_{i+1} - \mathbf{z}_{i+1} = h \sum_{j=1}^{6} (c_{4j} - c_{5j})\mathbf{K}_j \tag{7.9–3}$$

The functions $\mathbf{K}_j$ are formed according to (7.5–7)–(7.5–9) with the constants a_j and b_{jk}, where $2 \le j \le 6$, $1 \le k \le j - 1$. These constants are displayed in Table 7.7, while c_{4j} and c_{5j} are given in Table 7.8.

Appendix 7.2 contains a user's manual for a package of Pascal procedures based on the Runge-Kutta-Fehlberg method, called the RKF45 package for obvious reasons. There are two main integrator procedures in the package: the procedure Fehlberg which implements (7.9–1) and (7.9–2) and the main procedure RKF45 which handles input/output, error estimation, step-size changing, and other housekeeping things needed to make the integrator work. The order five

Table 7.7 THE COEFFICIENTS IN (7.5–7), (7.5–8), AND (7.5–9) FOR THE RUNGE-KUTTA-FEHLBERG METHOD

j	a_j	b_{jk}, $1 \le k \le j - 1$				
2	1/4	1/4				
3	3/8	3/32	9/32			
4	12/13	1932/2197	−7200/2197	7296/2197		
5	1	439/216	−8	3680/513	−845/4104	
6	1/2	−8/27	2	−3544/2565	1859/4104	−11/40

Table 7.8 THE CONSTANTS c_{4j} AND c_{5j} IN (7.9–1) AND (7.9–2) FOR THE RUNGE-KUTTA-FEHLBERG METHOD

j	1	2	3	4	5	6
c_{4j}	25/216	0	1408/2565	2197/4104	−1/5	0
c_{5j}	16/135	0	6656/12825	28561/56430	−9/50	2/55

method is used as the integrator; the estimated local discretization error (7.9–3) is used for step-size control, as follows. The user is required to supply two error estimates, r and a, to measure relative and absolute error, respectively. An estimated error tolerance is computed as

$$T = |h| \left[\underset{1 \le k \le n}{\text{maximum}} \left\{ \frac{|E_k|}{a + 0.5r(w_k + z_k)} \right\} \right] \tag{7.9–4}$$

where h is the current step-size, E_k is the kth component of the estimated error (7.9–3), w_k and z_k are the kth components of the computed solution at t_i and t_{i+1}, respectively. If $T > 1$ then the step-size is reduced to qh, where

$$q = \begin{cases} 0.1 & \text{if } T \ge 9^5 = 59049 \\ \dfrac{0.9}{T^{0.2}} & \text{if } T < 59049 \end{cases}$$

RKF45 does not require the user to supply an initial step-size. An initial step-size is determined by the procedure using the user-selected error tolerances r and a, the initial value of t, and the final value of t, "tout."

RKF45 can be used in two modes: interval or single-step. In the interval mode, the user specifies both the starting value of the independent variable, "t," and the stopping value, "tout"; RKF45 performs the integration from t to *tout* while trying to maintain the error criterion (7.9–4), with a = "AbsErr" and r = "RelErr." Control is returned to the calling program when *tout* is reached, or if an error condition is encountered before *tout* is attained. This mode of operation is most useful if one wants the values of the dependent variables at only specific independent variable points, but not at any other intermediate points. However, if we want to examine the behavior of the solution over a range of points, but we do not wish to specify these except for the end points, then we should use the single-step mode. The single-step mode operates in a similar fashion to the interval mode, but control is returned to the calling program after each integration step; to continue the integration, RKF45 must be executed over and over again until *tout* is reached. The single-step mode would be used if we wanted to plot graphs using the computed solution to the differential equation.

The program in Appendix 7.3 is used to integrate the differential equation in Example 7.1; the purpose is to show how RKF45 is used in the interval mode.

EXERCISE 7.7

Use RKF45 in the interval mode to integrate the differential equation in Exercise 7.1, and thus obtain the error at $t = 1$. Use RelErr $= 10^{-5}$ and AbsErr $= 0.0$. The program in Appendix 7.3 can be adapted for this problem.

Answer The magnitude of the error at $t = 1$ is 7.67×10^{-7}.

7.10 SOME EXAMPLE PROBLEMS

EXAMPLE 7.2

The famous van der Pol differential equation models the behavior of a triode oscillator. The equation is

$$\frac{d^2x}{dt^2} = -x + b(1 - x^2)\frac{dx}{dt} \tag{7.10–1}$$

where $b > 0$ is a constant. We take $b = 1$. The solutions of (7.10–1) have an interesting property. There is a closed curve in the *phase plane*, the plane whose coordinate axes are x and dx/dt, with the property that all solutions of (7.10–1) tend towards that curve as $t \to \infty$. Such a closed curve is called a (stable) *limit cycle*.

In order to solve (7.10–1) using RKF45, we first convert it into a system of two first-order ordinary differential equations by setting

$$y_1 = x, \qquad y_2 = \frac{dx}{dt}$$

to give the system

$$\frac{dy_1}{dt} = y_2 \qquad \frac{dy_2}{dt} = -y_1 + (1 - y_1^2)y_2$$

Figure 7.6 shows the limit cycle and two trajectories that tend towards it; the inner trajectory, starting at $(y_1, y_2) = (-0.5, -0.5)$, "spirals" outward toward the limit cycle while the outer trajectory, starting at $(y_1, y_2) = (0.0, 2.9)$, "spirals" inward.

EXAMPLE 7.2 (cont'd)

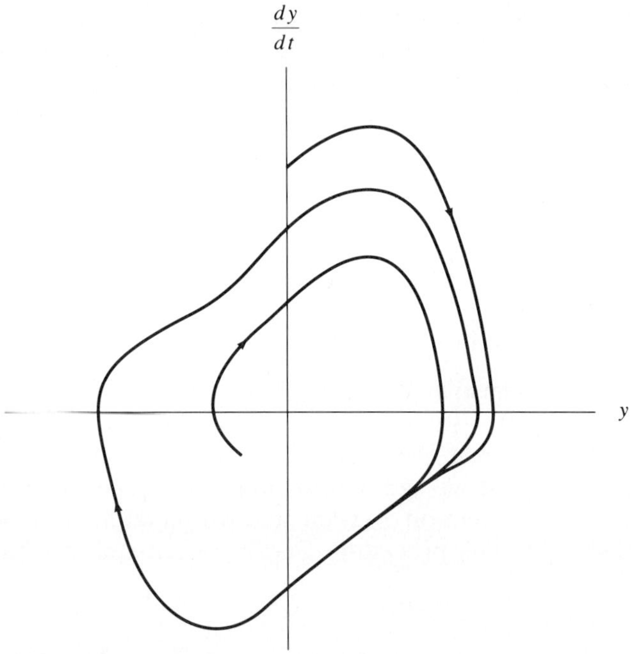

Figure 7.6 Example 7.2, limit cycle for van der Pol's equation.

EXAMPLE 7.3

In this problem we want to determine the period of the simple pendulum whose equation of motion is given by (7.1–1). If the pendulum is released from rest at the angle θ_0, the period is four times the least time taken for the position of the pendulum to change from θ_0 to 0. We want to obtain the period for different values of θ_0, given any value of g/ℓ. This means that for each pair of numbers, $(g/\ell, \theta_0)$, we must integrate (7.1–1). We can do this more efficiently if we make the following change-of-variables: if the independent variable in (7.1–1) is s, set $s = bt$, where b is a constant to be determined. Then, (7.1–1), with the imposed initial conditions, becomes

$$\frac{d^2\theta}{dt^2} = -b^2\left(\frac{g}{\ell}\right)\sin(\theta); \qquad \theta(0) = \theta_0, \qquad \theta'(0) = 0$$

If we now set $b = \sqrt{\ell/g}$, the transformed differential equation is independent of (g/ℓ). Hence, the transformed equation need only be integrated for each different θ_0. If t_* is the least "time" required for θ to change from θ_0 to 0, then the period of the pendulum is given by $S_* = 4t_*\sqrt{\ell/g}$. Because the differential equation and the initial conditions remain the same when θ is replaced by $-\theta$,

EXAMPLE 7.3 (cont'd)

we need only look at starting positions satisfying $0 < \theta_0 < \pi$. (We must exclude $\theta_0 = \pi$ because the period is infinite.) The new independent variable, t, is dimensionless, whereas the old variable, s, had units of time; θ is already dimensionless. By using dimensionless variables, we have saved a considerable amount of computation work.

The system of first-order differential equations to be integrated using RKF45 is

$$\frac{dy_1}{dt} = y_2, \qquad \frac{dy_2}{dt} = -\sin(y_1) \tag{7.10–2}$$

$$y_1(0) = \theta_0, \qquad y_2(0) = 0$$

There is one more item that we have yet to resolve. We want to obtain the least positive value of t, t_*, such that $y_1(t_*) = 0$. It is unreasonable to expect that an integration step should occur exactly at t_*. Therefore, we would need to monitor the solution so that we can determine when y_1 changes sign. We would then have the zero of the function $y_1(t)$ located within an interval determined by the last integration step. The procedure ZeroIn, discussed in Chapter 4, could

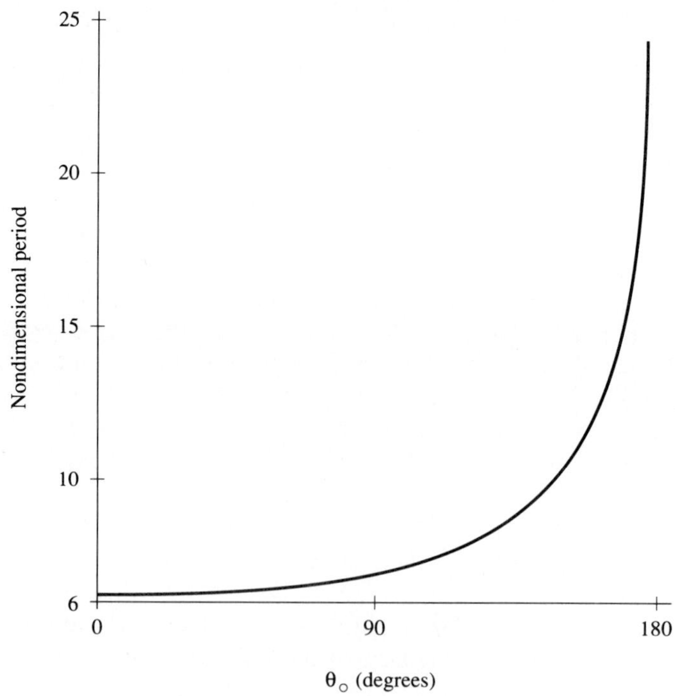

Figure 7.7 Example 7.3, the nondimensional period of the simple pendulum as a function of the starting position, θ_0.

EXAMPLE 7.3 (cont'd)

then be used to find the zero, t_*. We have adapted ZeroIn to be used in conjunction with RKF45, because there are many problems which require the zero of a function whose argument depends on the solution of a differential equation. This adaptation of ZeroIn, called ZeroRKF, uses the procedure Fehlberg from the RKF45 package; a user's manual for ZeroRKF is given in Appendix 7.2 as part of the RKF45 package.

We integrated (7.10–2) for θ_0 from 1 to 179 degrees in steps of one degree. For each initial value we determined the nondimensional period, $4t_*$; these periods are shown in Fig. 7.7. The program listing is given in Appendix 7.4.

EXAMPLE 7.4

A thin uniform cantilever beam of length L has one end horizontally embedded in concrete. The deflection of the beam from true horizontal, due to its own weight, is given by the differential equation

$$EIy'' = g\rho[1 + (y')^2]^{3/2}\left\{x\left(x - \frac{L}{2}\right) + \frac{L^2}{2}\right\} \tag{7.10–3}$$

where x measures length along the beam from the clamped end, y is the beam deflection measured positively downward, y' and y'' denote the first and second derivatives of y with respect to x, g is gravity, ρ is the linear mass density of the beam, I is the cross-sectional moment of inertia of the beam about its central axis, and E is Young's modulus for the material in the beam. Because the beam is clamped at $x = 0$, $y(0) = 0$ and $y'(0) = 0$, engineers frequently approximate (7.10–3) by assuming that $(y')^2 \ll 1$, giving the linearized equation

$$EIy'' = g\rho\left\{x\left(x - \frac{L}{2}\right) + \frac{L^2}{2}\right\}$$

This equation can be immediately integrated because the right-hand side does not contain y. Its solution is

$$y(x) = g\rho x^2\,\frac{6L^2 + x^2 - 4Lx}{24EI} \tag{7.10–4}$$

Equation (7.10–4) will be a good approximation to the solution of (7.10–3) provided the slope of the beam, y', remains small. We want to compare (7.10–4) with the solution of (7.10–3).

Taking $g\rho/(EI) = 0.1\ m^{-3}$ we integrated (7.10–3) using RKF45. The results shown in Table 7.9 were obtained.

EXAMPLE 7.4 (cont'd)

Table 7.9 EXAMPLE 7.4, BEAM DEFLECTION DIFFERENTIAL EQUATIONS

Length of the Beam (Meters)	Deflection at the End of the Beam (Meters)		Maximum Slope
	Eq. (7.10–3)	Eq. (7.10–4)	
0.5	0.0007813	0.0007813	0.00208
1.0	0.01250	0.01250	0.0167
1.5	0.06336	0.06328	0.0563
2.0	0.2014	0.2000	0.135
2.5	0.5016	0.4883	0.270
3.0	1.1039	1.0125	0.504

The approximation solution, (7.10–4), is very good provided the maximum slope remains small; as the maximum slope increases the error between (7.10–4) and the solution to (7.10–3) becomes greater.

7.11 LINEAR MULTISTEP METHODS

One of the disadvantages of the Runge-Kutta class of methods is the large number of evaluations of the derivative function required to advance the solution by one step. Also, information from previous steps is not utilized in taking a step. For this reason linear multistep methods are often used. A linear multistep method used to integrate Eq. (7.1–4) has the form

$$\mathbf{y}_i = \sum_{j=1}^{k} a_j \mathbf{y}_{i-j} + h \sum_{j=\alpha}^{k} b_j \mathbf{f}(t_{i-j}, \mathbf{y}_{i-j}) \tag{7.11–1}$$

where $\alpha = 1$ for an explicit method and $\alpha = 0$ for an implicit method. The single-step Euler's method, Section 7.3, is an explicit multistep method of order one. The coefficients a_j and b_j are determined by applying Eq. (7.11–1) to a single differential equation, such as (7.1–5). Each term on the right-hand side of the single-equation version of (7.11–1) is expanded in a Taylor series about the point (t_i, y_i). The resulting expressions are collected in powers of h, and the coefficient of each power of h is set to zero, yielding the system of equations

$$\left.\begin{aligned} &\sum_{j=1}^{k} a_j = 1 \\ &\sum_{j=1}^{k} a_j j^p = \sum_{j=\alpha}^{k} (pb_j) j^{p-1}; \qquad p = 1, 2, \ldots, m. \end{aligned}\right\} \tag{7.11–2}$$

The integer m is the order of the method. The set of equations in (7.11–2) is derived for constant step-size h. If the step-size is allowed to change from step to step (as could happen with a single-step method), then the coefficients a_j and b_j will depend on the step-sizes used in the previous k steps.

One commonly used family of multistep methods is the *Adams methods*, characterized by $a_1 = 1$ and $a_j = 0$ for $j > 1$. The Adams family of methods can also be obtained by using a polynomial to interpolate y' at various points. The degree of the interpolating polynomial is the order of the method. Another linear multistep family is the *backward differentiation formulas*, with $b_0 \neq 0$ and $b_j = 0$ for $j > 0$.

We shall need the following definitions in our discussion of linear multistep methods.

Definition 7.3

A linear multistep method is *consistent* if, in (7.11–2), the first equation and the second equation (with $p = 1$) are satisfied. This means that the local discretization error is $0(h^{m+1})$, with $m \geq 1$.

A consistent linear multistep method is one in which the local discretization error tends to zero as the step-size, h, tends to zero.

Definition 7.4

A linear multistep method is *convergent* provided the solution of (7.11–1) approaches the solution of (7.1–4) as the step-size tends to zero.

There are two polynomials associated with any multistep method, (7.11–1); these are

$$P(x) = x^k - \sum_{j=1}^{k} a_j x^{k-j} \quad \text{and} \quad Q(x) = \sum_{j=\alpha}^{k} b_j x^{k-j}$$

Definition 7.5

$P(x)$ is said to satisfy the *root condition* if all roots of $P(x)$ are located inside or on the unit circle, centered at the origin, in the complex plane, and all the roots of $P(x)$ on the unit circle are simple roots.

It can be shown that the consistency of a linear multistep method is equivalent to the condition that the associated polynomials, $P(x)$ and $Q(x)$, satisfy the equations $P(1) = 0$ and $P'(1) = Q(1)$.

Definition 7.6

A linear multistep method, (7.11–1), is said to be *stable* if its associated polynomial $P(x)$ satisfies the root condition. The method is *strongly stable* if it is stable and only one root lies on the unit circle. The method is called *weakly stable* if it is stable and more than one root lies on the unit circle.

Stability in a multistep method means that small perturbations in the values of the dependent variable(s) at one point produce only small perturbations in subsequent values. Furthermore, the stability and consistency of a multistep method are necessary and sufficient for the method to be convergent. The Adams methods are often used because they are all strongly stable; they have $P(x) = x^{k-1}(1 - x)$.

Eq. (7.11–2) determines the particular linear multistep method that we want to use. There are $2k + \alpha$ unknown coefficients and $m + 1$ equations to be solved. We might expect to set $m = 2k + \alpha - 1$, and solve the resulting system for the coefficients a_j and b_j. Unfortunately, (7.11–1) is an unstable method for $m > k + 2$ (k even) and $m > k + 1$ (k odd). Therefore, m is at most $k + 1$ or k. In general, we can fix some of the coefficients in advance. For the Adams methods, with $a_1 = 1$ and $a_j = 0$ for $j > 1$, there are $k + \alpha$ unknowns; this satisfies the stability requirement.

Linear multistep methods are often used in pairs; an explicit method, the *predictor*, and an implicit method, the *corrector*. The predictor is used to obtain a first estimate of $\mathbf{y}_i$, denoted by $\mathbf{y}_i^{(0)}$; the corrector is then used to update, or correct, this estimate. We shall denote the predictor coefficients by a_j and b_j, and the corrector coefficients by a_j^* and b_j^*. The predictor-corrector pair is

$$\mathbf{y}_i^{(0)} = \sum_{j=1}^{k} a_j \mathbf{y}_{i-j} + h \sum_{j=1}^{k} b_j \mathbf{f}(t_{i-j}, \mathbf{y}_{i-j}) \tag{7.11–3}$$

and for $m = 0, 1, 2, \ldots$

$$\mathbf{y}_i^{(m+1)} = hb_0^* \mathbf{f}(t_i, \mathbf{y}_i^{(m)}) + \mathbf{w} \tag{7.11–4}$$

where

$$\mathbf{w} = \sum_{j=1}^{q} [a_j^* \mathbf{y}_{i-j} + hb_j^* \mathbf{f}(t_{i-j}, \mathbf{y}_{i-j})]$$

and $\mathbf{y}_{i-j}$, $j \geq 1$, denotes the accepted corrector values at previous steps. For example, the two first-order multistep Adams methods are

$$\mathbf{y}_i = \mathbf{y}_{i-1} + h\mathbf{f}(t_{i-1}, \mathbf{y}_{i-1}) \qquad \text{(Predictor)}$$

$$\mathbf{y}_i = \mathbf{y}_{i-1} + h\mathbf{f}(t_i, \mathbf{y}_i) \qquad \text{(Corrector)}$$

The predictor is Euler's method, while the corrector is the backward Euler method.

Explicit Adams methods (predictors) are called *Adams-Bashforth* methods, while implicit Adams methods (correctors) are called *Adams-Moulton* methods. The order of the corrector is usually chosen so that it is at least the order of the predictor; often, the corrector order is chosen to be one greater than the predictor order. The corrector, (7.11–4), is written so that the terms in $\mathbf{w}$ only depend on information from previous steps.

A predictor-corrector pair is used as follows: (1) we *P*redict by (7.11–3), (2) we *E*valuate the function $\mathbf{f}$, and (3) we *C*orrect by using (7.11–4). In general, steps (2) and (3) can be repeated so that the predictor-corrector sequence is written as

$P(EC)^{\ell}$ for $\ell = 1, 2, \ldots$. Sometimes the predictor-corrector sequence uses a final evaluation stage, so that it looks like $P(EC)^{\ell}E$.

Equation (7.11–4) is organized so that the fixed-point iteration method can be used (see Section 4.7). The fixed-point method converges if $| hb_0^* | L < 1$, where L is the Lipschitz constant for the function in (7.1–4). The fixed-point iteration is usually terminated after one or two iterations. But, we are not limited to the fixed-point method for the corrector. Newton's method could be used to solve the corrector equation for $\mathbf{y}_i$. This technique requires much more computation than the fixed-point method, although the improved accuracy which results may more than compensate for the extra computational expense. No matter what method is chosen to determine the corrected value of $\mathbf{y}_i$, a multistep method generally requires fewer evaluations of $\mathbf{f}$ than a corresponding Runge-Kutta method.

In general, multistep methods are more efficient than Runge-Kutta methods, but they are more costly in overhead (extra computations required to keep track of the past data points). One disadvantage of a linear multistep method is the problem of starting the method at $t = a$. Here, we only know $\mathbf{y}(a)$; the multistep method requires values of $\mathbf{y}$ at previous values of t. One method of starting a multistep method is to use a Runge-Kutta method to generate the requisite number of steps so that a multistep method can begin. Another way to start a multistep method is to use a first-order method to generate the first point, then use a second-order method, and so on. This entails having a family of methods available.

It is a fairly straightforward procedure to change the step-size in a single-step method, because the step from t_{i-1} to t_i depends only on the information at t_{i-1}. This is not the case with multistep methods; consequently, some means of obtaining intermediate values of $\mathbf{y}$ is required with multistep methods. Usually, a form of interpolation is used when the step-size is changed. The interpolated values are computed at values of t corresponding to the new step-size. However, the interpolation process itself could introduce some additional roundoff errors in the solution, which could be serious if the solution is very sensitive to small perturbations.

If we assume that the orders of the predictor and corrector are equal to k, then the jth component of the local discretization errors for the predictor, $D_j^{(0)}$, and the corrector, $D_j^{(\ell)}$, $\ell \geq 1$, are

$$D_j^{(0)}(t_i) = C_p h^{k+1} \frac{d^{k+1} y_j(\xi)}{dt^{k+1}} = y_j^{(0)}(t_i) - u_j(t_i)$$

$$D_j^{(\ell)}(t_i) = C_c h^{k+1} \frac{d^{k+1} y_j(\upsilon)}{dt^{k+1}} = y_j^{(\ell)}(t_i) - u_j(t_i)$$

where $u_j(t)$ is the jth component of the local solution $\mathbf{u}(t)$, and $y_j(t)$ is the jth component of the computed solution, $\mathbf{y}(t)$, to the differential equation. If we assume that $| h |$ is sufficiently small, then the derivative terms can be taken to be approximately equal. Subtracting these error expressions gives the estimate for jth component of the local discretization error as

$$D_j^{(\ell)}(t_i) \simeq \frac{C_c[y_j^{(\ell)}(t_i) - y_j^{(0)}(t_i)]}{C_c - C_p}$$

This can be used to estimate appropriate changes in the step-size, given a tolerance, ϵ, on the norm of $\mathbf{D}^{(\ell)}(t_i)$.

Use of certain linear multistep methods can lead to a new kind of instability, called *numerical instability*. For example, we want to use the second-order method

$$y_i = y_{i-2} + 2hf(t_{i-1}, y_{i-1}) \tag{7.11–5}$$

to integrate the differential equation

$$y' = 1 - y, \qquad y(0) = 2 \tag{7.11–6}$$

with $h = 1/32$. The solution of the initial value problem is

$$y(t) = 1 + e^{-t}$$

In order to start the multistep method, we set $y_0 = 2$ and $y_1 = 1 + e^{-h}$. For comparison, we will also use Euler's method. The computational results are shown in Table 7.10. Although the second-order method is "better" than Euler's method, its error grows as t increases, whereas the Euler-method error decreases in magnitude. In fact, the second-order method solution becomes larger and larger as t

Table 7.10 NUMERICAL INTEGRATION OF $y' = 1 - y$, $y(0) = 2$, $h = 1/32$

	Order 2 Multistep Method		Euler's Method	
t	$y(t)$	Error	$y(t)$	Error
0.0	2.00000	0.00000	2.00000	0.00000
1.0	1.36795	0.00007	1.36224	−0.00564
2.0	1.13540	0.00006	1.13115	−0.00419
3.0	1.04986	0.00007	1.04748	−0.00230
4.0	1.01846	0.00015	1.01719	−0.00112
5.0	1.00711	0.00037	1.00622	−0.000514
6.0	1.00347	0.00099	1.00225	−0.000225
7.0	1.00361	0.00270	1.00082	-9.60×10^{-5}
8.0	1.00767	0.00733	1.00030	-4.01×10^{-5}
9.0	1.02005	0.01993	1.00011	-1.65×10^{-5}
10.0	1.05420	0.05415	1.00004	-6.68×10^{-5}
11.0	1.14720	0.14718	1.00001	-2.68×10^{-6}
12.0	1.40003	0.40002	1.00001	-1.07×10^{-6}
13.0	2.08719	1.08719	1.00000	-4.23×10^{-7}
14.0	3.95481	2.95481	1.00000	-1.66×10^{-7}
15.0	9.03069	8.03069	1.00000	-6.50×10^{-8}

increases; the Euler method solution tends toward 1. The second-order method exhibits this numerical instability because the associated polynomial for this multistep method,

$$P(x) = x^2 - 1$$

has roots at ± 1, so that the method is weakly stable. Also, the difference equation (the multistep method) has a higher order than the differential equation. This gives extraneous solutions for the difference equation. Unless these extraneous solutions are damped out, they could affect the desired computed solution. In the example given above, the magnitude of the extraneous solution becomes larger as t increases; eventually it dominates the desired solution. This is easy to see if we examine the exact solution to (7.11–5) for the differential equation (7.11–6). The difference equation solution is

$$y_k = a_1(r - h)^k + (-1)^k a_2(r + h)^k + 1$$

where

$$r = (1 + h^2)^{1/2}$$

$$a_1 = \frac{r + h + e^{-h}}{2r}$$

$$a_2 = \frac{r - h - e^{-h}}{2r}$$

For $h = 1/32$ and $k = 480$ ($t = 15$) we obtain $y_{480} = 9.03069079232814$. In the difference equation solution, the term $(r + h)^k$ becomes larger as k increases because $(r + h) > 1$, and therefore, the difference equation solution diverges from the true solution to the differential equation.

The preceding example shows the importance of using a numerical method that generates y_{i+1} without magnifying any error that might be present in y_i. The strongly stable methods, such as the Adams methods, have this desirable property.

7.12 A VARIABLE-STEP VARIABLE-ORDER ADAMS METHOD INTEGRATOR

Although Runge-Kutta methods, such as those in the Pascal procedure RKF45 (see Section 7.9), are extremely useful for integrating a system of first-order ordinary differential equations, they require a large number of derivative function evaluations for each integration step. This is especially noticeable in computer time if the derivative function involves many arithmetic operations or the evaluation of a substantial number of elementary functions, such as sine, cosine, exponential, etc. If the derivative function is quite complicated, then most of the computer time is required to evaluate the derivative function. It would be sensible to have a numerical integration method which required fewer derivative function evaluations per step. Linear multistep methods satisfy this requirement, although they do have larger overhead costs because derivatives at previous values of the independent variable must be stored. To complement the single-step Runge-Kutta method used in RKF45, we include the integrator package ODE, which consists of various Adams methods of orders 1 to 12, along with a root-finder procedure ZeroODE, similar to the one for RKF45. See Appendix 7.5 for a user's manual.

The ODE integrator is more demanding of the user than RKF45, because the user must supply a weighting function procedure, SetWeight, in addition to the procedure to evaluate the derivative functions. The weighting functions are used in the estimation of local discretization errors; errors are measured relative to the user-supplied weights. The error estimates are used not only to change the step-size, but also to adjust the order of the method, either by raising the order or lowering it. If the order is to be changed, then ODE switches to a higher or lower order method as indicated. The integrator can operate in both the single-step mode or the interval mode, similar to RKF45. The user must supply two error tolerances, "RelErr" and "AbsErr" (similar to those required by RKF45), the initial value of t, the final value of t, "tout," and the values of the dependent variables at the initial t. The procedure ODE determines the initial step-size given the error tolerances and the direction of integration from the initial t to "tout."

Appendices 7.6 and 7.7 contain listings of programs using ODE on the differential equations in Examples 7.4 and 7.3, respectively.

EXERCISE 7.8

Use ODE in the interval mode to integrate the differential equation in Exercise 7.1, and thus obtain the error at $t = 1$. Use RelErr $= 10^{-5}$ and AbsErr $= 0.0$, and select the weighting function in the procedure SetWeight as $|y| + 1$. The program in Appendix 7.6 can be adapted for this problem.

Answer The magnitude of the error at $t = 1$ is 7.79×10^{-7}.

7.13 STIFF DIFFERENTIAL EQUATIONS

There are many systems, such as chemical reactions, electronic networks, or control systems, that can be described by a system of first-order ordinary differential equations whose solution contains some components that are slowly varying and other components that are rapidly varying. Such systems arise because they contain time constants, or time scales, with widely different magnitudes. For example, a system might have a transient component that rapidly decays to zero. These different time scales would not cause any computational problems if we were only interested in the behavior of the solution very close to the startup time. If we want the solution "far" from the startup time, the time constant corresponding to the transient would force a numerical method to use very small integration steps in order to follow the transient, thereby increasing the amount of computation. In certain cases, the "extra" computation could be expensive. The following examples illustrate the behavior when different time scales are present.

EXAMPLE 7.5

$$\frac{dy}{dt} = -50(\cos(t) + y); \qquad y(0) = 0$$

The solution of this differential equation is

$$y(t) = \frac{e^{-50t} - \cos(t) - 0.02\sin(t)}{1.004}$$

We tried to integrate the differential equation from $t = 0$ to $t = 20$, using both RKF45 and ODE. RelErr and AbsErr were set to 10^{-5}. The weight function needed for ODE was taken as $|y| + 1$. Neither RKF45 nor ODE managed to complete the integration to $t = 20$. Both procedures returned an error flag indicating too much work was needed in order to continue the integration. RKF45 quit at $t = 17.761077$ after 495 integration steps and 3006 derivative function evaluations. ODE quit at $t = 16.043966$ after 588 integration steps and 1233 derivative function evaluations.

EXAMPLE 7.6

$$\frac{dy_1}{dt} = -1999y_1 - 1998y_2; \qquad y_1(0) = 2$$

$$\frac{dy_2}{dt} = 999y_1 + 998y_2; \qquad y_2(0) = -3$$

The solutions are $y_1(t) = 4e^{-t} - 2e^{-1000t}$ and $y_2(t) = -4e^{-t} + e^{-1000t}$. We tried to integrate this system on the interval [0, 3], with RelErr and AbsErr set to 10^{-5} and each component of the weighting function for ODE taken as in Example 7.5. Once again, both integrations were terminated because of the excessive amount of work. RKF45 stopped at $t = 1.786038$, after 3003 derivative function evaluations and 497 integration steps. ODE quit at $t = 0.402966$ after 569 derivative function evaluations and 261 integration steps.

In both of these examples, ODE stopped sooner because it contains a test specially designed to detect this type of problem; RKF45 automatically gives an indication of excessive work after every 3000 derivative function evaluations. We could have continued the integration to the final requested value of t in both examples. We chose not to do this because the computational effort was already excessive.

These two examples have solutions with very different time constants. The solution for Example 7.5 contains the term e^{-50t}, which rapidly approaches zero as t increases from 0. The other terms are changing much more slowly. Similarly, the solutions for Example 7.6 contain the term e^{-t} which is slowly changing when compared to the other term e^{-1000t}. In both cases, the "fast" terms, e^{-50t} in Example 7.5 and e^{-1000t} in Example 7.6, quickly approach zero, but the integrator must still account for them; hence, small steps are taken to satisfy the imposed error criterion, and therefore the amount of computational work required may become excessive.

Examples 7.5 and 7.6 belong to a somewhat "fuzzy" set of differential equations; they are called *stiff* equations. Stiff differential equations are characterized by solutions that have widely differing time constants. Consequently, a solution may be slowly varying, but has a transient part that is rapidly damped. We can make this more precise for systems of first-order differential equations, as follows.

Let $\mathbf{J}(t, \mathbf{y})$ denote the $n \times n$ Jacobian matrix of the right-hand side of (7.1–4). The (i, j)th entry of $\mathbf{J}$ is the partial derivative of f_i with respect to y_j. Let λ_j denote the jth eigenvalue of $\mathbf{J}$. Recall that the number λ, real or complex, is an *eigenvalue* of the $n \times n$ matrix $\mathbf{A}$ if there is a nonzero vector $\mathbf{x}$, called an *eigenvector*, such that $\mathbf{Ax} = \lambda\mathbf{x}$ (see Appendix A.4).

A system of the form (7.1–4) is often called stiff if the following conditions are satisfied:

(a) $\text{Re}(\lambda_j) < 0 \qquad 1 \le j \le n$

(b) $$S = \frac{\underset{1 \le j \le n}{\text{maximum}} \{ \,|\, \text{Re}(\lambda_j) \,|\, \}}{\underset{1 \le j \le n}{\text{minimum}} \{ \,|\, \text{Re}(\lambda_j) \,|\, \}} \gg 1, \qquad \text{where } n > 1$$

S is called the *stiffness ratio*.

(c) The interval of integration is sufficiently "large."

$\text{Re}(\lambda)$ denotes the real part of the complex number λ. Condition (a) ensures that the system is stable as the independent variable increases; condition (b) shows that there are widely differing time constants; condition (c) indicates that we want solutions well beyond the transient phase. We should note that the behavior of a system may change from stiff to nonstiff and vice versa if the Jacobian matrix depends on $\mathbf{y}$ and/or t. Note that condition (b) does not apply to the case where $n = 1$; we have seen in Example 7.5 that a single differential equation can be stiff. It is also possible to have $S = 1$ in condition (b) and still have a stiff differential equation, because there could be terms with rapidly varying time scales.

It has been found that certain types of methods are more suitable for stiff equations than others; for example, the backward differentiation formulas in which $b_0 \neq 0$ and $b_j = 0$ for $j > 0$ in (7.11–1). We used a Pascal procedure based on the backward differentiation formulas to integrate the equations in Examples 7.5 and 7.6. In Example 7.5 the integration to $t = 20$ was completed taking 158 integration steps and 289 derivative function evaluations; the integration for Example 7.6 was completed to $t = 3$ with 124 derivative function evaluations and 77 integration steps. The program was run on an IBM PC using Turbo Pascal with an 8087 coprocessor. Clearly, the specialized method is computationally more efficient than RKF45 and ODE for stiff problems.

Although RKF45 does not contain a test to detect stiffness, the presence of stiffness is usually indicated by an excessive number of derivative function evaluations (more than the built-in bound of 3000). If this occurs you should examine the differential equation for stiffness.

The procedure ODE contains a specific test to detect the presence of stiffness. Stiffness is indicated if 50 or more consecutive integration steps are taken by Adams methods whose orders are less than four. As soon as a method with order at least four is used, the stiffness-step-counter is reset to zero.

7.14 BOUNDARY-VALUE PROBLEMS FOR ORDINARY DIFFERENTIAL EQUATIONS

So far we have discussed only the initial value problem for ordinary differential equations, (7.1–4), where the dependent variables are completely specified at one value of the independent variable. We often obtain the solution to an engineering or science problem in terms of a differential equation in which some of the dependent variables are specified at the starting value, t_0, while others are specified at some later value, t_f. If not all the dependent variables are known at t_0, we have, in this case, a *two-point boundary-value problem*. In general, when one or more of the dependent variables are specified, or satisfy algebraic equations, at $t_1 < t_2 < \ldots < t_f$ then we have a *multipoint boundary-value problem*. For example,

$$\frac{dy_1}{dt} = y_1(y_2 - 1.5) \tag{7.14–1}$$

$$\frac{dy_2}{dt} = y_2(1.5 - 2y_1) \tag{7.14–2}$$

$$y_1(0) = 1 \tag{7.14–3}$$

$$y_1(1) = 0.5 \tag{7.14–4}$$

In this two-point boundary-value problem, $y_2(0)$ is unknown.

One method of solving two-point boundary-value problems is based on solving a sequence of initial value problems. For (7.14–1) to (7.14–4) we set $y_2(0) = b$, a constant. This gives the new initial condition

$$y_2(0) = b \tag{7.14–5}$$

We guess a value for b and integrate the initial value problem (7.14–1), (7.14–2),

(7.14–3), and (7.14–5) over the interval [0, 1]. Unless we were very lucky, (7.14–4) will not be satisfied. We must then adjust b so that (7.14–4) can be more nearly satisfied. This new value of b is then used to re-integrate the initial value problem. The process is repeated until the terminal condition (7.14–4) is satisfied to within the desired accuracy. This procedure is called the *shooting method* for two-point boundary-value problems. We shall restate the shooting method in a more general setting than this particular example.

Suppose we have a first-order system of ordinary differential equations (7.1–4). Furthermore, let

$$\mathbf{y}(t_0) = \mathbf{K}(\mathbf{b}) \tag{7.14–6}$$

$$\mathbf{M}[\mathbf{y}(t_f)] = \mathbf{0} \tag{7.14–7}$$

where $\mathbf{b}^T = [b_1, \ldots, b_m]$ and $\mathbf{M}^T = [M_1, \ldots, M_m]$. $\mathbf{b}$ is a vector of unknown parameters to be determined so that the vector $\mathbf{M}$ of terminal conditions is satisfied. Because $\mathbf{y}(t_0)$ depends on the parameter vector $\mathbf{b}$ the solution of (7.1–4), $\mathbf{y}(t)$, will also depend on $\mathbf{b}$. Therefore, the terminal conditions (7.14–7) can be considered as a system of algebraic equations in terms of the unknowns $\mathbf{b}$, although the explicit dependence on $\mathbf{b}$ is quite complicated. We can use any method for solving nonlinear algebraic equations in order to obtain a $\mathbf{b}$ that satisfies (7.14–7). For example, if the number of unknowns, m, is one then we could use the procedure ZeroIn; if $m > 1$ we could use Powell's method, Chapter 4. In either case we would guess a value for $\mathbf{b}$, then integrate (7.1–4) with the initial conditions (7.14–6) to obtain $\mathbf{y}(t_f)$ so that (7.14–7) can be computed.

When the shooting method was applied to the two-point boundary-value problem (7.14–1) to (7.14–5) using ZeroIn, with b assumed to lie in the interval [0.5, 1.5] and error tolerance 10^{-11}, the iterative method converged to b = 0.849272102865593 after seven evaluations of the terminal condition (7.14–4). (See Table 7.11 and Fig. 7.8). The differential equations (7.14–1) and (7.14–2) were integrated using RKF45 with RelErr and AbsErr set to 10^{-8}.

One can also solve multipoint boundary-value problems for ordinary differ-

Table 7.11 SHOOTING METHOD SOLUTION OF THE TWO-POINT BOUNDARY VALUE PROBLEM (7.14–1) TO (7.14–5)

k, Evaluation of Equation (7.14–4)	$y_2(0) = b_k$	$\lvert y_1(1) - 0.5 \rvert$
1	0.5	0.1279
2	1.5	0.2660
3	0.824755896678028	9.480×10^{-3}
4	0.849850099700738	2.243×10^{-4}
5	0.849270163958209	7.523×10^{-7}
6	0.849272102714874	5.848×10^{-11}
7	0.849272102865593	0.0

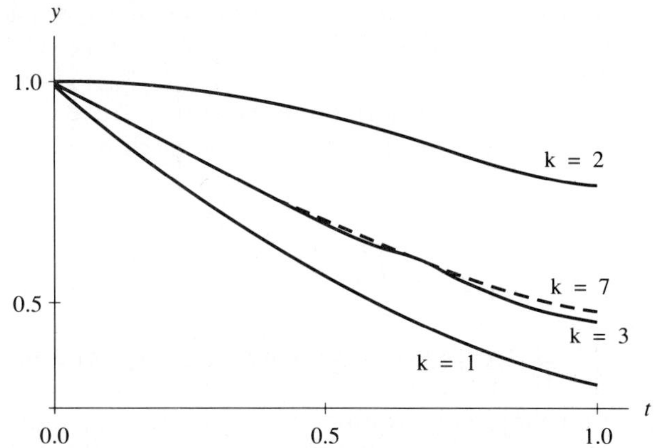

Figure 7.8 The first three iterations (k = 1, 2, 3) of the shooting method solution of the two-point boundary-value problem (7.14-1) to (7.14-5), and the final converged solution k = 7.

ential equations by a finite-difference method. Each derivative is replaced by an appropriate finite difference approximation at each point of a preselected set of mesh points on the interval $[t_0, t_f]$ (see Section 6.12). This gives a large nonlinear system of algebraic equations to be solved for the values of the dependent variables at the mesh points. The system of equations could be solved using Powell's method, Chapter 4.

7.15 ADDITIONAL TOPICS IN THE NUMERICAL SOLUTION OF ORDINARY DIFFERENTIAL EQUATIONS

We have stated in an earlier section that one is interested in obtaining the numerical solution of a differential equation with a given accuracy. In other words, we want to control the global error; unfortunately, this is often impractical because we can only estimate the local discretization error, not the global error. We could re-integrate the differential equation, using smaller error tolerances, in order to check the global accuracy. But it is possible that the numerical method might produce the same output values over a large range of error tolerances. This can happen because the error may not be a monotonic function of the error tolerance. A better method to test the global accuracy of the computed solution would be to integrate the differential equation using two different numerical methods, such as RKF45 and ODE; the solutions would "agree" if the imposed error tolerances are the same.

One is often faced with the choice of numerical method to use; should RKF45 or ODE be used for a nonstiff problem? Some general guidelines can be stated.

When the derivative functions are relatively simple in that they do not require many time-consuming machine operations, and if the error tolerance requested is approximately 10^{-4} to 10^{-6} then a Runge-Kutta method, such as RKF45, may be more efficient than ODE. If the derivative functions are complicated or the requested accuracy is high, then an Adams method, such as ODE is preferred. If one is solving a two-point boundary-value problem by the shooting method, then the differential equations are integrated many times; a method requiring few derivative function evaluations may be preferable. However, if the differential equation is to be integrated only once, then such considerations of efficiency may not be worthwhile; use the method with which you are most familiar.

If you encounter a stiff differential equation in your work, you may find that procedures such as RKF45 and ODE can handle the equation. If not, these procedures will indicate that there is a problem with the differential equation, which may be due to stiffness. In such a case, you should be aware of special procedures for dealing with stiff equations. The FORTRAN subroutine DGEAR, in the IMSL package of mathematical programs, can be used for stiff equations.

When you apply your favorite differential equation solver to some system of differential equations, you may find that the solver is not as efficient as you had expected. If the equations are not stiff, then perhaps a change in variables could help. For example, if one of the variables, y, is oscillating rapidly, the transformation $y = \sin(z)$ may help. In general, you might want to consider a transformation of the dependent variables $\mathbf{y} \to \mathbf{x}$, and the independent variable $s \to t$. The transformations $\mathbf{y} = \mathbf{N}(\mathbf{x})$ and $s = g(t)$ should be selected so that (1) $\mathbf{N}(\mathbf{x})$ and $g(t)$ are invertible functions, and (2) the differential equations are "simplified." The choice of the functions $\mathbf{N}(\mathbf{x})$ and $g(t)$ will depend on your ingenuity and insight into the particular problem that you are solving. See Problem 23 in Section 7.16 for an example.

7.16 PROBLEMS

1. Integrate each of the following differential equations over the given interval using one or more of the single-step methods discussed in Sections 7.3–7.7. For each method used, vary the given step-size from the value given with each equation. If m denotes either the number of terms retained in the Taylor series method or the number of levels for the Extrapolation method (with Euler's method), use different values of m when these methods are employed.

   (a) $\dfrac{dy}{dt} = t + y^2;\ y(0) = 1;\ 0 \le t \le 0.5;\ h = 0.05$

   (b) $\dfrac{dy}{dt} = \dfrac{y(t + y)}{t^2};\ y(1) = 1;\ 1 \le t \le 2;\ h = 0.05$

   (c) $\dfrac{dy}{dt} = \tan^{-1}(y);\ y(1) = 1;\ 1 \le t \le 2;\ h = 0.1$

(d) $\dfrac{dy}{dt} = e^{\sin(y)}$; $y(0) = 0$; $0 \le t \le 1$; $h = 0.1$

(e) $\dfrac{dy}{dt} = y(ty - 2)$; $y(0) = 2$; $0 \le t \le 1$; $h = 0.1$

2. **(a)** Write a Pascal procedure to implement the classical Runge-Kutta method (Section 7.5) for a system of n first-order ordinary differential equations.

(b) The differential equation

$$t^2y'' + ty' + (t^2 - b^2)y = 0,$$

where b is a constant, is called *Bessel's differential equation*. Convert this equation into a system of two first-order ordinary differential equations. Use your Pascal procedure from part (a) to obtain $y(1)$ and $y'(1)$ given $b = 0$, $y(0) = 1$ and $y'(0) = 0$ with the step-size $h = 0.05$.
Hint: In order to evaluate the derivatives at $t = 0$, you will need to compute $y''(0)$. This can be done by solving for y'' in Bessel's equation and then using l'Hôpital's rule to find $\lim_{t \to 0} y''(t)$.

(c) Repeat part (b) with the differential equation

$$y'' + ty^2y' + t^2 + y^2 = 4$$

$y(0) = 0$, $y'(0) = 0$, $0 \le t \le 1$, with $h = 0.1$.

(d) Integrate the differential equation in part (b) using RKF45 in the single-step mode, with RelErr and AbsErr equal to 10^{-5}. Compare the number of integration steps and the number of derivative function evaluations required for RKF45 and the fixed step method in part (b).

(e) Repeat part (d) using the differential equation in part (c).

3. The Adams method order 4 predictor has coefficients $b_1 = 55/24$, $b_2 = -59/24$, $b_3 = 37/24$, and $b_4 = -9/24$. The Adams method order 5 corrector has coefficients $b_0^* = 251/720$, $b_1^* = 646/720$, $b_2^* = -264/720$, $b_3^* = 106/720$, and $b_4^* = -19/720$.

(a) Write a Pascal procedure to implement the order 4/order 5 Adams predictor-corrector pair for a system of n first-order ordinary differential equations. Your procedure should be based upon the *PECE* scheme discussed in Section 7.11.

(b) Adapt your Runge-Kutta procedure from Problem 2(a) for a system of n first-order ordinary differential equations. This revised procedure can be used as the starter for the Adams method in part (a).

(c) Use the procedures developed in parts (a) and (b) to integrate each differential equation in Problem 2.

4. When a simple pendulum is subjected to viscous damping, the pendulum equation becomes $\theta'' + [\alpha^2 c/(m\ell^2)]\theta' + (g/\ell)\sin(\theta) = 0$, where c is the coefficient of the viscous damping mechanism and α is the distance from the point of suspension of the pendulum to the point at which the viscous damper is attached. The other symbols have the same meanings as in Example 7.3. By making the same change-of-variables as was done in Example 7.3, the nondimensionalized viscously damped pendulum satisfies the differential equation

$$\frac{d^2\theta}{dt^2} + \mu\frac{d\theta}{dt} + \sin(\theta) = 0; \qquad \theta(0) = \theta_0, \qquad \theta'(0) = 0,$$

where $\mu = (\alpha^2 c)/(m\ell^{1.5}\sqrt{g})$. Use RKF45 to integrate this differential equation for various values of μ and θ_0 to study the motion of the pendulum. The motion may be displayed graphically by a phase plane plot of $d\theta/dt$ vs. θ. Try $\mu = 0.1$.

5. One end of a nonlinear spring, with spring constant k, is attached to a wall and the other end is attached to a mass, m. The mass can slide along a rough surface subject to a Coulomb friction force, of magnitude cmg, that acts in the opposite direction to the velocity of the mass; c is the coefficient of friction and g is gravity. Let x denote the position of the mass as measured from the equilibrium position of the spring and let s denote time. The equation of motion of the mass-spring system is

$$mx'' + cmg\cdot\text{sgn}(x') + kx^3 = 0$$

where $$\text{sgn}(z) = \begin{cases} -1 & \text{if } z < 0 \\ 0 & \text{if } z = 0 \\ +1 & \text{if } z > 0. \end{cases}$$

When the mass is at rest it does not move if $k\,|\,x\,|^3 \le cmg$. The change-of-variables $x = (cmg/k)^{1/3}y$ and $s = [m/(kc^2g^2)]^{1/6}t$ transforms the differential equation into the nondimensionalized form

$$\frac{d^2y}{dt^2} + y^3 + \text{sgn}(dy/dt) = 0.$$

Use RKF45 to obtain the phase plane trajectory, the graph of dy/dt vs. y, starting at $(y, dy/dt) = (3, 0)$.
Note: The term involving the sign function, sgn, is piecewise constant. If the derivative of y is not zero on an interval, the last term of the differential equation is either 1 or -1. You will need to use the root-finder ZeroRKF to determine where dy/dt changes sign. Also, if $|\,y\,| \le 1$ when $dy/dt = 0$, there is no motion of the mass.

6. The study of predator-prey interactions has led to the development of the following system of differential equations:

$$\frac{dx}{dt} = a_1x - a_2xy \qquad x(0) = x_0$$

$$\frac{dy}{dt} = -b_1y + b_2xy \qquad y(0) = y_0$$

where a_1, a_2, b_1 and b_2 are positive constants. The variables x and y could represent rabbits and wolves, for example. Use RKF45 to obtain graphs of y vs. x for each of the following initial conditions if $a_1 = 1.7$, $a_2 = 0.03$, $b_1 = 0.6$, and $b_2 = 0.002$.

(a) $x_0 = 400$, $y_0 = 50$

(b) $x_0 = 400$, $y_0 = 100$

(c) $x_0 = 400$, $y_0 = 200$

(d) $x_0 = 400$, $y_0 = 250$

7. Suppose the Earth is considered to be a sphere, of radius r, rotating about its axis at the constant angular speed ω. The origin of the (x, y, z)-coordinate system is fixed to the surface of the sphere at the latitude angle θ, with the positive x-axis pointing "east," the positive y-axis pointing "north," and the positive z-axis indicating "up." The (x, y, z)-axis system is a local coordinate system that is rotating with respect to some inertial reference frame. The motion of a particle near the surface of the Earth, measured in the local coordinate system, is given by the differential equations

$$x'' = \omega^2 x + 2\omega\{y' \sin(\theta) - z' \cos(\theta)\}$$

$$y'' = \omega^2 y \sin^2(\theta) - 2\omega x' \sin(\theta) - (r + z)\omega^2 \sin(\theta)\cos(\theta)$$

$$z'' = 2\omega x' \cos(\theta) - \omega^2 y \sin(\theta)\cos(\theta) + (r + z)\omega^2 \cos^2(\theta) - g$$

where g is constant gravity. Set $g = 9.8$ m/sec^2, $r = 6.373 \times 10^6$ m, and $\omega = 2\pi/(24 \times 3600)$ sec^{-1}. The latitude angle θ is considered to be positive for the northern hemisphere.

In each of the following questions, use RKF45 to integrate the differential equations, and use ZeroRKF if required.

(a) This problem determines how far a particle that is thrown upwards can deviate in the rotating coordinate system due to the Coriolis forces. Set $x(0) = y(0) = z(0) = x'(0) = y'(0) = 0$, $z'(0) = 50$ m/sec. For each $\theta = 10i$ degrees, $-9 \leq i \leq 9$, integrate the differential equations to obtain $(x(t_*), y(t_*))$, where t_* is the least positive t such that $z(t_*) = 0$.

(b) This problem repeats (a), but for a particle that is dropped from rest. The initial conditions are $x(0) = y(0) = x'(0) = y'(0) = z'(0) = 0$, $z(0) = 100$ m.

Because $\omega \simeq 7.2 \times 10^{-5}$ the preceding system of equations is often approximated by neglecting the terms in ω^2 to give

$$x'' = 2\omega\{y' \sin(\theta) - z' \cos(\theta)\}$$

$$y'' = -2\omega x' \sin(\theta)$$

$$z'' = 2\omega x' \cos(\theta) - g.$$

(c) Repeat part (a) with the approximate differential equations, and compare your results with those of part (a).

(d) Repeat part (b) with the approximate differential equations, and compare your results with those of part (b).

8. The matrix Riccati differential equation

$$\frac{d\mathbf{K}}{dt} - \mathbf{A}(t) + \mathbf{B}(t)\mathbf{K} + \mathbf{K}\mathbf{B}^T(t) + \mathbf{K}\mathbf{C}(t)\mathbf{K} = \mathbf{0};\ \mathbf{K}(t_0) = \mathbf{K}_0$$

often arises in the solution of certain linear control systems. In the differential equation, $\mathbf{K}$, $\mathbf{A}(t)$, $\mathbf{B}(t)$, $\mathbf{C}(t)$, and $\mathbf{K}_0$ are $n \times n$ matrices with real entries; $\mathbf{A}(t)$, $\mathbf{C}(t)$, and $\mathbf{K}_0$ are symmetric matrices. When we write an $n \times n$ matrix as $\mathbf{S}(t)$ we mean that each entry in the matrix $\mathbf{S}$ is a function of t. It can be shown that the solution $\mathbf{K}$ is also a symmetric matrix. When the left-hand side of the matrix differential equation is combined into one $n \times n$ matrix, $\mathbf{M}$, and equated to the right-hand side, we see that each entry in $\mathbf{M}$ gives a differential equation. In general, the matrix differential equation would consist of n^2 scalar differential equations; because $\mathbf{K}$ is symmetric, there are only $n(n + 1)/2$ different equations.

If $$\mathbf{A}(t) = \begin{bmatrix} -1 & 0 \\ 0 & -1 \end{bmatrix}, \quad \mathbf{B}(t) = \begin{bmatrix} 0 & 0 \\ 1 & 0 \end{bmatrix},$$

$$\mathbf{C}(t) = \begin{bmatrix} 0 & 0 \\ 0 & -1 \end{bmatrix}, \quad \mathbf{K}_0 = \begin{bmatrix} 0.5 & 0 \\ 0 & 0 \end{bmatrix}, \quad t_0 = 0,$$

then use RKF45 to determine $\mathbf{K}$ at $t = 10$.

9. The motion of a certain controlled mechanism is governed by the differential equations

$$\frac{dy_1}{dt} = y_1(y_1 - 1.5) - \operatorname{sgn}(1 + y_2) \qquad y_1(0) = 1$$

$$\frac{dy_2}{dt} = y_2(1.5 - 2y_1) \qquad y_2(0) = b.$$

The sign function, sgn, is defined in Problem 5. Use RKF45 and ZeroRKF to obtain $y_1(t)$ and $y_2(t)$ for t in the interval $[0, 2]$, for the initial values $b = -0.5$ and $b = -1.5$.

10. The problem of launching a sounding rocket from the surface of the Earth so that the rocket reaches maximum altitude is known as the Goddard problem in honor of Dr. R. Goddard, the American rocket pioneer. The equations governing the rocket's motion are

$$\frac{dx}{dt} = v \qquad x(0) = 0$$

$$\frac{dv}{dt} = \frac{cu}{m} - g(x) - \frac{D(x, v)}{m} \qquad v(0) = 0$$

$$\frac{dm}{dt} = -u \qquad m(0) = m_0$$

where x measures the altitude of the rocket, $g(x)$ is the Earth's gravitational attraction at x, $D(x, v)$ is the atmospheric drag on the rocket, m is the mass of the rocket, c is the exhaust speed of the propellant, and u is the rate at which the propellant is burned. We shall define $g(x)$ and $D(x, v)$ as

$$g(x) = g_0R^2(x + R)^{-2} \quad \text{and} \quad D(x, v) = D_0v^2e^{-kx},$$

where g_0, R, D_0, and k are constants; R is the Earth's radius. Assume that $u(t) = b$, a constant, for $0 \le t \le t_1$, and $u(t) = 0$ for $t > t_1$. The time t_1 is the "burn-time."
Use RKF45 and ZeroRKF to find the maximum altitude that the rocket may attain if $t_1 = 5$ sec, $g_0 = 9.8$ m/sec^2, $m_0 = 1000$ Kg, $c = 1000$ m/sec, $b = 90$ Kg/sec, $D_0 = 100$ newtons, $k = 10^{-4}$/m, and $R = 6.373 \times 10^6$ m.

11. The Mathieu equation

$$\frac{d^2y}{dt^2} + (a + 2b\cos(t))y = 0$$

belongs to the large class of linear ordinary different equations with periodic coefficients. If $a = 3$ and $b = 1$ find the solution of Mathieu's equation, using RKF45, on [0, 20] if $y = 2$ and $dy/dt = 2$ at $t = 0$. The study of linear ordinary differential equations with periodic coefficients is called *Floquet theory*.

12. The differential equation

$$y'' - y'\left[a - \left(\frac{a}{3}\right)(y')^2\right] + ay = 0$$

is called the *Rayleigh oscillator*. Use RKF45 to obtain the phase plane trajectory, a graph of y' vs y, of this differential equation if $a = 21$, $y(0) = 0$, $y'(0) = 1$ for $0 \le t \le 3$. Also obtain the graph of $y(t)$ vs t.

13. Euler's equations of motion for a certain rotating body are

$$\frac{dy_1}{dt} = y_2y_3$$

$$\frac{dy_2}{dt} = -y_1y_3$$

$$\frac{dy_3}{dt} = -0.51\, y_1y_2$$

The solutions at multiples of $K = 1.86264080233273855\ldots$ are

j	$y_1(jK)$	$y_2(jK)$	$y_3(jK)$
0	0.0	1.0	1.0
1	1.0	0.0	0.7
2	0.0	−1.0	1.0
3	−1.0	0.0	0.7

The solution is periodic with period $4K$.

(a) Use RKF45 to integrate the differential equations on $[0, 4K]$, with RelErr and AbsErr taken as 10^{-8}. Obtain the solution at $t = jK$ for $j = 1, 2, 3, 4$. Also obtain the number of derivative function evaluations and integration steps required to complete the integration on $[0, 4K]$.

(b) Repeat part (a) using the integrator ODE. The weighting functions can be selected as $w_i(t) = |y_i(t)| + 1$. Compare the results with those of part (a).

(c) Repeat the integration on the interval $[0, 4K]$ but use ODE in the single-step mode. At each integration step print out the number of derivative function evaluations, the order of the Adams method used on that step, as well as t, $y_1(t)$, $y_2(t)$, and $y_3(t)$.

14. Obtain the phase plane trajectory, the graph of y' vs y of

$$y'' + y' \ell\text{og}[y^2 + (2y')^2] + y = 0$$

when

(a) $y(0) = 1.0$, $y'(0) = 0$.

(b) $y(0) = 0.1$, $y'(0) = 0$.

15. A jet of water streams vertically upward from a nozzle. The mass of water per unit time leaving the nozzle is r, a constant; the water leaves the nozzle at the speed v_0. The jet strikes a ball of mass m, and the vertical velocity of the water relative to the ball is destroyed on impact. If x denotes the displacement of the ball, s denotes time, and g denotes gravity, then the equation of motion is $x'' + g - r(v - x')^2/(mv) = 0$, where $v = \sqrt{v_0^2 - 2gx}$. The change-of-variables $x = v_0^2 y/(2g)$ and $s = v_0 t/(2g)$ puts the differential equation into the nondimensional form

$$\frac{d^2y}{dt^2} = -0.5 + \frac{\alpha}{(1-y)^{1/2}}\left[(1-y)^{1/2} - \frac{dy}{dt}\right]^2$$

$\alpha = rv_0/(2mg)$. Set $\alpha = 2$ and take $y = 0$, $dy/dt = 0$ at $t = 0$. Obtain the phase plane trajectory, the graph of dy/dt vs y.

16. The following differential equations occur in the study of a nonlinear system with feedback. Find $y_1(t)$, $y_2(t)$, and $y_3(t)$ for $0 \le t \le 10$, if $y_1(0) = 2$, $y_2(0) = 0$, and $y_3(0) = 1$.

$$\frac{dy_1}{dt} = y_1 - y_2 - y_1 y_3$$

$$\frac{dy_2}{dt} = y_1 + y_2 - y_2 y_3$$

$$\frac{dy_3}{dt} = -y_3 + y_1^2 + y_2^2$$

17. When a spherical gas-filled bubble in an inviscid compressible liquid is subjected to a tension wave, the resulting nondimensionalized equation of radial motion is

$$x\left(1 - \frac{x'}{v}\right)x'' + 1.5(x')^2\left(1 - \frac{x'}{3v}\right) = (1 + D)x^{-3\gamma}\left(1 - \frac{(3\gamma - 1)x'}{v}\right)$$
$$- \frac{D}{x} + \left(1 + \frac{x'}{v}\right)(Pf(t) - 1) + \frac{xPf'(t)}{v}$$

where $x = R(t)/R_0$, $R(t)$ is the bubble radius, R_0 is the bubble's equilibrium radius; $P = P_i/P_a$, P_i is the amplitude of the applied tension wave $f(t)$, P_a is the ambient liquid pressure; $v = c\sqrt{\rho/P_a}$, c is the speed of sound in the liquid, ρ is the liquid density; $D = 2\sigma/(R_0 P_a)$, σ is the liquid's surface tension; γ is the ratio of specific heats. The expansion and subsequent collapse of microbubbles is the mechanism that causes cavitation damage. Microbubbles also can form in blood; the existence of such bubbles can lead to the "bends," a disease due to rapid decompression.

(a) Set $R_0 = 0.001$ cm, $P_a = 10^6$ dyn/cm^2, $\sigma = 72.8$ dyn/cm, $\gamma = 1.4$, $c = 1.485 \times 10^5$ cm/sec, $P_i = 5P_a$, $\rho = 1$ gm/cm^3, $x(0) = 1$, and $x'(0) = 0$. The tension wave is modelled by the function $f(t) = \exp(-t/t_0)$, where $t_0 = 2.9 \times 10^{-5}(R_0)^{-1}\sqrt{P_a/\rho}$. This form of $f(t)$ represents a typical underwater explosion shock wave. Integrate the differential equation over the interval [0, 190] to obtain the values of x at each bubble-radius extremum.

(b) To study the free oscillations of the bubble, we set $P_i = 0$, and take $x(0) > 1$ with $x'(0) = 0$. Obtain a graph of $x(t)$ for $0 \le t \le 25$, when $x(0) = 3$; use the values of R_0, P_a, σ, γ, c, and ρ as given in part (a).

18. Integrate the following initial value problem on the interval [0, 10].

$$\frac{dy_1}{dt} = y_2 \qquad y_1(0) = 0$$

$$\frac{dy_2}{dt} = y_3 \qquad\qquad y_2(0) = 1$$

$$\frac{dy_3}{dt} = -\frac{y_1 y_3 + y_2^2}{3} \qquad\qquad y_3(0) = 0$$

19. The following differential equation occurs in the flow of an incompressible fluid over a semi-infinite flat plate. Such a flow is called a *boundary-layer flow* because the viscous effects are limited to a thin layer near the surface of the plate.

$$\frac{d^3y}{dt^3} + 0.5y\frac{d^2y}{dt^2} = 0; \qquad y(0) = 0, \qquad y'(0) = 0, \qquad y''(0) = 1.$$

Obtain $y(t)$, $y'(t)$, and $y''(t)$ at $t = 0.1k$, $k = 1, 2, 3, \ldots$ until $|y'(t)y''(t)| < 10^{-12}$.

20. When a simple pendulum is allowed to swing in any direction, it is called a *spherical pendulum*. In terms of nondimensionalized variables the equation of motion of a spherical pendulum is

$$y'' - \beta\cos(y)[\sin(y)]^{-3} + \sin(y) = 0,$$

where β is a constant. Obtain phase plane trajectories, graphs of y' vs y, for $\beta = 0.01$ and $\beta = 0.1$, given the initial conditions $y(0) = 1$ and $y'(0) = 0$. Compare the trajectories with those of the simple pendulum, $\beta = 0$. (See Example 7.3)

21. **(a)** In the analysis of the deflection angle of a cantilever beam, the following two-point boundary-value problem is obtained;

$$\frac{d^2y}{dt^2} + b\cos(y) = 0; \qquad y(0) = 0, \qquad y'(1) = 0.$$

Solve this two-point boundary-value problem for $b = 0.1k$, $k = 1, 2, 3, 4, 5$.

(b) The following two-point boundary-value problem arises in the solution of a control problem.

$$\frac{dy_1}{dt} = y_1(y_1 - 1.5) - \operatorname{sgn}(y_2 + 1)$$

$$\frac{dy_2}{dt} = y_2(1.5 - 2y_1)$$

$$y_1(0) = 1, \qquad y_1(1) = 0.5.$$

Solve the two-point boundary-value problem for $y_2(0) = \alpha$, where α can occur in each of the following intervals.

(i) $[-1.0, -0.8]$ (ii) $[-1.5, -1.4]$ (iii) $[-1.2, -1.0]$

22. Solve the two-point boundary-value problem

$$\frac{dy_1}{dt} = y_2 \qquad \frac{dy_3}{dt} = y_4$$

$$\frac{dy_2}{dt} = -y_1 - \text{sgn}(y_4) \qquad \frac{dy_4}{dt} = -y_3$$

$$y_1(0) = 3, \quad y_2(0) = 0.5, \quad y_1(t_f) = 0, \quad \text{and} \quad [y_3(t_f)]^2 = 1.$$

The unknowns are $y_3(0)$, $y_4(0)$, and t_f.

23. The differential equation

$$xx'' + \left(1.5x' + \frac{m}{x}\right)x' + \frac{D}{x} - \frac{1 + D}{x^{4.2}} + 1 = Pf(t)$$

$$x(0) = 1, \qquad x'(0) = 0$$

represents the radial motion of a spherical bubble in a viscous incompressible liquid, when the bubble is subjected to a tension wave $f(t)$. The variables x and t are the nondimensional bubble radius and time, respectively; the constants m, D, and P are the nondimensional viscosity, surface tension, and amplitude of the tension wave. The tension wave is selected so that $f(t) = e^{-t/50}$, and the constants are $P = 1.5$, $m = 0.05$, and $D = 0.15$.

(a) Use RKF45 to integrate the differential equation on the interval [0, 100], with RelErr = AbsErr = 10^{-6}. Record the number of derivative function evaluations and integration steps required to obtain the solution at $t = 100$.

(b) Consider the change-of-variable $t \rightarrow s$ defined by

$$\frac{dt}{ds} = x^r$$

where r is a constant to be determined. We can take $s = 0$ when $t = 0$. Use the change-of-variable to obtain the differential equation for x in terms of the new independent variable s. Find the value of r so that the term multiplying $(dx/ds)^2$ is zero.

(c) Repeat part (a) for the transformed differential equation and the equation for t, using the value of r found in part (b). Here, we must integrate three first-order differential equations; part (a) only required two equations to be integrated. Compare the number of derivative functions evaluations and integration steps for the transformed equation with those quantities found in part (a). Comment on the difference in the results.

24. Obtain the graph of $y(t)$ on the interval [0, 1] if

$$t\frac{d^2y}{dt^2} + (0.5 - t)\frac{dy}{dt} - 3y = 0$$

with $y = 1$ and $\dfrac{dy}{dt} = 6$ at $t = 0$.

25. A circuit containing a resistor of 1 ohm in series with a nonlinear inductor is driven by a voltage source of $v(t)$ volts, where t is time. The flux of the inductor is $\phi(i) = \tanh(i)$, where i is the current flowing through the circuit. This flux function is a fairly good approximation to the actual characteristic of many saturating devices. The voltage across the nonlinear inductor is the derivative of ϕ with respect to t.

    (a) Obtain the differential equation that describes the current in the circuit in the form

    $$\frac{di}{dt} = f(t, i).$$

    (b) Using the initial condition $i(0) = 0$ obtain the graph of i vs t for $v(t) = 0.1u(t)$, $v(t) = u(t)$, and $v(t) = 3u(t)$, on the interval $0 \le t \le 5$. The function $u(t)$ is the unit step function,

    $$u(t) = \begin{cases} 0 & \text{if } t < 0 \\ 1 & \text{if } t \geq 0 \end{cases}$$

7.17 REFERENCES

F. S. Acton. *Numerical Methods that Work.* New York: Harper & Row, 1970.

R. L. Burden, J. D. Faires, and A. C. Reynolds. *Numerical Analysis.* 2d Ed. Boston: Prindle, Weber & Schmidt, 1981.

S. D. Conte and C. de Boor. *Elementary Numerical Analysis.* 3d Ed. New York: McGraw-Hill, 1980.

G. Dahlquist, A. Bjorck, and N. Anderson. *Numerical Methods.* Englewood Cliffs, NJ: Prentice-Hall, 1974.

G. E. Forsythe, M. A. Malcolm, and C. B. Moler. *Computer Methods for Mathematical Computations.* Englewood Cliffs, NJ: Prentice-Hall, 1977.

P. Henrici. *Discrete Variable Methods in Ordinary Differential Equations.* New York: Wiley, 1962.

J. R. Rice. *Numerical Methods, Software, and Analysis.* New York: McGraw-Hill, 1983.

L. F. Shampine and R. C. Allen, Jr. *Numerical Computing: An Introduction.* Philadelphia: W.B. Saunders, 1973.

L. F. Shampine and M. K. Gordon. *Computer Solution of Ordinary Differential Equations: The Initial Value Problem.* San Francisco: W.H. Freeman, 1975.

APPENDIX 7.1

Listing of the Pascal Program for Example 7.1

The program illustrates several single-step methods.

```
PROGRAM OrdinaryDifferentialEquation;

CONST
  HighestDerivative = 4 ; MaxLevel = 4 ;
```

```
TYPE
  String80 = String[80];
  VectorL = ARRAY[ 0..MaxLevel ] OF REAL;
  VectorN = ARRAY[ 1..HighestDerivative ] OF REAL;

VAR
  i, steps, n, Level : INTEGER;
  t, ye, yme, yh, yrk, yt, yex : REAL;
  a, t0, tFinal, h : REAL;
  Equation : String80;

(*$I Response.Pas *)
(*$I ReadNum.Pas *)

PROCEDURE SetUp;
BEGIN
  Equation := 'dy/dt = y*cos(t)';
  t0 := 0.0 ; tFinal := 3.0 ; a := 1.0 ; h := 0.25 ;
  n := 3 ; Level := 3
END; (* SetUp *)

PROCEDURE F( t, y : REAL ; VAR yp : REAL );
BEGIN
  yp := y*cos(t)
END;

PROCEDURE Derivatives( terms : INTEGER ; t , y : REAL ; VAR yp : VectorN );
VAR
  c, s : REAL;
  k : INTEGER;
BEGIN
  c := cos(t) ; s := sin(t) ;
  FOR  k := 1  TO  terms  DO
    CASE  k  OF
      1 : yp[1] := y*c;
      2 : yp[2] := yp[1]*c - y*s;
      3 : yp[3] := yp[2]*c - 2.0*yp[1]*s - yp[1];
      4 : yp[4] := yp[3]*c - 3.0*yp[2]*s - 3.0*yp[1]*c + y*s
    END
END; (* Derivatives *)

PROCEDURE Euler( h, t : REAL ; VAR y : REAL );
VAR
  yp : REAL;
BEGIN
  F(t,y,yp);
  t := t + h;
  y := y + h*yp
END; (* Euler *)

PROCEDURE ModifiedEuler( h, t : REAL ; VAR y : REAL );
VAR
  yp, ym, tm, hBy2 : REAL;
BEGIN
  hBy2 := h/2.0;
  tm := t + hBy2;
  F(t,y,yp);
  t := t + h;
  ym := y + hBy2*yp;
```

```
  F(tm,ym,yp);
  y := y + h*yp
END; (* ModifiedEuler *)

PROCEDURE Heun( h, t : REAL ; VAR y : REAL );
VAR
  yp1, yp2, ym : REAL;
BEGIN
  F(t,y,yp1);
  t := t + h;
  ym := y + h*yp1;
  F(t,ym,yp2);
  y := y + h*(yp1 + yp2)/2.0
END; (* Heun *)

PROCEDURE Extrapolation( ExtLevel : INTEGER ; h, t : REAL ; VAR y : REAL );
(* Extrapolation method with Euler's method *)
VAR
  a : VectorL;
  ds, s, z : REAL;
  m, n, steps : INTEGER;
BEGIN
  ds := h;
  FOR  m := 0  TO  ExtLevel  DO
    BEGIN
      s := t ; z := y ; steps := Round(h/ds) ;
      FOR  n := 1  TO  steps  DO
        BEGIN
          Euler(ds,s,z) ; s := s + ds
        END;
      a[m] := z ; ds := ds/2.0
    END;
  z := 1.0;

  FOR  n := 1  TO  ExtLevel  DO
    BEGIN
      z := 2.0*z ; ds := z - 1.0 ;
      FOR  m := 0  TO  ExtLevel - n  DO
        a[m] := (z*a[m+1] - a[m])/ds
    END;
  y := a[0]
END; (* Extrapolation *)

PROCEDURE RungeKutta( h, t : REAL ; VAR y : REAL );
VAR
  ym, yp1, yp2, yp3, yp4, tm, hBy2 : REAL;
BEGIN
  hBy2 := h/2.0;
  tm := t + hBy2;
  F(t,y,yp1);
  t := t + h;
  ym := y + hBy2*yp1;
  F(tm,ym,yp2);
  ym := y + hBy2*yp2;
  F(tm,ym,yp3);
  ym := y + h*yp3;
  F(t,ym,yp4);
  y := y + h*(yp1 + 2.0*(yp2 + yp3) + yp4)/6.0
END; (* RungeKutta *)
```

```
PROCEDURE TaylorSeries( terms : INTEGER ; h , t : REAL ; VAR y : REAL );
VAR
  yp : VectorN;
  p : REAL;
  k : INTEGER;
BEGIN
  Derivatives(terms,t,y,yp);
  p := 1.0;
  FOR  k := 1  TO  terms  DO
    BEGIN
      p := k*p ; yp[k] := yp[k]/p
    END;
  p := yp[terms];
  FOR  k := terms - 1  DOWNTO  1  DO
    p := h*p + yp[k];
  y := y + h*p
END; (* TaylorSeries *)


PROCEDURE Message(x, y, offset, m : INTEGER ; name : String80 ; n : INTEGER);
BEGIN
  GoToXY(x + offset,y) ; HighVideo ;
  WRITE(m,' <= ',name,' <= ',n) ; LowVideo ;
  Delay(1500) ; GoToXY(x,y) ; ClrEol
END; (* Message *)


PROCEDURE Change;
VAR
  x, y : INTEGER;
  answer : CHAR;
BEGIN
  REPEAT
    ClrScr;
    WRITELN('The current initial values are:');
    WRITELN('  (1) initial  t  = ', t0:8:5);
    WRITELN('  (2)  final   t  = ', tFinal:8:5);
    WRITELN('  (3) initial  y  = ',  a:8:5);
    WRITELN('  (4) step-size   = ', h:8:5);
    WRITELN('  (5) highest order derivative in Taylor series method is ',n);
    WRITELN('  (6) maximum level for the extrapolation method is ',Level);
    WRITELN;
    WRITELN('Enter  0  if these are acceptable.');
    WRITE('Otherwise, enter the number of the one to be changed:  ');
    Echo(answer);
    IF  answer  IN  ['1'..'6']  THEN
      CASE  answer  OF
        '1' : BEGIN
                WRITE(' initial t = ') ; t0 := ReadReal
              END;
        '2' : BEGIN
                WRITE(' final t = ') ; tFinal := ReadReal
              END;
        '3' : BEGIN
                WRITE(' initial y = ') ; a := ReadReal
              END;
        '4' : BEGIN
                WRITE(' step-size = ');
                REPEAT
                  x := WhereX ; y := WhereY ; h := ReadReal ;
```

```
                IF  h = 0.0  THEN
                  ErrorHandler(x,y,15,'Invalid step-size')
              UNTIL  h <> 0.0
            END;
      '5' : BEGIN
              WRITE(' highest order derivative = ');
              REPEAT
                x := WhereX ; y := WhereY ;  n := ReadInteger ;
                IF  (n < 1) OR (n > HighestDerivative)  THEN
                  BEGIN
                    Errorhandler(x,y,7,'Invalid derivative order');
                    Message(x,y,7,1,'n',HighestDerivative)
                  END
              UNTIL  (1 <= n) AND (n <= HighestDerivative)
            END;
      '6' : BEGIN
              WRITE(' extrapolation level = ');
              REPEAT
                x := WhereX ; y := WhereY ; Level := ReadInteger ;
                IF  (Level < 0) OR (Level > MaxLevel)  THEN
                  BEGIN
                    ErrorHandler(x,y,7,'Invalid extrapolation level');
                    Message(x,y,7,0,'Level',MaxLevel)
                  END
              UNTIL  (0 <= Level) AND (Level <= MaxLevel)
            END
    END
  UNTIL  NOT (answer  IN  ['1'..'6']);
  WRITELN;
  WRITELN
END; (* Change *)

BEGIN (* main program *)
  SetUp ; LowVideo ; Change ;
  t := t0 ; steps := Round((tFinal - t0)/h);
  ye := a ; yme := a ; yh := a ; yrk := a ; yt := a ; yex := a ;
  WRITELN(' ':15,'Numerical integration of ',Equation);
  WRITELN;
  WRITELN(' ':30,'h = ',h:8:5);
  WRITELN;
  WRITELN(' ':20,'Using ',n+1,' terms in the Taylor series');
  WRITELN(' ':20,'Using ',Level,' levels in the extrapolation method');
  WRITELN;
  WRITELN('                       Modified               Taylor',
             '   Extrapolation   Runge'  );
  WRITELN('    t        Euler      Euler       Heun      series',
             '       method       Kutta');
  WRITELN;
  WRITELN(t:8:5,' ',ye:10:5,' ',yme:10:5,' ',yh:10:5,' ',
                             yt:10:5,' ',yex:10:5,' ':3,yrk:10:5);
  FOR  i := 1  TO  steps  DO
    BEGIN
      Euler(h,t,ye);
      ModifiedEuler(h,t,yme);
      Heun(h,t,yh);
      TaylorSeries(n,h,t,yt);
      Extrapolation(Level,h,t,yex);
      RungeKutta(h,t,yrk);
      t := t + h;
```

```
    WRITELN(t:8:5,' ',ye:10:5,' ',yme:10:5,' ',yh:10:5,' ',
                             yt:10:5,' ',yex:10:5,' ':3,yrk:10:5)
  END
END.
```

APPENDIX 7.2

User's Manual for the Pascal Package RKF45, which includes the RKF45 integrator and the root-finder ZeroRKF

1. User's manual for the Pascal package of procedures and functions, which is in the file RKF45.Pas. The main procedure is RKF45.

```
PROCEDURE RKF45( VAR n : INTEGER ; VAR y, yp : VectorN ; VAR t, tout, h,
                 RelErr, AbsErr : REAL ; VAR flag, nfe, steps : INTEGER ;
                      help : BOOLEAN ; VAR Data : DataRecord );
```

The following must be declared in the calling program

```
   TYPE
     String80 = String[80];
     VectorN = ARRAY[ 1..MaxN ] OF REAL;
     DataRecord = RECORD
                    kflag, jflag, MaxNfe, kop, init : INTEGER;
                    Eps, u26, saveRelErr, saveAbsErr, h : REAL
                  END;
```

RKF45.Pas uses the files Response.Pas and ReadNum.Pas, which must be declared before RKF45.Pas.

RKF45 is designed to solve non-stiff and mildly stiff differential equations when derivative evaluations are inexpensive. RKF45 should generally not be used when the user is demanding high accuracy. The RKF45 package can be used with the root-finder procedure ZeroRKF to locate a zero of a function F[t,**y**(t)], where **y** is the solution of a differential equation being integrated by RKF45.

RKF45 integrates a system of n first order ordinary differential equations of the form

$$\frac{d\mathbf{y}}{dt} = \mathbf{f}(t,\mathbf{y})$$

where **y** is known at t.

Typically the procedure is used to integrate from *t* to *tout* but it can be used as a single-step integrator to advance the solution one step in the direction of *tout*. On return form RKF45 the parameters in the calling list are set for continuing the integration. The user has only to call RKF45 again (and perhaps define a new value for *tout*). RKF45 contains an interactive error-handler facility to help the user recover from abnormal or error exits.

```
RKF45  uses the Runge-Kutta-Fehlberg (4,5)-method described in
  E. Fehlberg,
    "Low-order classical Runge-Kutta formulas with stepsize control",
    NASA TR R-315.

The performance of  RKF45  is illustrated in
  L.F. Shampine, H.A. Watts, and S. Davenport,
   "Solving non-stiff ordinary differential equations - the state of the art",
   S.I.A.M. Review, vol. 18, July 1976, pp. 376-411.

The parameters in the calling list of RKF45:

    n        The number of equations to be integrated  (1 ≤ n ≤ MaxN)
   y[*]      The solution vector at  t
   yp[*]     The derivative of the solution vector at  t
    t        The independent variable
   tout      The output point at which the solution is desired
    h        The step-size used on the last successful step
   RelErr,
    AbsErr   Relative and absolute error tolerances for the local error
               test.  At each step  RKF45  requires that
                   Abs(local error) ≤ RelErr*Abs(solution) + AbsErr
               for each component of the local error and solution vectors.
   flag      An indicator for the status of the integration
   nfe       A counter on the number of derivative function evaluations
   steps     A counter on the number of integration steps
   help      On/off indicator for the interactive error-handler
   Data      A work record containing variables required by  RKF45  between
               calls.  These variables must not be changed by the user.

User supplied procedure:

   PROCEDURE Frkf45( n : INTEGER ; t : REAL ; VAR y, yp : VectorN );
```

Evaluates the n derivatives $yp[i] = \frac{dy_i}{dt}$ for i = 1,2,...,n.

```
First call to RKF45:

   The following parameters must be initialized before the first call:

    n        The number of equations to be integrated.  (1 ≤ n ≤ MaxN)
   y[*]      The vector of initial conditions
    t        The starting point of the integration
   tout      The output point at which the solution is desired.
               t = tout  is allowed on the first call only, in which case
               RKF45  returns with  flag = 2  if continuation is possible.

RelErr,
 AbsErr   Relative and absolute local error tolerances which must be
            non-negative.  RKF45 should normally not be used with a
```

relative error control smaller than about 10^{-8}. To avoid limiting precision difficulties RKF45 requires *RelErr* to be larger than an internally computed relative error parameter which is machine dependent. In particular, pure absolute error is not permitted. If a smaller than allowable value of *RelErr* is attempted, RKF45 increases *RelErr* appropriately and returns control to the user before continuing the integration.

flag — An indicator to initialize RKF45 for each new problem.
Interval mode: *flag* = 1.
Single-step mode: *flag* = −1.

help — Set to TRUE if the interactive error-handler is to be used; otherwise, set to FALSE. The indicator can be reset by the user at any time control is returned from RKF45 to the calling program. If the interactive error-handler is invoked, then the output *flag* values of 3, 4, 5, 6, 7, 8, or 999 are intercepted by the error-handler within RKF45. The error-handler supplies information to the user about the error condition. The user may be requested to supply new values of some input variables.

Output from RKF45:

y[*] — The computed solution at *t*

yp[*] — The derivative of the computed solution at *t*

t — The last point reached in the integration

h — The step-size of the last successful step

nfe — The cumulative number of derivative function evaluations

steps — The cumulative number of integration steps

flag = 2 — The integration reached *tout*. This indicates a successful return from RKF45 and is the normal output value of *flag* for continuing the integration. To continue the integration, define a new *tout* (reset *flag* to −2 if in the single-step mode) and call RKF45 again.

−2 — A single successful step in the direction of *tout* has been taken. This is the normal output value of *flag* for continuing the integration in the single-step mode. To continue the integration towards *tout* just call RKF45 again.

3 — The integration was not completed because the relative error tolerance, *RelErr*, was too small. *RelErr* has been increased by RKF45. To continue the integration call RKF45 without changing the value of *flag*.

4 — The integration was not completed because more than 3000 derivative function evaluations were needed (approximately 500 integration steps). To continue the integration call RKF45 without changing the value of *flag*. An internal counter will be reset to allow another 3000 derivative fu

5 — The integration was not completed because the solution vanished making a pure relative error test impossible. A nonzero *AbsErr* must be used to continue. The single-step integration mode can be used for one step. The integration can be continued, with a positive value for *AbsErr*, by calling RKF45 again without changing the value of *flag*.

6 — The integration was not completed because the requested accuracy could not be achieved using the smallest allowable step-size. The user must increase the error tolerances, *RelErr* and/or *AbsErr*, before continuing the integration. Before calling RKF45, reset *flag* to 2 or −2. **Warning**: *flag* = 6 indicates a trouble spot (the solution is changing rapidly, or a singularity may be present) and it may not be advisable to continue.

7 — RKF45 may be inefficient for solving this problem. Too much output is restricting the natural step-size choice.

The single-step mode can be used or the Adams method procedure ODE can be used. To continue the integration with RKF45 reset *flag* to 2 or −2 before calling RKF45.

8 Invalid input parameters. This occurs the first time any of the following conditions are satisfied:

$n < 1$;
$(t = tout)$ and $(flag \neq \pm 1)$;
$(RelErr \leq 0.0)$ or $(AbsErr < 0.0)$;
flag not equal to one of ±1, ±2, 3, 4, 5.

The integration cannot be continued unless the invalid input parameters are corrected.

9 The interactive error-handler is being used, and the user has responded "no" to a request to continue the integration.

999 Fatal input error, probably caused by an illegal input value of *flag*, or the user has not responded to an output value of *flag* = 5, 6, 7, or 8. RKF45 cannot proceed.

Warnings:

1. The work variables in the variable *Data*, a record of type DataRecord, contain information required for subsequent integration steps. These variables **must not** be altered.
2. The interactive error-handler is designed to work in text mode only. Do **not** use the error-handler if you are running RKF45 in graphics mode.

Procedure RKF45 is based on the FORTRAN subroutine RKF45 that appeared in
G.E. Forsythe, M.A. Malcolm and C.B. Moler,
Computer Methods for Mathematical Computations
Prentice-Hall, 1977.

2. User's manual for the Pascal procedure ZeroRKF, which is in the file

ZeroRKF.Pas.

```
PROCEDURE ZeroRKF( VAR y, yp, ys : VectorN ; n : INTEGER ;
                   t, h, tol, ft, ftmh : REAL ; VAR root : REAL ;
                                   VAR Evaluations : INTEGER );
```

A zero of the function FzeroRKF(t-x*h) is computed in the interval [0,1]. It is assumed that FzeroRKF(t)*FzeroRKF(t-h) < 0.0 without a check. ZeroRKF returns a zero at *root*, and *Evaluations* equal to the number of times the function FzeroRKF was called. ZeroRKF is based on the Pascal procedure ZeroIn.

On entry to ZeroRKF:

- ft = FzeroRKF(t)
- ftmh = FzeroRKF(t-h)
- y[*], yp[*] These are the solution to the differential equation and its derivative at t, as determined by RKF45
- tol Convergence is achieved when Abs(FzeroRKF(s)) ≤ tol

The user supplies the following to evaluate the function whose zero is sought.

```
FUNCTION FzeroRKF( n : INTEGER ; s : REAL ; y : VectorN ) : REAL;
```

```
y[*] is the solution of the system of n differential equations at  s
as found by the procedure  Fehlberg  from RKF45.
```

APPENDIX 7.3

Listing of the Program using RKF45 to integrate the differential equation of Example 7.1

```
PROGRAM DifferentialEquation;

CONST
  MaxN = 1;

TYPE
  String80 = String[80];
  VectorN = ARRAY[ 1..MaxN ] OF REAL;
  DataRecord = RECORD
                 kflag, jflag, MaxNfe, kop, init : INTEGER;
                 Eps, u26, saveRelErr, saveAbsErr, h : REAL
               END;

VAR
  y, yp : VectorN;
  Data : DataRecord;
  tfinal, delta, t, tout, h, RelErr, AbsErr : REAL;
  j, intervals, n, flag, nfe, steps : INTEGER;

PROCEDURE Frkf45( n : INTEGER ; t : REAL ; VAR y, yp : VectorN );
BEGIN
  yp[1] := y[1]*cos(t)
END; (* Frkf45 *)

(*$I Response.Pas *)
(*$I ReadNum.Pas *)
(*$I RKF45.Pas *)

BEGIN (* main program *)
  LowVideo;
  RelErr := 1.0E-05 ; AbsErr := 0.0 ; n := 1 ; flag := 1 ; j := 0 ;
  t := 0.0 ; tout := 0.0 ; y[1] := 1.0 ; tfinal := 10.0 ; delta := 1.0 ;
  intervals := Round(tfinal/delta);
  WRITELN('             RKF45     actual');
  WRITELN('      t    solution     error') ; WRITELN ;
  WRITELN(t:8:2,' ',y[1]:10:5,'   ',y[1]-Exp(sin(t)):10);
  REPEAT
    j := j + 1 ; tout := tout + delta ;
    RKF45(n,y,yp,t,tout,h,RelErr,AbsErr,flag,nfe,steps,TRUE,Data);
    IF  flag = 2  THEN
      WRITELN(t:8:2,' ',y[1]:10:5,'   ',y[1]-Exp(sin(t)):10)
    ELSE
      WRITELN('flag = ',flag)
  UNTIL  (j = intervals) OR (flag > 2);
  WRITELN
END.
```

APPENDIX 7.4

Listing of the Program for Example 7.3

```
PROGRAM PeriodOfPendulum;

CONST
  MaxN = 2 ; Pi = 3.14159265358979323846264 3 ;

TYPE
  String80 = String[80];
  VectorN = ARRAY[ 1..MaxN ] OF REAL;
  DataRecord = RECORD
                 kflag, jflag, MaxNfe, kop, init : INTEGER;
                 Eps, u26, saveRelErr, saveAbsErr, h : REAL
               END;

VAR
  y, yp, ys : VectorN;
  t, tout, h, RelErr, AbsErr, ftmh, s : REAL;
  theta, period : REAL;
  Data : DataRecord;
  n, flag, nfe, steps, jj : INTEGER;

PROCEDURE Frkf45( n : INTEGER ; t : REAL ; VAR y, yp : VectorN );
BEGIN
  yp[1] :=  y[2];
  yp[2] := -sin(y[1])
END; (* Frkf45 *)

FUNCTION FzeroRKF( n : INTEGER ; s : REAL ; y : VectorN ) : REAL;
BEGIN
  FzeroRKF := y[1]
END; (* FzeroRKF *)

(*$I Response.Pas *)
(*$I ReadNum.Pas *)
(*$I RKF45.Pas *)
(*$I ZeroRKF.Pas *)

BEGIN (* main program *)
  LowVideo ; theta := 1.0 ;
  WRITELN(' theta       nondimensional period') ; WRITELN ;
  REPEAT
    RelErr := 1.0E-09 ; AbsErr := 0.0 ; n := 2 ; flag := -1;
    t := 0.0 ; y[2] := 0.0 ; tout := 1000.0 ;
    y[1] := Pi*theta/180.0 ; ftmh := FzeroRKF(n,t,y) ;
    REPEAT
      RKF45(n,y,yp,t,tout,h,relerr,abserr,flag,nfe,steps,TRUE,Data);
      IF  y[1] < 0.0  THEN
        BEGIN
          ZeroRKF(y,yp,ys,n,t,h,1.0E-11,y[1],ftmh,s,jj);
          flag := 2
        END;
      ftmh := y[1]
    UNTIL  flag >= 2;
    period := 4.0*s;
```

```
    WRITELN(theta:5:1,'   ',period:20:10);
    theta := theta + 1.0
  UNTIL  (theta > 179.5) OR (flag >= 9)
END.
```

APPENDIX 7.5

User's manual for the Pascal package ODE, which includes the ODE integrator and the root-finder ZeroODE

1. User's manual for the Pascal procedure ODE, which is in the file ODE.Pas.

```
PROCEDURE ODE( n : INTEGER ; VAR y : VectorN ; VAR t : REAL ; tout : REAL ;
                VAR RelErr, AbsErr : REAL ; VAR steps, nfe, flag : INTEGER ;
                  VAR Data : DataRecord );
```

ODE integrates a system of n first order ordinary differential equations of the form

$$\frac{dy}{dt} = f(t,y)$$

where y is given at t. ODE can be used with the root-finder procedure ZeroODE to locate a zero of a function F[t,y(t)], where y is the solution of a differential equation being integrated by ODE.

The following must be declared in the calling program

```
 TYPE
   Vector12    = ARRAY[ 1..12 ] OF REAL;
   VectorN     = ARRAY[ 1..MaxN ] OF REAL;
   MatrixN16   = ARRAY[ 1..MaxN , 1..16 ] OF REAL;
   DataRecord  = RECORD
                   first, notRnd, phase, start : BOOLEAN;
                   k, n, OldIsn, OldK, stiff : INTEGER;
                   ds, eps, fourU, h, OldH,
                         OldT, OldX, twoU, Usq, x : REAL;
                   p, wt, yp, z, zp : VectorN;
                   phi : MatrixN16;
                   alpha, beta, psi, v, w : Vector12;
                   g, gs, sigma, two : ARRAY[ 1..13 ] OF REAL
                 END;
```

Information required by ODE between calls is stored in a variable of type DataRecord; that variable must not be changed by the user. The following information in DataRecord may be of use to the user:

OldH The step-size used for the last successful step
OldK The order used for the last successful step.

The parameters in the calling list of ODE represent:

n	The number of equations to be integrated
y[*]	The solution vector at t
t	The independent variable
tout	The output point at which the solution is desired
RelErr, AbsErr	Relative and absolute error tolerances for the local error test
steps	A counter on the number of integration steps
nfe	A counter on the number of derivative function evaluations
flag	An indicator for the status of the integration
Data	A work record of containing varaibles required by ODE between calls. These variable **must not** be changed by the user.

User supplied procedures:

```
Fode( n : INTEGER ; t : REAL ; VAR y, yp : VectorN );
```

Evaluates the n derivatives yp[i] $= \dfrac{dy_i}{dt}$ for i = 1,2,...,n.

```
SetWeight( n : INTEGER ; t : REAL ; VAR y, yp, wt : VectorN ;
                       RelEps, AbsEps : REAL ; VAR first : BOOLEAN );
```

Computes the weight vector wt[*]. The variable *first* can be used to distinguish the initial point at which the integration starts. It is set TRUE in ODE before the first call to SetWeight. If the facility is to be used then the procedure fragment should be

```
IF  first  THEN
  BEGIN
    first := FALSE:
    (* statements to find wt[*] at the initial t *)
  END
 ELSE
   (* statements to find wt[*] at other values of t *)
```

The Euclidean norm of the vector, with i-th component LocalError[i]/wt[i], is required to be less than *eps* for a successful step. The array wt[*] allows the user to specify an error test appropriate for the problem being solved. Examples of possible choices for wt[i]:

- wt[i] = 1.0 specifies absolute error.
- wt[i] = Abs(y[i]) specifies error relative to the most recent value of the i-th component of the solution.
- wt[i] = Abs(yp[i]) specifies error relative to the most recent value of the i-th component of the derivative.
- wt[i] = Max(wt[i],Abs(y[i])) specifies error relative to the largest magnitude of the i-th component of the solution obtained so far.
- wt[i] = Abs(y[i])*RelEps + AbsEps specifies a mixed relative/absolute error test, where eps = Max(RelErr,AbsErr), RelEps = RelErr/eps and AbsEps = AbsErr/eps.

First call to ODE:

n	The number of equations to be integrated (1 ≤ n ≤ MaxN)
y[*]	The vector of initial conditions
t	The starting point of the integration
tout	The point at which the solution is desired

```
 RelErr,
  AbsErr  Relative and absolute error tolerances for the local error test
 flag     Interval mode: flag =  1 (normal input)
                         flag = -1 (the user should use this value only
                                    if it is impossible to continue the
                                    integration beyond tout)
          Single-step mode: flag = 0
```

Output from ODE:

```
 y[*]      The solution at  t
  t        The last point reached in the integration (see flag below)
              Interval mode: t = tout.
              Single-step mode:  the value of the independent variable at
                                 the end of one integration step.
RelErr,
 AbsErr    On a normal return the tolerances are unchanged.  For an
             abnormal return with Abs(flag) = 3, the tolerances have been
             increased.
 steps     The cumulative number of integration steps.
  nfe      The cumulative number of derivative function evaluations.
 flag = 2  The normal return for the interval mode, or tout was reached in
             the single-step mode.  To continue the integration set a new
             value of tout (reset flag to 5 if in the single-step mode)
             and call ODE again.
        3  The integration was not completed because the error tolerances
             were too small.  RelErr and AbsErr have been increased
             by ODE.  To continue the integration in the interval mode
             just call ODE again without changing flag.  In the single-
             step mode reset flag to 5 before calling ODE again.
        4  The differential equations appear to be stiff.  To continue the
             integration in the interval mode just call ODE again.  In the
             single-step mode, set flag to 5 before calling ODE.
        5  The normal return for the single-step mode if tout has not yet
             been reached.
      999  Invalid input parameters.
    The value of flag is returned negative when the input value is negative
    and the integration does not reach tout; that is, -3, -4.
```

Switching modes:

The user may switch between the interval mode and the single-step mode after the first call to ODE by changing *flag* to 2 for the interval mode, or to 5 for the single-step mode.

Warning:

The work variables in the variable Data, a record of type DataRecord, contain information required for the subsequent integration steps. These variables **must not** be altered.

Procedure ODE is based on the FORTRAN subroutines ODE/DE/STEP/INTRP that appeared in

L.F. Shampine and M.K. Gordon,
Computer Solution of Ordinary Differential Equations: The Initial Value Problem
W.H. Freeman, San Francisco, 1975.

2. User's manual for the Pascal procedure ZeroOde, which is in the file

   ZeroOde.Pas.

```
PROCEDURE ZeroODE( VAR y, z : VectorN ; n, OldK : INTEGER ;
                   t, OldH, tol, ft, ftmh : REAL ; VAR phi : MatrixN16 ;
                   VAR psi : Vector12 ; VAR root : REAL ;
                   VAR Evaluations : INTEGER );
```

A zero of the function FzeroODE(t-x*OldH) is computed in the interval [0,1]. It is assumed that FzeroODE(t)*FzeroODE(t-OldH) < 0.0 without a check. ZeroODE returns a zero at *root*, and *Evaluations* equal to the number of times the function FzeroODE was called. ZeroODE is based on the Pascal procedure ZeroIn.

On entry to ZeroODE:

```
ft      FzeroODE(t)
ftmh    FzeroODE(t-OldH)
y[*]    The solution to the differential equation at  t, as determined
           by  ODE.
tol     Convergence is achieved when  Abs(FzeroODE(s)) ≤ tol
OldK,OldH,phi,psi are variables from the variable Data of type
          DataRecord used in ODE.
```

The user supplies the following to evaluate the function whose zero is sought

```
FUNCTION FzeroODE( n : INTEGER ; s : REAL ; VAR z : VectorN ) : REAL;

   z[*] is the solution of the system of n differential equations at  s
   as found by the procedure Interpolate from the ODE package.
```

APPENDIX 7.6

Listing of a Program using the ODE integrator in the Interval mode, for the differential equation of Example 7.4

```
PROGRAM BeamDeflection;

CONST
  MaxN = 2;

TYPE
  Vector12    = ARRAY[ 1..12 ] OF REAL;
  VectorN     = ARRAY[ 1..MaxN ] OF REAL;
  MatrixN16   = ARRAY[ 1..MaxN , 1..16 ] OF REAL;
  DataRecord  = RECORD
                  first, notRnd, phase, start : BOOLEAN;
                  k, n, OldIsn, OldK, stiff : INTEGER;
                  ds, eps, fourU, h, OldH,
                    OldT, OldX, twoU, Usq, x : REAL;
```

```
                p, wt, yp, z, zp : VectorN;
                phi : MatrixN16;
                alpha, beta, psi, v, w : Vector12;
                g, gs, sigma, two : ARRAY[ 1..13 ] OF REAL
              END;

VAR
  y : VectorN;
  t, tout, RelErr, AbsErr, maxy2, L, L2, s : REAL;
  neqn, flag, nfe, steps : INTEGER;
  Data : DataRecord;

PROCEDURE Fode( neqn : INTEGER ; x : REAL ; VAR y, yp : VectorN );
VAR
  q : REAL;
BEGIN
  yp[1] := y[2];
  IF  Abs(y[2]) > maxy2  THEN
    maxy2 := Abs(y[2]);
  q := 1.0 + Sqr(y[2]);
  yp[2] := s*q*Sqrt(q)*(x*(x/2.0 - L) + L2)
END; (* Fode *)

PROCEDURE SetWeight( neqn : INTEGER ; t : REAL ; VAR y, yp, wt : VectorN ;
                     RelEps, AbsEps : REAL ; VAR first : BOOLEAN );
(* Set the weight vector, wt[*], for use in the procedure ODE.  See
   procedure ODE for various ways of defining  wt[*].
     y[*]   solution vector at  t
     yp[*]  derivative of the solution vector at  t
     RelEps = RelErr/Max(RelErr,AbsErr)
     AbsEps = AbsErr/Max(RelErr,AbsErr)
     first  is TRUE for initialization          *)
VAR
  i : INTEGER;
BEGIN
  FOR  i := 1  TO  neqn  DO
    IF  Abs(y[i]) < 1.0  THEN
      wt[i] := 1.0
    ELSE
      wt[i] := Abs(y[i])
END; (* SetWeight *)

(*$I ODE.Pas *)

BEGIN (* main program *)
  L := 0.0 ; s := 0.1 ;
  WRITELN('   Length of        Deflection at end of the beam');
  WRITELN('   the beam       nonlinear model     linear model',
        ' ':3,'max slope');
  REPEAT
    L := L + 0.5 ; L2 := L*L/2.0 ; maxy2 := 0.0 ;
    RelErr := 1.0E-08 ; AbsErr := 0.0 ; neqn := 2 ; flag := 1 ;
    t := 0.0 ; tout := L ;
    y[1] := 0.0 ; y[2] := 0.0 ;
    ODE(neqn,y,t,tout,RelErr,AbsErr,steps,nfe,flag,Data);
    WRITELN('    ',tout:5:1,' ':8,y[1]:15:8,
          '  ',s*Sqr(L2)/2.0:15:8,' ':3,maxy2:10:7)
  UNTIL  L > 2.6
END.
```

APPENDIX 7.7

Listing of a Program using the ODE integrator in the single-step mode, with the root-finder ZeroODE, for the differential equation in Example 7.3

```
PROGRAM PeriodOfPendulum;

CONST
  MaxN = 2 ; Pi = 3.14159265358979323846264 3 ;

TYPE
  Vector12    = ARRAY[ 1..12 ] OF REAL;
  VectorN     = ARRAY[ 1..MaxN ] OF REAL;
  MatrixN16   = ARRAY[ 1..MaxN , 1..16 ] OF REAL;
  DataRecord = RECORD
                 first, notRnd, phase, start : BOOLEAN;
                 k, n, OldIsn, OldK, stiff : INTEGER;
                 ds, eps, fourU, h, OldH,
                   OldT, OldX, twoU, Usq, x : REAL;
                 p, wt, yp, z, zp : VectorN;
                 phi : MatrixN16;
                 alpha, beta, psi, v, w : Vector12;
                 g, gs, sigma, two : ARRAY[ 1..13 ] OF REAL
               END;

VAR
  y, ys : VectorN;
  t, tout, RelErr, AbsErr, s, ftmh : REAL;
  theta, period : REAL;
  neqn, flag, nfe, steps, jj : INTEGER;
  Data : DataRecord;

PROCEDURE Fode( neqn : INTEGER ; t : REAL ; VAR y, yp : VectorN );
BEGIN
  yp[1] :=  y[2];
  yp[2] := -sin(y[1])
END; (* Fode *)

PROCEDURE SetWeight( neqn : INTEGER ; t : REAL ; VAR y, yp, wt : VectorN ;
                       RelEps, AbsEps : REAL ; VAR first : BOOLEAN );
(* Set the weight vector, wt[*], for use in the procedure ODE.  See
   procedure ODE for various ways of defining  wt[*].
     y[*]   solution vector at  t
     yp[*]  derivative of the solution vector at  t
     RelEps = RelErr/Max(RelErr,AbsErr)
     AbsEps = AbsErr/Max(RelErr,AbsErr)
     first  is TRUE for initialization         *)
VAR
  i : INTEGER;
BEGIN
  FOR  i := 1  TO  neqn  DO
    IF  Abs(y[i]) < 1.0  THEN
      wt[i] := 1.0
    ELSE
      wt[i] := Abs(y[i])
END; (* SetWeight *)
```

```
FUNCTION FzeroODE( neqn : INTEGER ; s : REAL ; VAR y : VectorN ) : REAL;
BEGIN
  FzeroODE := y[1]
END; (* FzeroODE *)

(*$I ODE.Pas *)
(*$I ZeroODE.Pas *)

BEGIN (* main program *)
  theta := 1.0;
  WRITELN(' theta          reduced period') ; WRITELN ;
  REPEAT
    RelErr := 1.0E-09 ; AbsErr := 0.0 ; neqn := 2 ; flag := 0 ;
    t := 0.0 ; tout := 1000.0 ;
    y[2] := 0.0 ; y[1] := Pi*theta/180.0 ; ftmh := FzeroODE(neqn,t,y) ;
    REPEAT
      ODE(neqn,y,t,tout,RelErr,AbsErr,steps,nfe,flag,Data);
      IF  y[1] < 0.0  THEN
        BEGIN
          ZeroODE(y,ys,neqn,Data.OldK,t,Data.OldH,1.0E-11,y[1],ftmh,
                       Data.phi,Data.psi,s,jj);
          flag := 2
        END
      ELSE
        ftmh := y[1]
    UNTIL  flag IN [2,3,4];
    period := 4.0*s;
    WRITELN(theta:5:1,'   ',period:20:10);
    theta := theta + 1.0
  UNTIL  theta > 179.5
END.
```

Chapter

8

Eigenvalues and Eigenvectors of Matrices

Numerical methods for obtaining the eigenvalues and eigenvectors of an $n \times n$ matrix are discussed. The power method and the inverse power method are introduced for finding single eigenvalues. Special methods for finding the characteristic polynomial are given. Orthogonal transformations and the QR method are examined for finding the eigenvalues of a real matrix.

8.1 INTRODUCTION

The problem of determining the eigenvalues and the eigenvectors of a square matrix is of great importance to engineers and scientists. It arises in many disciplines. Some common examples are the dynamic analysis of linear systems, structural analysis, determination of the natural frequencies of vibration in mechanical systems, and the identification of the principal normal stresses in stress analysis.

In this chapter, we shall study several methods for calculating the eigenvalues and eigenvectors of a matrix. These methods belong to two main groups. In the so-called "direct" methods, we first determine the characteristic polynomial of the matrix. The eigenvalues are then evaluated as the roots of this polynomial. The second group of methods attempts to obtain the eigenvalues by performing a series of transformations to the matrix which do not alter its eigenvalues but change its structure so that the eigenvalues may be obtained by inspection. In addition, we shall discuss some iterative methods which lead to the determination of the eigenvalue with either the largest or the smallest magnitude with much less effort, if certain conditions are satisfied.

Before going into these methods, it will be desirable to review some preliminary ideas from linear algebra that relate to the eigenvalue problem.

8.2 SOME MATRIX PRELIMINARIES

In this section we shall present some basic ideas from linear algebra which will enable the reader to understand the rest of the chapter.

8.2.1 Definition of Eigenvalues and Eigenvectors of a Matrix

We shall start by introducing a definition of the eigenvalues and eigenvectors of a matrix which provides a geometric interpretation.

Let $\mathbf{A}$ be an $n \times n$ matrix, and let $\mathbf{x}$ be any vector of dimension n. In general, the product $\mathbf{Ax}$ will be a different n-dimensional vector in the underlying n-dimensional Euclidean space, E. We now ask if there exists a nonzero vector $\mathbf{v}$ such that the product $\mathbf{Av}$ has the same direction as $\mathbf{v}$. Thus,

$$\mathbf{Av} = \lambda \mathbf{v} \qquad \textbf{(8.2–1)}$$

where λ is a scalar. The vector $\mathbf{v}$ defines a "normal" coordinate axis in E. The scalar λ is called an *eigenvalue* of $\mathbf{A}$ and the nonzero vector $\mathbf{v}$ is called an *eigenvector* of $\mathbf{A}$. We may rewrite equation (8.2–1) as

$$(\lambda \mathbf{I} - \mathbf{A})\mathbf{v} = \mathbf{0} \qquad \textbf{(8.2–2)}$$

Evidently one solution to equation (8.2–2) is the zero vector, but this trivial solution is not allowed because an eigenvector must be nonzero. A nontrivial solution of (8.2–2) will exist if, and only if, the rank of the matrix $(\lambda \mathbf{I} - \mathbf{A})$ is less than n; this will occur only if

$$\det(\lambda \mathbf{I} - \mathbf{A}) = 0 \qquad \textbf{(8.2–3)}$$

where $\det(\mathbf{Q})$ is the determinant of the matrix $\mathbf{Q}$. The expression $\det(\lambda \mathbf{I} - \mathbf{A})$ is a polynomial of degree n in λ, called the *characteristic polynomial* of $\mathbf{A}$. There will be n values of λ which will satisfy equation (8.2–3); these are the roots of the characteristic polynomial, although not all of these roots may be distinct. Equation (8.2–3) is called the *characteristic equation* of $\mathbf{A}$. Thus, the $n \times n$ matrix $\mathbf{A}$ will have n eigenvalues, which may be denoted by $\lambda_1, \lambda_2, \ldots, \lambda_n$. Generally, each eigenvalue λ will have one or more eigenvectors associated with it.

Because each eigenvalue has one or more eigenvectors associated with it, we might ask if an $n \times n$ matrix $\mathbf{A}$ can have a set of n linearly independent eigenvectors. If so, then these eigenvectors can form a *basis* for the underlying n-dimensional space, E. This means that any vector in E can be written as a unique linear combination of the basis eigenvectors. Thus, the eigenvectors will define a set of (normal) coordinate axes. When the matrix $\mathbf{A}$ is written in terms of these normal coordinates, its form is particularly simple: it is a diagonal matrix with the eigenvalues along the main diagonal. If all the eigenvalues of the $n \times n$ matrix **A** are distinct, then we are guaranteed that **A** will have a set of n linearly independent eigenvectors; however, **A** can have repeated eigenvalues and still have a set of n linearly independent eigenvectors.

If all the elements of **A** are real, then all the coefficients of the characteristic polynomial will also be real. For this case all the eigenvalues must either be real or occur in conjugate pairs if complex. It may be noted that if we multiply an eigenvector of **A** by a nonzero scalar the resulting vector will also satisfy equation (8.2–1). Hence, it is possible to scale the eigenvectors to obtain a unique set. For example, one may divide each element of an eigenvector by the square root of the sum of the squares of the elements to obtain an eigenvector that has unit Euclidean *norm*. Another choice is to divide each element by the element with the largest absolute value. The choice of how the eigenvectors should be scaled depends on their subsequent usage.

We illustrate the method of finding eigenvalues and eigenvectors through the following example.

EXAMPLE 8.1

Consider the matrix

$$\mathbf{A} = \begin{bmatrix} 0 & 1 \\ -2 & -3 \end{bmatrix}$$

whose characteristic polynomial is

$$\det(\lambda\mathbf{I} - \mathbf{A}) = \begin{vmatrix} \lambda & -1 \\ 2 & \lambda + 3 \end{vmatrix} = \lambda^2 + 3\lambda + 2$$

The roots of the above quadratic, the eigenvalues of **A**, are $\lambda_1 = -1$ and $\lambda_2 = -2$. If we substitute $\lambda = \lambda_1 = -1$ into equation (8.2–2) we get

$$\begin{bmatrix} -1 & -1 \\ 2 & 2 \end{bmatrix} \mathbf{v}_1 = \mathbf{0}$$

It will be seen that the two rows of the matrix are linearly dependent, as expected, due to the fact that $\det(\lambda_1\mathbf{I} - \mathbf{A}) = 0$. Hence, many solutions to $(\lambda_1\mathbf{I} - \mathbf{A})\mathbf{v}_1 = \mathbf{0}$ exist. One solution is $\mathbf{v}_1 = [1 \quad -1]^T$. Other nontrivial solutions may be obtained by multiplying the vector $\mathbf{v}_1$ by any nonzero constant. It may be noted that this merely scales the vector by changing its length without altering its direction. Hence, an eigenvector is uniquely determined up to a nonzero scalar. A unique solution for the eigenvector may be obtained by selecting the scale factor so that the norm of the vector is equal to one. Many different norms may be defined (see Section 2.5). When the Euclidean norm (the ℓ_2 norm) is used, the unit length eigenvector would be $[1/\sqrt{2} \quad -1/\sqrt{2}]^T$. The eigenvector $\mathbf{v}_1$ is, of course, already normalized with respect to the ℓ_∞ norm. To complete Example 8.1, an eigenvector corresponding to the eigenvalue $\lambda_2 = -2$ is $\mathbf{v}_2 = [1 \quad -2]^T$.

Recall that we shall always get a complete set of n eigenvectors if the n eigenvalues of the $n \times n$ matrix $\mathbf{A}$ are distinct. It is not as straightforward for repeated eigenvalues. If the multiplicity of a certain eigenvalue, λ, is r then we can get r linearly independent eigenvectors for this eigenvalue if, and only if, the rank of the matrix $(\lambda\mathbf{I} - \mathbf{A})$ is $(n - r)$. A matrix is said to be *defective* unless this rank condition is satisfied for all of its eigenvalues. For defective matrices we cannot obtain a full set of n eigenvectors. For example, the matrix

$$\mathbf{A} = \begin{bmatrix} 1 & 1 \\ 0 & 1 \end{bmatrix}$$

has eigenvalues 1, 1; the rank of $(\mathbf{I} - \mathbf{A})$ is one so that there is only one eigenvector, $\mathbf{v} = [1 \quad 0]^T$. In the case of defective matrices it is possible to define *generalized eigenvectors*, which together with the eigenvectors will form a basis for the underlying space, E. When a defective matrix is given in terms of a coordinate system whose axes are the directions of the generalized eigenvectors, the matrix will not have the nice property of being a diagonal matrix. This is the price we pay because, originally, there were not enough eigenvectors.

It should be noted that many $n \times n$ matrices arising in science and engineering problems have a complete set of n linearly independent eigenvectors (they are not defective). In particular, a real symmetric matrix cannot be defective. The interested reader is advised to refer to a suitable text on linear algebra for more information on generalized eigenvectors.

8.2.2 Some Useful Theorems

We shall now present some theorems, related to the eigenvalues of a matrix, which will be utilized in the subsequent sections.

Theorem 8.1

An $n \times n$ matrix $\mathbf{A}$ and its transpose $\mathbf{A}^T$ have the same eigenvalues.

This follows from the fact that the determinant of a matrix and its transpose are identical. Hence both $\mathbf{A}$ and $\mathbf{A}^T$ will have the same characteristic polynomial.

Definition 8.1

A set of vectors $\{\mathbf{v}_1, \mathbf{v}_2, \ldots, \mathbf{v}_n\}$ is called *orthogonal* if $\mathbf{v}_i^T\mathbf{v}_j = 0$ for all $i \neq j$. In addition, the set is called *orthonormal* if $\mathbf{v}_i^T\mathbf{v}_i = 1$ for $i = 1, 2, \ldots, n$.

Theorem 8.2

Let $\mathbf{v}$ be an eigenvector of the real $n \times n$ matrix $\mathbf{A}$ corresponding to the eigenvalue λ and let $\mathbf{w}$ be an eigenvector of $\mathbf{A}^T$ corresponding to the eigenvalue μ. If $\lambda \neq \mu$ then $\mathbf{v}$ and $\mathbf{w}$ are orthogonal, $\mathbf{w}^T\mathbf{v} = 0$.

Proof

From the definition of eigenvalues and eigenvectors, we have

$$\mathbf{A}\mathbf{v} = \lambda\mathbf{v}$$

$$\mathbf{A}^T\mathbf{w} = \mu\mathbf{w}.$$

Premultiply the first equation by $\mathbf{w}^T$, transpose both sides of the second equation and postmultiply by $\mathbf{v}$. The results of these operations are

$$\mathbf{w}^T\mathbf{A}\mathbf{v} = \lambda\mathbf{w}^T\mathbf{v}$$

$$\mathbf{w}^T\mathbf{A}\mathbf{v} = \mu\mathbf{w}^T\mathbf{v}$$

When these last two equations are subtracted we get

$$(\lambda - \mu)\mathbf{w}^T\mathbf{v} = 0$$

Because $\lambda \neq \mu$, this last equation requires that $\mathbf{w}$ and $\mathbf{v}$ be orthogonal.

Corollary 8.1

The eigenvectors of a real symmetric matrix corresponding to distinct eigenvalues are orthogonal.

Proof

This follows from the fact that $\mathbf{A}^T = \mathbf{A}$, so $\mathbf{A}$ and $\mathbf{A}^T$ have the same eigenvectors and, by Theorem 8.2, the eigenvectors corresponding to distinct eigenvalues must be orthogonal.

Theorem 8.3

All eigenvalues of a real symmetric matrix are real numbers.

Proof

We shall prove the theorem by contradiction. Let λ be a nonzero complex eigenvalue of $\mathbf{A}$. Then, since the elements of $\mathbf{A}$ are real, $\bar{\lambda}$, the complex conjugate of λ, must also be an eigenvalue of $\mathbf{A}$. Also, if $\mathbf{v}$ is an eigenvector corresponding to the eigenvalue λ, then an eigenvector corresponding to $\bar{\lambda}$ must be $\bar{\mathbf{v}}$, the complex conjugate of $\mathbf{v}$. But Corollary 8.1 requires that these eigenvectors be mutually orthogonal, $\mathbf{v}^T\bar{\mathbf{v}} = 0$. But $\mathbf{v}^T\bar{\mathbf{v}} > 0$ because $\mathbf{v} \neq \mathbf{0}$, it follows that a real symmetric matrix cannot have complex eigenvalues.

We must often deal with *polynomial matrices*, which are defined as follows. Given a polynomial

$$P(s) = \sum_{k=1}^{m} c_k s^k + c_0$$

we define the polynomial matrix $P(\mathbf{A})$ for the $n \times n$ matrix $\mathbf{A}$ as

$$P(\mathbf{A}) = \sum_{k=1}^{m} c_k \mathbf{A}^k + c_0\mathbf{I}$$

$P(\mathbf{A})$ is obtained from $P(s)$ by replacing s by $\mathbf{A}$ and by replacing c_0 by $c_0\mathbf{I}$. The next theorem relates the eigenvalues and eigenvectors of $P(\mathbf{A})$ to those of $\mathbf{A}$.

Theorem 8.4

Let λ be an eigenvalue of the $n \times n$ matrix $\mathbf{A}$ and $\mathbf{v}$ a corresponding eigenvector. If $P(s)$ is a polynomial, then $P(\lambda)$ and $\mathbf{v}$ are an eigenvalue and a corresponding eigenvector of the matrix $P(\mathbf{A})$.

Proof

Premultiplying equation (8.2–1) by $\mathbf{A}$ and using (8.2–1) gives $\mathbf{A}^2\mathbf{v} = \mathbf{A}(\lambda\mathbf{v}) = \lambda^2\mathbf{v}$. Continuing in this way, we get $\mathbf{A}^k\mathbf{v} = \lambda^k\mathbf{v}$ for any positive integer k. It follows that λ^k is an eigenvalue of $\mathbf{A}^k$ and $\mathbf{v}$, a corresponding eigenvector. Postmultiplying $P(\mathbf{A})$ by $\mathbf{v}$ and using the expression for $\mathbf{A}^k\mathbf{v}$ shows that $P(\mathbf{A})$ will have $P(\lambda)$ as an eigenvalue and $\mathbf{v}$ as a corresponding eigenvector.

Corollary 8.2

Let $\mathbf{A}$ be an $n \times n$ matrix with λ an eigenvalue and $\mathbf{v}$ an associated eigenvector. If the polynomial matrix $P(\mathbf{A})$ is nonsingular, then $[P(\lambda)]^{-1}$ is an eigenvalue and $\mathbf{v}$ is a corresponding eigenvector of $[P(\mathbf{A})]^{-1}$.

Theorem 8.5

Let $\mathbf{A}$ be an $n \times n$ matrix and let $\mathbf{S}$ be any nonsingular $n \times n$ matrix. Then the eigenvalues of $\mathbf{B} = \mathbf{S}^{-1}\mathbf{A}\mathbf{S}$ will be the same as those of $\mathbf{A}$. Furthermore, if $\mathbf{v}$ is an eigenvector of $\mathbf{A}$ corresponding to the eigenvalue λ, then $\mathbf{S}^{-1}\mathbf{v}$ is an eigenvector of $\mathbf{B}$, corresponding to λ.

Proof

Premultiplying both sides of equation (8.2–1) by $\mathbf{S}^{-1}$, we may write

$$\mathbf{S}^{-1}\mathbf{A}(\mathbf{S}\mathbf{S}^{-1})\mathbf{v} = \lambda\mathbf{S}^{-1}\mathbf{v}$$

which may be rewritten as

$$\mathbf{B}(\mathbf{S}^{-1}\mathbf{v}) = \lambda(\mathbf{S}^{-1}\mathbf{v})$$

This shows that λ is an eigenvalue of $\mathbf{B}$ and $\mathbf{S}^{-1}\mathbf{v}$ is a corresponding eigenvector.

The matrix $\mathbf{B}$, in Theorem 8.5, is said to be *similar* to $\mathbf{A}$; alternatively, $\mathbf{B}$ is called a *similarity transformation* of $\mathbf{A}$. In Section 8.5 we shall study methods based on similarity transformations.

Theorem 8.6 (Shift of axis)

Let λ be any eigenvalue of the $n \times n$ matrix $\mathbf{A}$ and $\mathbf{v}$ a corresponding eigenvector. Then $\mathbf{v}$ will be an eigenvector of the matrix $(\mathbf{A} - s\mathbf{I})$, but the corresponding eigenvalue will be $\lambda - s$.

Proof

From the definition of an eigenvalue and eigenvector we have $\mathbf{A}\mathbf{v} = \lambda\mathbf{v}$ and $s\mathbf{I}\mathbf{v} = s\mathbf{v}$. Subtracting gives $(\mathbf{A} - s\mathbf{I})\mathbf{v} = (\lambda - s)\mathbf{v}$.

This proves the theorem, which is very useful, as it allows all the eigenvalues to be displaced by the amount s. We shall see many applications in subsequent sections.

Definition 8.2

An $n \times n$ real matrix $\mathbf{P}$ is said to be *orthogonal* if $\mathbf{P}^{-1} = \mathbf{P}^T$.

Thus, an orthogonal matrix satisfies the condition $\mathbf{P}^T\mathbf{P} = \mathbf{I}$.

Definition 8.3

An $n \times n$ matrix $\mathbf{A}$ is said to be in *upper Hessenberg* form if all elements of $\mathbf{A}$ which are below the subdiagonal are zero.

Theorem 8.7

The elements on the main diagonal of an $n \times n$ upper triangular matrix are its eigenvalues.

The proof is straightforward and is left to the reader.

Definition 8.4

A *block* matrix is a matrix whose elements are matrices.

As a consequence of Definition 8.4, a block $n \times n$ matrix can be written so that its main diagonal consists of square matrices of various sizes.

Corollary 8.3

If an $n \times n$ matrix is in block upper triangular form, then its eigenvalues are those of the blocks on the main diagonal.

Definition 8.5

An $n \times n$ matrix $\mathbf{A}$ is said to be in *upper Frobenius normal* (or *companion*) *form* if it has the structure

$$\mathbf{A} = \begin{bmatrix} c_1 & c_2 & c_3 & \dots & c_{n-1} & c_n \\ 1 & 0 & 0 & \dots & 0 & 0 \\ 0 & 1 & 0 & \dots & 0 & 0 \\ \vdots & \vdots & \vdots & & \vdots & \vdots \\ 0 & 0 & 0 & \dots & 1 & 0 \end{bmatrix} \tag{8.2–4}$$

Theorem 8.8

The characteristic polynomial of the Frobenius matrix (8.2–4) is

$$\det(\lambda \mathbf{I} - \mathbf{A}) = \lambda^n - c_1\lambda^{n-1} - c_2\lambda^{n-2} - \dots - c_n \tag{8.2–5}$$

Proof

The proof is based on induction. First, we verify that it holds for a 3×3 matrix by Laplace expansion (expansion by minors) of the determinant of $(\lambda\mathbf{I} - \mathbf{A})$ along its first column. Next, we assume that it is valid for an $m \times m$ matrix and then show that the characteristic polynomial of an $(m + 1) \times (m + 1)$ matrix, obtained by Laplace expansion of the determinant about the first column, is of the form given by equation (8.2–5). The details will be left to the reader.

Corollary 8.4

If a matrix is in the upper Frobenius normal form (8.2–4) and λ is one of its eigenvalues, then a corresponding eigenvector is given by

$$\mathbf{v} = [\lambda^{n-1} \quad \lambda^{n-2} \quad \dots . \quad \lambda \quad 1]^T \tag{8.2–6}$$

Proof

Multiplying **A** in (8.2–4) by **v** we get

$$\mathbf{Av} = \begin{bmatrix} c_1\lambda^{n-1} + c_2\lambda^{n-2} + \cdots + c_n \\ \lambda^{n-1} \\ \lambda^{n-2} \\ \vdots \\ \lambda^2 \\ \lambda \end{bmatrix} \tag{8.2–7}$$

Since λ is an eigenvalue of **A**, it must be a root of the characteristic polynomial of **A**. Hence, the first row of the right-hand side of equation (8.2–7) may be replaced by λ^n. Consequently, we may write $\mathbf{Av} = \lambda\mathbf{v}$. This shows that **v**, given by equation (8.2–6), is an eigenvector of **A** corresponding to the eigenvalue λ.

Finally, we observe that the eigenvalues of a matrix **A** can be located in the complex plane by use of the Gerschgorin disk theorem (Theorem 2.5). This provides a check on any numerical method for finding eigenvalues.

The next corollary relates the eigenvalues of an $n \times n$ matrix to its determinant.

Corollary 8.5

If **A** is an $n \times n$ matrix with eigenvalues $\lambda_1, \lambda_2, \ldots, \lambda_n$ then

$$\det(\mathbf{A}) = \lambda_1 \lambda_2 \ldots \lambda_n.$$

This completes many of the mathematical results that we will need in subsequent sections of this chapter. We now return, briefly, to Example 8.1. Our purpose is to reexamine the steps we took to obtain the eigenvalues and eigenvectors in that simple example and look at them in light of a general $n \times n$ matrix.

There were three basic steps involved in the calculation of the eigenvalues and eigenvectors of the matrix in Example 8.1. The first step in the process was to find the characteristic polynomial of the matrix, $\det(\lambda\mathbf{I} - \mathbf{A})$. The second step involved the determination of the roots of the characteristic polynomial. This is called the "direct" approach; it is quite suitable for matrices of low order, but difficulties arise as the order is increased, especially in the case of repeated eigenvalues. The main problem is the high sensitivity of the roots of most polynomials to small changes in the values of the polynomial coefficients. When the characteristic polynomial of a matrix is determined numerically, the coefficients may be in error due to roundoff errors made during the various calculations needed for the computations. Other methods, based on similarity transformations, following Theorem 8.5, are more powerful for use with computers. In addition, there are iterative methods which can be utilized to determine a single eigenvalue, either the largest or the smallest in magnitude, for certain matrices. This may then be followed by deflation to determine the next largest or smallest magnitude eigenvalue, and so on. (These methods will be discussed in the next section; other methods are discussed in the following sections.) After we had found the eigenvalues in Example 8.1, we solved the system $(\lambda_k\mathbf{I} - \mathbf{A})\mathbf{v} = \mathbf{0}$ for the eigenvectors associated with each eigenvalue, λ_k. For some methods this is done; however, for methods based on similarity transformations the eigenvectors are obtained in terms of the transformation matrices.

Because the eigenvalues of the matrix in Example 8.1 were real, the eigenvectors were also real vectors because the entries in the matrix were real numbers. Consequently, the underlying two-dimensional space for this example is a two-dimensional real Euclidean space; that is, the components of the vectors in the space are real numbers. There are real matrices that have complex eigenvalues; for example, the matrix

$$\mathbf{A} = \begin{bmatrix} 1 & -1 \\ 2 & 1 \end{bmatrix}$$

with characteristic polynomial $P(\lambda) = \lambda^2 - 2\lambda + 3$, has eigenvalues $1 \pm j\sqrt{2}$ and eigenvectors $\mathbf{v} = [1 \quad \mp j\sqrt{2}]^T$, where $j = \sqrt{-1}$. Here, the underlying space is a two-dimensional complex Euclidean space; the vectors in the space have components which are complex numbers.

EXERCISE 8.1

Determine the eigenvalues of the matrices

(a) $\begin{bmatrix} -6 & -11 & -6 \\ 1 & 0 & 0 \\ 0 & 1 & 0 \end{bmatrix}$ (b) $\begin{bmatrix} -5 & 4 & 2 \\ 0 & 3 & 1 \\ 0 & 0 & 6 \end{bmatrix}$ (c) $\begin{bmatrix} -5 & 4 & 2 \\ 0 & 3 & 1 \\ 0 & -1 & 3 \end{bmatrix}$

Answer (a) $-1, -2, -3$, (b) $-5, 3, 6$, (c) $-5, 3 + j1, 3 - j1$.

EXERCISE 8.2

Show that the eigenvectors of the matrix given in Exercise 8.1(a) can be written as $[1 \quad -1 \quad 1]^T$, $[4 \quad -2 \quad 1]^T$ and $[9 \quad -3 \quad 1]^T$.

EXERCISE 8.3

Show that

$$\mathbf{P} = \begin{bmatrix} 1 & 0 & 0 \\ 0 & \cos(\theta) & \sin(\theta) \\ 0 & -\sin(\theta) & \cos(\theta) \end{bmatrix}$$

is an orthogonal matrix.

EXERCISE 8.4

Show that interchanging the ith row of a matrix with its jth row, followed by the interchange of the ith and jth columns does not alter its eigenvalues. *Hint*: First show that the interchange of rows i and j of a matrix is equivalent to pre-multiplying it by a permutation matrix $\mathbf{P}$ (see Chapter 2). Similarly, the interchange of columns i and j is equivalent to postmultiplication by $\mathbf{P}^T$. Finally, invoke Theorem 8.5 and show that the two interchanges constitute a similarity transformation; the two permutation matrices are inverses of each other.

8.3 METHODS FOR A SINGLE EIGENVALUE

In many problems (e.g., the stability analysis of linear discrete-time systems) it is sufficient to determine the eigenvalue with the largest magnitude, known as the *dominant* eigenvalue. The power method described in the next subsection can be used effectively if the dominant eigenvalue is simple (it has multiplicity one).

8.3.1 The Power Method

Suppose that the eigenvalues of the $n \times n$ matrix $\mathbf{A}$ are ordered so that

$$|\lambda_1| > |\lambda_2| \geq |\lambda_3| \geq \ldots \geq |\lambda_n| \tag{8.3–1}$$

and that $\mathbf{A}$ has a set of n linearly independent eigenvectors, $\{\mathbf{v}_i\}$. Our objective is to find the dominant eigenvalue, λ_1. We start by assuming an arbitrary vector $\mathbf{v}(0)$ as the initial guess for $\mathbf{v}_1$, an eigenvector corresponding to λ_1. The method is based on the fact that the vector $\mathbf{v}(0)$ can be expressed as a unique linear combination of the eigenvectors of $\mathbf{A}$. Therefore, we may write

$$\mathbf{v}(0) = \sum_{j=1}^{n} \alpha_j \mathbf{v}_j \tag{8.3–2}$$

We now postmultiply $\mathbf{A}$ by $\mathbf{v}(0)$ to get $\mathbf{v}(1) = \mathbf{A}\mathbf{v}(0)$. Substituting for $\mathbf{v}(0)$ from equation (8.3–2), we get

$$\mathbf{v}(1) = \sum_{j=1}^{n} \alpha_j \lambda_j \mathbf{v}_j$$

Similarly,

$$\mathbf{v}(2) = \sum_{j=1}^{n} \alpha_j (\lambda_j)^2 \mathbf{v}_j$$

and continuing the process we get

$$\mathbf{v}(k) = \sum_{j=1}^{n} \alpha_j (\lambda_j)^k \mathbf{v}_j \tag{8.3–3}$$

Equation (8.3–3) may be rewritten as

$$\mathbf{v}(k) = (\lambda_1)^k \sum_{j=1}^{n} \alpha_j \left(\frac{\lambda_j}{\lambda_1}\right)^k \mathbf{v}_j \simeq (\lambda_1)^k \alpha_1 \mathbf{v}_1$$

for large k, since $(\lambda_j/\lambda_1)^k \to 0$ for $j > 1$. Hence, $\mathbf{v}(k) \to \lambda_1 \mathbf{v}(k - 1)$ as $k \to \infty$. Thus, if two successive iteration vectors $\mathbf{v}(m)$ and $\mathbf{v}(m - 1)$ are produced that are related through a scalar ratio which is constant, or nearly constant, then we can terminate our search.

One problem that may occur in this procedure is that the magnitude of the vectors will keep increasing if $|\lambda_1| > 1$. This may lead to overflow in the computer. Overflow may be avoided by normalizing the vectors after each multiplication. We may have additional difficulty if the initial vector $\mathbf{v}(0)$ is selected such that $\alpha_1 = 0$; that is, $\mathbf{v}(0)$ is orthogonal to $\mathbf{v}_1$. In many practical cases, however, roundoff errors in the computations may cause the subsequent vectors to have a component in the direction of $\mathbf{v}_1$.

The power method requires the following steps:

Step 0. Select a nonzero vector $\mathbf{v}(0)$. Often a good choice is to make all the elements of this vector equal to one. Set $m = -1$.
Step 1. Replace m by $m + 1$.
Step 2. Calculate $\mathbf{y}(m + 1) = \mathbf{A}\mathbf{v}(m)$.
Step 3. Let μ_{m+1} be the value of that element of $\mathbf{y}(m + 1)$ which has the largest magnitude. Divide each element of $\mathbf{y}(m + 1)$ by μ_{m+1} to obtain a new vector $\mathbf{v}(m + 1)$.
Step 4. If $|\mu_{m+1} - \mu_m| \le \epsilon$, where ϵ is a specified tolerance, accept μ_{m+1} as the desired dominant eigenvalue; otherwise go to Step 1.

Notice that the power method not only gives the dominant eigenvalue, μ_{m+1}, but it also gives an associated eigenvector $\mathbf{v}(m + 1)$. In some (rare) cases, we may find that $\mathbf{y}(1)$ turns out to be the zero vector. This would happen if 0 is one of the eigenvalues of the matrix and our choice of $\mathbf{v}(0)$ turns out to be a corresponding eigenvector. In such cases, it will simply be necessary to start with another $\mathbf{v}(0)$. A computer program for this algorithm is given in Appendix 8.1.

EXAMPLE 8.2

Consider the matrix

$$\mathbf{A} = \begin{bmatrix} -15 & -54 & -40 \\ 1 & 0 & 0 \\ 0 & 1 & 0 \end{bmatrix}$$

EXAMPLE 8.2 (cont'd)

Starting with $\mathbf{v}(0) = [1 \quad 1 \quad 1]^T$, the program in Appendix 8.1 produced the vectors $\mathbf{v}(m)$ shown in Table 8.1. We used an error tolerance $\epsilon = 10^{-6}$ (Step 4 of the algorithm). In this example, the algorithm has converged in 19 iterations to very close to λ_1, the dominant eigenvalue. This happened because the ratio $|\lambda_2/\lambda_1|$ is 0.4. The convergence will be slower if $|\lambda_2/\lambda_1|$ is closer to one. In particular, the method will fail if a complex conjugate pair of eigenvalues happen to be dominant, but this violates condition (8.3–1).

Table 8.1 EXAMPLE 8.2, THE POWER METHOD FOR THE DOMINANT EIGENVALUE

m		$[\mathbf{v}(m+1)]^T$		μ_{m+1}
0	1.000000	−0.009174	−0.009174	−109.000000
1	1.000000	−0.070733	0.000649	−14.137615
2	1.000000	−0.089235	0.006312	−11.206360
3	1.000000	−0.095843	0.008553	−10.433783
4	1.000000	−0.098361	0.009427	−10.166605
5	1.000000	−0.099348	0.009772	−10.065580
6	1.000000	−0.099740	0.009909	−10.026064
7	1.000000	−0.099896	0.009964	−10.010399
8	1.000000	−0.099958	0.009985	−10.004155
9	1.000000	−0.099983	0.009994	−10.001661
10	1.000000	−0.099993	0.009998	−10.000664
11	1.000000	−0.099997	0.009999	−10.000266
12	1.000000	−0.099999	0.010000	−10.000106
13	1.000000	−0.100000	0.010000	−10.000043
14	1.000000	−0.100000	0.010000	−10.000017
15	1.000000	−0.100000	0.010000	−10.000007
16	1.000000	−0.100000	0.010000	−10.000003
17	1.000000	−0.100000	0.010000	−10.000001
18	1.000000	−0.100000	0.010000	−10.000000

The convergence of the power method to the dominant eigenvalue of a real symmetric matrix can be accelerated if we use the *Rayleigh quotient* of $\mathbf{v}$,

$$\mu_{m+1} = \frac{[\mathbf{v}(m)]^T \mathbf{A}\mathbf{v}(m)}{[\mathbf{v}(m)]^T \mathbf{v}(m)} \tag{8.3–4}$$

A computer program using (8.3–4) is in Appendix 8.1. It may be noted that although this method accelerates the convergence to the dominant eigenvalue, it does not affect the convergence to the eigenvector.

EXERCISE 8.5

Find the dominant eigenvalue of

$$\mathbf{A} = \begin{bmatrix} 9 & 8 & 2 & 2 \\ 8 & 9 & 2 & 2 \\ 2 & 2 & 7 & 4 \\ 2 & 2 & 4 & 7 \end{bmatrix}$$

using the program in Appendix 8.1. How many iterations does it take for the algorithm to converge within a tolerance of 10^{-7}?

Answer $\lambda_1 = 19$, $\mathbf{v}_1 = [1 \quad 1 \quad 0.5 \quad 0.5]^T$, 24 iterations.

EXERCISE 8.6

Repeat Exercise 8.5 using the program in Appendix 8.1. How do the rates of convergence compare in the two cases?

Answer $\lambda_1 = 19$, $\mathbf{v}_1 = [1 \quad 1 \quad 0.500025 \quad 0.500025]^T$, 13 iterations.

8.3.2 The Inverse Power Method

In many problems it is desired to find the eigenvalue which has the least magnitude. The power method can be easily modified for this case, provided that $\mathbf{A}$ is nonsingular, which we assume. Since the eigenvalues of $\mathbf{A}^{-1}$ are the reciprocals of the eigenvalues of $\mathbf{A}$, it follows immediately that estimating the dominant eigenvalue of $\mathbf{A}^{-1}$ will give the eigenvalue of $\mathbf{A}$ which has the smallest magnitude. To apply the power method to $\mathbf{A}^{-1}$ means that the eigenvalues of $\mathbf{A}^{-1}$ must satisfy (8.3–1); we assume that (8.3–1) holds for $\mathbf{A}^{-1}$.

Thus, starting with an arbitrary vector $\mathbf{v}(0)$, we calculate

$$\mathbf{v}(1) = \mathbf{A}^{-1}\mathbf{v}(0) \tag{8.3–5}$$

and continue as in Section 8.3.1. Furthermore, it is not necessary to invert $\mathbf{A}$ to proceed with this algorithm. Rewrite (8.3–5) as $\mathbf{A}\mathbf{v}(1) = \mathbf{v}(0)$ and solve for $\mathbf{v}(1)$. This solution can be carried out most conveniently by the L-U factorization of $\mathbf{A}$ and storing these factors for use in computing the other vectors. Therefore, we obtain $\mathbf{A} = \mathbf{LU}$ (see Chapter 2) so that

$$\mathbf{LU}\mathbf{v}(k + 1) = \mathbf{v}(k), \qquad k = 1, 2, 3, \ldots \tag{8.3–6}$$

After the dominant eigenvalue of $\mathbf{A}^{-1}$ is obtained by this method, its reciprocal will give the eigenvalue of $\mathbf{A}$ with the smallest magnitude. Because the inverse power method finds the dominant eigenvalue of $\mathbf{A}^{-1}$ it also finds an associated eigenvector. Hence, the smallest magnitude eigenvalue of **A** and an associated eigenvector are found.

The convergence of the inverse power method can be accelerated by a shift of origin. Using Theorem 8.6, we replace **A** by $(\mathbf{A} - s\mathbf{I})$ in (8.3–6); the inverse power method will give the eigenvalue of **A** closest to s. If the estimated eigenvalue of $\mathbf{A}^{-1}$ is μ_m then the eigenvalue of **A** that is closest to s is $s + 1/\mu_m$.

Appendix 8.1 contains a listing of a computer program for the inverse power method with the shift of origin.

EXAMPLE 8.3

We shall use the inverse power method to estimate that eigenvalue of the matrix **A**, given in Example 8.2, which has the smallest magnitude. Starting with the vector $[1 \quad 1 \quad 1]^T$, the eigenvalue and eigenvector were found in 13 iterations without shifting the origin. The results are shown in Table 8.2.

Table 8.2 EXAMPLE 8.3. THE INVERSE POWER METHOD

m	$[\mathbf{v}(m+1)]^T$			μ_{m+1}
0	−0.571429	−0.571429	1.000000	−0.571429
1	0.509554	−0.891720	1.000000	−0.891720
2	0.867142	−0.972437	1.000000	−0.972437
3	0.965664	−0.993034	1.000000	−0.993034
4	0.991296	−0.998249	1.000000	−0.998249
5	0.997811	−0.999561	1.000000	−0.999561
6	0.999452	−0.999890	1.000000	−0.999890
7	0.999863	−0.999973	1.000000	−0.999973
8	0.999966	−0.999993	1.000000	−0.999993
9	0.999991	−0.999998	1.000000	−0.999998
10	0.999998	−1.000000	1.000000	−1.000000
11	0.999999	−1.000000	1.000000	−1.000000
12	1.000000	−1.000000	1.000000	−1.000000

EXAMPLE 8.4

Repeat Example 8.3 but with the origin shifted to −0.9. In this case, convergence was reached in 6 iterations. The results are shown in Table 8.3.

EXAMPLE 8.4 (cont'd)

Table 8.3 EXAMPLE 8.4. THE INVERSE POWER METHOD WITH ORIGIN SHIFT OF −0.9

m	$[\mathbf{v}(m+1)]^T$			μ_{m+1}
0	0.804999	−0.950009	1.000000	−0.950009
1	0.991716	−0.998225	1.000000	−0.998225
2	0.999710	−0.999941	1.000000	−0.999941
3	0.999990	−0.999998	1.000000	−0.999998
4	1.000000	−1.000000	1.000000	−1.000000
5	1.000000	−1.000000	1.000000	−1.000000

EXERCISE 8.7

Use the inverse power method to determine the smallest eigenvalue of the matrix **A** given in Exercise 8.5, without using any shift of the origin. Use $\mathbf{v}(0) = [1 \quad 1 \quad 1 \quad 1]^T$ as the initial vector.

Answer $\lambda_4 = 1.0$, $\mathbf{v}_4 = [1 \quad -1 \quad 0 \quad 0]^T$, requiring 26 iterations because $\mathbf{v}(0)$ is orthogonal to $\mathbf{v}_4$.

EXERCISE 8.8

Repeat Exercise 8.7 with the initial vector $\mathbf{v}(0) = [1 \quad 0 \quad 0 \quad 0]^T$.

Answer Same as for Exercise 8.7, with only 9 iterations.

EXERCISE 8.9

Repeat Exercise 8.7 with a shift in origin, $s = 0.9$.

Answer Same as for Exercise 8.7, with only 15 iterations.

EXERCISE 8.10

Repeat Exercise 8.8 with a shift in origin, $s = 0.9$.

Answer Same as for Exercise 8.7, with only 5 iterations.

8.3.3 Deflation

After obtaining a good estimate of the dominant eigenvalue and a corresponding eigenvector of a matrix, it may be of interest to determine other eigenvalues and eigenvectors. The power method can be utilized for this purpose after the matrix is suitably *deflated*. This is based on the following theorem.

Theorem 8.9

Let λ_1 be an eigenvalue of the $n \times n$ matrix $\mathbf{A}$ and $\mathbf{v}_1$ be a corresponding eigenvector. Let the other eigenvalues of $\mathbf{A}$ be $\lambda_2, \lambda_3, \ldots, \lambda_n$ (with $\lambda_i \neq \lambda_1$, $i \geq 2$) and let $\mathbf{v}_2, \mathbf{v}_3, \ldots, \mathbf{v}_n$ be the corresponding eigenvectors. Let $\mathbf{z}$ be any vector such that $\mathbf{z}^T\mathbf{v}_1 = 1$. Then the matrix

$$\mathbf{B} = \mathbf{A} - \lambda_1 \mathbf{v}_1 \mathbf{z}^T \tag{8.3–7}$$

has eigenvalues $0, \lambda_2, \lambda_3, \ldots, \lambda_n$ and associated eigenvectors $\mathbf{v}_1, \mathbf{w}_2, \mathbf{w}_3, \mathbf{w}_4, \ldots, \mathbf{w}_n$, where $\mathbf{v}_i$ and $\mathbf{w}_i$ are related through the equation

$$\mathbf{v}_i = (\lambda_i - \lambda_1)\mathbf{w}_i + \lambda_1(\mathbf{z}^T\mathbf{w}_i)\mathbf{v}_1 \tag{8.3–8}$$

for $i = 2, 3, \ldots, n$.

Proof

First we shall prove that $\mathbf{B}$ has an eigenvalue of zero with $\mathbf{v}_1$ as an associated eigenvector. This follows by postmultiplying both sides of (8.3–7) by $\mathbf{v}_1$ and noting that the right-hand side is reduced to zero in view of the equalities $\mathbf{A}\mathbf{v}_1 = \lambda_1\mathbf{v}_1$ and $\mathbf{z}^T\mathbf{v}_1 = 1$. Hence, zero must be an eigenvalue of $\mathbf{B}$ and $\mathbf{v}_1$ a corresponding eigenvector. To prove the rest of the theorem, we shall postmultiply both sides of equation (8.3–7) by $\mathbf{w}_i$ and use (8.3–8) to obtain

$$\begin{aligned}
\mathbf{B}\mathbf{w}_i &= \mathbf{A}\mathbf{w}_i - \lambda_1\mathbf{v}_1\mathbf{z}^T\mathbf{w}_i \\
&= \mathbf{A}\frac{1}{\lambda_i - \lambda_1}(\mathbf{v}_i - \lambda_1\mathbf{z}^T\mathbf{w}_i\mathbf{v}_1) - \lambda_1\mathbf{v}_1\mathbf{z}^T\mathbf{w}_i \\
&= \frac{1}{\lambda_i - \lambda_1}[\mathbf{A}\mathbf{v}_i - \lambda_1(\mathbf{z}^T\mathbf{w}_i)\mathbf{A}\mathbf{v}_1] - \lambda_1\mathbf{v}_1\mathbf{z}^T\mathbf{w}_i \\
&= \frac{1}{\lambda_i - \lambda_1}[\lambda_i\mathbf{v}_i - \lambda_1(\mathbf{z}^T\mathbf{w}_i)\lambda_1\mathbf{v}_1] - \lambda_1\mathbf{v}_1\mathbf{z}^T\mathbf{w}_i \\
&= \frac{\lambda_i\mathbf{v}_i}{\lambda_i - \lambda_1} - \lambda_1(\mathbf{z}^T\mathbf{w}_i)\mathbf{v}_1\left[\frac{\lambda_1}{\lambda_i - \lambda_1} + 1\right] \\
&= \frac{\lambda_i}{\lambda_i - \lambda_1}[\mathbf{v}_i - \lambda_1(\mathbf{z}^T\mathbf{w}_i)\mathbf{v}_1] \\
&= \lambda_i\mathbf{w}_i
\end{aligned}$$

This proves that λ_i is an eigenvalue of $\mathbf{B}$ and $\mathbf{w}_i$ is a corresponding eigenvector.

Wielandt's deflation method, which is based on Theorem 8.9, will now be described. The vector $\mathbf{z}$ is defined as

$$\mathbf{z} = \frac{[a_{r1} \quad a_{r2} \ldots a_{rn}]^T}{\lambda_1 v_{1r}}$$

where v_{1r} is the rth component of the vector $\mathbf{v}_1$; v_{1r} must be nonzero. For better numerical accuracy, v_{1r} should be the element of $\mathbf{v}_1$ that has the largest absolute value. The main advantage of this choice is that the rth row of the resulting matrix $\mathbf{B}$ consists entirely of zeros. As a result, the rth component of $\mathbf{w}_i$ must also be zero. Therefore, we can omit the rth row and column of $\mathbf{B}$ to obtain the matrix $\tilde{\mathbf{B}}$, which has the eigenvalues $\lambda_2, \lambda_3, \ldots, \lambda_n$. From an eigenvector $\tilde{\mathbf{w}}_i$ of $\tilde{\mathbf{B}}$, we can obtain $\mathbf{w}_i$ by simply adding a zero rth row. The eigenvector $\mathbf{v}_i$ can be calculated by using (8.3–8).

EXAMPLE 8.5

In Example 8.2, it was found that $\mathbf{A}$ had an eigenvalue $\lambda_1 = -10$ and a corresponding eigenvector was found to be $\mathbf{v}_1 = [1.0 \quad -0.1 \quad 0.01]^T$. Using Wielandt's method, we have

$$\mathbf{z} = \frac{[-15 \quad -54 \quad -40]^T}{-10 \times 1} = [1.5 \qquad 5.4 \qquad 4.0]^T$$

It can be verified that $\mathbf{z}^T\mathbf{v}_1 = 1$. We now obtain

$$\mathbf{B} = \mathbf{A} - \lambda_1 \mathbf{v}_1 \mathbf{z}^T = \begin{bmatrix} 0.0 & 0.0 & 0.0 \\ -0.5 & -5.4 & -4.0 \\ 0.15 & 1.54 & 0.4 \end{bmatrix}$$

We must now find the eigenvalues of the 2×2 matrix $\tilde{\mathbf{B}}$, given by

$$\tilde{\mathbf{B}} = \begin{bmatrix} -5.4 & -4.0 \\ 1.54 & 0.4 \end{bmatrix}$$

Using the power method, we obtain $\lambda_2 = -4.0$ and $\tilde{\mathbf{w}}_2 = [1 \quad -0.35]^T$, so that

$$\mathbf{w}_2 = [0 \qquad 1 \qquad -0.35]^T$$

and

$$\mathbf{v}_2 = (\lambda_2 - \lambda_1)\mathbf{w}_2 + \lambda_1(\mathbf{z}^T\mathbf{w}_2)\mathbf{v}_1 = [-40 \qquad 10 \qquad -2.5]^T$$

Although it is possible to find all real eigenvalues by repeated use of the power (or inverse power) method together with deflation, it is not very suitable due to the cumulative effects of roundoff errors that may be caused during the deflation process. Nevertheless, these iterative methods are specially suitable for finding the dominant (or the smallest modulus) eigenvalue of a real symmetric matrix. In the next section we shall discuss direct methods for calculating all the eigenvalues of a matrix.

EXERCISE 8.11

Consider the matrix **A** in Exercise 8.5. Assuming that the values of λ_1 and $\mathbf{v}_1$ are known, as given in the answer to that problem, use deflation to determine the next largest eigenvalue, λ_2, and the corresponding eigenvector.

Answer $\lambda_2 = 9.0$, $\mathbf{v}_2 = [1 \quad 1 \quad -2 \quad -2]^T$.

8.4 DIRECT METHODS

As discussed in Section 8.2, all the eigenvalues of the $n \times n$ matrix **A** are the roots of the characteristic polynomial

$$P(\lambda) = \det(\lambda\mathbf{I} - \mathbf{A}) \tag{8.4–1}$$

To use (8.4–1) involves (1) determination of the characteristic polynomial of the matrix and (2) computation of the roots of the characteristic polynomial. We have studied the latter problem in Chapter 3. Except for the case when the degree of the polynomial was four or less, the roots have to be found by using search or iterative techniques. Both the Laguerre and the Jenkins-Traub methods are excellent procedures for finding the roots of high-order polynomials (see Chapter 3). It was seen, however, that in many cases the roots are very sensitive to small variations in the values of the coefficients of the polynomial. This means that it is necessary to calculate the coefficients of the characteristic polynomial with as high precision as possible. In this section we shall discuss two popular methods for finding the characteristic polynomial which are suitable for use on computers.

8.4.1 Leverrier's Method

The characteristic polynomial of an $n \times n$ matrix will be a polynomial of degree n in λ. Hence, it may be written as

$$P(\lambda) = \det(\lambda\mathbf{I} - \mathbf{A}) = \lambda^n - c_1\lambda^{n-1} - c_2\lambda^{n-2} - \dots - c_{n-1}\lambda - c_n \tag{8.4–2}$$

We shall now present Leverrier's method, as modified by Fadeev, to find the coefficients, c_i, of the polynomial. Although it requires a large amount of computation, it is fairly insensitive to the individual peculiarities of the matrix **A**. The method has an added advantage that the inverse of **A**, if it exists, is also obtained in the process of determining the coefficients, c_i. One may then utilize both **A** and $\mathbf{A}^{-1}$ for checking the accuracy of individual eigenvalues using the power or the inverse power method. We shall start with the definition of the trace of a matrix.

Definition 8.6

The *trace* of an $n \times n$ matrix $\mathbf{A}$, denoted as $tr(\mathbf{A})$, is the sum of the elements on the main diagonal of $\mathbf{A}$,

$$tr(\mathbf{A}) = \sum_{i=1}^{n} a_{ii}$$

The trace is related to the eigenvalues of $\mathbf{A}$ by

Corollary 8.6

If $\mathbf{A}$ is an $n \times n$ matrix with eigenvalues $\lambda_1, \lambda_2, \ldots, \lambda_n$ then

$$tr(\mathbf{A}) = \sum_{i=1}^{n} \lambda_i$$

The Leverrier-Fadeev method determines the coefficients, c_i, of $P(\lambda)$ by obtaining the trace of each of the matrices, $\mathbf{B}_1, \mathbf{B}_2, \ldots, \mathbf{B}_n$, generated as follows. Set $\mathbf{B}_1 = \mathbf{A}$ and compute $c_1 = tr(\mathbf{B}_1)$. Then we compute

$$\left.\begin{aligned} \mathbf{B}_k &= \mathbf{A}(\mathbf{B}_{k-1} - c_{k-1}\mathbf{I}) \\ c_k &= \left(\frac{1}{k}\right) tr(\mathbf{B}_k) \\ k &= 2, 3, \ldots, n \end{aligned}\right\} \qquad \textbf{(8.4–3)}$$

Fadeev has also shown that the inverse of a nonsingular matrix $\mathbf{A}$ can be obtained from the relationship

$$\mathbf{A}^{-1} = \left(\frac{1}{c_n}\right)(\mathbf{B}_{n-1} - c_{n-1}\mathbf{I})$$

The following example illustrates the procedure.

EXAMPLE 8.6

We use Leverrier's algorithm to determine the characteristic polynomial and the inverse of the matrix

$$\mathbf{A} = \begin{bmatrix} 5 & 4 & 1 & 1 \\ 4 & 5 & 1 & 1 \\ 1 & 1 & 4 & 2 \\ 1 & 1 & 2 & 4 \end{bmatrix}$$

Since $\mathbf{B}_1 = \mathbf{A}$, we have $c_1 = tr(\mathbf{B}_1) = 18$, so that

$$\mathbf{B}_2 = \mathbf{A}(\mathbf{B}_1 - c_1\mathbf{I}) = \begin{bmatrix} -47 & -30 & -3 & -3 \\ -30 & -47 & -3 & -3 \\ -3 & -3 & -50 & -18 \\ -3 & -3 & -18 & -50 \end{bmatrix}, \qquad c_2 = -97$$

EXAMPLE 8.6 (cont'd)

$$\mathbf{B}_3 = \mathbf{A}(\mathbf{B}_2 - c_2\mathbf{I}) = \begin{bmatrix} 124 & 44 & 2 & 2 \\ 44 & 124 & 2 & 2 \\ 2 & 2 & 146 & 16 \\ 2 & 2 & 16 & 146 \end{bmatrix}, \qquad c_3 = 180$$

$$\mathbf{B}_4 = \mathbf{A}(\mathbf{B}_3 - c_3\mathbf{I}) = \begin{bmatrix} -100 & 0 & 0 & 0 \\ 0 & -100 & 0 & 0 \\ 0 & 0 & -100 & 0 \\ 0 & 0 & 0 & -100 \end{bmatrix}, \qquad c_4 = -100$$

Hence, $P(\lambda) = \lambda^4 - 18\lambda^3 + 97\lambda^2 - 180\lambda + 100$. Also, the inverse of **A** is obtained as

$$\mathbf{A}^{-1} = \left(\frac{1}{c_4}\right)[\mathbf{B}_3 - c_3\mathbf{I}] = \begin{bmatrix} 0.56 & -0.44 & -0.02 & -0.02 \\ -0.44 & 0.56 & -0.02 & -0.02 \\ -0.02 & -0.02 & 0.34 & -0.16 \\ -0.02 & -0.02 & -0.16 & 0.34 \end{bmatrix}$$

A computer program for finding the characteristic polynomial of a matrix using Leverrier's algorithm is given in Appendix 8.2. It also has the provision for obtaining the inverse of the matrix.

EXERCISE 8.12

Use the program in Appendix 8.2 to determine the characteristic polynomial of the matrix **A** in Exercise 8.5.

Answer $\lambda^4 - 32\lambda^3 + 286\lambda^2 - 768\lambda + 513$.

The following theorem can be of help in checking the accuracy of the coefficients of a characteristic polynomial, if they were determined numerically by Leverrier's method, or the method to be discussed in the next section. In addition, the theorem is useful if large integer powers of a matrix are to be computed and the coefficients of the characteristic polynomial are known.

Theorem 8.10 (Cayley-Hamilton theorem)

Let **A** be an $n \times n$ matrix whose characteristic polynomial is $P(\lambda)$. Then $P(\mathbf{A}) = \mathbf{0}$.

The Cayley-Hamilton theorem says that **A** satisfies its own characteristic equation. If $P(\lambda)$ is given by (8.2–5), then we have

$$\mathbf{A}^n = \sum_{k=1}^{n-1} c_{n-k}\mathbf{A}^k + c_n\mathbf{I}$$

8.4.2 Danilevsky's Method

This is a very elegant and efficient method for obtaining the characteristic polynomial of a matrix, requiring much less computation than Leverrier's method. The basic idea is to perform a sequence of similarity transformations to the matrix **A**, so that it is reduced to the Frobenius normal form defined in Section 8.2 (see Definition 8.5). As stated in Theorem 8.8, the characteristic polynomial is then obtained immediately from the topmost row of the resulting matrix. The procedure requires $(n - 1)$ similarity transformations, but these can be carried out without matrix multiplications. As a result it requires only $(n - 1)(n^2 + n - 1)$ multiplications and divisions as compared with $(n - 1)n^3$ multiplications for Leverrier's method. We shall now study the procedure for transforming **A** to Frobenius normal form.

Starting with the last row of the matrix **A**, which is

$$a_{n1} \quad a_{n2} \quad a_{n3} \quad \cdots \quad a_{n,n-1} \quad a_{nn}$$

we would like to change it to

$$0 \quad 0 \quad 0 \quad \cdots \quad 1 \quad 0$$

We shall assume that $a_{n,n-1}$ is not zero. The special case when $a_{n,n-1}$ is zero will be considered later. First, we divide all elements of column $n - 1$ by $a_{n,n-1}$. This is followed by subtracting, from the rest of the columns, this $(n - 1)$th column, multiplied by $a_{n1}, a_{n2}, \ldots, a_{nn}$, respectively. The result of these operations will be that the nth row of the resulting matrix will be transformed to the desired form. This entire procedure is equivalent to postmultiplying **A** by the matrix

$$[\mathbf{P}(n - 1)]^{-1} = \begin{bmatrix} 1 & 0 & \cdots & 0 & 0 \\ 0 & 1 & \cdots & 0 & 0 \\ \vdots & \vdots & & \vdots & \vdots \\ \dfrac{-a_{n1}}{a_{n,n-1}} & \dfrac{-a_{n2}}{a_{n,n-1}} & \cdots & \dfrac{1}{a_{n,n-1}} & \dfrac{-a_{nn}}{a_{n,n-1}} \\ 0 & 0 & \cdots & 0 & 1 \end{bmatrix}$$

In order that the matrix $\mathbf{A}[\mathbf{P}(n - 1)]^{-1}$ may have the same eigenvalues as **A**, we must premultiply it by $\mathbf{P}(n - 1)$. It is easily shown that

$$\mathbf{P}(n - 1) = \begin{bmatrix} 1 & 0 & \cdots & 0 & 0 \\ 0 & 1 & \cdots & 0 & 0 \\ \vdots & \vdots & & \vdots & \vdots \\ a_{n1} & a_{n2} & \cdots & a_{n,n-1} & a_{nn} \\ 0 & 0 & \cdots & 0 & 1 \end{bmatrix}$$

This premultiplication does not affect the last row, so that the resulting matrix $[\mathbf{P}(n-1)]\mathbf{A}[\mathbf{P}(n-1)]^{-1}$ has its last row of the desired form. It may be noted that we do not have to do the matrix multiplication, as only the elements in the $(n-1)$th row are changed; the element in the kth column of the nth row can be found as the scalar product of vectors formed from the nth row of **A** and the kth column of $\mathbf{A}[\mathbf{P}(n-1)]^{-1}$.

The next step in the algorithm is to change the $(n-1)$th row to the desired form by using the element in the $(n-1)$th row and the $(n-2)$th column as the pivot, provided the element is nonzero. This process is continued, moving up the rows, until finally, the second row of the transformed matrix is of the form $[1 \;\; 0 \;\; 0 \;\; \ldots \;\; 0]$. The following example will help in understanding the procedure.

EXAMPLE 8.7

Consider the matrix

$$\mathbf{A} = \begin{bmatrix} 3 & 2 & 1 \\ 2 & 1 & 1 \\ 6 & 2 & 2 \end{bmatrix}$$

To transform **A** to Frobenius normal form, we start with the element in the third row and second column. First we store the third row for later use. Since $a_{32} = 2$, we divide each element of the second column by 2. This is followed by subtracting 6 times the resulting second column from the first column and 2 times the same resultant second column from the third. The result is

$$\mathbf{A}[\mathbf{P}(2)]^{-1} = \begin{bmatrix} -3 & 1 & -1 \\ -1 & 1/2 & 0 \\ 0 & 1 & 0 \end{bmatrix}$$

The next step of premultiplication by $\mathbf{P}(2)$ is taken into account by replacing elements in the second row by the scalar products of the third row of **A** with each of the corresponding columns of the matrix $\mathbf{A}[\mathbf{P}(2)]^{-1}$. The result is

$$\mathbf{A}(2) = \mathbf{P}(2)\mathbf{A}[\mathbf{P}(2)]^{-1} = \begin{bmatrix} -3 & 1 & -1 \\ -20 & 9 & -6 \\ 0 & 1 & 0 \end{bmatrix}$$

The matrix $\mathbf{A}(2)$ has the same eigenvalues as **A** and has its third row in the desired form. Now we have to transform the second row. Again, we first store this row. This time we take the element in the first column of the second row and divide every element in the first column by this element, $a_{21} = -20$. Now subtract 9 times the resulting first column from the second and add 6 times the resultant first column to the third, thereby obtaining

$$\mathbf{A}(2)[\mathbf{P}(1)]^{-1} = \begin{bmatrix} 3/20 & -7/20 & -1/10 \\ 1 & 0 & 0 \\ 0 & 1 & 0 \end{bmatrix}$$

EXAMPLE 8.7 (cont'd)

Finally, to make $\mathbf{A}(2)[\mathbf{P}(1)]^{-1}$ similar to $\mathbf{A}$, we must premultiply by $\mathbf{P}(1)$. This is done by replacing elements in the first row by the scalar products of the second row of $\mathbf{A}(2)$ with each of the corresponding columns of the matrix $\mathbf{A}(2)[\mathbf{P}(1)]^{-1}$, giving

$$\mathbf{A}(3) = \mathbf{P}(1)\mathbf{A}(2)[\mathbf{P}(1)]^{-1} = \begin{bmatrix} 6 & 1 & 2 \\ 1 & 0 & 0 \\ 0 & 1 & 0 \end{bmatrix}$$

Since the matrix $\mathbf{A}(3)$ is similar to $\mathbf{A}$ and in the Frobenius form (8.2–4), we obtain the characteristic polynomial of $\mathbf{A}$, $P(\lambda) = \lambda^3 - 6\lambda^2 - \lambda - 2$.

So far we have assumed that the pivot element $a_{k,k-1}$ is nonzero at each step of the transformation. We shall now discuss the situation when this is not so. If $a_{k,k-1} = 0$ for some k, we can interchange the $(k - 1)$th column of $\mathbf{A}$ with another column, say the jth, where $j < (k - 1)$, so that a_{kj} is nonzero. It can be shown that if this interchange is followed by the interchange of the jth row and the $(k - 1)$th row, the eigenvalues of the matrix are unchanged (see Exercise 8.4). In fact, like pivoting in Gaussian elimination, the accuracy of the numerical computations can be improved if the column interchange is carried out to get the largest magnitude element in the kth row, to the left of the $(k - 1)$th column, into the $(k - 1)$th column.

Sometimes we may also run into the situation where $a_{k,k-1}$ is zero and there are only zero elements to its left in the kth row. In this case we are not able to continue any further. However, this implies that the matrix $\mathbf{A}$ has been reduced by similarity transformations to the form

$$\mathbf{A}(n - k) = \begin{bmatrix} \mathbf{A}_{11} & \mathbf{A}_{12} \\ \mathbf{0} & \mathbf{A}_{22} \end{bmatrix}$$

$\mathbf{A}(n - k)$ is seen to be a block triangular matrix. Hence, it follows that the characteristic polynomial of $\mathbf{A}$ is the product of the characteristic polynomials of $\mathbf{A}_{11}$ and $\mathbf{A}_{22}$. Moreover, $\mathbf{A}_{22}$ is already in Frobenius form so that its characteristic polynomial can be obtained by inspection. Thus, our problem is reduced to transforming $\mathbf{A}_{11}$ to Frobenius form by following the same procedure as described earlier in this section.

EXAMPLE 8.8

This example illustrates both aspects discussed above. We consider

$$\mathbf{A} = \begin{bmatrix} 3 & 2 & 4 \\ 2 & 0 & 2 \\ 4 & 2 & 3 \end{bmatrix}$$

EXAMPLE 8.8 (cont'd)

Since the element in the second column of the third row is not the element with largest absolute value in that row, we shall interchange the second column with the first and follow it by interchanging the first and the second rows of the resulting matrix. This leads to

$$\mathbf{A}(1) = \begin{bmatrix} 0 & 2 & 2 \\ 2 & 3 & 4 \\ 2 & 4 & 3 \end{bmatrix}$$

Now that we have done our pivoting, we shall divide the second column of $\mathbf{A}(1)$ by 4 and then subtract 2 times the resulting second column from the first column and 3 times the resultant second column from the third, giving

$$\mathbf{A}(1)[\mathbf{P}(2)]^{-1} = \begin{bmatrix} -1 & 1/2 & 1/2 \\ 1/2 & 3/4 & 7/4 \\ 0 & 1 & 0 \end{bmatrix}$$

Next, we replace the elements in the second row of $\mathbf{A}(1)[\mathbf{P}(2)]^{-1}$ with the scalar product of the third row of $\mathbf{A}(1)$ and the corresponding columns of $\mathbf{A}(1)[\mathbf{P}(2)]^{-1}$, obtaining

$$\mathbf{A}(2) = \mathbf{P}(2)\mathbf{A}(1)[\mathbf{P}(2)]^{-1} = \begin{bmatrix} -1 & 1/2 & 1/2 \\ 0 & 7 & 8 \\ 0 & 1 & 0 \end{bmatrix}$$

We now observe that we cannot proceed any further since the element in the first column and second row of $\mathbf{A}(2)$ is zero. Hence, it is not possible to transform this matrix into Frobenius form. However, noting the block triangular structure of $\mathbf{A}(2)$, we can obtain the characteristic polynomial as $P(\lambda) = (\lambda + 1)(\lambda^2 - 7\lambda - 8)$. Because we are interested in obtaining the eigenvalues of $\mathbf{A}$, we would not multiply out the factors in $P(\lambda)$. Rather, we find the roots of each factor separately. This shows that if we obtain a block upper triangular form, then the characteristic polynomial of $\mathbf{A}$ is obtained as the product of two lower degree polynomials.

A computer program for finding the characteristic polynomial of a matrix using Danilevsky's method is in Appendix 8.2.

EXERCISE 8.13

Repeat Exercise 8.12 using the computer program from Appendix 8.2.

Danilevsky's method also provides a simple method for computing the set of eigenvectors of a matrix after the eigenvalues have been obtained as the roots of the characteristic polynomial. As shown in Corollary 8.4, that follows Theorem 8.8, the elements of the eigenvectors of a matrix in the Frobenius form (8.2–4) are obtained from the various powers of the eigenvalues. We can, therefore, invoke Theorem 8.5 to obtain the eigenvectors of the original matrix if we keep track of the transformations. This is not a difficult problem since the matrix for premultiplication in each of the transformations differs from the identity matrix in only one row. Hence, if we store these rows, we can reconstruct the complete matrix that was used to transform the given matrix to the Frobenius form. This will be illustrated through the following example.

EXAMPLE 8.9

Consider the matrix

$$\mathbf{A} = \begin{bmatrix} -9 & -38 & -42 \\ 2 & -2 & -7 \\ -1 & 2 & 5 \end{bmatrix}$$

Here, the first matrix for premultiplication is

$$\mathbf{P}(2) = \begin{bmatrix} 1 & 0 & 0 \\ -1 & 2 & 5 \\ 0 & 0 & 1 \end{bmatrix}$$

Carrying out the first transformation we get

$$\mathbf{A}(2) = \begin{bmatrix} -28 & -19 & 53 \\ 30 & 22 & -57 \\ 0 & 1 & 0 \end{bmatrix}$$

The second matrix for transformation is now obtained as

$$\mathbf{P}(1) = \begin{bmatrix} 30 & 22 & -57 \\ 0 & 1 & 0 \\ 0 & 0 & 1 \end{bmatrix}$$

and the final matrix, in Frobenius form, is

$$\mathbf{A}(3) = \begin{bmatrix} -6 & -11 & -6 \\ 1 & 0 & 0 \\ 0 & 1 & 0 \end{bmatrix}$$

The characteristic polynomial is, therefore, $P(\lambda) = \lambda^3 + 6\lambda^2 + 11\lambda + 6$ so that the eigenvalues are determined as $\lambda_1 = -1$, $\lambda_2 = -2$ and $\lambda_3 = -3$. The eigenvectors of **A**(3), corresponding to these eigenvalues are

$$\mathbf{v}_1 = \begin{bmatrix} 1 \\ -1 \\ 1 \end{bmatrix}, \quad \mathbf{v}_2 = \begin{bmatrix} 4 \\ -2 \\ 1 \end{bmatrix} \quad \text{and} \quad \mathbf{v}_3 = \begin{bmatrix} 9 \\ -3 \\ 1 \end{bmatrix}$$

EXAMPLE 8.9 (cont'd)

Because $\mathbf{A}(3) = \mathbf{P}(1)\mathbf{P}(2)\mathbf{A}[\mathbf{P}(2)]^{-1}[\mathbf{P}(1)]^{-1}$ it follows that the corresponding eigenvectors of $\mathbf{A}$, denoted by $\mathbf{w}_1$, $\mathbf{w}_2$, $\mathbf{w}_3$ are given by $\mathbf{P}(1)\mathbf{P}(2)\mathbf{w}_i = \mathbf{v}_i$, for $i = 1$, 2, and 3. It should noted that this equation for $\mathbf{w}_i$ does not require matrix inversion, since each of the transformations changes only one row of the vector. For example, to solve for $\mathbf{w}_1$, we first let $\mathbf{P}(2)\mathbf{w}_1 = \mathbf{z}_1$ and solve the equation $\mathbf{P}(1)\mathbf{z}_1 = \mathbf{v}_1$. Using the values of $\mathbf{P}(1)$ and $\mathbf{v}_1$ we readily obtain

$$\mathbf{z}_1 = \begin{bmatrix} 8/3 \\ -1 \\ 1 \end{bmatrix}$$

Solving the equation $\mathbf{P}(2)\mathbf{w}_1 = \mathbf{z}_1$ we get

$$\mathbf{w}_1 = \begin{bmatrix} 8/3 \\ -5/3 \\ 1 \end{bmatrix}$$

It can be verified that $\mathbf{A}\mathbf{w}_1 = \lambda_1\mathbf{w}_1$.

8.5 METHODS BASED ON SIMILARITY TRANSFORMATIONS

The methods discussed in the previous section do not always give very accurate results due to roundoff errors introduced during the computation of the coefficients of the characteristic polynomial. The problem is compounded by the fact that in most cases, especially for polynomials of high degree, the roots are very sensitive to small variations in the values of the coefficients. Hence, it is often more desirable to find all the eigenvalues of a matrix simultaneously by transforming it to a form from which the eigenvalues can be obtained by inspection. This is always the case if the matrix is transformed to a diagonal, block diagonal, triangular or block triangular form, with the maximum size of any block not exceeding two. If these similarity transformations are carried out using orthogonal matrices, errors in computation are considerably reduced, as there is no need for computing matrix inverses.

Two main types of orthogonal matrices are commonly utilized for these transformations. The first type is called a *rotation matrix*, as it causes a rotation by the angle θ in the plane formed by the ith and jth coordinate axes. We call this plane the (i, j) *plane*. The rotation matrix for the (i, j) plane, denoted by $\mathbf{P}_{ij}(\theta)$, is the identity matrix except for the following entries in the ith and jth rows,

$$p_{ii} = p_{jj} = \cos(\theta), \qquad p_{ij} = -p_{ji} = -\sin(\theta). \tag{8.5–1}$$

For example, a 4×4 rotation matrix, $\mathbf{P}_{13}(\theta)$, will have the form

$$\mathbf{P}_{13}(\theta) = \begin{bmatrix} \cos(\theta) & 0 & -\sin(\theta) & 0 \\ 0 & 1 & 0 & 0 \\ \sin(\theta) & 0 & \cos(\theta) & 0 \\ 0 & 0 & 0 & 1 \end{bmatrix}$$

It is easily verified that $\mathbf{P}_{ij}(\theta)$ is an orthogonal matrix, and

$$[\mathbf{P}_{ij}(\theta)]^T = \mathbf{P}_{ij}(-\theta) \qquad \textbf{(8.5–2)}$$

The effect of premultiplication of a matrix $\mathbf{A}$ by $\mathbf{P}_{ij}(\theta)$ will be to change only the elements in rows i and j, whereas postmultiplication by $[\mathbf{P}_{ij}(\theta)]^T$, to complete the similarity transformation, will change only the elements in columns i and j.

The other type of orthogonal matrix is a *reflection matrix*. Here, we use the matrix

$$\mathbf{R} = \mathbf{I} - 2\mathbf{w}\mathbf{w}^T \qquad \textbf{(8.5–3)}$$

where $\mathbf{w}$ is any vector with unit Euclidean norm, $\mathbf{w}^T\mathbf{w} = 1$. It will be seen that $\mathbf{R}$ is a symmetric matrix and that

$$\mathbf{R}^2 = (\mathbf{I} - 2\mathbf{w}\mathbf{w}^T)(\mathbf{I} - 2\mathbf{w}\mathbf{w}^T) = \mathbf{I} - 4\mathbf{w}\mathbf{w}^T + 4\mathbf{w}\mathbf{w}^T\mathbf{w}\mathbf{w}^T = \mathbf{I}$$

If we premultiply $\mathbf{A}$ by $\mathbf{R}$, each column of the product $\tilde{\mathbf{A}}$ will be changed and will have the form

$$\tilde{\mathbf{a}}_k = (\mathbf{I} - 2\mathbf{w}\mathbf{w}^T)\mathbf{a}_k = \mathbf{a}_k - 2(\mathbf{w}^T\mathbf{a}_k)\mathbf{w}$$

where $\mathbf{a}_k$ is the kth column of $\mathbf{A}$ and $\tilde{\mathbf{a}}_k$ is the kth column of $\tilde{\mathbf{A}}$. Similarly, postmultiplication by $\mathbf{R}$ changes each row of the product.

8.5.1 Jacobi's Method for Symmetric Matrices

This method, proposed by Jacobi in 1846, is based on performing a sequence of similarity transformations, on a *symmetric* matrix $\mathbf{A}$, so that it tends to a diagonal matrix. These transformations, plane rotations of the type described in (8.5–1), attempt to reduce the largest-magnitude off-diagonal element to zero. If this element is a_{ij} then the transformation matrix is given by $\mathbf{P}_{ij}(\theta)$, where the angle θ satisfies

$$\tan(2\theta) = \frac{2a_{ij}}{a_{ii} - a_{jj}} \qquad \textbf{(8.5–4)}$$

This follows from the fact that the element in the ith row and jth column of the matrix $\tilde{\mathbf{A}} = [\mathbf{P}_{ij}(-\theta)]\mathbf{A}[\mathbf{P}_{ij}(\theta)]$ is given by

$$\tilde{a}_{ij} = (a_{jj} - a_{ii})\cos(\theta)\sin(\theta) + a_{ij}(\cos^2(\theta) - \sin^2(\theta))$$

$\tilde{a}_{ij}$ will be zero only if equation (8.5–4) is satisfied. This transformation changes only the elements in rows i and j and columns i and j. Since $\tilde{\mathbf{A}}$ is also symmetric, it is necessary only to calculate the elements in the upper triangle of the affected rows and columns. There is an efficient technique for obtaining the transformation matrix which does not require the evaluation of θ explicitly. If we let $t = \tan(\theta)$, then

$$\tan(2\theta) = \frac{2t}{1 - t^2} \tag{8.5–5}$$

Setting $a = \cot(2\theta)$, equation (8.5–5) can be solved for t by obtaining the roots of the quadratic $t^2 + 2at - 1 = 0$. The root closer to zero is selected in order that θ may lie between $-\pi/4$ and $\pi/4$. Then it follows that

$$\cos(\theta) = c = \frac{1}{(1 + t^2)^{1/2}}$$

and $\sin(\theta) = s = ct$. The procedure is continued until all off-diagonal elements are reduced to zero. It should be noted that any transformation is likely to change the values of the off-diagonal elements that were reduced to zero by earlier transformations. Hence, several iterations are required and the process may be very time-consuming for some cases.

EXAMPLE 8.10

We want to apply Jacobi's method to the matrix

$$\mathbf{A} = \begin{bmatrix} 5 & 4 & 1 \\ 4 & 6 & 2 \\ 1 & 2 & 3 \end{bmatrix}$$

Here, the largest magnitude off-diagonal element is in the second column of the first row (we look at the upper triangular part only). Hence, we obtain $\tan(2\theta) = (2 \times 4)/(5 - 6) = -8$. Using the procedure following equation (8.5–5), we get $\tan(\theta) = -0.8827822$, $c = 0.7496782$ and $s = -0.6618026$. As a result, the transformation matrix

$$\mathbf{P} = \begin{bmatrix} 0.7496782 & 0.6618026 & 0 \\ -0.6618026 & 0.7496782 & 0 \\ 0 & 0 & 1 \end{bmatrix}$$

is obtained. Using the transformation $\mathbf{A}(1) = \mathbf{P}^T\mathbf{A}\mathbf{P}$, we obtain

$$\mathbf{A}(1) = \begin{bmatrix} 1.4688711 & 0.0000000 & -0.5739270 \\ 0.0000000 & 9.5311289 & 2.1611589 \\ -0.5739270 & 2.1611589 & 3.0000000 \end{bmatrix}$$

The largest magnitude off-diagonal element of $\mathbf{A}(1)$ is in the third column of the second row. Proceeding as before, we get the matrix

$$\mathbf{A}(2) = \begin{bmatrix} 1.4688711 & -0.1653877 & -0.5495808 \\ -0.1653877 & 10.1814959 & 0.0000000 \\ -0.5495808 & 0.0000000 & 2.3496330 \end{bmatrix}$$

EXAMPLE 8.10 (cont'd)

It will be seen that in this process we have made the element in the second column of the first row nonzero, but it has a smaller magnitude than the corresponding entry in **A**. The next step is to reduce the element in the third column of the first row as it is the largest magnitude off-diagonal element in **A**(2). The result is

$$\mathbf{A}(3) = \begin{bmatrix} 1.2049976 & -0.1490929 & 0.0000000 \\ -0.1490929 & 10.1814959 & 0.0715849 \\ 0.0000000 & 0.0715849 & 2.6135066 \end{bmatrix}$$

The largest magnitude off-diagonal element in **A**(3) is $a_{12}(3)$. The next transformation yields

$$\mathbf{A}(4) = \begin{bmatrix} 1.2025219 & 0.0000000 & 0.0011885 \\ 0.0000000 & 10.1839715 & 0.0715750 \\ 0.0011885 & 0.0715750 & 2.6135066 \end{bmatrix}$$

We can see that the magnitudes of the off-diagonal elements getting smaller, but we need to do more iterations. The transformation to reduce $a_{23}(4)$ to zero leads to

$$\mathbf{A}(5) = \begin{bmatrix} 1.2025219 & 0.0000112 & 0.0011884 \\ 0.0000112 & 10.1846482 & 0.0000000 \\ 0.0011884 & 0.0000000 & 2.6128299 \end{bmatrix}$$

We now reduce the element $a_{13}(5)$ to zero, giving

$$\mathbf{A}(6) = \begin{bmatrix} 1.2025209 & 0.0000112 & 0.0000000 \\ 0.0000112 & 10.1846482 & 0.0000000 \\ 0.0000000 & 0.0000000 & 2.6128309 \end{bmatrix}$$

which is now almost diagonal except for $a_{12}(6)$. The result of the next transformation is

$$\mathbf{A}(7) = \begin{bmatrix} 1.2025209 & 0.0000000 & 0.0000000 \\ 0.0000000 & 10.1846482 & 0.0000000 \\ 0.0000000 & 0.0000000 & 2.6128309 \end{bmatrix}$$

As **A**(7) is diagonal to seven figures after the decimal point, we can stop and conclude that the eigenvalues of matrix **A** are 1.2025209, 10.1846482, and 2.6128309.

If the algorithm succeeds in transforming **A** to diagonal form, all the eigenvalues are obtained as the elements on the main diagonal. We may also easily determine all the eigenvectors as they are the columns of the matrix that is obtained by multiplying all the transformation matrices; that is, from the product $\mathbf{P} = \mathbf{P}_1(\theta_1)\mathbf{P}_2(\theta_2) \ldots \mathbf{P}_m(\theta_m)$, where m transformations were required for obtaining the diagonal form and $\mathbf{P}_k(\theta_k)$ is the matrix used for the kth transformation.

A computer program for eliminating off-diagonal elements of a symmetric matrix using Jacobi's method is in Appendix 8.3.

EXERCISE 8.14

Use Jacobi's method to determine the eigenvalues of the matrix $\mathbf{A}$ in Example 8.6. How many transformations are required to get the matrix in the diagonal form?

Answer 1, 2, 5, 10. The number of transformations required was 9.

8.5.2 Givens' Method

This is an attempt to improve the efficiency of Jacobi's method by first transforming a symmetric matrix to tridiagonal form. With such a matrix, further rotations to eliminate off-diagonal terms do not change the values of elements which were annihilated earlier. Hence, the number of iterations required does not exceed $(n - 1)(n - 2)/2$ for an $n \times n$ matrix.

Transformation of a matrix to tridiagonal form is carried out by applying a sequence of rotations in a strict order. Starting with the first column, the elements in positions (3, 1) to $(n, 1)$ are made zero, in that order, by rotations in planes (2, 3) to $(2, n)$. The process is then repeated for columns 2, 3, . . . , $(n - 2)$ in that order. For the jth column one must reduce the elements in the order $(j + 2, j)$, $(j + 3, j)$, . . . , (n, j) by rotation in the planes $(j + 1, j + 2)$, $(j + 1, j + 3)$, $(j + 1, j + 4)$, . . . , $(j + 1, n)$. It can be easily verified that if we proceed in this order then these transformations do not affect elements that were made zero earlier.

The angle θ must be chosen so that a rotation in the plane (i, j) reduces the element $a_{j,i-1}$ to zero. We can verify that the corresponding element of the matrix $\tilde{\mathbf{A}} = [\mathbf{P}_{ij}(-\theta)]\mathbf{A}[\mathbf{P}_{ij}(\theta)]$ is

$$\tilde{a}_{j,i-1} = -a_{i,i-1}\sin(\theta) + a_{j,i-1}\cos(\theta)$$

Hence, $\tan(\theta) = a_{j,i-1}/a_{i,i-1}$.

Appendix 8.3 contains a listing of a computer program for eliminating off-diagonal elements of a symmetric matrix using Givens' method. It was used for the following example.

EXAMPLE 8.11

We want to transform the matrix

$$\mathbf{A} = \begin{bmatrix} 5 & 4 & 1 & 1 \\ 4 & 5 & 1 & 1 \\ 1 & 1 & 8 & 4 \\ 1 & 1 & 4 & 8 \end{bmatrix}$$

to tridiagonal form using Givens' method. We must first eliminate the element a_{31}; this means that $\tan(\theta) = a_{31}/a_{21} = 0.25$ and the resulting transformation matrix is

$$\mathbf{P}_{23}(\theta) = \begin{bmatrix} 1 & 0 & 0 & 0 \\ 0 & \cos(\theta) & -\sin(\theta) & 0 \\ 0 & \sin(\theta) & \cos(\theta) & 0 \\ 0 & 0 & 0 & 1 \end{bmatrix}$$

The transformed matrix $\mathbf{A}(1) = [\mathbf{P}_{23}(-\theta)]\mathbf{A}[\mathbf{P}_{23}(\theta)]$ is

$$\mathbf{A}(1) = \begin{bmatrix} 5.0000000 & 4.1231056 & 0.0000000 & 1.0000000 \\ 4.1231056 & 5.6470588 & 1.5882353 & 1.9402850 \\ 0.0000000 & 1.5882353 & 7.3529412 & 3.6380344 \\ 1.0000000 & 1.9402850 & 3.6380344 & 8.0000000 \end{bmatrix}$$

It will be seen that both $a_{31}(1)$ and $a_{13}(1)$ have been reduced to zero. The next step is to eliminate the element $a_{41}(1)$. Proceeding as before, we obtain the matrix

$$\mathbf{A}(2) = \begin{bmatrix} 5.0000000 & 4.2426407 & 0.0000000 & 0.0000000 \\ 4.2426407 & 6.6666667 & 2.4009802 & 2.2636658 \\ 0.0000000 & 2.4009802 & 7.3529412 & 3.1611833 \\ 0.0000000 & 2.2636658 & 3.1611633 & 6.9803992 \end{bmatrix}$$

The final stage is to eliminate $a_{42}(2)$ giving

$$\mathbf{A}(3) = \begin{bmatrix} 5.0000000 & 4.2426407 & 0.0000000 & 0.0000000 \\ 4.2426407 & 6.6666667 & 3.2998316 & 0.0000000 \\ 0.0000000 & 3.2998316 & 10.3333333 & 0.0000000 \\ 0.0000000 & 0.0000000 & 0.0000000 & 4.0000000 \end{bmatrix}$$

The matrix $\mathbf{A}(3)$ is now tridiagonal. At this stage, we can use Jacobi's method to transform it to diagonal form and determine the eigenvalues.

Givens' method requires a total of approximately $4n^3/3$ multiplications as compared with approximately $2n^3$ multiplications for each cycle of the Jacobi method. Both methods, however, suffer from the disadvantage that square roots have to be evaluated in order to calculate the sines and cosines of the rotation.

These introduce errors in computation. However, it is possible to obtain a form of Givens' method that is square root-free, although we shall not discuss it here.

EXERCISE 8.15

Use Givens' method to transform the matrix

$$\begin{bmatrix} 8 & 6 & 4 & 2 & 1 \\ 6 & 5 & 3 & 1 & 1 \\ 4 & 3 & 7 & 3 & 2 \\ 2 & 1 & 3 & 9 & 1 \\ 1 & 1 & 2 & 1 & 4 \end{bmatrix}$$

to tridiagonal form.

Answer

$$\begin{bmatrix} 8.0000000 & 7.5498344 & 0.0000000 & 0.0000000 & 0.0000000 \\ 7.5498344 & 10.1754386 & 3.7656006 & 0.0000000 & 0.0000000 \\ 0.0000000 & 3.7656006 & 6.7486335 & 1.9692690 & 0.0000000 \\ 0.0000000 & 0.0000000 & 1.9692690 & 5.0970202 & -0.6882416 \\ 0.0000000 & 0.0000000 & 0.0000000 & -0.6882416 & 2.9789077 \end{bmatrix}$$

8.5.3 Householder's Method

This method transforms symmetric matrices to tridiagonal form by annihilating a column of elements in one operation. This is done by using a reflection-type transformation defined in equation (8.5–3). The following theorem will enable us to understand the method.

Theorem 8.11

Let $\mathbf{x}$ and $\mathbf{y}$ be two vectors ($\mathbf{x} \neq \mathbf{y}$) with the property that

$$\mathbf{x}^T\mathbf{x} = \mathbf{y}^T\mathbf{y} \tag{8.5–6}$$

Then there exists a vector $\mathbf{w}$ with the property $\mathbf{w}^T\mathbf{w} = 1$, such that

$$\mathbf{Rx} = \mathbf{y} \tag{8.5–7}$$

where $\mathbf{R} = \mathbf{I} - 2\mathbf{ww}^T$ is both orthogonal and symmetric; hence, $\mathbf{R} = \mathbf{R}^T = \mathbf{R}^{-1}$.

Proof

From equation (8.5–7), we have

$$\mathbf{y} = (\mathbf{I} - 2\mathbf{ww}^T)\mathbf{x} = \mathbf{x} - (2\mathbf{w}^T\mathbf{x})\mathbf{w}$$

so that

$$\mathbf{w} = \frac{\mathbf{x} - \mathbf{y}}{\alpha} \tag{8.5–8}$$

where the scalar α is selected to make $\mathbf{w}^T\mathbf{w} = 1$; thus,

$$\alpha = [(\mathbf{x} - \mathbf{y})^T(\mathbf{x} - \mathbf{y})]^{1/2}.$$

This shows the existence of a unique $\mathbf{w}$ which satisfies (8.5–7).

The Householder procedure for transforming a symmetric matrix to tridiagonal form uses a sequence of reflection matrices, $\mathbf{R}(k)$, in which the vector $\mathbf{w}(k)$ has the form

$$\mathbf{w}(k) = [0 \quad 0 \quad \ldots \quad 0 \quad w_{k+1}(k) \quad \ldots \quad w_n(k)]^T$$

The choice of $\mathbf{w}(k)$ follows from equation (8.5–8). Let the vector

$$\mathbf{x} = [x_1 \quad x_2 \quad \ldots \quad x_k \quad x_{k+1} \quad \ldots \quad x_n]^T$$

represent the kth column of $\mathbf{A}(k) = \mathbf{R}(k)\mathbf{A}(k-1)\mathbf{R}(k)$. Our objective is to select $\mathbf{w}(k)$ in such a way that premultiplication of $\mathbf{A}(k-1)$ by $\mathbf{R}(k)$ reduces the elements $x_{k+2}, x_{k+3}, \ldots, x_n$ to zero. Let

$$\sum_{i=k+1}^{n} x_i^2 = \sigma^2 \tag{8.5–9}$$

Then the vector

$$\mathbf{y} = [x_1 \quad x_2 \quad \ldots \quad x_k \quad \pm\sigma \quad 0 \quad \ldots \quad 0]^T$$

satisfies the condition $\mathbf{x}^T\mathbf{x} = \mathbf{y}^T\mathbf{y}$. Hence,

$$\mathbf{w} = \left(\frac{1}{\alpha}\right) [0 \quad 0 \quad \ldots \quad 0 \quad (x_{k+1} \mp \sigma) \quad x_{k+2} \quad \ldots \quad x_n]^T \tag{8.5–10}$$

where α is selected so that the Euclidean norm of $\mathbf{w}$ is one. To reduce roundoff error, the sign of σ is chosen to be the same as that of x_{k+1}:

$$\alpha^2 = (x_{k+1} + \sigma)^2 + \sum_{i=k+2}^{n} x_i^2 = 2\sigma(\sigma + x_{k+1}) \tag{8.5–11}$$

A computer program for eliminating all the elements below row $(k + 1)$ in column k of a symmetric matrix, using Householder's method, is given in Appendix 8.3.

For a general k, the similarity transformation has the form

$$\begin{aligned}\mathbf{R}(k)\mathbf{A}(k-1)\mathbf{R}(k) &= \{\mathbf{I} - 2\mathbf{w}(k)[\mathbf{w}(k)]^T\}\mathbf{A}(k-1)\{\mathbf{I} - 2\mathbf{w}(k)[\mathbf{w}(k)]^T\} \\ &= \mathbf{A}(k-1) - 2\mathbf{w}(k)[\mathbf{w}(k)]^T\mathbf{A}(k-1) \\ &\quad - 2\mathbf{A}(k-1)\mathbf{w}(k)[\mathbf{w}(k)]^T \\ &\quad + 4\mathbf{w}(k)[\mathbf{w}(k)]^T\mathbf{A}(k-1)\mathbf{w}(k)[\mathbf{w}(k)]^T\end{aligned} \tag{8.5–12}$$

Many elements of $\mathbf{w}$ are zero; we can reduce the number of arithmetic operations by simplifying (8.5–12). Let $\mathbf{v}(k) = \mathbf{A}(k-1)\mathbf{w}(k)$. Then $\mathbf{A}(k-1)\mathbf{R}(k) = \mathbf{A}(k-1) - 2\mathbf{v}(k)[\mathbf{w}(k)]^T$. With $\mathbf{q}(k) = \mathbf{v}(k) - \{[\mathbf{w}(k)]^T\mathbf{v}(k)\}\mathbf{w}(k)$, the transformation (8.5–12) can be written as

$$\mathbf{R}(k)\mathbf{A}(k-1)\mathbf{R}(k) = \mathbf{A}(k-1) - 2\mathbf{w}(k)[\mathbf{q}(k)]^T - 2\mathbf{q}(k)[\mathbf{w}(k)]^T$$

The various steps in this transformation, for each k, are

Step 1. Form the vector $\mathbf{w}(k)$ from the elements $a_{k,k+1}$ to a_{kn}, as indicated in (8.5–10).
Step 2. Update $a_{k,k+1}$ and $a_{k+1,k}$ as $-\sigma$, given by (8.5–9), with the sign of σ selected as that of x_{k+1}.
Step 3. Obtain $\mathbf{v}(k) = \mathbf{A}(k-1)\mathbf{w}(k)$. (Only the elements $k+1$ to n are nonzero.)
Step 4. Calculate the scalar product $u = [\mathbf{w}(k)]^T\mathbf{v}(k)$. (Note that the elements from 1 to k are zero.)
Step 5. Obtain $\mathbf{q}(k) = \mathbf{v}(k) - u\mathbf{w}(k)$.
Step 6. Calculate $\mathbf{A}(k) = \mathbf{A}(k-1) - 2\mathbf{w}(k)[\mathbf{q}(k)]^T - 2\mathbf{q}(k)[\mathbf{w}(k)]^T$.

A computer program using these steps will be more efficient than the program in Appendix 8.3. The interested reader is encouraged to write such a program.

EXAMPLE 8.12

We shall reconsider the matrix used in Example 8.11. In this case, we transform the matrix to tridiagonal form using Householder's method. Starting with the first column, we follow the procedure of equation (8.5–10) to obtain $\mathbf{w}(1) = [0.0000000 \quad 0.9855986 \quad 0.1195732 \quad 0.1195732]^T$. The transformation matrix is then determined from (8.5–3) as

$$\mathbf{R}(1) = \begin{bmatrix} 1.0000000 & 0.0000000 & 0.0000000 & 0.0000000 \\ 0.0000000 & -0.9428090 & -0.2357023 & -0.2357023 \\ 0.0000000 & -0.2357023 & 0.9714045 & -0.0285955 \\ 0.0000000 & -0.2357023 & -0.0285955 & 0.9714045 \end{bmatrix}$$

and the transformed matrix $\mathbf{A}(1) = \mathbf{R}(1)\mathbf{A}\mathbf{R}(1)$ is

$$\mathbf{A}(1) = \begin{bmatrix} 5.0000000 & -4.2426407 & 0.0000000 & 0.0000000 \\ -4.2426407 & 6.6666667 & -2.3333333 & -2.3333333 \\ 0.0000000 & -2.3333333 & 7.1666667 & 3.1666667 \\ 0.0000000 & -2.3333333 & 3.1666667 & 7.1666667 \end{bmatrix}$$

EXAMPLE 8.12 (cont'd)

The second stage of transformation requires the second column to be reduced so that the last element is zero. Proceeding as before, we have

$$\mathbf{w}(2) = [0.0000000 \quad 0.0000000 \quad -0.9238975 \quad -0.3826834]^T$$

$$\mathbf{R}(2) = \begin{bmatrix} 1.0000000 & 0.0000000 & 0.0000000 & 0.0000000 \\ 0.0000000 & 1.0000000 & 0.0000000 & 0.0000000 \\ 0.0000000 & 0.0000000 & -0.7071068 & -0.7071068 \\ 0.0000000 & 0.0000000 & -0.7071068 & 0.7071068 \end{bmatrix}$$

and

$$\mathbf{A}(2) = \begin{bmatrix} 5.0000000 & -4.2426407 & 0.0000000 & 0.0000000 \\ -4.2426407 & 6.6666667 & 3.2998316 & 0.0000000 \\ 0.0000000 & 3.2998316 & 10.3333333 & 0.0000000 \\ 0.0000000 & 0.0000000 & 0.0000000 & 4.0000000 \end{bmatrix}$$

It will be seen that **A**(2) is almost identical to the matrix **A**(3) obtained in Example 8.11 using Givens' method. There is only a difference in the sign of the elements a_{12} and a_{21} in the two cases, but it can be verified that the two matrices have the same eigenvalues.

EXERCISE 8.16

Repeat Exercise 8.15 using Householder's method.

Answer

$$\begin{bmatrix} 8.0000000 & -7.5498344 & 0.0000000 & 0.0000000 & 0.0000000 \\ -7.5498344 & 10.1754386 & 3.7656006 & 0.0000000 & 0.0000000 \\ 0.0000000 & 3.7656006 & 6.7486335 & 1.9692690 & 0.0000000 \\ 0.0000000 & 0.0000000 & 1.9692690 & 5.0970202 & 0.6882416 \\ 0.0000000 & 0.0000000 & 0.0000000 & 0.6882416 & 2.9789077 \end{bmatrix}$$

8.5.4 The QR Algorithm

This algorithm is considered as an excellent method for finding all the eigenvalues of a real matrix. It is suitable for both symmetric and nonsymmetric matrices. The basic idea of the QR method is to factor a given matrix **A**(k) as the product of an orthogonal matrix **Q**(k) and an upper triangular matrix **R**(k) as

$$\mathbf{A}(k) = \mathbf{Q}(k)\mathbf{R}(k) \tag{8.5–13}$$

which may be rewritten as

$$\mathbf{R}(k) = [\mathbf{Q}(k)]^T\mathbf{A}(k) \tag{8.5–14}$$

The procedure for obtaining $\mathbf{Q}(k)$ that will give us the upper triangular matrix $\mathbf{R}(k)$ will be discussed later. We note that the matrix $\mathbf{R}(k)$ will not have the same eigenvalues as $\mathbf{A}(k)$. Hence, we must postmultiply $\mathbf{R}(k)$ by $\mathbf{Q}(k)$ to obtain the matrix

$$\mathbf{A}(k + 1) = \mathbf{R}(k)\mathbf{Q}(k) = [\mathbf{Q}(k)]^T\mathbf{A}(k)\mathbf{Q}(k) \tag{8.5–15}$$

which will have the same eigenvalues as $\mathbf{A}(k)$.

Although $\mathbf{R}(k)$ is an upper triangular matrix, the product $\mathbf{R}(k)\mathbf{Q}(k)$ will not necessarily be upper triangular. The QR algorithm is based on the assumption that repeated applications of this procedure will cause $\mathbf{A}(k + 1)$ to approach triangular form for large k. In some cases, this convergence may be accelerated by a shift of axis. Here, the steps

$$\mathbf{A}(k) - s_k\mathbf{I} = \mathbf{Q}(k)\mathbf{R}(k) \tag{8.5–16}$$

$$\mathbf{A}(k + 1) = s_k\mathbf{I} + \mathbf{R}(k)\mathbf{Q}(k) \tag{8.5–17}$$

are taken. It can be easily shown that $\mathbf{A}(k)$ and $\mathbf{A}(k + 1)$ have the same eigenvalues. (See Theorem 8.6 in Section 8.2.) The value of the shift s_k is usually taken as that eigenvalue of the 2×2 matrix

$$\begin{bmatrix} a_{n-1,n-1}(k) & a_{n-1,n}(k) \\ a_{n,n-1}(k) & a_{nn}(k) \end{bmatrix}$$

which is closer to $a_{nn}(k)$ because this generally gives the fastest rate of convergence. For further discussion and proof of convergence in the case of the shifted matrix, the reader is referred to the book by Wilkinson.

For a full matrix, the QR algorithm requires $4n^3/3$ arithmetic operations. This can be considerably reduced if a nonsymmetric matrix is in Hessenberg form ($4n^2$ arithmetic operations) or if a symmetric matrix is tridiagonal ($12n$ operations). We have already studied two methods for transforming a symmetric matrix to tridiagonal form. Methods for transforming a nonsymmetric matrix to upper Hessenberg form will be discussed in the next section.

We shall now study the procedure for factoring a given matrix, $\mathbf{A}$, into the product $\mathbf{QR}$, where $\mathbf{Q}$ is orthogonal and $\mathbf{R}$ is upper triangular. First we note that equation (8.5–13) may be rewritten as

$$\mathbf{R} = \mathbf{Q}^{-1}\mathbf{A} = \mathbf{Q}^T\mathbf{A}$$

Thus, the problem of QR factorization can be viewed as that of performing a sequence of orthogonal transformations, $\{\mathbf{Q}_k{}^T\}$, such that the matrix $\mathbf{A}$ is reduced to upper triangular form. In the particular case when the matrix is in upper Hessenberg form, these transformations can be of the rotation type and will not affect each other if done in a proper sequence. Hence, for an $n \times n$ Hessenberg matrix, we shall require $(n - 1)$ transformations. The order of reductions must be in the following form

Column 1	$(2, 1), (3, 1), \ldots, (n, 1)$
Column 2	$(3, 2), (4, 2), \ldots, (n, 2)$
$\vdots$	$\vdots$
Column $(n - 1)$	$(n, n - 1)$

In order that the (i, j)th element of $\mathbf{Q}^T\mathbf{A}$ may be zero (for $i > j$), we make

$$\mathbf{Q}^T = \begin{bmatrix} \mathbf{I}_{j-1} & & & & \\ & c & & s & \\ & & \mathbf{I}_{i-j} & & \\ & -s & & c & \\ & & & & \mathbf{I}_{n-i-1} \end{bmatrix} \tag{8.5–18}$$

where $\mathbf{I}_m$ is an $m \times m$ identity matrix and

$$c = \frac{a_{jj}}{(a_{jj}^2 + a_{ij}^2)^{1/2}}, \qquad s = \frac{a_{ij}}{(a_{jj}^2 + a_{ij}^2)^{1/2}}.$$

The procedure will be illustrated through an example.

EXAMPLE 8.13

Consider the matrix

$$\mathbf{A} = \begin{bmatrix} -6 & -11 & -6 \\ 1 & 0 & 0 \\ 0 & 1 & 0 \end{bmatrix}$$

We see that $\mathbf{A}$ is in the upper Frobenius form (8.2–4), which is also an upper Hessenberg matrix. It is known that the eigenvalues of this matrix are -1, -2, and -3. We shall perform a few iterations of the QR algorithm for a better understanding of the procedure. First, we shall determine the matrix $\mathbf{Q}_1(1)$, so that the element in the second row and first column of $[\mathbf{Q}_1(1)]^T\mathbf{A}$ is zero. Using (8.5–18) we get

$$\mathbf{Q}_1(1) = \begin{bmatrix} -0.9863939 & -0.1643990 & 0.0000000 \\ 0.1643990 & -0.9863939 & 0.0000000 \\ 0.0000000 & 0.0000000 & 1.0000000 \end{bmatrix}$$

and

$$[\mathbf{Q}_1(1)]^T\mathbf{A} = \begin{bmatrix} 6.0827625 & 10.8503332 & 5.9183635 \\ 0.0000000 & 1.8083889 & 0.9863939 \\ 0.0000000 & 1.0000000 & 0.0000000 \end{bmatrix}$$

Next we find a matrix $\mathbf{Q}_2(1)$ such that $[\mathbf{Q}_2(1)]^T[\mathbf{Q}_1(1)]^T\mathbf{A}$ is upper triangular. We now annihilate the element in the second column of the third row. Using the same

EXAMPLE 8.13 (cont'd)

procedure as before, we have

$$\mathbf{Q}_2(1) = \begin{bmatrix} 1.0000000 & 0.0000000 & 0.0000000 \\ 0.0000000 & 0.8751130 & -0.4839186 \\ 0.0000000 & 0.4839186 & 0.8751130 \end{bmatrix}$$

and

$$[\mathbf{Q}_2(1)]^T[\mathbf{Q}_1(1)]^T\mathbf{A} = \begin{bmatrix} 6.0827625 & 10.8503332 & 5.9183635 \\ 0.0000000 & 2.0664632 & 0.8632062 \\ 0.0000000 & 1.0000000 & -0.4773344 \end{bmatrix}$$

The complete transformation matrix is

$$\mathbf{Q}(1) = \mathbf{Q}_1(1)\mathbf{Q}_2(1) = \begin{bmatrix} -0.9863939 & -0.1438677 & 0.0795557 \\ 0.1643990 & -0.8632062 & 0.4773444 \\ 0.0000000 & 0.4839186 & 0.8751130 \end{bmatrix}$$

We now postmultiply $[\mathbf{Q}_2(1)]^T[\mathbf{Q}_1(1)]^T\mathbf{A}$ by $\mathbf{Q}(1)$ to obtain

$$\mathbf{A}(2) = \begin{bmatrix} -4.2162162 & -7.3771812 & 10.8423925 \\ 0.3397242 & -1.3660623 & 1.7417969 \\ 0.0000000 & -0.2309910 & -0.4177215 \end{bmatrix}$$

This completes one cycle of the QR iteration. We note that $\mathbf{A}(2)$ is in upper Hessenberg form and the magnitude of each element on the subdiagonal is getting smaller. The second cycle will be started with the matrix $\mathbf{A}(2)$ and the transformation matrix $\mathbf{Q}_1(2)$ will be determined so that the element in the first column of the second row of $[\mathbf{Q}_1(2)]^T\mathbf{A}(2)$ will be reduced to zero. At the end of the second cycle of the QR iteration we have

$$\mathbf{A}(3) = \begin{bmatrix} -3.6344411 & -6.2554546 & -11.4811683 \\ 0.1580410 & -1.6496810 & -2.7526050 \\ 0.0000000 & -0.0846206 & -0.7158779 \end{bmatrix}$$

We can see further reduction in the magnitudes of the elements on the subdiagonal. Also, the elements on the main diagonal are approaching -1, -2, and -3, the actual eigenvalues of this matrix. Continuing in this manner, after 15 iterations we obtain

$$\mathbf{A}(16) = \begin{bmatrix} -3.0021627 & -5.4607174 & 11.5973910 \\ 0.0003969 & -1.9978757 & -4.0837087 \\ 0.0000000 & -0.0000094 & -0.9999617 \end{bmatrix}$$

It may be seen that $\mathbf{A}(16)$ is getting closer to the desired upper triangular form. After carrying out another 10 iterations, we get

$$\mathbf{A}(26) = \begin{bmatrix} -3.0000374 & -5.4609984 & 11.5958521 \\ 0.0000034 & -1.9999626 & -4.0882315 \\ 0.0000000 & 0.0000000 & -1.0000001 \end{bmatrix}$$

EXAMPLE 8.13 (cont'd)

which is almost triangular. This shows that the rate of convergence is not fast at this stage. We shall now examine the effect of a shift of axis, as given in equations (8.5–16) and (8.5–17). From the matrix **A**(16), we may estimate the shift as -0.9999617. Using this value and carrying out 10 iterations of the algorithm, we obtain

$$\tilde{\mathbf{A}}(26) = \begin{bmatrix} -3.0000021 & -5.4610048 & 11.5958252 \\ 0.0000004 & -1.9999978 & -4.0883065 \\ 0.0000000 & 0.0000000 & -1.0000001 \end{bmatrix}$$

which is seen to be much closer to the upper Hessenberg form. It is even more spectacular to see the effect of an axis shift in the earlier stages. If we start with the matrix **A** and use the shift of -1, after only 10 iterations we get

$$\tilde{\mathbf{A}}(10) = \begin{bmatrix} -3.0009237 & -5.4608362 & 11.5965154 \\ 0.0001693 & -1.9990763 & -4.0863493 \\ 0.0000000 & 0.0000000 & -1.0000000 \end{bmatrix}$$

At this stage, since the element in the second column of the third row has been reduced to zero, we can take advantage of the block triangular form of the matrix $\tilde{\mathbf{A}}(10)$ to determine one eigenvalue and then proceed with only the upper 2×2 part. This practice of splitting the matrix, when one element of the subdiagonal becomes zero, is very useful, especially in the case of larger matrices when it introduces a significant reduction in the number of arithmetic operations.

A computer program for performing a given number of iterations of the QR algorithm is in Appendix 8.4. The matrix is displayed after each complete transformation; in addition, any desired shift of the axis can be performed.

EXERCISE 8.17

Use the program in Appendix 8.4 to determine the eigenvalues of the symmetric matrix given in Exercise 8.15 after first bringing it to tridiagonal form using Givens' method.

Answer 17.5192394, 8.0745436, 4.3991403, 2.7291619, 0.2779148.

The procedure described above is sufficient for real matrices having real eigenvalues. Since a general real matrix may have eigenvalues that are complex conjugate pairs, it is seldom possible to effectively accelerate the rate of convergence by shifting while using only real arithmetic. In such cases, one may use the *double QR transformation*, which performs the equivalent of two basic QR transformations in one step. Not only does it accelerate convergence to a pair of complex conjugate eigenvalues, using only real arithmetic, but it also reduces the amount of computation because only $5n^2$ multiplications are required, compared with $8n^2$ multiplications for a double application of the QR algorithm. For further details of this method, the interested reader is referred to the book by Golub and Van Loan.

8.5.5 Transformation to Upper Hessenberg Form

A nonsymmetric matrix must be transformed to upper Hessenberg form before we can use the QR algorithm to find its eigenvalues, otherwise the number of arithmetic operations required is excessive. In this section, we shall discuss several methods for transforming a matrix to Hessenberg form.

We have already seen that Danilevsky's method (Section 8.4.2) transforms a matrix to the upper Frobenius form (8.2–4), which has the same structure as the Hessenberg form with some additional zeros above the subdiagonal. This is often not very practical, as it requires more arithmetic operations. However, the technique can be utilized for finding the roots of a polynomial of high degree by using the QR algorithm which is more robust than various methods for finding the roots of a polynomial. The idea is to write an $n \times n$ matrix in the Frobenius form which has the same characteristic polynomial as the given polynomial and then find all of its eigenvalues.

An efficient approach for transformation to Hessenberg form is to use a sequence of orthogonal matrices of the same type as used in Givens' method for transforming a symmetric matrix to tridiagonal form. The sequence of zeroing elements is the same as used with Givens' method; one starts with the element in the third row of the first column and proceeds down this column before moving to the fourth element in the second column, and so on. The following example will illustrate this.

EXAMPLE 8.14

The matrix

$$\mathbf{A} = \begin{bmatrix} 2 & 1 & 4 & 1 \\ 3 & 3 & 2 & -3 \\ 4 & 2 & -1 & 3 \\ 5 & -1 & 4 & 2 \end{bmatrix}$$

is to be transformed to upper Hessenberg form. We start with the element in the third row of the first column. The transformation matrix for zeroing this element is

$$\mathbf{Q}_1 = \begin{bmatrix} 1 & 0 & 0 & 0 \\ 0 & 0.6 & -0.8 & 0 \\ 0 & 0.8 & 0.6 & 0 \\ 0 & 0 & 0 & 1 \end{bmatrix}$$

EXAMPLE 8.14 (cont'd)

and the resulting matrix, $\mathbf{A}(2) = [\mathbf{Q}_1]^T\mathbf{A}\mathbf{Q}_1$, is

$$\mathbf{A}(2) = \begin{bmatrix} 2 & 3.8 & 1.6 & 1 \\ 5 & 2.36 & -2.48 & 1.2 \\ 0 & -2.48 & -0.36 & 3.4 \\ 5 & 2.6 & 3.2 & 2 \end{bmatrix}$$

The next step is to zero the element in the fourth row of the first column of $\mathbf{A}(2)$. The transformation matrix is

$$\mathbf{Q}_2 = \begin{bmatrix} 1.00000 & 0.00000 & 0.00000 & 0.00000 \\ 0.00000 & 0.70711 & 0.00000 & -0.70711 \\ 0.00000 & 0.00000 & 1.00000 & 0.00000 \\ 0.00000 & 0.70711 & 1.00000 & 0.70711 \end{bmatrix}$$

with the corresponding transformed matrix as

$$\mathbf{A}(3) = \begin{bmatrix} 2.00000 & 3.39411 & 1.60000 & -1.97990 \\ 7.07107 & 3.78000 & 0.50912 & -1.18000 \\ 0.00000 & 1.21622 & -0.36000 & 4.72347 \\ 0.00000 & 0.82000 & 4.01637 & 0.58000 \end{bmatrix}$$

The last step is to reduce the element in the fourth row of the second column of $\mathbf{A}(3)$ to zero. This will be left as an exercise for the student.

The listing of a Pascal program for transforming a nonsymmetric matrix to upper Hessenberg form using this method is given in Appendix 8.4. If the matrix is symmetric, it will be transformed to tridiagonal form.

8.6 FINDING EIGENVECTORS

So far in our examination of the matrix eigenvalue-eigenvector problem, we have been concerned mostly with finding the eigenvalues of a matrix. The problem of finding the eigenvectors is much simpler after the eigenvalues have been calculated. We have already seen that one may utilize the various transformation matrices for this purpose. In Section 8.4.2, while discussing Danilevsky's method, we had found that it was quite straightforward to determine the eigenvectors from the eigenvalues and the transformation matrices. A similar observation was made while studying Jacobi's method. In fact, any nondefective matrix can be transformed to diagonal form, as the following theorem shows. In addition, the theorem reveals some important features of eigenvectors.

Theorem 8.12

If $\lambda_1, \lambda_2, \ldots, \lambda_n$ are the eigenvalues of the $n \times n$ matrix **A**, which is not defective, and $\mathbf{v}_1, \mathbf{v}_2, \ldots, \mathbf{v}_n$ are corresponding eigenvectors, then **A** is similar to the diagonal matrix

$$\mathbf{D} = \mathbf{M}^{-1}\mathbf{A}\mathbf{M} = \text{diag}(\lambda_1, \lambda_2, \ldots, \lambda_n) \quad \textbf{(8.6–1)}$$

where

$$\mathbf{M} = [\mathbf{v}_1 \quad \mathbf{v}_2 \quad \ldots \quad \mathbf{v}_n] \quad \textbf{(8.6–2)}$$

Proof

The proof follows from the definition of eigenvalues and eigenvectors of a matrix. If we use equation (8.2–1) for each of the n eigenvalues and the corresponding eigenvectors of **A**, we get the set of equations

$$\mathbf{A}\mathbf{v}_k = \lambda_k\mathbf{v}_k \qquad k = 1, 2, \ldots, n \quad \textbf{(8.6–3)}$$

These equations can be concatenated to give

$$\mathbf{A}[\mathbf{v}_1 \quad \mathbf{v}_2 \quad \ldots \quad \mathbf{v}_n] = \mathbf{M}\cdot\text{diag}(\lambda_1, \lambda_2, \ldots, \lambda_n) \quad \textbf{(8.6–4)}$$

Equation (8.6–4) leads immediately to (8.6–1) because **M** is nonsingular, which is always the case if **A** is not defective.

The matrix **M** is called the *modal matrix* for **A**. If **A** is not defective, then we can diagonalize the matrix using a sequence of similarity transformations. It also follows that the eigenvectors are the columns of the modal matrix, **M**, which is used for the diagonalizing transformation given by equation (8.6–1).

In many engineering and science applications, the matrix **A** has only real entries, but it could have complex eigenvalues. If **A** satisfies the conditions of Theorem 8.12 then it is similar to a diagonal matrix, but in this case, both **D** and **M** are complex matrices. It is possible to obtain a factorization of **A** into a form where **D** and **M** have only real entries; however, **D** will be block diagonal with each diagonal block a 1×1 matrix corresponding to a real eigenvalue of **A** or a 2×2 matrix of the form

$$\begin{bmatrix} a & -b \\ b & a \end{bmatrix}$$

corresponding to a complex conjugate pair of eigenvalues $a \pm jb$. A column of **M** corresponding to a real eigenvalue will be the appropriate eigenvector. The situation for a pair of complex conjugate eigenvalues is handled as follows. If the eigenvector corresponding to $a + jb$ is $\mathbf{v}$ and the eigenvector corresponding to $a - jb$ is $\overline{\mathbf{v}}$ (the complex conjugate of $\mathbf{v}$), then the two corresponding columns in **M** are $(\mathbf{v} + \overline{\mathbf{v}})/2$ and $(\mathbf{v} - \overline{\mathbf{v}})/(2j)$.

If a similarity transformation reduces **A** to triangular form, we can determine the eigenvectors of the triangular matrix by solving a set of linear simultaneous equations. This is easily done due to the triangular nature of the matrix. We can then use the transformation matrix to find the eigenvectors of the original matrix. The procedure will be shown through the following example.

EXAMPLE 8.15

Consider the case when the matrix

$$\mathbf{P}^{-1}\mathbf{A}\mathbf{P} = \begin{bmatrix} 3 & 2 & 2 \\ 0 & 2 & 3 \\ 0 & 0 & 4 \end{bmatrix}$$

with the orthogonal transformation matrix **P** given by

$$\mathbf{P} = \begin{bmatrix} 0.60 & -0.80 & 0.00 \\ 0.64 & -0.48 & 0.60 \\ -0.48 & 0.36 & 0.80 \end{bmatrix}$$

From the matrix $\mathbf{P}^{-1}\mathbf{A}\mathbf{P}$ it is evident that the eigenvalues are $\lambda_1 = 3$, $\lambda_2 = 2$, and $\lambda_3 = 4$. We shall, therefore, first determine the eigenvectors of the matrix $\mathbf{P}^{-1}\mathbf{A}\mathbf{P}$. Let these be denoted by $\mathbf{w}_1$, $\mathbf{w}_2$, and $\mathbf{w}_3$, respectively. If the elements of $\mathbf{w}_1$ are x, y, and z, we get the system of equations

$$\begin{bmatrix} 3 & 2 & 2 \\ 0 & 2 & 3 \\ 0 & 0 & 4 \end{bmatrix} \begin{bmatrix} x \\ y \\ z \end{bmatrix} = 3 \begin{bmatrix} x \\ y \\ z \end{bmatrix}$$

As expected, the three equations are not independent. The last equation, $4z = 3z$, requires $z = 0$. The second equation, $2y + 3z = 3y$, gives $y = 0$. Lastly, the first equation, $3x + 2y + 2z = 3x$, is an identity for all x so we can use any nonzero real number for x; setting $x = 1$ we get

$$\mathbf{w}_1 = [1 \quad 0 \quad 0]^T$$

Proceeding in the same manner gives

$$\mathbf{w}_2 = [-2 \quad 1 \quad 0]^T$$

and

$$\mathbf{w}_3 = [5 \quad 1.5 \quad 1]^T$$

The eigenvectors of the original matrix, **A**, are now found from the relationship $\mathbf{v}_i = \mathbf{P}\mathbf{w}_i$, $i = 1, 2, 3$.

8.7 PASCAL PROCEDURES FOR EIGENVALUES AND EIGENVECTORS OF REAL MATRICES

Appendix 8.5 contains a user's manual for the Pascal procedure, EigenR, which finds the eigenvalues and eigenvectors of a general real matrix. EigenR includes balancing of matrices to increase numerical accuracy (see Problem 17 in Section 8.9), transformation of a matrix to Hessenberg form, the double QR shift for improved acceleration in the case of complex eigenvalues, and also provides a performance index that is a measure of the accuracy of the computations. The performance index is computed as follows. Let $\mathbf{v}_k$ be a computed eigenvector of the $n \times n$ matrix $\mathbf{A}$ associated with the computed eigenvalue λ_k. The performance index is

$$P = \underset{1 \le k \le n}{\text{maximum}} \left\{ \frac{\| \mathbf{A}\mathbf{v}_k - \lambda_k \mathbf{v}_k \|_1}{\| \mathbf{A} \|_1 \| \mathbf{v}_k \|_1 (10nu)} \right\}$$

where u is the relative machine precision, and the vector norm $\| \mathbf{z} \|_1$ is defined as

$$\| \mathbf{z} \|_1 = \sum_{i=1}^{n} | z_i |$$

for the complex vector $\mathbf{z} = [z_1, z_2, \ldots, z_n]^T$. This performance index was developed at the Argonne National Laboratory as a part of a software package for eigenvalues and eigenvectors.

Appendix 8.6 contains a main program using EigenR. Although the procedure EigenR can be used for real symmetric matrices, it is not as efficient as a specialized procedure would be. Therefore, we have included the Pascal procedure EigenRS.

The Pascal procedure EigenRS, whose user's manual is in Appendix 8.7 and a main program in Appendix 8.8, is designed to find the eigenvalues and eigenvectors of a real symmetric matrix. EigenRS uses Householder transformations and the QR method; like EigenR it returns a performance index on the accuracy of the computations.

The procedures EigenR and EigenRS and their accompanying main programs should be useful for solving a large variety of problems occurring in many science and engineering disciplines. They are highly recommended for these applications.

EXAMPLE 8.16

The general position of a point in space $\mathbf{X} = [X_1, X_2, X_3]^T$, measured with respect to the fixed axis system denoted by $\langle X_1, X_2, X_3 \rangle$, also can be located with respect to a rotated axis system $\langle x_1, x_2, x_3 \rangle$ in which the position vector is $\mathbf{x} = [x_1, x_2, x_3]^T$. The vector $\mathbf{x}$ is related to $\mathbf{X}$ by the matrix $\mathbf{A}$,

$$\mathbf{x} = \mathbf{A}\mathbf{X}$$

where $\mathbf{A}$ is the product of three 3×3 orthogonal matrices,

$$\mathbf{A} = \mathbf{A}_3\mathbf{A}_2\mathbf{A}_1$$

with

$$\mathbf{A}_3 = \begin{bmatrix} \cos(\psi) & \sin(\psi) & 0 \\ -\sin(\psi) & \cos(\psi) & 0 \\ 0 & 0 & 1 \end{bmatrix}$$

EXAMPLE 8.16 (cont'd)

$$\mathbf{A}_2 = \begin{bmatrix} 1 & 0 & 0 \\ 0 & \cos(\theta) & \sin(\theta) \\ 0 & -\sin(\theta) & \cos(\theta) \end{bmatrix}$$

$$\mathbf{A}_1 = \begin{bmatrix} \cos(\phi) & \sin(\phi) & 0 \\ -\sin(\phi) & \cos(\phi) & 0 \\ 0 & 0 & 1 \end{bmatrix}$$

The angles ψ, θ, and ϕ are called *Euler angles*. Each transformation is obtained by a rigid rotation of an axis system as follows.

1. The $\langle X_1, X_2, X_3\rangle$ axis system is rotated about the X_3 axis through an angle ϕ to give a new coordinate system $\langle Y_1, Y_2, Y_3\rangle$. Thus $\mathbf{Y} = \mathbf{A}_1\mathbf{X}$.
2. The $\langle Y_1, Y_2, Y_3\rangle$ axis system is rotated about the Y_1 axis through an angle θ to give a new coordinate system $\langle Z_1, Z_2, Z_3\rangle$. Thus $\mathbf{Z} = \mathbf{A}_2\mathbf{Y}$.
3. The $\langle Z_1, Z_2, Z_3\rangle$ axis system is rotated about the Z_3 axis through an angle ψ to give the coordinate system $\langle x_1, x_2, x_3\rangle$. Thus $\mathbf{x} = \mathbf{A}_3\mathbf{Z}$.

Therefore, $\mathbf{x} = \mathbf{A}_3\mathbf{Z} = \mathbf{A}_3\mathbf{A}_2\mathbf{Y} = \mathbf{A}_3\mathbf{A}_2\mathbf{A}_1\mathbf{X} = \mathbf{AX}$.

We now want to find the eigenvalues of **A** for $\psi = 22$ degrees, $\theta = 15$ degrees and $\phi = 17$ degrees. Remembering to convert the given angle data to radians, we find that **A** is

$$\mathbf{A} = \begin{bmatrix} 0.78087791380860 & 0.61711372492846 & 0.09695532079701 \\ -0.62008348484675 & 0.74693340261591 & 0.23997283987345 \\ 0.07567136543133 & -0.24750988376535 & 0.96592582628907 \end{bmatrix}$$

EigenR found the eigenvalues of **A** to be

$$\lambda_1 = 1.0, \quad \text{and} \quad \lambda_{2,3} = 0.74686857135679 \pm 0.66497168144175j$$

and the eigenvectors

$$\mathbf{v}_1 = \begin{bmatrix} 0.18327198632800 \\ -0.00800182773180 \\ 0.46513154029838 \end{bmatrix}$$

$$\mathbf{v}_{2,3} = \begin{bmatrix} -0.05378894367661 \mp 0.46208021875595j \\ 0.49620847276196 \mp 0.06093539457654j \\ 0.02973047422096 \pm 0.18102140514938j \end{bmatrix}$$

with a performance index of 0.064.

The matrix **A** represents a rotation of the $\langle X_1, X_2, X_3\rangle$ axis system to the $\langle x_1, x_2, x_3\rangle$ axis system. The axis of rotation is in the direction of the eigenvector $\mathbf{v}_1$ because $\mathbf{A}\mathbf{v}_1 = \mathbf{v}_1$.

EXAMPLE 8.17

The stresses at a point in an elastic body are completely determined by the six stress components $\sigma_{11}, \sigma_{12}, \sigma_{13}, \sigma_{22}, \sigma_{23}, \sigma_{33}$. These components form a 3×3 real symmetric matrix, called a *stress tensor* in the Theory of Elasticity,

$$\mathbf{S} = \begin{bmatrix} \sigma_{11} & \sigma_{12} & \sigma_{13} \\ \sigma_{12} & \sigma_{22} & \sigma_{23} \\ \sigma_{13} & \sigma_{23} & \sigma_{33} \end{bmatrix}$$

The eigenvalues of **S** are the principal stress components $\sigma_1, \sigma_2, \sigma_3$ and the corresponding eigenvectors form the principal stress axes. The characteristic polynomial of **S** is given in Example 3.8. Using the data from Example 3.8 we want to determine the principal stress components by finding the eigenvalues of **S**. From Example 3.8 we have

$$\mathbf{S} = \begin{bmatrix} 3.6 & 2.5 & -1.5 \\ 2.5 & 5.7 & 3.9 \\ -1.5 & 3.9 & -1.0 \end{bmatrix}$$

where the stresses are in units of 10^9 dyn/cm^2. Using EigenRS we found that the eigenvalues of **S** were

$$\lambda_1 = 3.75858130301091, \quad \lambda_2 = -3.62126929615344, \text{ and}$$

$$\lambda_3 = 8.16268799314253.$$

Thus, the principal stress components are $\sigma_k = \lambda_k$ for $k = 1, 2, 3$ in units of 10^9 dyn/cm^2.

To conclude this section we mention one problem in computing the eigenvalues of a matrix whose maximum modulus eigenvalue is very much greater than the moduli of other eigenvalues. If the maximum modulus eigenvalue, λ, can be determined to an accuracy of t decimal digits, then, in general, the error in all the other eigenvalues will be approximately of the order of $10^{-t}\,|\,\lambda\,|$. If another eigenvalue has a modulus of about $10^{-t}\,|\,\lambda\,|$ then it is likely that none of its computed digits are significant. For example, the matrix

$$\begin{bmatrix} 1.00000000000004 & 1.99999999999998 \\ 1.99999999999998 & 4.00000000000001 \end{bmatrix}$$

has eigenvalues 5.0 and 5×10^{-14}. When the eigenvalues were obtained using EigenR and EigenRS the following results were obtained.

EigenR	EigenRS
5.0	5.0
$4.97379915032070 \times 10^{-14}$	$4.99600361081320 \times 10^{-14}$

As you can see, none of the digits of the smaller eigenvalue is correct.

8.8 PROBLEMS

1. Determine the characteristic polynomial of each of the following matrices by transforming them to upper Frobenius form.

$$\text{(a)} \begin{bmatrix} 3 & 3 & 2 \\ 4 & 2 & -1 \\ 2 & 1 & 4 \end{bmatrix} \qquad \text{(b)} \begin{bmatrix} 2 & 1 & 4 & 1 \\ 3 & 3 & 2 & -3 \\ 4 & 2 & -1 & 3 \\ 5 & -1 & 4 & 2 \end{bmatrix}$$

2. Use Leverrier's method to determine the characteristic polynomial as well as the inverse of each of the matrices given in Problem 1.

3. Use the power method to determine that eigenvalue of the matrix **A** which has the largest magnitude.

$$\mathbf{A} = \begin{bmatrix} 10 & 6 & 2 & 1 \\ 6 & 8 & 3 & 2 \\ 2 & 3 & 5 & 1 \\ 1 & 2 & 1 & 3 \end{bmatrix}$$

4. Use the inverse power method to determine the smallest magnitude eigenvalue of the matrix in Problem 3.

5. Transform the matrix in Problem 3 to tridiagonal form. Give the transformed matrix as well as the matrix required for the transformation.

6. Determine all eigenvalues and eigenvectors of the matrix in Problem 3. Verify that the matrix **M**, formed from the eigenvectors of **A**, satisfies the relationship **AM** = **MD**, where **D** is a diagonal matrix with the eigenvalues of **A** on its main diagonal.

7. Transform the matrix

$$\mathbf{A} = \begin{bmatrix} -2.6 & 0.25 & -38 & 0 \\ -0.075 & -0.27 & 4.4 & 0 \\ 0.078 & -0.99 & -0.23 & 0.052 \\ 1 & 0.078 & 0 & 0 \end{bmatrix}$$

to upper Hessenberg form.

8. In the study of linear systems, one often uses state equations of the form

$$\frac{d\mathbf{x}}{dt} = \mathbf{Ax} + \mathbf{Bu}, \qquad \textbf{(8.8–1)}$$

where **A** is an $n \times n$ matrix, **B** is an $n \times m$ matrix, **x** is an n-dimensional vector of variables describing the state of the system (state variables) and **u** is an m-dimensional vector of input (or control) functions. The eigenvalues of the matrix **A** are called the *natural frequencies* of the system. For stability, all the eigenvalues must have negative real parts. Determine if the system with the **A** matrix as in Problem 7 will be stable. (Use the eigenvalue procedure EigenR in Appendix 8.5 for this purpose.)

9. The state equations for a helicopter near hover are (8.8–1), with

$$\mathbf{A} = \begin{bmatrix} -0.002 & -1.4 & 9.8 \\ -0.01 & -0.4 & 0 \\ 0 & 1 & 2 \end{bmatrix}, \qquad \mathbf{B} = \begin{bmatrix} 9.8 \\ 6.3 \\ 0 \end{bmatrix}$$

Determine the eigenvalues of the system matrix. Is it a stable system?

10. The linearized dynamic equations for a satellite in equatorial orbit are given by (8.8–1), where

$$\mathbf{A} = \begin{bmatrix} 0 & 1 & 0 & 0 \\ 3 & 0 & 0 & 2 \\ 0 & 0 & 0 & 1 \\ 0 & -2 & -3 & 0 \end{bmatrix}$$

Find the eigenvalues of **A** and hence determine if the system is stable.

11. The linearized dynamic equations for an automatic ship steering system are given by (8.8–1), where

$$\mathbf{A} = \begin{bmatrix} -0.05 & 6.00 & 0 & 0 \\ -0.001 & -0.15 & 0 & 0 \\ 1 & 0 & 0 & 13 \\ 0 & 1 & 0 & 0 \end{bmatrix}$$

Investigate the stability of the system by finding the eigenvalues of **A**.

12. A discrete-time system is stable if and only if the eigenvalues of the **A**-matrix of the state equation $\mathbf{x}(k + 1) = \mathbf{Ax}(k)$ are all smaller than one in magnitude. Determine if the system described by the matrix

$$\mathbf{A} = \begin{bmatrix} -0.095830473 & 0.08784725 & -3.227109600 & -11.9740360 \\ -0.011197641 & -0.03831703 & 15.763176000 & -1.2545593 \\ 0.001670204 & -0.01133869 & 0.497682210 & 0.0187607 \\ 0.024718663 & 0.00430285 & -0.057271536 & 0.0803222 \end{bmatrix}$$

is stable.

13. Repeated applications of the double QR algorithm transformed a matrix to the form shown below. Determine all the eigenvalues and eigenvectors of this matrix.

$$\begin{bmatrix} 4 & 3 & 1 & 0 \\ -3 & 4 & 2 & -1 \\ 0 & 0 & 3 & 4 \\ 0 & 0 & 0 & 2 \end{bmatrix}$$

14. As stated in Chapter 2, the condition number of a matrix can be defined as the ratio of the eigenvalues with the largest and the smallest magnitude. Calculate the condition number of Wilson's matrix

$$\begin{bmatrix} 10 & 7 & 8 & 7 \\ 7 & 5 & 6 & 5 \\ 8 & 6 & 10 & 9 \\ 7 & 5 & 9 & 10 \end{bmatrix}$$

by using the power and the inverse power methods.

15. The 5×5 Hilbert matrix, $\mathbf{H}$, is rather ill-conditioned. Determine its condition number by finding all of its eigenvalues. (To improve the numerical accuracy, you may multiply all elements of the matrix by 2520, as this will produce a matrix with integer entries without changing the condition number. Prove that the eigenvalues of $k\mathbf{A}$ are k times the eigenvalues of $\mathbf{A}$.)

$$\mathbf{H} = \begin{bmatrix} 1 & \frac{1}{2} & \frac{1}{3} & \frac{1}{4} & \frac{1}{5} \\ \frac{1}{2} & \frac{1}{3} & \frac{1}{4} & \frac{1}{5} & \frac{1}{6} \\ \frac{1}{3} & \frac{1}{4} & \frac{1}{5} & \frac{1}{6} & \frac{1}{7} \\ \frac{1}{4} & \frac{1}{5} & \frac{1}{6} & \frac{1}{7} & \frac{1}{8} \\ \frac{1}{5} & \frac{1}{6} & \frac{1}{7} & \frac{1}{8} & \frac{1}{9} \end{bmatrix}$$

16. The vector differential equation (8.8–1), where $\mathbf{A}$ and $\mathbf{B}$ are matrices whose entries are real numbers, can be solved quite conveniently by *decoupling* the state variables. This is easily accomplished through the transformation $\mathbf{y} = \mathbf{Px}$, so that the transformed equation takes the form

$$\frac{d\mathbf{y}}{dt} = \tilde{\mathbf{A}}\mathbf{y} + \tilde{\mathbf{B}}\mathbf{u},$$

where $\tilde{\mathbf{A}} = \mathbf{PAP}^{-1}$ and $\tilde{\mathbf{B}} = \mathbf{PB}$. $\tilde{\mathbf{A}}$ is a diagonal matrix if $\mathbf{P}^{-1}$ is properly selected; the columns of $\mathbf{P}^{-1}$ are the eigenvectors of $\mathbf{A}$. This will always work if $\mathbf{A}$ is not defective. If the eigenvalues of $\mathbf{A}$ are all real, then $\mathbf{P}$ and $\tilde{\mathbf{A}}$ are real matrices; otherwise, they may be complex. The effect of the decoupling is that the n original differential equations are replaced by another set of n differential equations, each of which contains only one state variable. Hence, they can be solved easily. The solution in terms of the original variables is then obtained by utilizing the inverse transformation

$$\mathbf{x} = \mathbf{P}^{-1}\mathbf{y}.$$

Use this method to solve the differential equations if

$$\mathbf{A} = \begin{bmatrix} -9 & -38 & -42 \\ 2 & -2 & -7 \\ -1 & 2 & 5 \end{bmatrix}, \quad \mathbf{B} = \begin{bmatrix} 1 \\ 0 \\ 1 \end{bmatrix}, \quad \text{and} \quad u(t) = 1$$

with the initial condition $\mathbf{x}(0) = [2 \quad 3 \quad 1]^T$.

17. The accuracy of the numerical procedures for calculating the eigenvalues of an $n \times n$ matrix $\mathbf{A} = (a_{ij})$ can be improved considerably if it is *balanced*; that is, we try to reduce $\|\mathbf{A}\|$ for some matrix norm. This is usually done by using a diagonal similarity transformation matrix $\mathbf{D}$, chosen so that

$$\sum_{\substack{j=1 \\ j\neq i}}^{n} |a_{ij}| \simeq \sum_{\substack{j=1 \\ j\neq i}}^{n} |a_{ji}| \quad \text{for each } i = 1, 2, \ldots, n.$$

This is the scheme used in the procedure Balance in the general eigenvalue/eigenvector procedure EigenR in Appendix 8.5. Use this technique to balance the matrix

$$\mathbf{A} = \begin{bmatrix} 1.5 & 2000 & 3 \\ 0.001 & 3 & -0.001 \\ 2 & 1000 & 5 \end{bmatrix}$$

Verify that the eigenvalues are unchanged.

18. Show that if k times the ith row of an $n \times n$ matrix, $\mathbf{A}$, is added to its jth row and this is followed by subtracting k times the jth column from the ith column, then the eigenvalues will be unchanged.

19. Verify the result of Problem 18 by determining the characteristic polynomials of the matrices $\mathbf{A}$ and $\mathbf{B}$, given below. (*Note*: $\mathbf{B}$ has been obtained from $\mathbf{A}$ by subtracting 0.5 times the first row from the second row, followed by adding 0.5 times the second column to the first. This result can be utilized for developing a systematic procedure for transforming a matrix to upper Frobenius form.)

$$\mathbf{A} = \begin{bmatrix} 2 & 2 & 4 \\ 3 & 5 & 2 \\ 1 & 4 & 5 \end{bmatrix} \qquad \mathbf{B} = \begin{bmatrix} 3 & 2 & 4 \\ 4 & 4 & 0 \\ 3 & 4 & 5 \end{bmatrix}$$

20. Use the Frobenius normal form (8.2–4) to find the roots of the polynomials in Problem 13 of Chapter 3, except (d) and (i). Compare your results with the roots obtained using either of the root-finding programs in Chapter 3.

21. Use the procedure EigenRS in Appendix 8.7 to find the eigenvalues and eigenvectors of the matrix in

(a) Problem 3 **(b)** Problem 14 **(c)** Problem 15

(d) Chapter 2, Problem 9 **(e)** Chapter 2, Problem 11.

22. The matrix

$$\mathbf{A} = \begin{bmatrix} a & b & c \\ -b & -a & c \\ c & -c & \sqrt{c} \end{bmatrix},$$

where $a = \dfrac{\sqrt{2}-1}{2\sqrt{2}}$, $b = \dfrac{\sqrt{2}+1}{2\sqrt{2}}$, $c = 0.5$, represents a rotation. Find a vector that determines the axis of rotation.

23. Prove that every $n \times n$ orthogonal matrix $\mathbf{P}$, with n odd and $\det(\mathbf{P}) = 1$, has an eigenvalue $\lambda = 1$.

24. The 12×12 Frank matrix is

$$\mathbf{F} = \begin{bmatrix} 12 & 11 & 10 & 9 & 8 & 7 & 6 & 5 & 4 & 3 & 2 & 1 \\ 11 & 11 & 10 & 9 & 8 & 7 & 6 & 5 & 4 & 3 & 2 & 1 \\ 0 & 10 & 10 & 9 & 8 & 7 & 6 & 5 & 4 & 3 & 2 & 1 \\ 0 & 0 & 9 & 9 & 8 & 7 & 6 & 5 & 4 & 3 & 2 & 1 \\ 0 & 0 & 0 & 8 & 8 & 7 & 6 & 5 & 4 & 3 & 2 & 1 \\ 0 & 0 & 0 & 0 & 7 & 7 & 6 & 5 & 4 & 3 & 2 & 1 \\ 0 & 0 & 0 & 0 & 0 & 6 & 6 & 5 & 4 & 3 & 2 & 1 \\ 0 & 0 & 0 & 0 & 0 & 0 & 5 & 5 & 4 & 3 & 2 & 1 \\ 0 & 0 & 0 & 0 & 0 & 0 & 0 & 4 & 4 & 3 & 2 & 1 \\ 0 & 0 & 0 & 0 & 0 & 0 & 0 & 0 & 3 & 3 & 2 & 1 \\ 0 & 0 & 0 & 0 & 0 & 0 & 0 & 0 & 0 & 2 & 2 & 1 \\ 0 & 0 & 0 & 0 & 0 & 0 & 0 & 0 & 0 & 0 & 1 & 1 \end{bmatrix}$$

(a) Find the eigenvalues of $\mathbf{F}$. (They are all real, positive and lie in the interval [0.03, 32.3]).

(b) Find the eigenvalues of the matrix obtained from $\mathbf{F}$ by replacing the (1, 12) entry in $\mathbf{F}$ by $(1 + 10^{-7})$. Compare your results with part (a).

25. Let $\mathbf{A}$ denote an $n \times n$ matrix and let λ be an eigenvalue of $\mathbf{A}$ with associated eigenvector $\mathbf{v}$,

$$\mathbf{A}\mathbf{v} = \lambda\mathbf{v}.$$

We sometimes refer to $\mathbf{v}$ as a *right eigenvector* of $\mathbf{A}$ because we can define another eigenvalue problem of $\mathbf{A}$ as finding a number μ and a nonzero vector $\mathbf{w}$ such that

$$\mathbf{w}^T\mathbf{A} = \mu\mathbf{w}^T.$$

$\mathbf{w}$ is called a *left eigenvector* of $\mathbf{A}$. Transposing this last equation gives

$$\mathbf{A}^T\mathbf{w} = \mu\mathbf{w}.$$

Because $\mathbf{A}$ and $\mathbf{A}^T$ have the same eigenvalues, we can equate λ and μ. Thus, if $\lambda = \mu$, $\mathbf{v}$ is a right eigenvector of $\mathbf{A}$ and $\mathbf{w}$ is a left eigenvector of $\mathbf{A}$ (and a right eigenvector of $\mathbf{A}^T$).

Wilkinson defines a condition number for each eigenvalue of an $n \times n$ diagonalizable matrix $\mathbf{A}$ as follows. Let $\mathbf{x}_k$ and $\mathbf{y}_k$ denote right and left eigenvectors of $\mathbf{A}$ associated with the eigenvalue λ_k. The eigenvectors are normalized so that

$$[\mathbf{x}_k]^T \mathbf{x}_k = [\mathbf{y}_k]^T \mathbf{y}_k = 1.$$

The number s_k is defined as

$$s_k = [\mathbf{y}_k]^T \mathbf{x}_k.$$

Note that $|s_k| \leq 1$ because s_k is the cosine of the angle between the vectors $\mathbf{x}_k$ and $\mathbf{y}_k$. Then, the *Wilkinson condition number* of the eigenvalue λ_k is defined as $1/|s_k|$. (Real, symmetric matrices have $\mathbf{y}_k = \mathbf{x}_k$, so that $s_k = 1$ for each k.)

(a) $\mathbf{P}$ is an $n \times n$ matrix whose columns are the right eigenvectors of the $n \times n$ diagonalizable matrix $\mathbf{A}$. Prove that the left eigenvectors of $\mathbf{A}$ are the rows of $\mathbf{P}^{-1}$.

(b) Determine the condition number for each eigenvalue of the matrix in part (a) of Problem 24.

26. The eigenvalues of a real symmetric matrix are well-conditioned as Problem 25 has shown. That is, small changes in the elements of the matrix lead to small changes in the eigenvalues. The same is not true of the eigenvectors. The following matrix (due to Givens) illustrates this.

$$\mathbf{A} = \begin{bmatrix} 1 + \epsilon \cos(2/\epsilon) & -\epsilon \sin(2/\epsilon) \\ -\epsilon \sin(2/\epsilon) & 1 - \epsilon \cos(2/\epsilon) \end{bmatrix}.$$

(a) By hand computation, determine the eigenvalues and eigenvectors of $\mathbf{A}$.

(b) Examine the behavior of the eigenvalues and eigenvectors of $\mathbf{A}$ as $\epsilon \to 0$.

27. Let $\mathbf{C}$ be an $n \times n$ real symmetric tridiagonal matrix,

$$\mathbf{C} = \begin{bmatrix} a_1 & b_1 & 0 & 0 & 0 & \cdots & & & & 0 \\ b_1 & a_2 & b_2 & 0 & 0 & \cdots & & & & 0 \\ 0 & b_2 & a_3 & b_3 & 0 & \cdots & & & & 0 \\ \vdots & & & & & & & & & \vdots \\ 0 & \cdots & & & & 0 & b_{n-3} & a_{n-2} & b_{n-2} & 0 \\ 0 & \cdots & & & & & 0 & b_{n-2} & a_{n-1} & b_{n-1} \\ 0 & \cdots & & & & & & 0 & b_{n-1} & a_n \end{bmatrix}$$

For any real number x we can form the sequence of polynomials

$$p_0(x) = 1, \qquad p_1(x) = a_1 - x,$$

$$p_k(x) = (a_k - x)p_{k-1}(x) - (b_k)^2 p_{k-2}(x) \quad \text{for} \quad k = 2, 3, \ldots, n.$$

The sequence $\{p_j(x)\}$ gives the principal minors of

$$\det(\mathbf{C} - x\mathbf{I}) = p_n(x).$$

Note that $p_n(x)$ is $(-1)^n$ times the characteristic polynomial of **C** as defined by the left-hand side of (8.2–3). We are interested in obtaining the eigenvalues of **C**. We assume that $b_k \neq 0$ for any k. If b_j were zero for some value of j, then **C** can be partitioned into two smaller order matrices, each of which is tridiagonal. In this case, the problem of finding the eigenvalues of the $n \times n$ matrix **C** is reduced to finding the eigenvalues of two other real symmetric tridiagonal matrices whose dimensions are less than n.

The polynomials $p_0(x), p_1(x), \ldots, p_n(x)$ form a *Sturm sequence*. If $x = \lambda$ the sequence of numbers $\{p_j(\lambda)\}$ has an interesting property. The number of sign changes between successive values $p_i(\lambda)$ and $p_{i+1}(\lambda)$ for $i = 0, 1, \ldots, n - 1$ gives the number of roots of $p_n(x)$ that are less than λ, and hence, the number of eigenvalues of **C** less than λ. If any value $p_i(\lambda) = 0$ then it is omitted from the sequence of values whose signs are to be determined.

(a) Write a Pascal program that determines n intervals, each containing one eigenvalue of an $n \times n$ real symmetric tridiagonal matrix **C**. Test your program by using the tridiagonal matrix found in Exercise 8.15.

(b) Modify your program in part (a) so that the procedure ZeroIn, from Chapter 4, can be used to find the eigenvalues.

8.9 REFERENCES

L. V. Atkinson and P. J. Harley. *An Introduction to Numerical Methods with Pascal*. London: Addison-Wesley, 1983.

R. L. Burden, J. D. Faires, and A. C. Reynolds. *Numerical Analysis*. 2d Ed. Boston: Prindle, Weber and Schmidt, 1981.

V. N. Fadeeva. *Computational Methods of Linear Algebra*. New York: Dover, 1959.

G. H. Golub and C. F. Van Loan. *Matrix Computations*. Baltimore, MD: The Johns Hopkins University Press, 1983.

A. Jennings. *Matrix Computation for Engineers and Scientists*. Chichester: Wiley, 1977.

G. Strang. *Linear Algebra and Its Applications*. New York: Academic Press, 1976.

J. H. Wilkinson. *The Algebraic Eigenvalue Problem*. Oxford: Oxford University Press, 1965.

APPENDIX 8.1

Listing of a Pascal Program that finds single eigenvalues by either the Power Method, the Power Method with Rayleigh Quotients, or the Inverse Power Method

```
PROGRAM FindSingleEigenvalue;

CONST
  MaxN = 25 ; MaxIterations = 100 ; tolerance = 1.0E-07 ;
  Field = 10  (* Fieldwidth *) ;
  Decimal = 6 (* Number of digits after decimal point *) ;

TYPE
  VectorN = ARRAY[ 1..MaxN ] OF REAL;
  VectorM = ARRAY[ 1..1 ] OF REAL;
  MatrixNN = ARRAY[ 1..MaxN , 1..MaxN ] OF REAL;
  MatrixNM = MatrixNN;
  IntVectorN = ARRAY[ 1..MaxN ] OF INTEGER;
  String14 = String[14];
  String80 = String[80];
  MatrixData = Record
                 LU : MatrixNN;
                 pivot : IntVectorN;
                 ConditionNumber : REAL;
                 residuals : VectorM
               END;

VAR
  A : MatrixNN;
  v : VectorN;
  n, iteration : INTEGER;
  answer : CHAR;
  new, old : REAL;
  OutFile : TEXT;
  display, Printing, converged : BOOLEAN;
  Data : MatrixData;

(*$I Response.Pas *)
(*$I Files.Pas *)
(*$I ReadNum.Pas *)
(*$I MatrixIO.Pas *)
(*$I MatrixUt.Pas *)
(*$I LinearSys.Pas *)


(****************************************************************************)
(*       Module 0   Utility procedures and functions                        *)
(****************************************************************************)

PROCEDURE QuerryPrinting;
VAR
  answer : CHAR;
BEGIN
  WRITE('Do you want the vectors printed?  (Y or N) ') ; Echo(answer) ;
  IF  answer IN ['Y','y']  THEN
```

```
    BEGIN
      ASSIGN(Outfile,'Prn') ; Printing := TRUE
    END
  ELSE
    BEGIN
      ASSIGN(OutFile,'Con') ; Printing := FALSE
    END;
  REWRITE(OutFile)
END; (* QuerryPrinting *)

PROCEDURE QuerryDisplay;
VAR
  answer : CHAR;
BEGIN
  WRITE('Do you want the vectors displayed?  (Y or N) ') ; Echo(answer) ;
  IF  answer IN ['Y','y']  THEN
    display := TRUE
  ELSE
    display := FALSE
END; (* QuerryDisplay *)

FUNCTION Max( v : VectorN ) : REAL;
VAR
  i : INTEGER;
  temp : REAL;
BEGIN
  temp := v[1];
  FOR  i := 2  TO  n  DO
    IF  Abs(v[i]) > Abs(temp)  THEN
      temp := v[i];
  Max := temp
END; (* Max *)

PROCEDURE Scale( VAR y : VectorN );
VAR
  i : INTEGER;
  t : REAL;
BEGIN
  t := Max(y);
  FOR  i := 1  TO  n  DO
    y[i] := y[i]/t
END; (* Scale *)

PROCEDURE Greeting;
BEGIN
  ClrScr;
  WRITELN('This program can be used to find either the largest-magnitude or');
  WRITELN('or the smallest-magnitude eigenvalue of a matrix, A.') ; WRITELN ;
  WRITELN('The methods available are:') ; WRITELN ;
  WRITELN('   (a) For the largest-magnitude eigenvalue') ; WRITELN ;
  WRITELN('            (1) The power method');
  WRITELN('            (2) The power method with Rayleigh quotients');
  WRITELN;
  WRITELN('   (b) For the smallest-magnitude eigenvalue of a nonsingular ',
                      'matrix') ; WRITELN ;
  WRITELN('            (3) The inverse power method') ; WRITELN ;
  WRITELN('   The default is the power method (1)') ; WRITELN
END; (* Greeting *)
```

```
(**************************************************************************)
(*       Module 1   Input and Output procedures                           *)
(**************************************************************************)

PROCEDURE DataIn( message1, message2 : String80 );
VAR
  answer : CHAR;
BEGIN
  WRITELN(message1) ; WRITELN(message2) ; WRITELN ;
  WRITELN('Get matrix A.');
  GetMatrix(A,n,n,TRUE,MaxN,MaxN,Field,Decimal);
  WRITE('Do you want to print the matrix A?  (Y or N)  ') ; Echo(answer) ;
  IF  answer IN ['Y','y']  THEN
    PutMatrix(A,n,n,Field,Decimal,'The matrix  A:','P')
END; (* DataIn *)

PROCEDURE WriteVector( v : VectorN ); (* This procedure outputs the vector v *
VAR
  i : INTEGER;
BEGIN
  WRITELN(OutFile,'The vector v(',iteration,')');
  FOR  i := 1  TO  n  DO
    WRITELN(OutFile,v[i]:Field:decimal);
  WRITELN(OutFile,'Estimated eigenvalue is ',new:Field:Decimal);
  IF  NOT Printing  THEN
    Pause
END; (* WriteVector *)


(**************************************************************************)
(*       Module 2   Initial setup and final answer                        *)
(**************************************************************************)

PROCEDURE Initialize;
VAR
  zero : BOOLEAN;
  i : INTEGER;
  answer : CHAR;
BEGIN
  zero := TRUE;
  WRITELN('Each component of v has been initially set to  1.');
  WRITE('Do you want to change it?  (Y or N) ') ; Echo(answer) ;
  IF  answer IN ['Y','y']  THEN
    FOR  i := 1  TO  n  DO
      BEGIN
        WRITE('enter the value of v[',i,'] :   ') ; v[i] := ReadReal ;
        IF  v[i] <> 0.0  THEN
          zero := FALSE
      END;
  IF  zero  THEN
    FOR  i := 1  TO  n  DO
      v[i] := 1.0;
  QuerryDisplay ; QuerryPrinting ; ClrScr ;
  iteration := 0 ; new := 1.0
END; (* Initialize *)
```

```
PROCEDURE ShowAnswer;
VAR
  i : INTEGER;
BEGIN
  WRITELN(OutFile,'The estimated eigenvalue is ',new:Field:Decimal);
  WRITELN(OutFile,'The estimated eigenvector is ');
  FOR  i := 1  TO  n  DO
    WRITELN(OutFile,v[i]:Field:Decimal);
 WRITELN(OutFile,'The number of iterations required was  ',iteration)
END; (* ShowAnswer *)

PROCEDURE Finalize;
BEGIN
  ShowAnswer;
  IF  NOT converged  THEN
    WRITELN(OutFile,'The algorithm did not converge after ',MaxIterations,
                ' iterations.')
END; (* Finalize *)

(****************************************************************************)
(*       Module 3   Power Method                                            *)
(****************************************************************************)

PROCEDURE PowerMethod;
CONST
  title =
    'The largest-magnitude eigenvalue of the matrix A, by the power method';
VAR
  y : vectorN;
BEGIN
  DataIn(title,' ');
  Initialize;
  REPEAT
    iteration := iteration + 1;
    VectorMultiply(n,n,A,v,y);
    v := y ; old := new ;
    new := Max(y);
    Scale(v);
    IF  display  THEN
      WriteVector(v);
    converged := Abs(new - old) <= tolerance
  UNTIL  converged OR (iteration = MaxIterations);
  Finalize
END; (* PowerMethod *)

(****************************************************************************)
(*       Module 4   Power Method using Rayleigh quotients                   *)
(****************************************************************************)

PROCEDURE PowerMethodWithRayleighQuotients;
CONST
  title1 =
    'The largest-magnitude eigenvalue of the matrix A, by the power method';
  title2 = 'using Rayleigh quotients';
VAR
  y : VectorN;
BEGIN (* main program *)
  DataIn(title1,title2);
  Initialize;
  REPEAT
    iteration := iteration + 1;
```

```
    VectorMultiply(n,n,A,v,y);
    old := new;
    new := DotProduct(n,v,y)/DotProduct(n,v,v);
    v := y;
    Scale(v);
    IF  display  THEN
      WriteVector(v);
    converged := Abs(new - old) <= tolerance
  UNTIL  converged OR (iteration = MaxIterations);
  Finalize
END; (* PowerMethodWithRayleighQuotients *)



(****************************************************************************)
(*       Module 5   Inverse Power Method                                    *)
(****************************************************************************)



PROCEDURE InversePowerMethod;
CONST
  title1 = 'The smallest-magnitude eigenvalue of the nonsingular matrix A,';
  title2 = 'by the inverse power method';
VAR
  flag, i : INTEGER;
  FirstTime : BOOLEAN;
  shift : REAL;


PROCEDURE QuerryShift;
VAR
  i : INTEGER;
  answer : CHAR;
BEGIN
  WRITE('Do you want to shift the axis?  (Y or N)    ') ; Echo(answer) ;
  IF  answer IN ['Y','y']  THEN
    BEGIN
      WRITE('Enter the desired value of the shift :    ') ; shift := ReadReal
    END
  ELSE
    shift := 0.0;
  FOR  i := 1  TO  n  DO
    A[i,i] := A[i,i] - shift
END; (* QuerryShift *)


PROCEDURE Solve;
(* This procedure solves the equation  Ay = v, using L-U decomposition *)
VAR
  b : MatrixNM;
  digits, i : INTEGER;
BEGIN
  ClrScr ; digits := 7 ;
  FOR  i := 1  TO  n  DO
    b[i,1] := v[i];
  LinearSystemSolver(n,A,1,b,digits,NOT FirstTime,flag,Data);
  IF  FirstTime  THEN
    FirstTime := FALSE;
```

```
  FOR  i := 1  TO  n  DO
    v[i] := b[i,1];
  IF  flag = 2  THEN
    WRITELN('The matrix  A  is singular')
END; (* Solve *)


BEGIN (* InversePowerMethod *)
  DataIn(title1,title2);
  QuerryShift;
  Initialize;
  FirstTime := TRUE;
  REPEAT
    iteration := iteration + 1;
    old := new;
    Solve;
    IF  flag <> 2  THEN
      BEGIN
        new := shift + 1.0/Max(v);
        Scale(v);
        IF  display  THEN
          BEGIN
            WRITELN(OutFile,'The vector v(',iteration,') is ');
            FOR  i := 1  TO  n  DO
              WRITELN(OutFile,v[i]:Field:Decimal);
            WRITELN(OutFile,'The estimated eigenvalue is ',
                                              new:Field:Decimal);
            Pause
          END
      END;
    converged := Abs(new - old) < tolerance
  UNTIL  converged OR (iteration = MaxIterations) OR (flag = 2);
  IF  flag <> 2  THEN
    Finalize
END; (* InversePowerMethod *)


(****************************************************************************)
(*       Module 6   Main program                                            *)
(****************************************************************************)


BEGIN
  LowVideo;
  REPEAT
    Greeting;
    WRITE('Select  1, 2, or 3  ') ; Echo(answer) ;
    IF  NOT (answer IN ['1','2','3'])  THEN
      answer := '1';
    ClrScr;
    CASE  answer  OF
           '1' : PowerMethod;
           '2' : PowerMethodWithRayleighQuotients;
           '3' : InversePowerMethod
    END;
    CLOSE(OutFile) ; WRITELN ;
    WRITE('Do you want to do another matrix?  (Y or N)   ') ; Echo(answer)
  UNTIL  NOT (answer IN ['Y','y'])
END.
```

APPENDIX 8.2

Listing of a Pascal Program for finding the Characteristic Polynomial of a matrix by either Leverrier's Method or Danilevsky's Method

```
PROGRAM CharacteristicPolynomial;

CONST
  MaxN = 9;
  Field = 14  (* Fieldwidth *) ;
  Decimal = 7 (* Number of digits after decimal point *) ;

TYPE
  MatrixNM = ARRAY[ 1..MaxN , 1..MaxN ] OF REAL;
  VectorN = ARRAY[ 1..MaxN ] OF REAL;
  String14 = String[14];
  String80 = String[80];

VAR
  A : MatrixNM;
  n : INTEGER;
  answer : CHAR;

(*$I Response.Pas *)
(*$I Files.Pas *)
(*$I ReadNum.Pas *)
(*$I MatrixIO.Pas *)
(*$I MatrixUt.Pas *)

(****************************************************************************)
(*     Module 0    Utility procedure                                        *)
(****************************************************************************)

PROCEDURE Greeting;
BEGIN
  ClrScr;
  WRITELN('This program finds the coefficients of the characteristic',
         ' polynomial, P(s),');
  WRITELN('of a matrix A.');
  WRITELN;
  WRITELN('The following methods are available:') ; WRITELN ;
  WRITELN('    (1) Leverrier''s method');
  WRITELN('    (2) Danilevsky''s method');
  WRITELN;
  WRITELN('The default is Leverrier''s method');
  WRITELN
END; (* Greeting *)


(****************************************************************************)
(*     Module 1    Input/output procedures                                  *)
(****************************************************************************)

PROCEDURE DataIn;
VAR
  answer : CHAR;
```

```
BEGIN
  WRITELN('Get matrix A.');
  GetMatrix(A,n,n,TRUE,MaxN,MaxN,Field,Decimal);
  WRITE('Do you want to print A?  (Y or N)  ') ; Echo(answer) ;
  IF  answer IN ['Y','y']  THEN
    PutMatrix(A,n,n,Field,Decimal,'The matrix A:','P')
END; (* DataIn *)


PROCEDURE WriteVector( VAR w : VectorN ; d : CHAR ; switch : BOOLEAN );
VAR
  OutFile : TEXT;
  i : INTEGER;
BEGIN
  IF  d IN ['P','p']  THEN
    ASSIGN(OutFile,'Prn')
  ELSE
    ASSIGN(OutFile,'Con');
  REWRITE(OutFile) ; WRITELN(OutFile) ;
  IF  switch  THEN  (* Leverrier's method *)
    FOR  i := 1  TO  n  DO
      WRITELN(OutFile,'c[',i,'] = ',-w[i]:Field:Decimal)
  ELSE
    FOR  i := 1  TO  n  DO (* Danilevsky's method *)
      WRITELN(OutFile,'c[',n-i,'] = ',w[i]:Field:Decimal);
  CLOSE(OutFile)
END; (* WriteVector *)


PROCEDURE OutputMatrix( VAR A : MATRIXNM ; title : String80 );
VAR
  answer : CHAR;
BEGIN
  PutMatrix(A,n,n,Field,Decimal,Title,'C');
  WRITELN;
  WRITE('Do you want to print the matrix?  (Y or N)  ') ; Echo(answer) ;
  IF  answer IN ['Y','y']  THEN
    PutMatrix(A,n,n,Field,Decimal,Title,'P');
  WRITE('Do you want to store the matrix in a file ?   (Y or N)  ');
  Echo(answer);
  IF  answer IN ['Y','y']  THEN
    PutMatrix(A,n,n,Field,Decimal,Title,'F')
END; (* OutputMatrix *)


PROCEDURE OutputVector( VAR w : VectorN ; message1, message2 : String80 ;
                                                         switch : BOOLEAN )
VAR
  answer : CHAR;
BEGIN
  ClrScr;
  WRITELN('The coefficients of the characteristic polynomial, P(s),',
          ' of degree  n = ',n) ;  WRITELN ;
  WRITELN(message1) ; WRITELN(message2) ;
  WriteVector(w,'C',switch) ; WRITELN ;
  WRITE('Print the coefficients of the characteristic polynomial?');
  WRITE('   (Y or N)   ') ; Echo(answer) ;
  IF  answer IN ['Y','y']  THEN
    WriteVector(w,'P',switch)
END; (* OutputVector *)
```

```
(****************************************************************************)
(*     Module 2    Leverrier's Method                                       *)
(****************************************************************************)


PROCEDURE LeverrierMethod;
CONST
  message1 = '         n             n-1';
  message2 = 'P(s) = s   +  c[1]*s     +  ... + c[n-1]*s + c[n]';
VAR
  Ai : MatrixNM;
  c : VectorN;


PROCEDURE MainAlgorithm;
VAR
  i, j, k, m : INTEGER;
  temp, B : MatrixNM;
  t : REAL;
BEGIN
  B := A ; c[1] := Trace(n,B) ;
  FOR  k := 2  TO  n-1  DO
    BEGIN
      FOR  i := 1  TO  n  DO
        FOR  j := 1  TO  n  DO
          BEGIN
            t := 0.0;
            FOR  m := 1  TO  n  DO
              IF  m = j  THEN
                t := t + A[i,j]*(B[j,j] - c[k-1])
              ELSE
                t := t + A[i,m]*B[m,j];
            temp[i,j] := t
          END;
      B := temp ; c[k] := Trace(n,B)/k
    END;
  Ai := B;
  FOR  i := 1  TO  n  DO
    Ai[i,i] := Ai[i,i] - c[n-1];
  MatrixMultiply(n,n,n,A,Ai,B);
  c[n] := Trace(n,B)/n;
  IF  c[n] <> 0.0  THEN
    FOR  i := 1  TO n  DO
      FOR  j := 1  TO  n  DO
        Ai[i,j] := Ai[i,j]/c[n]
END; (* MainAlgorithm *)


BEGIN (* LeverrierMethod *)
  DataIn;
  MainAlgorithm;
  IF  c[n] = 0.0  THEN
    WRITELN('The matrix  A  is singular')
  ELSE
    OutputMatrix(Ai,'A-inverse:');
  OutputVector(c,message1,message2,TRUE)
END; (* LeverrierMethod *)
```

```
(*****************************************************************************)
(*     Module 3    Danilevsky's Method                                       *)
(*****************************************************************************)

PROCEDURE DanilevskyMethod;
CONST
  message1 = '         n              n-1';
  message2 = 'P(s) = s   +  c[n-1]*s     +  ... + c[1]*s + c[0]';
VAR
  v, w : VectorN;
  largest : INTEGER;
  pivot : REAL;
  flag : BOOLEAN;

PROCEDURE RowInterchange( j, k : INTEGER );
(* This procedure interchanges rows j and k of the matrix A *)
VAR
  i : INTEGER;
  temp : REAL;
BEGIN
  FOR  i := 1  TO  n  DO
    BEGIN
      temp := A[j,i] ; A[j,i] := A[k,i] ; A[k,i] := temp
    END
END; (* RowInterchange *)

PROCEDURE ColumnInterchange( j, k : INTEGER );
(* This procedure interchanges columns j and k of the matrix A *)
VAR
  i : INTEGER;
  temp : REAL;
BEGIN
  FOR  i := 1  TO  n  DO
    BEGIN
      temp := A[i,j] ; A[i,j] := A[i,k] ; A[i,k] := temp
    END
END; (* ColumnInterchange *)

PROCEDURE Max( k : INTEGER );
(* Returns the (column) location of the largest element in
   row k, before the k-th column and sets the pivot as this element *)
VAR
  i : INTEGER;
  temp : REAL;
BEGIN
  largest := 1 ; temp := Abs(A[k,largest]) ;
  FOR  i := 2  TO  k-1  DO
    BEGIN
      IF temp < Abs(A[k,i]) THEN
        BEGIN
          temp := Abs(A[k,i]) ; largest := i
        END
    END;
  pivot := A[k,largest]
END; (* Max *)
```

```
PROCEDURE CopyRow( k : INTEGER );
(* Copies the k-th row of matrix A as vector v *)
VAR
  j : INTEGER;
BEGIN
  FOR  j := 1  TO  n  DO
    v[j] := A[k,j]
END; (* CopyRow *)



PROCEDURE ScaleColumn( k : INTEGER );
(* Divides the k-th column of A by the pivot *)
VAR
  i : INTEGER;
BEGIN
  FOR  i := 1  TO  n  DO
    A[i,k] := A[i,k]/pivot
END; (* ScaleColumn *)



PROCEDURE TransformOneRow( k :INTEGER );
(* Transforms the k-th row to the (k-1)-th unit vector *)
VAR
  i, j : INTEGER;
  temp : REAL;
BEGIN
  IF  largest <> k-1  THEN
  (* Rows and columns are interchanged to make the largest element
     the pivot *)
    BEGIN
      ColumnInterchange(largest,k-1);
      RowInterchange(largest,k-1)
    END;
  CopyRow(k); (* The k-th row is copied as vector v *)
  ScaleColumn(k-1); (* The (k-1)-th column of A is divided by the pivot *)
  (* From the remaining columns subtract v[j] times the (k-1)-th column *)
  FOR  j := 1  TO  k-2  DO
    FOR  i := 1  TO  n  DO
      A[i,j] := A[i,j] - v[j]*A[i,k-1];
  FOR  j := k  TO  n  DO
    FOR  i := 1  TO  n  DO
      A[i,j] := A[i,j] - v[j]*A[i,k-1];
 (* Complete the transformation by replacing the elements in the
    (k-1)-th row by the dot product of v and the corresponding column *)
  FOR  j := 1  TO  n  DO
    BEGIN
      temp := 0.0;
      FOR  i := 1  TO  n  DO
        temp := temp + v[i]*A[i,j];
      A[k-1,j] := temp
    END
END; (* TransformOneRow *)
```

```
PROCEDURE MainAlgorithm;
VAR
  j, k : INTEGER;
BEGIN
  flag := FALSE;
  FOR  k := n  DOWNTO  2  DO
    BEGIN
      Max(k);
      IF  pivot <> 0.0  THEN
        TransformOneRow(k)
      ELSE
        BEGIN
          flag := TRUE;
          WRITELN('The matrix A cannot be put in the Frobenius form.');
          OutputMatrix(A,'The matrix A:')
        END
    END;
  (* If successful then get the vector w from the first row of the
     transformed matrix.  w gives the coefficients of the characteristic
     polynomial *)
  IF  NOT flag  THEN
    FOR  j := 1  TO  n  DO
      w[j] := -A[1,j]
END; (* MainAlgorithm *)



BEGIN (* DanilevskyMethod *)
  DataIn;
  MainAlgorithm;
  IF  NOT flag  THEN
    BEGIN
      OutputMatrix(A,'The matrix in the companion form:');
      OutputVector(w,message1,message2,FALSE)
    END
END; (* DanilevskyMethod *)


(***************************************************************************)
(*     Module 4    Main Program                                           *)
(***************************************************************************)


BEGIN
  LowVideo;
  REPEAT
    Greeting;
    WRITE('Select 1 or 2    ') ; Echo(answer) ;
    IF  NOT (answer IN ['1','2'])  THEN
      answer := '1';
    IF  answer = '1'  THEN
      LeverrierMethod
    ELSE
      DanilevskyMethod;
    WRITELN;
    WRITE('Do another matrix?  (Y or N)    ') ; Echo(answer)
  UNTIL  NOT (answer IN ['Y','y'])
END.
```

APPENDIX 8.3

Listing of a Pascal Program that performs Similarity Transformations on Real Symmetric Matrices using Jacobi's Method, Givens' Method, or Householder's Method

```
PROGRAM SimilarityTransformationsOnRealSymmetricMatrices;

CONST
  MaxN = 10;
  Field = 14  (* Fieldwidth *) ;
  Decimal = 7 (* Number of digits after decimal point *) ;

TYPE
  MatrixNM = ARRAY[ 1..MaxN , 1..MaxN ] OF REAL;
  VectorN = ARRAY[ 1..MaxN ] OF REAL;
  String14 = String[14];
  String80 = String[80];

VAR
  A : MatrixNM;
  n : INTEGER;
  answer : CHAR;

(*$I Response.Pas *)
(*$I Files.Pas *)
(*$I ReadNum.Pas *)
(*$I MatrixIO.Pas *)
(*$I MatrixUt.Pas *)

(****************************************************************************)
(*     Module 0    Utility procedure                                        *)
(****************************************************************************)

PROCEDURE Greeting;
BEGIN
  ClrScr;
  WRITELN('This program performs similarity transformations on a real ',
          'symmetric matrix A.');
  WRITELN;
  WRITELN('The following methods are available:') ; WRITELN ;
  WRITELN('    (1) Jacobi''s method');
  WRITELN('    (2) Givens'' method');
  WRITELN('    (3) Householder''s method');
  WRITELN;
  WRITELN('The default is Householder''s method');
  WRITELN
END; (* Greeting *)

(****************************************************************************)
(*     Module 1    Input/output procedures                                  *)
(****************************************************************************)

PROCEDURE DataIn( message1, message2 : String80 );
VAR
  answer : CHAR;
```

```
BEGIN
  ClrScr;
  WRITELN(message1) ; WRITELN(message2) ; WRITELN ;
  WRITELN('Get the real symmetric matrix A.');
  GetMatrix(A,n,n,TRUE,MaxN,MaxN,Field,Decimal);
  WRITE('Do you want to print A?  (Y or N)  ') ; Echo(answer) ;
  IF  answer IN ['Y','y']  THEN
    PutMatrix(A,n,n,Field,Decimal,'The matrix A:','P')
END; (* DataIn *)

PROCEDURE WriteVector( VAR w : VectorN ; d : CHAR );
VAR
  OutFile : TEXT;
  i : INTEGER;
BEGIN
  IF  d IN ['P','p']  THEN
    ASSIGN(OutFile,'Prn')
  ELSE
    ASSIGN(OutFile,'Con');
  REWRITE(OutFile) ; WRITELN(OutFile) ;
  FOR  i := 1  TO  n  DO
    WRITELN(OutFile,w[i]:Field:Decimal);
  CLOSE(OutFile)
END; (* WriteVector *)

PROCEDURE OutputMatrix( title : String80 );
VAR
  answer : CHAR;
BEGIN
  PutMatrix(A,n,n,Field,Decimal,title,'C');
  WRITELN;
  WRITE('Do you want to print the matrix?  (Y or N)  ') ; Echo(answer) ;
  IF  answer IN ['Y','y']  THEN
    PutMatrix(A,n,n,Field,Decimal,title,'P');
  WRITE('Do you want to store the matrix in a file ?  (Y or N)  ');
  Echo(answer);
  IF  answer IN ['Y','y']  THEN
    PutMatrix(A,n,n,24,-1,title,'F')
END; (* OutputMatrix *)

(***************************************************************************)
(*     Module 2    Jacobi's method                                        *)
(***************************************************************************)

PROCEDURE JacobiMethod;
CONST
  title =
 'Rotation transformations are performed as required in the Jacobi algorithm';
  name = 'The transformed matrix eliminating element ';
VAR
  P, Ptranspose, Temp : MatrixNM;
  k, m : INTEGER;
  answer : CHAR;
  continue : BOOLEAN;

PROCEDURE GetLargestMagnitudeOffDiagonalElement( VAR k, m : INTEGER );
VAR
  i, j : INTEGER;
  max, t : REAL;
```

```
BEGIN
  max := 0.0;
  FOR  i := 1  TO  n-1  DO
    FOR  j := i+1  TO  n  DO
      BEGIN
        t := Abs(A[i,j]);
        IF  t > max  THEN
          BEGIN
            max := t ; k := i ; m := j
          END
      END;
  IF  max = 0.0  THEN
    continue := FALSE
  ELSE
    continue := TRUE
END; (* GetLargestMagnitudeOffDiagonalElement *)


PROCEDURE GetRotationMatrix;
(* Get the location of the element to be annihilated and obtain the
   transformation matrix *)
VAR
  cot2theta, TanTheta, LargeRoot, c, s : REAL;
BEGIN
  GetLargestMagnitudeOffDiagonalElement(k,m);
  IF  continue  THEN
    BEGIN
      cot2theta := (A[k,k] - A[m,m])/(2.0*A[k,m]);
      IF  cot2theta > 0.0  THEN
        LargeRoot := -cot2theta - Sqrt(1.0 + Sqr(cot2theta))
      ELSE
        LargeRoot := -cot2theta + Sqrt(1.0 + Sqr(cot2theta));
      TanTheta := -1.0/LargeRoot;
      c := 1.0/Sqrt(1.0 + Sqr(TanTheta)) ; s := c*TanTheta ;
      P[m,m] := c ; P[k,k] := c ; P[m,k] := s ; P[k,m] := -s
    END
END; (* GetRotationMatrix *)

BEGIN (* JacobiMethod *)
  DataIn(title,' ');
  REPEAT
    IdentityMatrix(n,P);
    GetRotationMatrix;
    IF  continue  THEN
      BEGIN
        Transpose(n,n,P,Ptranspose);
        MatrixMultiply(n,n,n,Ptranspose,A,Temp);
        MatrixMultiply(n,n,n,Temp,P,A);
        OutputMatrix(Concat(name,chr(k+48),', ',chr(m+48)));
        IF  continue  THEN
          BEGIN
            WRITE('Do you want to do more rotations?  (Y or N)  ');
            Echo(answer)
          END
        ELSE
          answer := 'n'
      END
    ELSE
      answer := 'n'
  UNTIL  NOT (answer IN ['Y','y'])
END; (* JacobiMethod *)
```

```
(*****************************************************************************)
(*     Module 3    Givens' method                                            *)
(*****************************************************************************)

PROCEDURE GivensMethod;
CONST
  title1 = 'Rotation transformations are performed as required in the';
  title2 = 'Givens'' algorithm';
  name = 'The transformed matrix eliminating element ';
VAR
  P, Ptranspose, Temp : MatrixNM;
  k, m : INTEGER;
  answer : CHAR;

PROCEDURE GetRotationMatrix( k, m : INTEGER );
(* Obtain the transformation matrix annihilating the (k,m) element of  A *)
VAR
  temp, c, s : REAL;
BEGIN
  temp := A[k,m]/A[m+1,m] ; c := 1.0/Sqrt(1.0 + Sqr(temp)) ; s := c*temp ;
  P[m+1,m+1] := c ; P[k,k] := c ; P[m+1,k] := -s ; P[k,m+1] := s
END; (* GetRotationMatrix *)

BEGIN (* GivensMethod *)
  DataIn(title1,title2);
  m := 0;
  REPEAT
    m := m + 1;
    FOR  k := m+2  TO  n  DO
      BEGIN
        IdentityMatrix(n,P);
        GetRotationMatrix(k,m);
        Transpose(n,n,P,Ptranspose);
        MatrixMultiply(n,n,n,Ptranspose,A,Temp);
        MatrixMultiply(n,n,n,Temp,P,A);
        OutputMatrix(Concat(name,chr(k+48),', ',chr(m+48)))
      END;
    IF  m < n-2  THEN
      BEGIN
        WRITE('Do you want to do more rotations?  (Y or N)  ');
        Echo(answer)
      END
    ELSE
      answer := 'n'
  UNTIL  answer IN ['N','n']
END; (* GivensMethod *)

(*****************************************************************************)
(*     Module 4    Householder's method                                      *)
(*****************************************************************************)

PROCEDURE HouseholderMethod;
CONST
  title =
    'Reflection transformations as required in the Householder algorithm';
  name = 'The transformed matrix eliminating column ';
VAR
  R : MatrixNM;
  w : VectorN;
  k : INTEGER;
  answer : CHAR;
```

```
PROCEDURE FormW;
(* This procedure forms the vector w for the reflection matrix *)
VAR
  j : INTEGER;
  s, alpha, SumOfSquares : REAL;
BEGIN
  FOR  j := 1  TO  k  DO
    w[j] := 0.0;
  SumOfSquares := 0.0;
  FOR  j := k+1  TO  n  DO
    SumOfSquares := SumOfSquares + Sqr(A[k,j]);
  IF  A[k,k+1] < 0.0  THEN
    s := -Sqrt(SumOfSquares)
  ELSE
    s := Sqrt(SumOfSquares);
  alpha := Sqrt(2.0*s*(s + A[k,k+1]));
  w[k+1] := (s + A[k,k+1])/alpha;
  FOR  j := k+2  TO  n  DO
    w[j] := A[k,j]/alpha
END; (* FormW *)

PROCEDURE MainAlgorithm;
VAR
  i, j : INTEGER;
  Temp1, Temp2 : MatrixNM;
BEGIN
  ClrScr ; FormW ;
  WRITELN('The vector w') ; WRITELN ; WriteVector(w,'C') ;
  IdentityMatrix(n,R);
  OuterProduct(n,n,w,w,Temp1);
  FOR  i := 1  TO  n  DO
    FOR  j := 1  TO  n  DO
      R[i,j] := R[i,j] - 2.0*Temp1[i,j];
  PutMatrix(R,n,n,Field,Decimal,'The transformation matrix ','C');
  MatrixMultiply(n,n,n,R,A,Temp2);
  MatrixMultiply(n,n,n,Temp2,R,A);
  OutputMatrix(Concat(name,chr(k+48)))
END; (* MainAlgorithm *)

BEGIN (* HouseholderMethod *)
  DataIn(title,' ') ; k := 0 ;
  REPEAT
    k := k + 1;
    MainAlgorithm;
    IF  k < n-2  THEN
      BEGIN
        WRITE('Do you want to reduce more columns?  (Y or N)  ');
        Echo(answer)
      END
    ELSE
      answer := 'n'
  UNTIL  answer IN ['N','n']
END; (* HouseholderMethod *)

(****************************************************************************)
(*     Module 5    Main Program                                            *)
(****************************************************************************)

BEGIN
  LowVideo;
```

```
  REPEAT
    Greeting;
    WRITE('Select 1, 2, or 3   ') ; Echo(answer) ;
    IF  NOT (answer IN ['1','2','3'])  THEN
      answer := '3';
    CASE  answer  OF
           '1' : JacobiMethod;
           '2' : GivensMethod;
           '3' : HouseholderMethod
    END;
    OutputMatrix('The final transformed matrix:');
    WRITELN;
    WRITE('Do another matrix?  (Y or N)   ') ; Echo(answer)
  UNTIL  NOT (answer IN ['Y','y'])
END.
```

APPENDIX 8.4

Listing of a Pascal Program that performs Similarity Transformations on Real Matrices using either the QR Algorithm or a reduction to Upper Hessenberg Form

```
PROGRAM SimilarityTransformationsOnRealMatrices;

CONST
  MaxN = 10;
  Field = 14  (* Fieldwidth *) ;
  Decimal = 7 (* Number of digits after decimal point *) ;

TYPE
  MatrixNM = ARRAY[ 1..MaxN , 1..MaxN ] OF REAL;
  VectorN = ARRAY[ 1..MaxN ] OF REAL;
  String14 = String[14];
  String80 = String[80];

VAR
  A : MatrixNM;
  n : INTEGER;
  answer : CHAR;

(*$I Response.Pas *)
(*$I Files.Pas *)
(*$I ReadNum.Pas *)
(*$I MatrixIO.Pas *)
(*$I MatrixUt.Pas *)

(***************************************************************************)
(*     Module 0    Utility procedure                                      *)
(***************************************************************************)

PROCEDURE Greeting;
BEGIN
  ClrScr;
  WRITELN('This program performs similarity transformations on a real',
        ' matrix A.');
  WRITELN;
  WRITELN('The following methods are available:') ; WRITELN ;
  WRITELN('    (1) The QR method');
```

```
  WRITELN('    (2) Reduction to upper Hessenberg form');
  WRITELN;
  WRITELN('The default is reduction to upper Hessenberg form');
  WRITELN
END; (* Greeting *)


(****************************************************************************)
(*     Module 1    Input/output procedures                                  *)
(****************************************************************************)

PROCEDURE DataIn( message1, message2 : String80 );
VAR
  answer : CHAR;
BEGIN
  ClrScr;
  WRITELN(message1) ; WRITELN(message2) ; WRITELN ;
  WRITELN('Get matrix A.');
  GetMatrix(A,n,n,TRUE,MaxN,MaxN,Field,Decimal);
  WRITE('Do you want to print A?  (Y or N)  ') ; Echo(answer) ;
  IF  answer IN ['Y','y']  THEN
    PutMatrix(A,n,n,Field,Decimal,'The matrix A:','P')
END; (* DataIn *)

PROCEDURE OutputMatrix;
CONST
  title = 'The transformed matrix:';
VAR
  answer : CHAR;
BEGIN
  PutMatrix(A,n,n,Field,Decimal,title,'C');
  WRITELN;
  WRITE('Do you want to print the matrix?  (Y or N)  ') ; Echo(answer) ;
  IF  answer IN ['Y','y']  THEN
    PutMatrix(A,n,n,Field,Decimal,title,'P');
  WRITE('Do you want to store the matrix in a file ?   (Y or N)  ');
  Echo(answer);
  IF  answer IN ['Y','y']  THEN
    PutMatrix(A,n,n,24,-1,title,'F')
END; (* OutputMatrix *)

PROCEDURE Questions( VAR NumberOfIterations : INTEGER ; VAR shift : REAL );
VAR
  x, y : INTEGER;
  answer : CHAR;
BEGIN
  WRITE('How many iterations do you want?  ');
  REPEAT
    x := WhereX ; y := WhereY ; NumberOfIterations := ReadInteger ;
    IF  NumberOfIterations < 1  THEN
      ErrorHandler(x,y,7,'Invalid number of iterations')
  UNTIL  NumberOfIterations >= 1;
  WRITE('Do you want a shift of the axis?  (Y or N)  ');
  Echo(answer) ; shift := 0.0 ;
  IF  answer IN ['Y','y']  THEN
    BEGIN
      WRITE('Enter the desired shift:  ') ; shift := ReadReal
    END
END; (* Questions *)
```

```
(*****************************************************************************)
(*     Module 2    QR method                                                   *)
(*****************************************************************************)

PROCEDURE QRmethod;
CONST
  title1 =
  'A specified number of iterations of the QR algorithm will be performed';
  title2 = 'on an upper Hessenberg form matrix A.';
  tolerance = 1.0E-16;
VAR
  P, B, NewA, C : MatrixNM;
  x, y, NumberOfIterations, k, m, count : INTEGER;
  shift : REAL;

PROCEDURE ShiftAxis( VAR M : MatrixNM );
VAR
  i : INTEGER;
BEGIN
  FOR  i := 1  TO  n  DO
    M[i,i] := M[i,i] - shift
END; (* ShiftAxis *)

PROCEDURE ShiftBack( VAR M : MatrixNM );
VAR
  i : INTEGER;
BEGIN
  FOR  i := 1  TO  n  DO
    M[i,i] := M[i,i] + shift
END; (* ShiftBack *)

PROCEDURE GetRotationMatrix;
VAR
  temp, c, s : REAL;
BEGIN
  IdentityMatrix(n,P);
  IF  Abs(NewA[m,m]) < tolerance  THEN
    NewA[m,m] := 0.0;
  IF  Abs(NewA[k,m]) < tolerance  THEN
    NewA[k,m] := 0.0;
  temp := Sqrt(Sqr(NewA[m,m]) + Sqr(NewA[k,m]));
  IF  temp <> 0.0  THEN
    BEGIN
      c := NewA[m,m]/temp ; s := NewA[k,m]/temp
    END
  ELSE
    BEGIN
      c := 1.0 ; s := 0.0
    END;
  P[m,m] := c ; P[k,k] := c ; P[m,k] := -s ; P[k,m] := s
END; (* GetRotationMatrix *)

PROCEDURE MainAlgorithm;
VAR
  Temp, Ptr : MatrixNM;
BEGIN
  Transpose(n,n,P,Ptr);
  MatrixMultiply(n,n,n,PTr,NewA,Temp);
  NewA := Temp;
  MatrixMultiply(n,n,n,C,P,B);
```

```
  IF  Abs(NewA[k,m]) < tolerance  THEN
    NewA[k,m] := 0.0;
  C := B
END; (* MainAlgorithm *)

BEGIN (* QRmethod *)
  DataIn(title1,title2) ; Questions(NumberOfIterations,shift) ;
  WRITE('Iteration ') ; x := WhereX ; y := WhereY ;
  FOR  count := 1  TO  NumberOfIterations  DO
    BEGIN
      GoToXY(x,y) ; ClrEol ; WRITE(count) ;
      IF  (count MOD 50) = 0  THEN
        WRITELN;
      IdentityMatrix(n,C);
      NewA := A;
      ShiftAxis(NewA);
      FOR  m := 1  TO  n-1  DO
        BEGIN
          k := m + 1;
          GetRotationMatrix;
          MainAlgorithm
        END;
      MatrixMultiply(n,n,n,NewA,B,A);
      ShiftBack(A)
    END
END; (* QRmethod *)

(****************************************************************************)
(*     Module 3    Reduction to upper Hessenberg form                       *)
(****************************************************************************)

PROCEDURE HessenbergForm;
CONST
  title1 = 'Transforms a real matrix to upper Hessenberg form.';
  title2 = 'Symmetric matrices are made tridiagonal.';
VAR
  P, C, Ptranspose : MatrixNM;
  k, m : INTEGER;

PROCEDURE GetRotationMatrix( k, m : INTEGER );
(* Obtain the transformation matrix annihilating the element (k,m) of A *)
VAR
  temp, c, s : REAL;
BEGIN
  temp := A[k,m]/A[m+1,m] ; c := 1.0/Sqrt(1.0 + Sqr(temp)) ; s := c*temp ;
  P[m+1,m+1] := c ; P[k,k] := c ; P[m+1,k] := -s ; P[k,m+1] := s
END; (* GetRotationMatrix *)

BEGIN (* HessenbergForm *)
  DataIn(title1,title2);
  FOR  m := 1  TO  n-2  DO
    FOR  k := m+2  TO  n  DO
      BEGIN
        IdentityMatrix(n,P);
        GetRotationMatrix(k,m);
        Transpose(n,n,P,Ptranspose);
        MatrixMultiply(n,n,n,Ptranspose,A,C);
        MatrixMultiply(n,n,n,C,P,A)
      END
END; (* HessenbergForm *)
```

```
(****************************************************************************)
(*     Module 4    Main Program                                             *)
(****************************************************************************)

BEGIN
  LowVideo;
  REPEAT
    Greeting;
    WRITE('Select 1 or 2   ') ; Echo(answer) ;
    IF  NOT (answer IN ['1','2'])  THEN
      answer := '2';
    IF  answer = '1'  THEN
      QRmethod
    ELSE
      HessenbergForm;
    OutputMatrix;
    WRITELN;
    WRITE('Do another matrix?  (Y or N)   ') ; Echo(answer)
  UNTIL  NOT (answer IN ['Y','y'])
END.
```

APPENDIX 8.5

User's Manual for the Pascal Procedure EigenR, which is in the file EigenR.Pas

```
PROCEDURE EigenR( VAR A : MatrixNM ; n : INTEGER ; VAR EigenValue : Cvector ;
       VAR EigenVector : Cmatrix ; VAR Index : REAL ; VAR flag : INTEGER );
```

The procedure EigenR finds the eigenvalues and eigenvectors of a real n×n matrix A. The following must be declared in the calling program

```
TYPE
  VectorN   = ARRAY[ 1..MaxN ] OF REAL;
  MatrixNM  = ARRAY[ 1..MaxN , 1..MaxM ] OF REAL;
  Cvector  = RECORD
               Re, Im : VectorN
             END;
  Cmatrix = RECORD
              Re, Im : MatrixNM
            END;
```

Input:

A   The matrix whose eigenvalues and eigenvectors are to be found
n   The order of the matrix

Output:

EigenValue.Re[k]   The real part of eigenvalue k
EigenValue.Im[k]   The imaginary part of eigenvalue k
EigenVector.Re[j,k]   The real part of the j-th component of the k-th eigenvector
EigenVector.Im[j,k]   The imaginary part of the j-th component of the k-th eigenvector
Index   The performance index.
If Index < 1.0 then the performance of the routines is considered to be excellent, in the sense that the residuals

```
              Az - wz  are as small as can be expected.  When
                              1.0 < Index < 100.0
              the performance is good.  When  Index > 100.0  the performance
              is considered poor.
      flag = n  Normal termination; all eigenvalues/eigenvectors have been
                  successfully found.
      flag < n  Error termination; procedure QRmethod failed to converge.
                  Eigenvalues  1,2,...,flag  have been computed correctly.
                  Eigenvalues  flag+1,flag+2,...,n  are set to  zero and
                  Index  is set to 1000.0 .
 The eigenvalues and eigenvectors are sorted so that
        EigenValue.Re[k+1] ≤ EigenValue.Re[k]  for  1 ≤ k ≤ (flag - 1).


Procedure EigenR is based on the IMSL FORTRAN subroutine EIGRF
```

APPENDIX 8.6

Listing of a Pascal Program for finding Eigenvalues and Eigenvectors of a Real Matrix using the Procedure EigenR

```
PROGRAM EigenvaluesEigenvectors; (* of an  n by n  real matrix *)

CONST
  MaxN = 12;
  Field = 12;  (* Fieldwidth *)
  Decimal = 8; (* Number of digits after the decimal point *)

TYPE
  VectorN   = ARRAY[ 1..MaxN ] OF REAL;
  MatrixNM  = ARRAY[ 1..MaxN , 1..MaxN ] OF REAL;
  Cvector  = RECORD
               Re, Im : VectorN
             END;
  Cmatrix = RECORD
               Re, Im : MatrixNM
             END;
  String14 = String[14];
  String80 = String[80];

VAR
  A : MatrixNM;
  EigenValue : Cvector;
  EigenVector : Cmatrix;
  Index : REAL;
  flag, n : INTEGER;
  OutFile : TEXT;
  answer  : CHAR;
  FileName : String14;

(*$I Response.Pas *)
(*$I Files.Pas *)
(*$I ReadNum.Pas *)
```

```
(*$I MatrixIO.Pas *)
(*$I EigenR.Pas *)

PROCEDURE PrintComplex( a, b : REAL );
(* print the complex number  a + jb  *)
BEGIN
  WRITE(OutFile,a:Field:Decimal);
  IF  b >= 0.0  THEN
    WRITE(OutFile,'  + j ')
  ELSE
    WRITE(OutFile,'  - j ');
  WRITE(OutFile,Abs(b):Field:Decimal)
END; (* PrintComplex *)


PROCEDURE DataIn;
VAR
  answer : CHAR;
BEGIN
  ClrScr ; LowVideo ;
  WRITE('This program finds the eigenvalues and eigenvectors');
  WRITELN(' of a real matrix A.');
  GetMatrix(A,n,n,TRUE,MaxN,MaxN,Field,Decimal);
  WRITE('Do you want to print the matrix A?  (Y or N)   ') ; Echo(answer) ;
  IF  answer IN ['Y','y']  THEN
    PutMatrix(A,n,n,Field,Decimal,'The matrix A:','P')
END; (* DataIn *)

PROCEDURE DataOut;
VAR
  i, k : INTEGER;
BEGIN
  WRITELN(OutFile);
  FOR  k := 1  TO  flag  DO
    BEGIN
      WRITE(OutFile,'Eigenvalue[',k,'] = ');
      PrintComplex(Eigenvalue.Re[k],Eigenvalue.Im[k]);
      WRITELN(OutFile) ; WRITELN(OutFile) ;
      WRITE(OutFile,'Eigenvector[',k,'] :');
      FOR  i := 1  TO  n  DO
        BEGIN
          IF  i > 1  THEN
            WRITE(OutFile,' ':16);
          PrintComplex(EigenVector.Re[i,k],EigenVector.Im[i,k]);
          WRITELN(OutFile)
        END;
      WRITELN(OutFile)
    END;
  IF  flag = n  THEN
    WRITELN(OutFile,'performance index = ',Index:8:3);
  CLOSE(OutFile)
END; (* DataOut *)

BEGIN (* main program *)
  DataIn;
  WRITE('Do you want the output on the Console or the Printer?');
  WRITE(' (C or P)  ') ; Echo(answer) ;
  IF  answer IN ['P','p']  THEN
    ASSIGN(OutFile,'Prn')
```

```
  ELSE
    ASSIGN(OutFile,'Con');
  REWRITE(OutFile);
  EigenR(A,n,Eigenvalue,EigenVector,Index,Flag);
  DataOut;
  WRITELN;
  WRITE('Do you want to store the eigenvalues and ',
        'eigenvectors in a file?  (Y or N) ');
  Echo(answer);
  IF  answer in ['Y','y']  THEN
    BEGIN
      CreateFile(OutFile,FileName);
      DataOut
    END
END.
```

APPENDIX 8.7

User's Manual for the Pascal Procedure EigenRS, which is in the file EigenRS.Pas

```
PROCEDURE EigenRS( VAR c : MatrixNM ; n : INTEGER ; VAR EigenValue : VectorN ;
        VAR EigenVector : MatrixNM ; VAR Index : REAL ; VAR flag : INTEGER );

This procedure finds the eigenvalues and eigenvectors of the real, symmetric
nxn matrix  c.

The following must be declared in the calling program

     TYPE
       VectorN  = ARRAY[ 1..MaxN ] OF REAL;
       MatrixNM = ARRAY[ 1..MaxN , 1..MaxN ] OF REAL;
       LongVector = ARRAY[ 1..n2 ] OF REAL;

     where  n2 = MaxN*(MaxN + 1)/2

   Input:
        c     The real symmetric matrix whose eigenvalues and eigenvectors
               are to be found.
        n     The order of the matrix  c.

   Output:
        EigenValue[k]      The k-th eigenvalue of  c.
        EigenVector[j,k]   The j-th component of the k-th eigenvector of  c.
                            The eigenvectors are returned are unit vectors.
        Index   The performance index.
                  If  Index < 1.0  then the performance of the routines is
                  considered to be excellent, in the sense that the residuals
                  c*z - w*z  are as small as can be expected. (Here, z  is an
                  eigenvector corresponding to the eigenvalue  w.)  When
                               1.0 < Index < 100.0
                  the performance is good.  When  Index > 100.0  the
                  performance is considered poor.
        flag = n  Normal termination; all eigenvalues/eigenvectors have been
                   successfully found.
```

```
flag < n  Error termination; procedure QRmethod failed to converge.
           Eigenvalues  1,2,...,flag  have been computed correctly.
           Eigenvalues  flag+1,flag+2,...,n  are set to zero, and
           Index  is set to 1000.0 .

The eigenvalues and eigenvectors are sorted so that
       EigenValue[k+1] ≤ EigenValue[k]  for  1 ≤ k ≤ (flag - 1).

Procedure EigenRS is based on the IMSL FORTRAN subroutine EIGRS
```

APPENDIX 8.8

Listing of a Pascal Program for finding Eigenvalues and Eigenvectors of a Real Symmetric Matrix using the Procedure EigenRS

```
PROGRAM EigenvaluesEigenvectors; (* of a real symmetric matrix *)

CONST
  MaxN = 12;
  n2 = 78 ; (* n2 = MaxN*(MaxN + 1) div 2  *)
  Field = 12;  (* Fieldwidth *)
  Decimal = 8; (* Number of digits after the decimal point *)

TYPE
  VectorN   = ARRAY[ 1..MaxN ] OF REAL;
  MatrixNM = ARRAY[ 1..MaxN , 1..MaxN ] OF REAL;
  LongVector = ARRAY[ 1..n2 ] OF REAL;
  String14 = String[14];
  String80 = String[80];

VAR
  A, EigenVector : MatrixNM;
  EigenValue : VectorN;
  Index : REAL;
  flag, n : INTEGER;
  OutFile : TEXT;
  FileName : String14;
  answer : CHAR;

(*$I Response.Pas *)
(*$I Files.Pas *)
(*$I ReadNum.Pas *)
(*$I MatrixIO.Pas *)
(*$I EigenRS.Pas *)

PROCEDURE DataIn;
VAR
  answer : CHAR;
BEGIN
  ClrScr ; LowVideo ;
  WRITE('This program finds the eigenvalues and eigenvectors');
  WRITELN(' of a real symmetric matrix A');
  GetMatrix(A,n,n,TRUE,MaxN,MaxN,Field,Decimal);
  WRITE('Do you want to print the matrix A?  (Y or N)   ') ; Echo(answer) ;
```

```
  IF  answer IN ['Y','y']  THEN
    PutMatrix(A,n,n,Field,Decimal,'The matrix A:','P')
END; (* DataIn *)

PROCEDURE DataOut;
VAR
  i, k : INTEGER;
BEGIN
  WRITELN(OutFile);
  FOR  k := 1  TO  flag  DO
    BEGIN
      WRITE(OutFile,'Eigenvalue[',k,'] = ');
      WRITELN(OutFile,Eigenvalue[k]:Field:Decimal) ; WRITELN(OutFile) ;
      WRITE(OutFile,'Eigenvector[',k,'] :');
      FOR  i := 1  TO  n  DO
        BEGIN
          IF  i > 1  THEN
            WRITE(OutFile,' ':16);
          WRITELN(OutFile,Eigenvector[i,k]:Field:Decimal)
        END;
      WRITELN(OutFile)
    END;
  IF  flag = n  THEN
    WRITELN(OutFile,'performance index = ',Index:8:3);
  CLOSE(OutFile)
END; (* DataOut *)

BEGIN (* main program *)
  DataIn;
  WRITE('Do you want the output on the Console or the Printer?');
  WRITE(' (C or P)  ') ; Echo(answer) ;
  IF  answer IN ['P','p']  THEN
    ASSIGN(OutFile,'Prn')
  ELSE
    ASSIGN(OutFile,'Con');
  REWRITE(OutFile);
  EigenRS(A,n,EigenValue,EigenVector,Index,flag);
  DataOut;
  WRITELN;
  WRITE('Do you want to store the eigenvalues and ',
        'eigenvectors in a file?  (Y or N) ');
  Echo(answer);
  IF answer in ['Y','y'] THEN
    BEGIN
      CreateFile(OutFile,FileName);
      DataOut
    END
END.
```

Chapter

9

Optimization: Minimization or Maximization of a Function

Various methods are examined for determining the minimum of a function of one variable on an interval. These include Golden Section Search, Fibonacci Search, and Brent's method. Methods for minimization of a function of n variables are also discussed. These include the method of Steepest Descent, the Conjugate Gradient method, and quasi-Newton methods.

9.1 INTRODUCTION

Optimization is concerned with doing things in a "best" possible manner. For example, in many problems of engineering design one is required to select the values of certain components from an admissible set in such a way that a specified performance criterion is minimized or maximized. Because of the importance of such problems, a great deal of attention has been given to this topic during recent years. At present, several excellent algorithms are available for computer-aided optimization.

One may state an optimization problem as follows: Given a scalar real-valued function, which is a function of one or more independent variables, it is desired to find the values of these variables where the function takes on a maximum or minimum value. One possible direct approach to optimization is the use of the calculus of extrema to obtain the points where an extremum (maximum or minimum) of the function is reached. If the function is differentiable we can set the derivative of the function (the partial derivatives in the case of optimization in many dimensions), to zero and find the zeros of the derivative function. In most cases determination of these zeros will not be very straightforward, but we can use the algorithms presented in Chapter 4 for obtaining a numerical solution. However, the vanishing of the derivative is only a necessary condition for the existence of an extremum; it is not a sufficient condition. To be certain that a maximum or minimum exists at a point, x_*, where the derivative is zero, we must examine the second derivative of the function, which must be nonzero at x_* in order for us to tell the type of extremum obtained. If the second derivative is also zero, then it is necessary to examine higher order derivatives or to look at the function in a neighborhood of x_*.

The problem of finding extrema is often more complicated due to the fact that sometimes the function may only be piecewise smooth, or even piecewise continuous, with respect to the independent variables. Thus, one cannot use the calculus of extrema. Hence, often the only possible approach to optimization is the use of a suitable search scheme which may be combined with the evaluations of derivatives in order to obtain the best direction for search. This may often require considerable computational effort because the search must be made in one or more dimensions. From the computational point of view, a good algorithm for optimization should do this quickly and cheaply, without requiring a large amount of computer memory. The computational effort may require repeated evaluation of the function as well as its partial derivatives with respect to the various independent variables, and this may be time-consuming. Hence, a good algorithm should require a small number of these evaluations. Another difficulty is caused by the fact that an extremum, x_*, can be either *local* (an extremum in a small neighborhood of x_*) or *global* (an extremum for all values of the independent variables). All search methods that are commonly used at present lead only to local extrema.

Optimization problems may be either *unconstrained* or *constrained*. In the former, there are no limits on the values of the independent variables, whereas in the latter these limits are clearly specified. Most practical problems require constrained optimization. For example, while designing an electrical circuit in which the values of resistors and capacitors have to be determined for optimum performance, we must keep in mind that these values cannot be negative. Furthermore, there are often only a discrete number of component values from which we may choose. All of these factors must be taken into account while developing a suitable algorithm for optimization under constraints.

In this chapter we shall study the basic principles of optimization of a function of one or more independent variables. In addition, some of the well-known algorithms that are suitable for use with personal computers will be given. First we shall discuss the case when there is only one independent variable, so that we need optimization in one dimension. This will be followed by the study of optimization in several dimensions. In all of our discussions, we shall be considering the case when the function is to be minimized and there are no constraints on the independent variables. Problems involving maximization of a function can be easily handled by the minimization of the negative of the function.

9.2 OPTIMIZATION IN ONE DIMENSION

In this case, we are concerned with finding the value of the independent variable, x, for which a continuous function $f(x)$ will retain its minimum value over the interval $0 \leq x \leq 1$. It should be noted that restricting x to the interval [0, 1] does not reduce the generality of the discussion. If we have a function $F(y)$, where y is in the interval $[a, b]$, then we can use the transformation

$$x = \frac{y - a}{b - a} \tag{9.2–1}$$

to transform $F(y)$ so that it becomes $F(a + (b - a)x) = f(x)$. Figure 9.1 shows a typical continuous function whose global minimum is sought on the interval $[a, b]$. It will also be assumed that the function $f(x)$ to be minimized (or maximized) is *unimodal* on the interval [0, 1]. A unimodal function has the property of having only one global minimum (or maximum) in the interval [0, 1].

Let us assume that there is a unique value of x, denoted by x_*, for which the minimum of $f(x)$ is attained and that the function increases on either side of this point. Our aim is to find x_*. This can be done by using a search method, which successively reduces the *interval of uncertainty*, which is the interval containing x_*. For example, we may consider the use of a procedure similar to the Bisection Method described in Chapter 4, where we attempted to locate a zero of $F(x)$ by successively reducing the interval of uncertainty by a factor of two. In order to

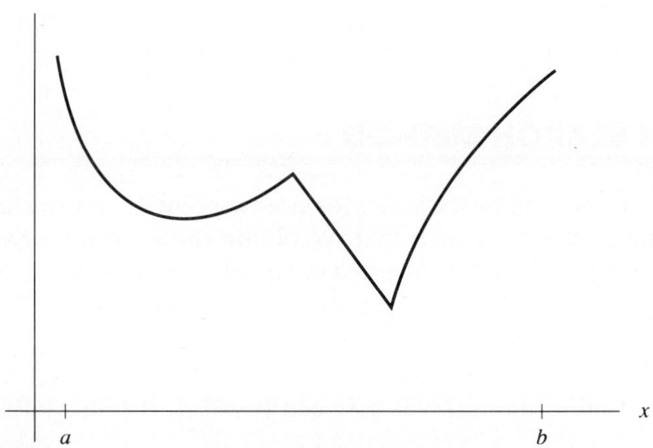

Figure 9.1 A typical function for minimization.

start the Bisection Method, it was required that the function $F(x)$ have opposite signs at the two endpoints of the interval, as this guaranteed the existence of at least one zero within the interval. The optimization problem is slightly more complicated, as we must evaluate $f(x)$ at three values of x in order to select the region in which x_* exists. Denoting these three values of x by a, b, and c, where $a < b < c$, a minimum is indicated for $a \leq x \leq c$ only when $f(b)$ is less than both $f(a)$ and $f(c)$. The strategy would be to continue until the interval between the two outer points a and c is reduced to an accepted tolerance level. The minimum value of this tolerance level depends upon the machine episilon (see Chapter 1) as well as some properties of the function to be minimized, as will be shown in the next section.

9.3 MINIMUM INTERVAL OF UNCERTAINTY

It is desirable to know the minimum possible interval of uncertainty in a given optimization problem, as this knowledge may save a large number of unnecessary function evaluations. To do this, we can use the Taylor's series expansion of $f(x)$ about $x = x_*$,

$$f(x) - f(x_*) \simeq \tfrac{1}{2} f''(x_*)\cdot(x - x_*)^2 \tag{9.3–1}$$

where $f'(x_*) = 0$, because $f(x)$ attains a minimum at x_*. In order to distinguish the two terms on the left-hand side of (9.3–1) we must have

$$| f(x) - f(x_*) | \geq u | f(x_*) |$$

where u is the machine epsilon. Thus,

$$2u | f(x_*) | \simeq | f''(x_*) |\cdot(x - x_*)^2 \tag{9.3–2}$$

The minimum interval of uncertainty can be obtained by solving (9.3–2) for $| x - x_* |$. For most functions, it would be of the order of $\sqrt{u}$.

In the next section we shall study a method which attempts to reach the optimum with a minimum number of function evaluations.

9.4 GOLDEN SECTION SEARCH METHOD

Let us now determine the best choice for a new point, x, given the three points a, b, and c, with $a < b < c$, such that we obtain the maximum reduction in the interval of uncertainty, which is at present $[a, c]$. Let us define

$$w = \frac{b - a}{c - a} \tag{9.4–1}$$

which is the fraction of the interval between b and a. It follows that

$$1 - w = \frac{c - b}{c - a} \tag{9.4–2}$$

Let the next trial point x be an additional fraction z beyond b,

$$z = \frac{x - b}{c - a} \tag{9.4–3}$$

Then the next interval will be either of length $(w + z)$ relative to the current interval, or of length $(1 - w)$. The worst case possibility will be minimized if we make these equal, by setting

$$w + z = 1 - w \tag{9.4–4}$$

It will be seen that the new point x is selected such that

$$| b - a | = | x - c |$$

It also means that the point x lies in the larger of the two intervals $[b, c]$ and $[a, b]$. This follows from the fact that z is positive only if $w < 0.5$. If we assume that every new point is chosen with the same strategy, the relative reduction in the interval of uncertainty should be the same at each stage. Hence,

$$w = \frac{z}{1 - w} \tag{9.4–5}$$

Eliminating z from (9.4–4) and (9.4–5) we obtain the quadratic equation

$$w^2 - 3w + 1 = 0 \tag{9.4–6}$$

Solving (9.4–6) for the root w_* which is less than 0.5 gives

$$w_* = \frac{3 - \sqrt{5}}{2} \simeq 0.3819660115 \tag{9.4–7}$$

and

$$1 - w_* = \frac{\sqrt{5} - 1}{2} \simeq 0.6180339885 \tag{9.4–8}$$

Thus, we get the following rule for an optimal search method. Given, at any stage, a set of three points $\{a, b, c\}$ such that $a < b < c$ and the optimum point, x_*, is located within the interval $[a, c]$, the next point to be tried is that which is a fraction w_* of the larger of the two intervals, measured from the central point b. This procedure is called the *Golden Section Search Method*; the name has been attributed to the ancient Greeks.

To start the Golden Section Search Method, given the endpoints 0 and 1, we first locate two points, at w_* and $(1 - w_*)$, respectively. After evaluating $f(x)$ at these four points, we decide on the set of three points which bracket the minimum. We now select the next point using the rule stated above and continue in this manner. It may be noted that each iteration reduces the interval of uncertainty by a factor of

$$(1 - w_*) \simeq 0.61803$$

This is comparable to, but not as good as, the factor of 0.5 which is obtained when the Bisection Method is used for finding a zero of a function. The Golden Section Search Method is guaranteed to work in the worst case. The price we pay is the slow rate of convergence. In order to terminate the method we use an error tolerance, τ. If the current interval of uncertainty, $[a, c]$, is such that

$$|c - a| \leq \tau(|x_1| + |x_2|)$$

where x_1 and x_2 are any two points in the interval $[a, c]$ at which we have computed function values, then we terminate the Golden Section Search Method. The Golden Section Search procedure is illustrated in Figure 9.2. The points w_* and $(1 - w_*)$ are shown in the figure. A Pascal procedure for the Golden Section Search Method is given in Appendix 9.1.

EXAMPLE 9.1

Consider the function

$$f(x) = 6 - 11x + 6x^2 - x^3$$

It is desired to locate the minimum of $f(x)$ for x in the interval $[1, 2]$. In this case it is possible to locate the minimum of the function by finding the roots of the derivative of $f(x)$, but we used the Golden Section Search Method. The complete program is given in Appendix 9.2. With an error tolerance of 10^{-7}, the algorithm required 62 function evaluations to reach the minimum at $x_* = 1.4226497$. See Table 9.1.

Table 9.1 GOLDEN SECTION SEARCH, EXAMPLE 9.1

Number of Function Evaluations	Length of the Interval of Uncertainty
2	1.0
3	0.382
4	0.236
5	0.236
6	0.146
7	0.146
8	0.0902
9	0.0344
10	0.0558
20	5.03×10^{-3}
30	4.53×10^{-4}
40	4.09×10^{-5}
50	3.68×10^{-6}
60	3.32×10^{-7}
62	2.05×10^{-7}

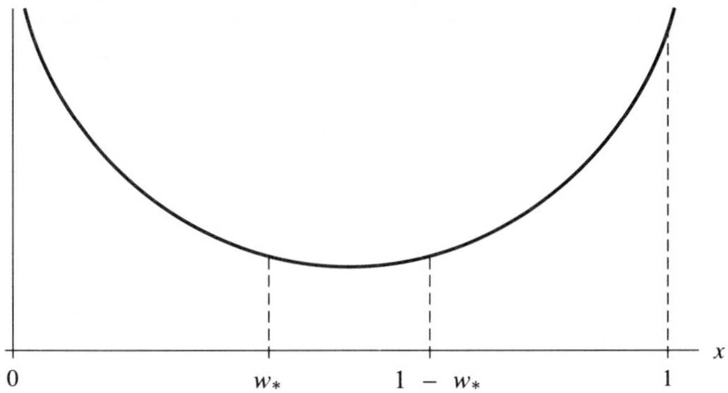

Figure 9.2 Golden Section Search.

EXERCISE 9.1

Find the location of the minimum of the function

$$f(x) = e^x - 3x^2 - 2e^{-2x}$$

for x in the interval [0, 5] using a tolerance of 10^{-7}, with the program in Appendix 9.2.

Answer $x_* = 2.8318850$, 66 function evaluations.

9.5 FIBONACCI SEARCH METHOD

In the particular case when we want to get the best result for a given number of evaluations of the function $f(x)$, it is preferable to use the Fibonacci Search Method. For example, if we are allowed to make only two function evaluations, the Golden Section Search Method will reduce the length of the interval of uncertainty from 1 to $(1 - w_*) \simeq 0.61803$. It is possible to obtain an interval of uncertainty of length 0.5 if the function is evaluated at the points $(0.5 + \Delta)$ and $(0.5 - \Delta)$, instead of w_* and $(1 - w_*)$ for the Golden Section Search Method, where Δ is a suitable tolerance level. Depending upon the values of the function at these two points, the next interval for search will be either $[0, 0.5 - \Delta]$ or $[0.5 + \Delta, 1]$, as shown in Figure 9.3. Let us now try to extend this idea to the case when m function evaluations are allowed. Let w_i be the width of the interval of uncertainty after the ith trial. Then the requirement of symmetry for evaluations yields

$$w_{i-2} = w_{i-1} + w_i \qquad \text{for} \quad 1 < i < m \tag{9.5–1}$$

Also we require the last evaluation to be such that

$$w_{m-1} = 2w_m - 2\Delta \tag{9.5–2}$$

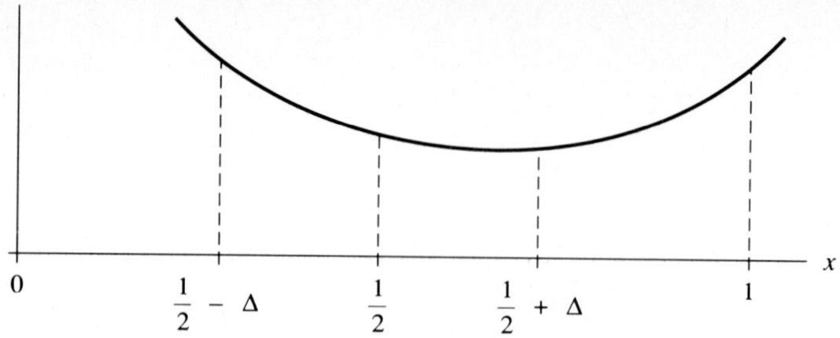

Figure 9.3 Search with only two function evaluations.

It is possible to work backwards from these two relationships to determine the locations for the various function evaluations for a given m. For example,

$$w_{m-3} = w_{m-2} + w_{m-1} = (w_{m-1} + w_m) + w_{m-1} = 5w_m - 4\Delta$$

Similarly, it can be shown that

$$w_{m-4} = 8w_m - 6\Delta$$

The general relationship for the ith stage is given by

$$w_{m-i} = F_{i+1}w_m - 2\Delta F_{i-1} \tag{9.5-3}$$

where the coefficients $\{F_j\}$ are the Fibonacci numbers, which are defined as

$$F_0 = 1 \qquad F_1 = 1$$

$$F_j = F_{j-1} + F_{j-2} \qquad \text{for } j = 2, 3, \ldots \tag{9.5-4}$$

Table 9.2 shows some Fibonacci numbers. Returning to the Fibonacci Search method, we see that the size of the final interval of uncertainty is given by

$$w_m = \frac{1}{F_m} + 2\Delta\frac{F_{m-2}}{F_m} \tag{9.5-5}$$

In order to use this method we must first select the value of m, the number of function evaluations. The spacing for the function evaluations is then planned using the information about w_i. Since $w_1 = w_0 = 1$, the process must start with two function evaluations placed at a distance w_2 from the opposite ends of the initial interval, [0, 1]. In this case,

$$w_2 = \frac{F_{m-1}}{F_m} + \Delta\frac{(-1)^m}{F_m} \tag{9.5-6}$$

Table 9.2 THE FIRST FOURTEEN FIBONACCI NUMBERS

i	0	1	2	3	4	5	6	7	8	9	10	11	12	13
F_i	1	1	2	3	5	8	13	21	34	55	89	144	233	377

where Δ is the smallest distance by which two evaluations may be separated and still be distinguishable from each other. (Usually $\Delta = \sqrt{u}$, where u is the machine epsilon.) Suppose the points where the function is to be evaluated are located at t and s. The next step is to locate another evaluation point, z, so that z is w_3 units from either t or s, depending on whether $f(t)$ is greater or less than $f(s)$. The appropriate interval is retained and the procedure is repeated until m function evaluations have been carried out. A Pascal procedure for the Fibonacci Search Method is given in Appendix 9.3.

EXAMPLE 9.2

The optimization problem in Example 9.1 was solved using the Fibonacci Search Method described above. The Pascal program for this case is given in Appendix 9.4. Table 9.3 gives the location of the lower and upper limits, a and b, of the interval of uncertainty for some specified number of function evaluations, m. It will be seen that we reach fairly close to the optimum with only 20 function evaluations. Table 9.4 gives the length of the interval of uncertainty for a given number of function evaluations.

Table 9.3 RESULTS WITH FIBONACCI SEARCH, EXAMPLE 9.2

m	a	b
15	1.4224924	1.4235056
20	1.4226201	1.4227115
30	1.4226496	1.4226503

Table 9.4 FIBONACCI SEARCH, EXAMPLE 9.2

Number of Function Evaluations	Length of the Interval of Uncertainty
1	1.0
2	0.618
3	0.382
4	0.236
5	0.146
6	0.0902
7	0.0557
8	0.0344
9	0.0213
10	0.0132
15	1.19×10^{-3}
20	1.07×10^{-4}
25	9.66×10^{-6}
30	7.49×10^{-7}

EXERCISE 9.2

Repeat Exercise 9.1 using the Fibonacci Search Method with 25 function evaluations.

Answer $2.8318766 < x_* < 2.8319178$.

9.6 BRENT'S METHOD

Another approach to accelerating the convergence to the optimum is to use some kind of interpolation to determine the next trial point. If the function is smooth near the optimum, it is possible to fit a parabola to the triplet of points $\{(a, f(a)), (b, f(b)), (c, f(c))\}$ and determine the abscissa corresponding to the apex of the parabola. This could take us to the optimum in a single leap, or at least very close to the optimum. We may now use this new point as the next trial point. This interpolation process is called *inverse parabolic interpolation* (see Chapter 5).

The formula for the abscissa x which is the minimum of a parabola passing through the three points $(a, f(a))$, $(b, f(b))$, and $(c, f(c))$ is

$$x = b + \frac{1}{2}\frac{(c - b)^2[f(a) - f(b)] - (b - a)^2[f(c) - f(b)]}{(c - b)[f(a) - f(b)] + (b - a)[f(c) - f(b)]} \tag{9.6–1}$$

The denominator of the second term on the right of (9.6–1) is zero in the special case when the three points are on a straight line. This implies that the optimum point is at infinity, or outside the interval under consideration. A set of typical steps using this scheme is shown in Fig. 9.4.

It should be noted that the formula of equation (9.6–1) would lead to either a minimum or a maximum of $f(x)$. Hence, any procedure depending entirely on this formula cannot be very reliable. It appears more logical to combine it with

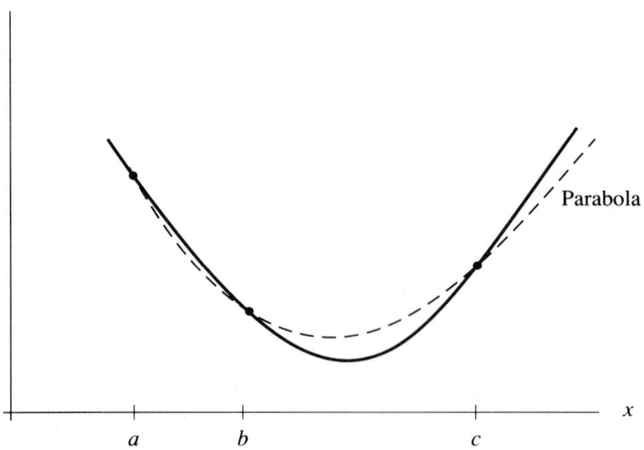

Figure 9.4 Convergence to a minimum with inverse parabolic interpolation.

the Golden Section Search Method that would guarantee convergence. The idea is to switch to inverse parabolic interpolation whenever the function is smooth, but use the Golden Section Search Method when it is not. In order that this may work, we must have a very robust scheme for deciding if the function is suitable for inverse parabolic interpolation in a given region.

Brent's method attempts to combine these ideas. At any particular stage, first it must be insured that the optimum occurs between the points, a and c by evaluating the function at an intermediate point, b, where its value must be less than those at a or c. Inverse parabolic interpolation is tried for obtaining the next point. To be acceptable, it is required that the new point lie within the bounding interval $[a, c]$ and it implies a movement from b which further decreases the value of $f(x)$. This interpolation step may be repeated several times, provided that the size of the last step is less than half the movement in the previous step. The last condition insures that the inverse parabolic interpolation scheme converges to some point. A Pascal procedure based on Brent's method is given in Appendix 9.5.

EXAMPLE 9.3

The optimization problem described in Example 9.1 was solved using Brent's method with the program in Appendix 9.6. The tolerance required by Brent's method was set at 10^{-7}. The minimum was located at $x_* = 1.4226497$ in 10 function evaluations. See Table 9.5.

Table 9.5 BRENT'S METHOD, EXAMPLE 9.3

Number of Function Evaluations	Length of the Interval of Uncertainty
1	1.0
2	0.618
3	0.382
4	0.236
5	0.0511
6	0.0427
7	0.0408
8	1.45×10^{-4}
9	1.38×10^{-4}
10	1.31×10^{-7}

EXERCISE 9.3

Repeat Exercise 9.1 using Brent's method with a tolerance of 10^{-7}.

Answer $x_* = 2.8318849$, 12 function evaluations.

9.7 SEARCH USING THE FIRST DERIVATIVE

The methods that we have considered so far are based on evaluating only the function $f(x)$ at suitably selected points. In some cases the search for the optimum can be accelerated by using the derivative of $f(x)$. The derivative is evaluated at the central point b of the triplet $\{a, b, c\}$ which brackets the optimum. The sign of the derivative at this point can be utilized to determine whether the next test point should be taken in the interval $[a, b]$ or the interval $[b, c]$. Furthermore, we can use the Secant Method (see Chapter 4) to extrapolate to the point where the derivative will be zero from the values of the derivative at this point and at another point, which may be the second best among the points determined so far. This should bring us fairly close to the optimum if the function is smooth in the interval. Since the convergence rate of the Secant Method is known to be superlinear (see Chapter 4), we shall obtain a significant improvement over the Golden Section Search Method. This is Brent's derivative method. The price for this better convergence is the extra work required in calculating the value of the derivative. A Pascal procedure based on this method is given in Appendix 9.7. In this procedure we compute the derivatives internally by a finite difference method (see Section 6.11) so that the user only has to supply a Pascal function to evaluate $f(x)$.

EXAMPLE 9.4

The optimization problem described in Example 9.1 was solved using the program in Appendix 9.8. With the tolerance set at 10^{-7}, the minimum was located at $x_* = 1.4226497$ in 49 function evaluations (see Example 9.1).

EXERCISE 9.4

Repeat Exercise 9.1 using Brent's derivative method. Set the tolerance at 10^{-7}.

Answer $x_* = 2.8318849$, 51 function evaluations.

9.8 OPTIMIZATION IN MANY DIMENSIONS

In many problems in science and engineering, the function which is to be optimized depends upon several independent variables. These problems of multidimensional optimization are much more complicated but, because of their importance, a great deal of work has been done during the last twenty-five years to solve them. The earliest attempts were based on using the gradient of the function. Such methods, called *Steepest Descent Methods* for function minimization, or *Steepest Ascent Methods* for function maximization, date back to the work of Cauchy in 1845.

The Method of Steepest Descent starts by finding the direction in which the function is decreasing at the maximum rate. This is the direction of the steepest descent and is given by the negative of the gradient of the function at any point. Let the vector

$$\mathbf{x} = [x_1 \quad x_2 \ldots x_n]^T$$

denote any point in the n-dimensional space of the independent variables. The gradient of $f(\mathbf{x})$ is denoted by the column vector $\mathbf{g}(\mathbf{x})$, whose ith component is the partial derivative of $f(\mathbf{x})$ with respect to x_i,

$$g_i = \frac{\partial f}{\partial x_i}$$

After determining the direction of steepest descent, $-\mathbf{g}(\mathbf{x})$, at the starting point, $\mathbf{x}_0$, we utilize a one-dimensional search technique to determine a value of the parameter t so that the function

$$F(t) = f(\mathbf{x}_0 - t\mathbf{g}(\mathbf{x}_0)) \tag{9.8–1}$$

is minimized. This means we are looking for a minimum of $f(\mathbf{x})$ on the line through the point $\mathbf{x}_0$ in the direction of the vector $-\mathbf{g}(\mathbf{x}_0)$. If t_0 denotes the value of t that minimizes (9.8–1), then we set

$$\mathbf{x}_1 = \mathbf{x}_0 - t_0\mathbf{g}(\mathbf{x}_0)$$

At $\mathbf{x}_1$, we again determine the gradient of $f(\mathbf{x})$ and proceed along the new direction of steepest descent. The process is repeated until a local minimum is obtained, at a point where the gradient is zero.

Although the Method of Steepest Descent is conceptually straightforward, in many cases it may take a very large number of evaluations of the function and the gradient for the method to converge to a local minimum of $f(\mathbf{x})$. For example, consider the function of two independent variables in the following example.

EXAMPLE 9.5

$$f(x, y) = (x^2 - y)^2 + (1 - x)^2$$

The Steepest Descent Method was used to minimize $f(x, y)$, starting at the initial point $(-4, 16)$, with $f(-4, 16) = 25$. After 40 iterations the method had reached $f(-1.535, 3.181) = 7.107$. After 70 iterations, the method had reached $f(1.012, 1.031) = 1.88 \times 10^{-4}$. Figure 9.5 shows several contours of constant values of $f(x, y)$. In Figure 9.5, the points shown on each contour were those obtained at every tenth iteration of the Steepest Descent Method. The minimum at (1, 1) is also shown. It will be seen that the path to the minimum proceeds very slowly. This is due to the fact that the local gradient does not point towards the minimum at any point along the path. See Table 9.6.

EXAMPLE 9.5 (cont'd)

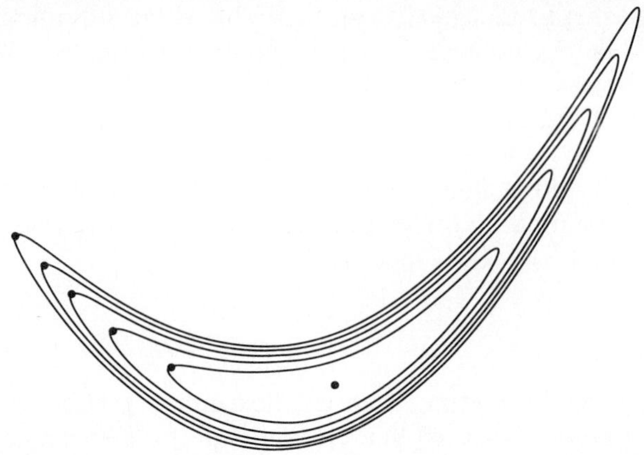

Figure 9.5 Slow convergence with the Method of Steepest Descent for the function $f(x, y) = (x^2 - y)^2 + (1 - x)^2$. The points shown on the contours are at every tenth iteration.

Table 9.6 STEEPEST DESCENT METHOD, EXAMPLE 9.5

Iteration k	x_k	y_k	$f(x_k, y_k)$
1	−4.00000	16.00000	25.0
2	−3.92077	16.00000	24.6
3	−3.92077	15.37247	24.2
4	−3.83956	15.37247	23.8
5	−3.83956	14.74225	23.4
6	−3.75621	14.74225	23.0
7	−3.75621	14.10914	22.6
8	−3.67055	14.10914	22.2
9	−3.67055	13.47292	21.8
10	−3.58237	13.47292	21.4
20	−3.09412	10.23518	17.2
30	−2.48141	6.85888	12.6
40	−1.53474	3.18121	7.11
50	1.09835	1.25114	0.0117
60	1.03597	1.09060	1.60×10^{-3}
70	1.01228	1.03077	1.88×10^{-4}

At each iteration of the Steepest Descent Method, $\mathbf{g}(\mathbf{x}_{k+1})$ is orthogonal to $\mathbf{g}(\mathbf{x}_k)$. Example 9.5 shows that, in spite of the orthogonality of the descent directions, the convergence can be very slow. Perhaps a suitable linear combination of the descent directions will give a direction pointing more closely towards the actual minimum. The Conjugate Gradient Method improves the convergence of the Steepest Descent Method by taking its first step identical to a Steepest Descent step, but each subsequent step is a linear combination of the current gradient vector and the preceding direction vector.

Let $\mathbf{x}_*$ denote the point at which $f(\mathbf{x})$ attains a local minimum and let $\mathbf{x}_k$ denote the kth approximation to $\mathbf{x}_*$. We can denote $\mathbf{g}(\mathbf{x}_k)$ as $\mathbf{g}_k$ and the direction of search at $\mathbf{x}_k$ by $\mathbf{b}_k$. We first determine the value of t, t_k, so that $f(\mathbf{x}_k - t\mathbf{b}_k)$ is a minimum; we define $\mathbf{x}_{k+1}$ as

$$\mathbf{x}_{k+1} = \mathbf{x}_k - t_k\mathbf{b}_k$$

We now calculate the gradient $\mathbf{g}_{k+1}$. The next direction of search is given by

$$\mathbf{b}_{k+1} = \frac{[\mathbf{g}_{k+1}]^T\mathbf{g}_{k+1}}{[\mathbf{g}_k]^T\mathbf{g}_k}\mathbf{b}_k + \mathbf{g}_{k+1}$$

Note that $\mathbf{g}_{k+1}$ is orthogonal to $\mathbf{b}_k$ so that $[\mathbf{g}_{k+1}]^T\mathbf{b}_{k+1} = [\mathbf{g}_{k+1}]^T\mathbf{g}_{k+1}$

EXAMPLE 9.6

The Conjugate Gradient Method was used to minimize the function in Example 9.5, starting at the same initial point, $(-4, 16)$. The method reached $f(1.031, 0.9641) = 0.0107$ after 40 iterations; this is considerably better than the Steepest Descent Method. See Table 9.7.

Table 9.7 CONJUGATE GRADIENT METHOD, EXAMPLE 9.6

Iteration k	x_k	y_k	$f(x_k, y_k)$
1	−4.00000	16.00000	25.0
2	−3.92077	16.00000	24.6
3	−2.59582	5.44304	14.6
4	−1.90575	1.74830	12.0
5	−1.44046	−0.11153	10.7
6	−1.09795	−1.16658	10.0
7	−0.83426	−1.80029	9.60
8	−0.62528	−2.19308	9.32
9	−0.45580	−2.44074	9.13
10	−0.31537	−2.59780	9.01
20	0.66617	−2.42644	8.35
30	2.77946	7.92575	3.21
40	1.03096	0.96407	0.0107
50	0.99572	0.98202	1.08×10^{-4}
60	0.99936	0.99824	6.40×10^{-7}
70	0.99995	0.99988	2.97×10^{-9}

Another approach to minimizing $f(\mathbf{x})$ is to try to find a zero of the gradient $\mathbf{g}(\mathbf{x})$. We can use Newton's Method, Chapter 4, to try to solve the equation

$$\mathbf{g}(\mathbf{x}) = \mathbf{0} \tag{9.8–2}$$

This means that we must compute the Jacobian matrix of $\mathbf{g}$; this is the $n \times n$ real symmetric matrix $\mathbf{H}(\mathbf{x})$, called the *hessian* of $f(\mathbf{x})$, whose (i, j)th entry is

$$h_{ij} = \frac{\partial^2 f}{\partial x_i \partial x_j}.$$

Then Newton's Method, equation (4.10–1), for solving equation (9.8–2) is

$$\mathbf{x}_{k+1} = \mathbf{x}_k - [\mathbf{H}(\mathbf{x}_k)]^{-1}\mathbf{g}(\mathbf{x}_k) \tag{9.8–3}$$

If (9.8–3) is used to try to find a minimum of $f(\mathbf{x})$ then care must be taken that the sequence of iterates $\{\mathbf{x}_k\}$ converges to a local minimum rather than a local maximum or a saddle point of $f(\mathbf{x})$.

Recall from Chapter 4 that Newton's Method has quadratic convergence if the initial guess $\mathbf{x}_0$ is sufficiently close to a solution $\mathbf{x}_*$ of (9.8–2). However, for many problems the initial guess $\mathbf{x}_0$ may not be close to $\mathbf{x}_*$. Also, there is the requirement that both the gradient, $\mathbf{g}$, and the hessian, $\mathbf{H}$, must be computed at each iteration. This may be a formidable computational task for many problems. Consequently, a modification to (9.8–3) is required. We modify (9.8–3) by using a quasi-Newton method, see equation (9.8–4) below, in which $\mathbf{H}(\mathbf{x}_k)$ is not computed at each iteration; rather, an estimate of $\mathbf{H}(\mathbf{x})$ is obtained at each iteration. The estimated hessian, denoted by $\mathbf{H}_k$, is updated at each iteration so that as k increases and $\mathbf{x}_k$ tends towards the sought-for minimum, $\mathbf{x}_*$, $\mathbf{H}_k$ tends towards $\mathbf{H}(\mathbf{x}_*)$. Usually one starts with $\mathbf{H}_0 = \mathbf{I}$, the $n \times n$ identity matrix. Because $\mathbf{H}(\mathbf{x})$ is a real symmetric matrix, one can build up $\mathbf{H}_k$ by storing the iterates as $\mathbf{L}_k\mathbf{L}_k^T$, a Cholesky factorization (see Problem 7, Chapter 2).

The quasi-Newton Method is given by

$$\mathbf{x}_{k+1} = \mathbf{x}_k - \alpha_k[\mathbf{H}_k]^{-1}\mathbf{g}(\mathbf{x}_k) \tag{9.8–4}$$

The numbers, α_k, are chosen to insure convergence of the method. A user's manual for the Pascal procedure QNM, for obtaining a local minimum of a function using the quasi-Newton method, is given in Appendix 9.9 along with a main program in Appendix 9.10. The procedure QNM only requires the user to supply the function $f(\mathbf{x})$; the gradient $\mathbf{g}$ is computed internally within the procedure by finite differences (see Section 6.11).

EXAMPLE 9.7

When QNM was used to minimize the function in Example 9.5, starting at $(-4, 16)$, 98 function evaluations were required to reach the minimum. See Table 9.8.

EXAMPLE 9.7 (cont'd)

Table 9.8 PROCEDURE QNM, EXAMPLE 9.7

Iteration k	x_k	y_k	$f(x_k, y_k)$
1	−4.00000	16.00000	25.0
2	−3.92427	16.00000	24.6
3	−2.64543	6.02017	14.2
4	−2.13645	2.76358	13.1
5	−1.85030	3.68127	8.17
6	−1.01359	0.19327	4.75
7	−0.55685	−0.74078	3.53
8	-8.17471×10^{-3}	0.161281	1.04
9	0.35143	−0.18190	0.514
10	0.77014	0.11145	0.285
15	1.00003	1.00007	1.03×10^{-9}
20	1.00000	1.00000	3.12×20^{-24}
22	1.00000	1.00000	6.16×10^{-32}

EXERCISE 9.5

Use the procedure QNM to find the minimum of the function

$$f(x_1, x_2) = w^2 + z^2$$

where

$$w = x_1 + x_2[x_2(5 - x_2) - 2] - 13$$

$$z = x_1 + x_2[x_2(1 + x_2) - 14] - 29$$

with the initial guess $\mathbf{x}_0 = [15.0, -2.0]^T$.

Answer $\mathbf{x}_* = [5.0, 4.0]^T$, 89 function evaluations.

EXAMPLE 9.8

Example 4.6 was repeated by using QNM to minimize the function $E(x_1, x_2, x_3)$, starting from $(x_1, x_2, x_3) = (1, 2, 10)$. This required 94 evaluations of the function $E(x_1, x_2, x_3)$ to converge to the same solution obtained in Example 4.6. The final value of E was 0.1521029.

9.9 ADDITIONAL TOPICS IN THE MINIMIZATION OF FUNCTIONS

In this chapter, we have discussed numerical methods for finding a minimum of a function $f(\mathbf{x})$. In the case where $\mathbf{x}$ is a scalar, we have a one-dimensional optimization problem. Several methods have been presented for optimization in one dimension, and computer programs have been given in the appendices to this chapter. The case of multidimensional optimization is more complicated and is often solved by first optimizing in a certain direction and then changing the direction of search until a local minimum is obtained. The Steepest Descent Method is the most direct approach to this problem, but convergence may often be very slow. Using a quasi-Newton method, such as QNM in Appendix 9.9 or a conjugate gradient method, can overcome the slow convergence.

Minimization methods such as QNM may be useful in problems involving the solution of systems of equations, $\mathbf{f}(\mathbf{x}) = \mathbf{0}$, as studied in Chapter 4. It may be helpful to use QNM to get "close" to a solution before using a procedure, such as Powell in Chapter 4, to find zeros of functions. For example, the procedure QNM could be used to minimize

$$\sum_{i=1}^{n} [f_i(\mathbf{x})]^2 \qquad \textbf{(9.9–1)}$$

where $\mathbf{f} = [f_1 \quad f_2 \ldots f_n]^T$. This would be helpful in the cases where the initial guess $\mathbf{x}_0$ supplied to the procedure Powell leads to a local minimum of (9.9–1) but the value of the expression (9.9–1) is "far" from zero.

When using a multidimensional procedure, such as QNM, to minimize a function $f(\mathbf{x})$, it is advisable to scale $\mathbf{x}$ so that each component of the solution $\mathbf{x}_*$ has approximately the same magnitude (near 1, say). If the components of $\mathbf{x}_*$ differ widely in magnitude, then the smallest-magnitude component may not have as many significant digits of accuracy as the largest-magnitude component. This scaling may have to be done after a first, preliminary, solution, $\mathbf{y}$, has been found. The scaling can be obtained by examining the components of $\mathbf{y}$.

9.10 PROBLEMS

1. It is desired to manufacture cylindrical tin cans having a volume of 1000 cubic centimeters and covered with circular lids at both ends. What should be the radius and height of the can in order that the surface area of tin plate may be minimized?
Hint: Use the volume constraint to eliminate one of the two independent variables and then utilize the minimization routine in Appendix 9.5 to determine the optimum parameters.

2. Consider the case when the can is shaped like a circular cone with a circular lid at the bottom. What will be the height of the cone and the radius of the base that will require minimum area of the tin plate?

3. Consider the case when two power stations are required to share the total load in a given area in the most economical manner. The total cost of operation at the two stations is given by

$$f(x, y) = 0.1 + 0.25x + 0.1y + 0.00125y^2$$

where x is the power supplied by the first station and y is the power supplied by the second station. (The values are in megawatts.) Determine the most economical values of x and y if the total power supplied is

(a) 100 megawatts

(b) 200 megawatts.

Hint: The sum of x and y must equal the total power supplied.

4. In the electrical circuit shown below, the power output to the load resistor, R, is given by

$$P = \frac{10^4 R}{(R + 20)^2}$$

The resistance R can be varied between 5 and 35 ohms. Find the value of R which will give maximum power output.

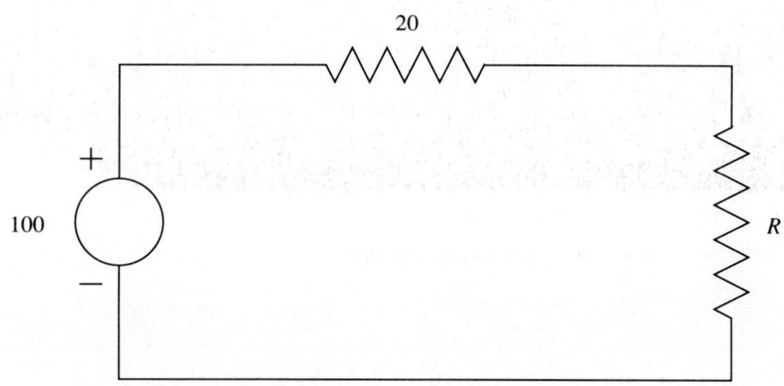

5. In the electrical network shown below, the power source provides an a.c. voltage of effective value 100 volts. The average power output to the load resistor, R, is given by

$$P = \frac{10^4 R}{(R + 20)^2 + 15^2}$$

The resistance R can be varied between 5 and 35 ohms. Find the value of R which will give maximum power output.

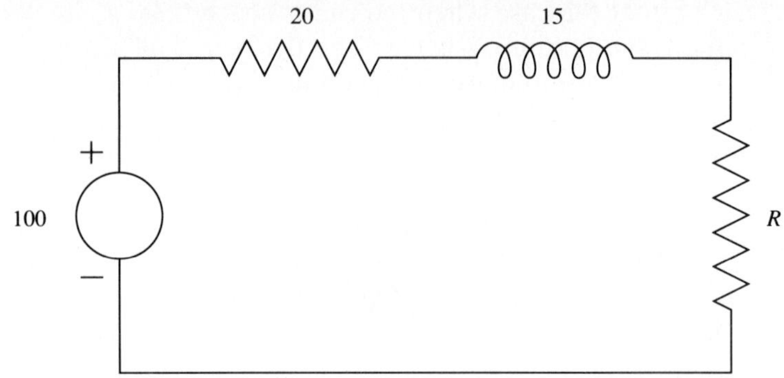

6. Consider the electrical circuit of the previous problem with the difference that both the resistive and the reactive components of the load can be varied. In this case, the average power output to the load is given by

$$P = \frac{10^4 R}{(R + 20)^2 + (X + 15)^2}$$

Determine the values of R and X for which maximum power output will be obtained. Use the quasi-Newton procedure given in Appendix 9.9 and take the initial guess as $R = 10$ and $X = 5$.

7. Determine the values of x and y which will minimize the function

$$f(x, y) = 2x^2 - 4xy + 4y^2 - 6y.$$

Start with the initial guess $(x, y) = (1.0, 2.0)$.

8. The following data were obtained as the result of an experiment.

i	1	2	3	4	5	6
x_i	0	1	2	3	4	5
y_i	3.1	5.92	7.05	8.96	11.11	12.88

It is desired to fit a straight line $y = ax + b$ to the data. Find the values of a and b which will minimize the function

$$f(a, b) = \sum_{i=1}^{6} [ax_i + b - y_i]^2.$$

This is the method of least squares that was discussed in Chapter 5.

9. Repeat Problem 8 for the following data if it is desired to fit the data to t quadratic $y = ax^2 + bx + c$.

i	1	2	3	4	5	6
x_i	0.2	0.5	0.8	1.0	1.5	2.0
y_i	14.1	16.4	18.8	20.1	22.4	24.2

10. Consider again the problem of the most economical load sharing that was described in Problem 3. In this problem there are three power stations and the individual costs are given by

$$f_1 = 0.1 + 0.25x$$

$$f_2 = 0.08 + 0.12y + 0.00125y^2$$

$$f_3 = 0.05 + 0.09z + 0.001z^2 + 0.0001z^3$$

where x, y, and z are the loads (in megawatts) supplied by the three power stations, respectively. Determine the values of these loads which will minimize the total cost if the total power to be supplied is

(a) 100 megawatts

(b) 200 megawatts.

11. The path of a comet can be described by the hyperbola

$$\frac{x^2}{64} - \frac{y^2}{49} = 1$$

where the coordinates are normalized and the origin is the center of the Earth. Find the minimum distance between the comet and a man-made satellite in a synchronous orbit that can be denoted by the point with fixed coordinates (1, 1).

12. Repeat Problem 11 if, instead of the comet, a second artificial satellite is in an elliptical orbit that can be described by the equation

$$\frac{x^2}{12} + \frac{y^2}{9} = 1.$$

13. Find the minimum of each of the following functions, using the given initial guesses.

(a) $f(x, y) = w^2 + z^2$,

where

$$w = \frac{\cos(x^2 - \sqrt{\sin(xy) + 3})}{4 + (xy)^2} + \sin(xy - 1) - 0.5$$

$$z = \exp(\cos[(xy)^3 - 3]) + \tan(x[0.08 + \cos(x)]/y) - 1.5.$$

Initial guess: $(x, y) = (1.0, 1.0)$.

(b) Rosenbrock's function

$$f(x, y) = 100(x^2 - y)^2 + (1 - x)^2$$

is well known as a difficult function for minimization methods to handle. It is often used as a test example for new methods that have been developed for function minimization.
Initial guess: $(x, y) = (-1.2, 1.0)$.

(c) Wood's function is a four-dimensional version of the function in part (b).

$$f(x, y, w, z) = 100(y - x^2)^2 + (1 - x)^2 + 90(z - w^2)^2$$
$$+ (1 - w)^2 + 10.1(y - 1)^2 + (z - 1)^2 + 19.8(y - 1)(z - 1).$$

Initial guess: $(x, y, w, z) = (-1.2, 1.0, -1.2, 1.0)$.

(d) Fletcher-Powell function.

$$f(x, y, z) = 100[(z - 10h)^2 + (w - 1)^2] + z^2,$$

where

$$w = (x^2 + y^2)^2$$

$$2\pi h = \begin{cases} \tan^{-1}(y/x) & \text{if } x > 0 \\ \pi + \tan^{-1}(y/x) & \text{if } x < 0 \end{cases}$$

Initial guess: $(x, y, z) = (-1.0, 0.0, 0.0)$.

14. Repeat Problem 12 of Chapter 5 by minimizing the function

$$f(a, b) = \sum_{i=1}^{n} [y_i - ax_i \exp(-x_i/b)]^2$$

where $n = 11$.

15. Repeat Problem 15 of Chapter 5 by minimizing an appropriate function. In part (a) there will be three unknowns to be determined; in part (b), six unknowns are to be found.

16. At time $t = 0$ the comet mentioned in Problem 11 is at the point (8, 0) while the artificial satellite mentioned in Problem 12 is at the point (0, 3). Find the time, $t_* > 0$, at which the comet and satellite are closest together.
Hint: Parameterize the equation describing an orbit by expressing each variable as a function of t. For example, the satellite orbit can be expressed as $x(t) = 2\sqrt{3}\sin(t)$ and $y(t) = 3\cos(t)$.

17. Repeat Problem 8 if the function to be minimized is

$$f(a, b) = \sum_{i=1}^{6} | ax_i + b - y_i |^k,$$

(a) $k = 1$ **(b)** $k = 4$ **(c)** $k = 8$.

18. Use the procedure QNM to solve the systems of equations in Chapter 4, Problems 15(a), 15(b), and 20(a). In each case, write the equations as $f(x, y) = 0$, $g(x, y) = 0$ and minimize the function $f^2 + g^2$. Compare your results to those obtained using the equation-solver procedure Powell in Chapter 4.

19. The function

$$f(x, y) = 100(x^2 - y)^8 + (1 - x)^8$$

can be considered as a generalization of Rosenbrock's function, Problem 13(b). Use the procedure QNM to minimize $f(x, y)$, given the initial point $(x, y) = (-1.2, 1.0)$.

20. The operation of a certain mechanism is controlled by a parameter, b, which can take values in the interval $[-10, 10]$. The cost of operation is given by the function

$$f(b) = \begin{cases} 0.25(1 + b^2)(e^{2(b-1)} - 1)/(b - 1) & \text{if } b \neq 1 \\ 1 & \text{if } b = 1 \end{cases}$$

Find the value of b for which the cost function is minimized.

21. Consider the following function

$$f(x) = \begin{cases} x(bx + c)/(2a) & \text{if } 0 \le x \le 2a \\ -a(x - 1)/(2a - 1) & \text{if } 2a \le x \le 1 \end{cases}$$

where $b = (2 - a)/a$ and $c = a - 4$. Suppose that the machine epsilon of your computer has the form $p \times 10^{-t}$, where $1 \le p < 10$. Choose the parameter a in $f(x)$ as $a = 10^{k-t}$ for $k = 5, 0, -5$. For each value of k, find the minimum of $f(x)$ using the Golden Section Search Method, the Fibonacci Search Method, and Brent's Method. If any method fails to find the correct minimum of $f(x)$, explain why.

22. Find the minimum value of

$$f(x) = x^2 - \sinh(\sin(x))$$

on the interval $[-1, 1]$.

23. For x in the interval $[0.5, 1.5]$ find the minimum value of

$$f(x) = \int_{-1}^{2} g(x, t)\, dt$$

where

$$g(x, t) = \frac{\cosh(2xt) - \cos(2t) + t}{\sinh(xt)}.$$

24. A gasoline storage tank, similar to the one in Problem 22, Chapter 4, is to be constructed. The cost of building either the cylindrical body or the two spherical caps is proportional to the surface area of that part of the tank. In addition, the cost of building the two spherical caps is three times the cost of the cylindrical body. The cost function (in nondimensional units) is given by

$$C(a) = v/a - 2[1 - a^2]^{1/2}(4af/3 + 3\pi) + 6\pi, \qquad 0 < a \le 1,$$

$$f = \int_0^1 [a^{-2} - 1 + t^2]^{-1}[1 - t^2]^{3/2}\, dt$$

v is the nondimensional volume of the tank.

The parameter a is the ratio of the cylinder radius to the radius of the sphere that forms one of the identical caps at each end of the tank. Find the value of a that minimizes $C(a)$ when $v = 2.5$.

9.11 REFERENCES

A. V. Fiacco and G. P. McCormick. *Nonlinear Sequential Unconstrained Minimization Techniques*. New York: Wiley, 1968.

G. E. Forsythe, M. A. Malcolm, and C. B. Moler. *Computer Methods for Mathematical Computations*. Englewood Cliffs, NJ: Prentice-Hall, 1977.

R. L. Fox. *Optimization Methods for Engineering Design*. Reading, MA: Addison-Wesley, 1971.

H. P. Kuenzi, H. G. Tzschach, and C. A. Zehnder. *Numerical Methods of Mathematical Optimization*. New York: Academic Press, 1971.

E. Polak. *Computational Methods in Optimization*. New York: Academic Press, 1971.

W. H. Press, B. P. Flannery, S. A. Teukolsky, and W. T. Vettering. *Numerical Recipes: The Art of Scientific Computing*. New York: Cambridge University Press, 1986.

T. E. Shoup. *A Practical Guide to Computer Methods for Engineers*. Englewood Cliffs, NJ: Prentice-Hall, 1979.

J. Stoer and R. Burlisch. *Introduction to Numerical Analysis*. New York: Springer-Verlag, 1980.

D. J. Wilde and C. S. Beightler. *Foundations of Optimization*. Englewood Cliffs, NJ: Prentice-Hall, 1967.

APPENDIX 9.1

Listing of a Pascal Procedure for the Golden Section Search Method, which is in the file Golden.Pas

```
PROCEDURE GoldenSectionSearch( VAR Left, Right : REAL ; tol : REAL ;
                               VAR xmin : REAL ; VAR evaluations : INTEGER );

(* Golden Section Method of minimization of a function on the interval,
   [Left,Right], to within a user-supplied tolerance of  tol.  The user
   must also supply the Pascal function
                FUNCTION F( x : REAL ) : REAL;
   to evaluate the function, whose minimum is desired, at any  x.
   The minimum is found at  xmin, and the interval of uncertainty is
   returned as  [Left, Right]. *)
```

```
VAR
  golden, x0, x1, x2, x3, f1, f2 : REAL;

PROCEDURE Initialize;
(* From the given end points Left and Right, determine
   intermediate points  x1  and  x2. *)
BEGIN
  golden := (3.0 - Sqrt(5.0))/2.0;
  x0 := Left;
  x3 := Right;
  x1 := Left + golden*(Right - Left);
  x2 := Right - golden*(Right - Left);
END; (* Initialize *)

PROCEDURE UseRightPart; (* Select the interval between  x1  and  x3  *)
BEGIN
  x0 := x1;
  x1 := x2;
  x2 := x3 - golden*(x3 - x1);
  f1 := f2;
  f2 := F(x2);
  evaluations := evaluations + 1
END; (* UseRightPart *)

PROCEDURE UseLeftPart; (* Select the interval between  x0  and  x2  *)
BEGIN
  x3 := x2;
  x2 := x1;
  x1 := x0 + golden*(x2 - x0);
  f2 := f1;
  f1 := F(x1);
  evaluations := evaluations + 1
END; (* UseLeftPart *)

PROCEDURE GetLimits;
BEGIN
  IF  f1 < f2  THEN
    BEGIN
      xmin := x1 ; Left := x0 ; Right := x1
    END
  ELSE
    BEGIN
      xmin := x2 ; Left := x2 ; Right := x3
    END
END; (* GetLimits *)

BEGIN (* GoldenSectionSearch *)
  Initialize;
  f1 := F(x1);
  f2 := F(x2);
  evaluations := 2;
  WHILE  (Abs(x3 - x0) > tol*(Abs(x1) + Abs(x2)))  DO
    BEGIN
      IF  f2 < f1  THEN
        UseRightPart
      ELSE
        UseLeftPart
    END;
  GetLimits
END; (* GoldenSectionSearch *)
```

APPENDIX 9.2

Listing of a Pascal Program for Example 9.1

```
PROGRAM GoldenSearchInOneDimension;

CONST
  tol = 1.0E-07;  (* tolerance *)

VAR
  xmin, Left, Right : REAL;
  evaluations : INTEGER;

FUNCTION F( x : REAL ) : REAL;
BEGIN
  F := 6.0 - ((x - 6.0)*x + 11.0)*x
END;

(*$I Golden.Pas *)

BEGIN
  Left := 1.0 ; Right := 2.0 ; (* Initial interval, chosen by the user *)
  GoldenSectionSearch(Left,Right,tol,xmin,evaluations);
  ClrScr;
  WRITELN('The minimum is located at  ',xmin);
  WRITELN('F(xmin) = ',F(xmin));
  WRITELN('Interval of uncertainty: [',Left,' ,',Right,'  ]');
  WRITELN('Number of function evaluations = ',evaluations)
END.
```

APPENDIX 9.3

Listing of a Pascal Procedure for the Fibonacci Search Method, which is in the file Fibon.Pas

```
PROCEDURE FibonacciSearch( Var Left, Right : REAL ;
                              VAR xmin : REAL ; evaluations : INTEGER );

(* Fibonacci search for the minimum of a function on the interval
   [Left, Right]  using "evaluations" function evaluations.  The user
   supplies the Pascal function
                 FUNCTION F( x : REAL ) : REAL;
   to evaluate the function at any point.  The minimum of the function is
   found at  xmin, and the interval of uncertainty is returned as
   [Left, Right] .   *)

CONST
  base = 2.0; (* Machine number base.  This is for Turbo Pascal. *)
  MaxEvaluations = 40;

VAR
  count : INTEGER;
  delta, x0, x1, x2, x3, f1, f2, ScaleFactor : REAL;
  fibonacci, w : ARRAY[ 0..MaxEvaluations ] OF REAL;
```

```
FUNCTION MachineEpsilon : REAL;
VAR
  a, b : REAL;
BEGIN
  a := 1.0;
  REPEAT
    a := a/base ; b := 1.0 + a
  UNTIL  b = 1.0;
  MachineEpsilon := a*base
END; (* MachineEpsilon *)

PROCEDURE GenerateFibonacciNumbers;
VAR
  i : INTEGER;
BEGIN
  fibonacci[0] := 1.0 ; fibonacci[1] := 1.0 ;
  FOR  i := 2  TO  evaluations  DO
    fibonacci[i] := fibonacci[i-1] + fibonacci[i-2]
END; (* GenerateFibonacciNumbers *)

PROCEDURE GetIntervals;
VAR
  i : INTEGER;
BEGIN
  w[evaluations] := (1.0 + delta*fibonacci[evaluations-2])/
                                           fibonacci[evaluations];
  FOR  i := 1  TO  evaluations-2  DO
    w[evaluations-i] := fibonacci[i+1]*w[evaluations] - delta*fibonacci[i-1];
  w[0] := 1.0 ; w[1] := 1.0
END; (* GetIntervals *)

PROCEDURE Initialize;
BEGIN
  IF  evaluations <= 0  THEN
    evaluations := 20
  ELSE
    IF  evaluations > MaxEvaluations  THEN
      evaluations := MaxEvaluations;
  delta := Sqrt(MachineEpsilon);
  x0 := Left;
  x3 := Right;
  ScaleFactor := Right - Left;
  GenerateFibonacciNumbers;
  GetIntervals
END; (* Initialize *)

PROCEDURE Select; (* Select the larger value as x2 *)
VAR
  temp : REAL;
BEGIN
  IF  Abs(x2) < Abs(x1)  THEN
    BEGIN
      temp := x1;
      x1 := x2;
      x2 := temp
    END
END; (* Select *)
```

```
PROCEDURE GetLimits;
BEGIN
  IF  f1 < f2  THEN
    BEGIN
      xmin := x1 ; Left := x0 ; Right := x1
    END
  ELSE
    BEGIN
      xmin := x2 ; Left := x2 ; Right := x3
    END
END; (* GetLimits *)

BEGIN (* FibonacciSearch *)
  Initialize;
  FOR  count := 1  TO  evaluations  DO
    BEGIN
      x1 := x3 - w[count]*ScaleFactor;
      x2 := x0 + w[count]*ScaleFactor;
      Select;
      f1 := F(x1);
      f2 := F(x2);
      IF  f1 < f2  THEN
        x3 := x2  (* Select left part *)
      ELSE
        x0 := x1  (* Select right part *)
    END;
  GetLimits
END; (* FibonacciSearch *)
```

APPENDIX 9.4

Listing of a Pascal Program for Example 9.2

```
PROGRAM FibonacciSearchInOneDimension;

VAR
  xmin, Left, Right : REAL;
  evaluations : INTEGER;

FUNCTION F( x : REAL ) : REAL;
BEGIN
  F := 6.0 - ((x - 6.0)*x + 11.0)*x
END;

(*$I Fibon.Pas *)

BEGIN (* Main program *);
  Left := 1.0 ; Right := 2.0 ; (* Initial interval, chosen by the user *)
  evaluations := 30;           (* Chosen by the user.  Maximum of 40 *)
  FibonacciSearch(Left,Right,xmin,evaluations);
  ClrScr;
  WRITELN('Number of function evaluations = ',evaluations);
  WRITELN('The minimum is located at  ',xmin);
  WRITELN('F(xmin) = ',F(xmin));
  WRITELN('Interval of uncertainty: [',Left,' ,',Right,'  ]')
END.
```

APPENDIX 9.5

Listing of a Pascal Procedure for Brent's Method, which is in the file Brent.Pas

```
PROCEDURE Brent( Left, Right, tol : REAL ; VAR xmin : REAL ;
                                    VAR Evaluations : INTEGER );

(* An approximation to the point where the function  f(x) attains a minimum
   on  the  interval  [Left,Right]  is  determined.   The  algorithm  is  a
   combination of the Golden Section Search method  and  inverse  parabolic
   interpolation.  The function  f(x)  is  never  evaluated at  two  points
   closer together than
                   r*Abs(minimum  of  f) + (tol/3),
   where  r  is the square root of the relative machine precision.  If f(x)
   is  unimodal  and the computed values of  f(x)  are always unimodal when
   separated by at least
                        r*Abs(x) + (tol/3),
   then  the  procedure  approximates the abscissa of the global minimum of
   f(x) on the interval [Left,Right] with an error less than
                     3*r*Abs(minimum of f) + tol.
   If f(x) is not unimodal, then the procedure may approximate a local, but
   perhaps nonglobal, minimum to the same accuracy.


   The user supplies the following Pascal function to evaluate  f(x)
         FUNCTION F( x : REAL ) : REAL;


   Input:

      Left    Left endpoint of the initial interval
      Right   Right endpoint of the initial interval
       tol    Desired length of the interval of uncertainty of the final
                result (tol >= 0.0)


   Output:

         xmin      Abscissa approximating the point where  f(x)  attains
                     a minimum
      Evaluations  The number of function evaluations required to obtain
                     the result, xmin

Procedure Brent is based on the FORTRAN function FMIN that appeared in
       G.E Forsythe, M.A. Malcolm and C.B. Moler
       "Computer Methods for Mathematical Computations"
       Prentice-Hall, Englewood Cliffs, N.J.  1977, pp. 185-187    *)


CONST
  base = 2.0; (* Machine number base.  This is for Turbo Pascal *)

VAR
  golden, d, e, eps, fu, fv, fw, fx, p : REAL;
  q, r, tol1, tol2, u, v, w, x, xm : REAL;
  gSection : BOOLEAN;
```

```
FUNCTION MachineEpsilon : REAL;
VAR
  a, b : REAL;
BEGIN
  a := 1.0;
  REPEAT
    a := a/base ; b := 1.0 + a
  UNTIL  b = 1.0;
  MachineEpsilon := a*base
END; (* MachineEpsilon *)

FUNCTION Sign( x, y : REAL ) : REAL;
BEGIN
  IF  y >= 0.0  THEN
    Sign := Abs(x)
  ELSE
    Sign := -Abs(x)
END; (* Sign *)

PROCEDURE Initialization;
BEGIN
  golden := (3.0 - Sqrt(5.0))/2.0;
  eps := Sqrt(MachineEpsilon);
  v := Left + golden*(Right - Left);
  w := v ; x := v ; e := 0.0 ; fx := F(x) ;
  fv := fx ; fw := fx ; Evaluations := 1
END; (* Initialization *)

PROCEDURE Start;
BEGIN
  xm := (Left + Right)/2.0;
  tol1 := eps*Abs(x) + tol/3.0;
  tol2 := 2.0*tol1;
  gSection := FALSE
END; (* Start *)

PROCEDURE GoldenSectionStep;
BEGIN
  gSection := TRUE;
  IF  x >= xm  THEN
    e := Left - x
  ELSE
    e := Right - x;
  d := e*golden
END; (* GoldenSectionStep *)

PROCEDURE Parabola;
BEGIN
  r := (x - w)*(fx - fv);
  q := (x - v)*(fx - fw);
  p := (x - v)*q - (x - w)*r;
  q := 2.0*(q - r);
  IF  q > 0.0  THEN
    p := -p
  ELSE
    q := Abs(q);
  r := e ; e := d
END; (* Parabola *)
```

```
PROCEDURE UpDate;
VAR
  exit : BOOLEAN;
BEGIN
  exit := FALSE;
  IF  fu > fx  THEN
    BEGIN
      IF  u < x  THEN
        Left := u
      ELSE
        Right := u;
      IF  (fu <= fw) OR (w = x)  THEN
        BEGIN
          v := w ; fv := fw ; w := u ; fw := fu ; exit := TRUE
        END;
      IF  NOT exit  THEN
        IF  (fu <= fv) OR (v = x) OR (v = w)  THEN
          BEGIN
            v := u ; fv := fu
          END;
      exit := TRUE
    END;
  IF  NOT exit  THEN
    BEGIN
      IF  u >= x  THEN
        Left := x
      ELSE
        Right := x;
      v := w ; fv := fw ; w := x ; fw := fx ; x := u ; fx := fu
    END
END; (* Update *)


BEGIN (* Brent *)
  Initialization;
  Start;
  WHILE  Abs(x - xm) > (tol2 - (Right - Left)/2.0)  DO
    BEGIN
      IF  Abs(e) <= toll  THEN
        GoldenSectionStep
      ELSE
        BEGIN
          Parabola; (* fit parabola, then see if it is acceptable *)
          IF  Abs(p) >= Abs(q*r/2.0)  THEN
            GoldenSectionStep
          ELSE
            IF  p <= q*(Left - x)  THEN
              GoldenSectionStep
            ELSE
              IF  p >= q*(Right - x)  THEN
                GoldenSectionStep;
          IF  NOT gSection  THEN
            BEGIN (* parabolic interpolation step *)
              d := p/q ; u := x + d ;
              IF  ((u - Left) < tol2) OR ((Right - u) < tol2)  THEN
                d := Sign(toll,xm - x)
            END
        END;
      IF  Abs(d) >= toll  THEN
        u := x + d
```

```
      ELSE
        u := x + Sign(toll,d);
      fu := F(u) ; Evaluations := Evaluations + 1 ;
      UpDate;
      Start
    END; (* while *)
  xmin := x
END; (* Brent *)
```

APPENDIX 9.6

Listing of a Pascal Program for Example 9.3

```
PROGRAM MinimizationInOneDimension;

CONST
  Left = 1.0 ; Right = 2.0 ; tol = 1.0E-07 ;

VAR
  xmin : REAL;
  evaluations : INTEGER;

FUNCTION F( x : REAL ) : REAL;
BEGIN
  F := 6.0 - ((x - 6.0)*x + 11.0)*x
END;

(*$I Brent.Pas *)

BEGIN
  Brent(Left,Right,tol,xmin,evaluations);
  ClrScr;
  WRITELN('The minimum is located at  ',xmin);
  WRITELN('F(xmin) = ',F(xmin));
  WRITELN('Number of function evaluations = ',evaluations')
END.
```

APPENDIX 9.7

Listing of a Pascal Procedure for the Modified Brent's Method, which is in the file dBrent.Pas

```
PROCEDURE dBrent( Left, Right, tol : REAL; VAR xmin : REAL ;
                                     VAR evaluations : INTEGER );

(* This procedure is a modification of Brent's method that uses derivative
   information with the secant method to determine the minimum of a
   unimodal function on the interval [Left, Right].
```

```
   The user must supply the Pascal function
            FUNCTION F( x : REAL ) : REAL;
   to evaluate the function whose minimum is sought.
   Derivatives are computed internally by a finite difference method.  *)

CONST
  base = 2.0; (* Machine number base.  This is for Turbo Pascal *)

VAR
  a, b, u, du, fu, v, dv, fv, x, dx, fx, w, dw, fw, xm : REAL;
  e, olde, d, eps, tol1, tol2, d1, d2, u1, u2 : REAL;
  done : BOOLEAN;

FUNCTION MachineEpsilon : REAL;
VAR
  a, b : REAL;
BEGIN
  a := 1.0;
  REPEAT
    a := a/base ; b := 1.0 + a
  UNTIL  b = 1.0;
  MachineEpsilon := a*base
END; (* MachineEpsilon *)

FUNCTION Sign( x, y : REAL ) : REAL;
BEGIN
  IF  y >= 0.0  THEN
    Sign := Abs(x)
  ELSE
    Sign := -Abs(x)
END; (* Sign *)

FUNCTION Derivative( x, fx : REAL ) : REAL;
VAR
  h, t : REAL;
BEGIN
  t := Abs(x);
  IF  t < 0.1  THEN
    h := 0.1*eps
  ELSE
    h := t*eps;
  IF  x = Right  THEN
    h := -h;
  Derivative := (F(x+h) - fx)/h;
  evaluations := evaluations + 1
END; (* Derivative *)

PROCEDURE Initialization;
BEGIN
  done := FALSE;
  eps := Sqrt(MachineEpsilon);
  IF  Left > Right  THEN
    BEGIN
      a := Right ; b := Left ; Left := a ; Right := b
    END
  ELSE
    BEGIN
      a := Left ; b := Right
    END;
```

```
  v := (a + b)/2.0 ; w := v ; x := v ; e := 0.0 ;
  fx := F(x) ; evaluations := 1 ;
  fv := fx ; fw := fx ;
  dx := Derivative(x,fx);
  dv := dx ; dw := dx
END; (* Initialization *)

FUNCTION Secant( y, dy : REAL ) : REAL;
BEGIN
  Secant := (y*dx - x*dy)/(dx - dy)
END; (* Secant *)

FUNCTION Flag( u, d : REAL ) : BOOLEAN;
BEGIN
  Flag := (a < u) AND (u < b) AND ((dx*d) <= 0.0)
END; (* Flag *)


PROCEDURE CheckConvergence;
BEGIN
  IF  Abs(d) >= toll  THEN
    BEGIN
      u := x + d ; fu := F(u) ; evaluations := evaluations + 1
    END
  ELSE
    BEGIN
      u := x + Sign(toll,d) ; fu := F(u) ; evaluations := evaluations + 1 ;
      IF  fu > fx  THEN
        done := TRUE
    END
END; (* CheckConvergence *)

PROCEDURE PickSegment;
BEGIN
  IF  dx >= 0.0  THEN
    e := a - x
  ELSE
    e := b - x;
  d := e/2.0
END; (* PickSegment *)

PROCEDURE Move( a, fa, da : REAL );
BEGIN
  v := w ; fv := fw ; dv := dw ;
  w := a ; fw := fa ; dw := da
END; (* Move *)

PROCEDURE NewValues( VAR a, fa, da : REAL );
BEGIN
  a := u ; fa := fu ; da := du
END; (* NewValues *)

PROCEDURE Test;
VAR
  flag1, flag2 : BOOLEAN;
BEGIN
  flag1 := Flag(u1,d1) ; flag2 := Flag(u2,d2) ;
  olde := e ; e := d ;
```

```
IF  NOT (flag1 OR flag2)  THEN
  BEGIN
    PickSegment;
    CheckConvergence
  END
    ELSE
      BEGIN
        IF  flag1 AND flag2  THEN
          IF  Abs(d1) < Abs(d2)  THEN
            d := d1
          ELSE
            d := d2
        ELSE
         IF  flag1  THEN
           d := d1
         ELSE
           d := d2;
       IF  Abs(d) > Abs(olde/2.0)  THEN
         BEGIN
           PickSegment;
           CheckConvergence
         END
       ELSE
         BEGIN
           u := x + d;
           IF  ((u - a) < tol2) OR ((b - u) < tol2)  THEN
             d := Sign(tol1,xm - x);
           CheckConvergence
         END
     END
 END; (* Test *)


 PROCEDURE UpDate;
 BEGIN
   du := Derivative(u,fu);
   IF  fu <= fx  THEN
     BEGIN
       IF  u >= x  THEN
         a := x
       ELSE
         b := x;
       Move(x,fx,dx);
       NewValues(x,fx,dx)
     END
   ELSE
     BEGIN
       IF  u < x  THEN
         a := u
       ELSE
         b := u;
       IF  (fu <= fw) OR (w = x)  THEN
         Move(u,fu,du)
       ELSE
         IF  (fu <= fv) OR (v = x) OR (v = w)  THEN
           NewValues(v,fv,dv)
     END
  END; (* Update *)
```

```
BEGIN (* dBrent *)
  Initialization;
  REPEAT
    xm := (a + b)/2.0;
    tol1 := eps*Abs(x) + tol/3.0;
    tol2 := 2.0*tol1;
    IF  Abs(x - xm) <= (tol2 - (b - a)/2.0)  THEN
      done := TRUE
    ELSE
      BEGIN
        IF  Abs(e) > tol1  THEN
          BEGIN
            d1 := 2.0*(b - a) ; d2 := d1 ;
            IF  dw <> dx  THEN
              d1 := Secant(w,dw);
            IF  dv <> dx  THEN
              d2 := Secant(v,dv);
            u1 := x + d1 ; u2 := x + d2 ;
            Test
          END
        ELSE
          BEGIN
            PickSegment;
            CheckConvergence
          END;
        IF  NOT done  THEN
          Update
      END
  UNTIL  done;
  xmin := x
END; (* dBrent *)
```

APPENDIX 9.8

Listing of a Pascal Program for Example 9.4

```
PROGRAM BrentDerivativeMethod;

CONST
  tol = 1.0E-07;  (* tolerance *)

VAR
  Left, Right, xmin : REAL;
  evaluations : INTEGER;

FUNCTION F( x : REAL ) : REAL;
BEGIN
  F := 6.0 - ( 11.0 + (x - 6.0)*x )*x
END;

(*$I dBrent.Pas *)

BEGIN
  Left := 1.0 ; Right := 2.0 ; (* Initial interval, chosen by the user *)
  dBrent(Left,Right,tol,xmin,evaluations);
```

```
  ClrScr;
  WRITELN('The minimum is located at  ',xmin);
  WRITELN('F(xmin) = ',F(xmin));
  WRITELN('Number of function evaluations = ',evaluations)
END.
```

APPENDIX 9.9

User's Manual for the Pascal Procedure QNM, which is in the file QNM.Pas

```
PROCEDURE QNM( n : INTEGER ; VAR x, g : VectorN ; VAR f : REAL ;
                  VAR digits : INTEGER ;  MaxEvaluations, print : INTEGER ;
                       VAR flag : INTEGER ; VAR OutFile : TEXT );
```

The procedure QNM finds the minimum of a function, f(x), of n variables using a quasi-Newton method. It is assumed that the gradient of f(x) and the Hessian matrix of f(x) exist, but the user is not required to supply procedures to calculate them. The gradient is computed by finite difference methods. The Hessian matrix is initially taken as the identity matrix and is updated as the computations proceed.

The following must be declared in the calling program

```
        TYPE
          VectorN = ARRAY[ 1..MaxN ] OF REAL;
          LongVector = ARRAY[ 1..n2 ] OF REAL;

   where   n2 = MaxN*(MaxN + 1)/2 .
```

The user supplies the following to evaluate the function, whose minimum is sought, at the point x = (x[1], x[2], ..., x[n]).

```
        FUNCTION Fqnm( n : INTEGER ; VAR x : VectorN ) : REAL;
```

Input:

n	The number of unknowns. (1 ≤ n ≤ MaxN)
x	The initial guess at the unknowns
digits	The number of digits of accuracy required in the answer. Note that f(x) may have to be rewritten by scaling the components of x because leading zeros to the right of the decimal point are counted in the significant digit test. The scaling should be done so that all components x[i] have approximately the same magnitude at the solution.
MaxEvaluations	Maximum allowable number of function evaluations. The procedure imposes a minimum number of max(200,2n).
print	Output indicator = 0 No output execept for the the initial guess and the "converged" solution. = m Output each m iterations, including the last one.
OutFile	Output is written to the file OutFile. Before calling QNM, the user should ensure that OutFile has been properly opened.

```
Output:

     x           The "converged" solution
     g           The gradient of f(x) at the "converged" solution
     f           The value of f(x) at the "converged" solution
   digits        An estimate of the number of significant digits in the
                  "converged" solution
    flag         A status indicator
                      0   Convergence achieved, and no errors occurred.
                      1   Could not find answer to the requested accuracy
                            of digits.  It is possible that rounding
                            errors have become dominant.
                      2   Termination because MaxEvaluations exceeded.
```

Procedure QNM contains the following machine dependent constant which is set for Turbo Pascal with 8087 support.

giant = 1.0E+307

Procedure QNM is based on the IMSL FORTRAN subroutine ZXMIN

APPENDIX 9.10

Listing of a Program using the Procedure QNM for Example 9.5

```
PROGRAM QuasiNewtonMethod;

CONST
  d = 'Con'; (* Output to the console *)
  MaxN = 2 ; n2 = 3 ; (* n2 = MaxN*(MaxN+1) DIV 2  *)
  MaxEvaluations = 500 ; print = 1 ;

TYPE
  VectorN = ARRAY[ 1..MaxN ] OF REAL;
  LongVector = ARRAY[ 1..n2 ] OF REAL;

VAR
  OutFile : TEXT;
  x, g : VectorN;
  f : REAL;
  n, digits, flag : INTEGER;

FUNCTION Fqnm( n : INTEGER ; VAR x : VectorN ) : REAL;
BEGIN
  Fqnm := Sqr(x[2] - Sqr(x[1])) + Sqr(1.0 - x[1])
END;

(*$I QNM.Pas *)

BEGIN
  n := 2;
  digits := 4;  (* Number of digits in solution.  Supplied by user *)
  x[1] := -4.0 ; x[2] := 16.0 ;  (* Initial guess *)
  ASSIGN(OutFile,d) ; REWRITE(OutFile) ;
  QNM(n,x,g,f,digits,MaxEvaluations,print,flag,OutFile);
  CLOSE(OutFile)
END.
```

Appendix

A

A Brief Review of Linear Algebra

A.1 VECTORS

Let $\mathbf{x}$ denote the column array of numbers

$$\mathbf{x} = \begin{bmatrix} x_1 \\ x_2 \\ \vdots \\ x_n \end{bmatrix} \quad \text{(A.1–1)}$$

If each x_i is a real number, then $\mathbf{x}$ is a real vector; if one or more x_i are complex numbers, then $\mathbf{x}$ is a complex vector. Let E denote the set of all vectors $\mathbf{x}$ of the form (A.1–1). The ith component of the vector $(\mathbf{x} + \mathbf{y})$ is $(x_i + y_i)$, where $\mathbf{x}$ and $\mathbf{y}$ are in E. The vector $t\mathbf{x}$, where t is a scalar, has tx_i as its ith component.

The set of vectors $\{\mathbf{x}_1, \mathbf{x}_2, \ldots, \mathbf{x}_k\}$ in E is said to be *linearly independent* if the only solution to the equation

$$\sum_{i=1}^{k} a_i\mathbf{x}_i = \mathbf{0}$$

is $a_i = 0$ for $i = 1, 2, \ldots, k$; otherwise, the set of vectors is said to be *linearly dependent*. A linearly independent set of vectors

$$\{\mathbf{x}_1, \mathbf{x}_2, \ldots, \mathbf{x}_n\}$$

in E is said to be a *basis* for E if any vector $\mathbf{y}$ in E can be written as a unique linear combination

$$\mathbf{y} = \sum_{i=1}^{n} c_i \mathbf{x}_i$$

This means that when $\{\mathbf{x}_1, \mathbf{x}_2, \ldots, \mathbf{x}_n\}$ and $\mathbf{y}$ are given, the n numbers c_i are uniquely determined.

If all vectors in E are real, we can define a real inner product, denoted by $\langle \cdot, \cdot \rangle$, as a function that satisfies the conditions

(a) $\langle \mathbf{x}, \mathbf{x} \rangle > 0$ for all $\mathbf{x} \neq \mathbf{0}$ in E

(b) $\langle \mathbf{x}, \mathbf{y} \rangle = \langle \mathbf{y}, \mathbf{x} \rangle$ for all $\mathbf{x}$ and $\mathbf{y}$ in E

(c) $\langle t\mathbf{x}, \mathbf{y} \rangle = t \langle \mathbf{x}, \mathbf{y} \rangle$ for all real numbers t and all $\mathbf{x}$ and $\mathbf{y}$ in E

(d) $\langle \mathbf{x} + \mathbf{y}, \mathbf{z} \rangle = \langle \mathbf{x}, \mathbf{z} \rangle + \langle \mathbf{y}, \mathbf{z} \rangle$ for all $\mathbf{x}$, $\mathbf{y}$, and $\mathbf{z}$ in E.

If $\langle \mathbf{x}, \mathbf{y} \rangle = 0$ then $\mathbf{x}$ and $\mathbf{y}$ are said to be *orthogonal*.

If $\{\mathbf{x}_1, \mathbf{x}_2, \ldots, \mathbf{x}_k\}$ is a linearly independent set of vectors in E, we can construct a mutually orthogonal set of vectors $\{\mathbf{y}_1, \mathbf{y}_2, \ldots, \mathbf{y}_k\}$ from the $\mathbf{x}$-vectors by using the *Gram-Schmidt* process:

(1) Set $\mathbf{y}_1 = \mathbf{x}_1$.

(2) For $j = 2, 3, \ldots, k$ we compute $\mathbf{y}_j$ as

$$\mathbf{y}_j = \mathbf{x}_j - \sum_{i=1}^{j-1} c_{ij} \mathbf{y}_i$$

where $c_{ij} = \langle \mathbf{x}_j, \mathbf{y}_i \rangle / \langle \mathbf{y}_i, \mathbf{y}_i \rangle$ for $i = 1, 2, \ldots, j - 1$.

If $\{\mathbf{y}_1, \mathbf{y}_2, \ldots, \mathbf{y}_k\}$ is a mutually orthogonal set of vectors with

$$\langle \mathbf{y}_i, \mathbf{y}_i \rangle = 1 \quad \text{for all } i$$

then $\{\mathbf{y}_1, \mathbf{y}_2, \ldots, \mathbf{y}_k\}$ is an *orthonormal* set of vectors.

A.2 MATRICES

If n and m are positive integers, the $n \times m$ matrix $\mathbf{A} = (a_{ij})$ is a rectangular array (if $n \neq m$), or square array ($n = m$), of the nm numbers a_{ij}. If all a_{ij} are real numbers, then $\mathbf{A}$ is called a *real* matrix; if some of the a_{ij} are complex numbers, then $\mathbf{A}$ is a *complex* matrix. The entry in the ith row and jth column of $\mathbf{A}$ is a_{ij}. The transpose of the $n \times m$ matrix $\mathbf{A}$, denoted by $\mathbf{A}^T$, is the $m \times n$ array whose (i, j)th entry is a_{ji}. If $\mathbf{A} = (a_{ij})$ and $\mathbf{B} = (b_{ij})$ are $n \times m$ matrices, then the (i, j)th entry of the matrix $(\mathbf{A} + \mathbf{B})$ is $(a_{ij} + b_{ij})$. If $\mathbf{A} = (a_{ij})$ is $n \times m$ and $\mathbf{C} = (c_{jk})$ is $m \times p$, then the (i, k)th entry of the $n \times p$ product matrix $\mathbf{AC}$ is

$\sum_{j=1}^{m} a_{ij}c_{jk}$. The vector **x**, with m components, can be considered as an $m \times 1$ matrix. Thus, the product **Ax** is a vector with n components, if **A** is $n \times m$.

If $\mathbf{A} = (a_{ij})$ is an $n \times n$ matrix, then the *main diagonal* of **A** consists of the entries a_{ii} for $i = 1, 2, \ldots, n$. The *subdiagonal* of **A** contains the entries $a_{i,i-1}$ for $i = 2, 3, \ldots, n$; this is the diagonal immediately below the main diagonal. The *superdiagonal* of **A** is made up of the entries $a_{i-1,i}$ for $i = 2, 3, \ldots, n$; this is the diagonal immediately above the main diagonal. An *upper triangular* matrix is one in which $a_{ij} = 0$ for all $j < i$; a *lower triangular* matrix is one in which $a_{ij} = 0$ for all $j > i$. A diagonal matrix has $a_{ij} = 0$ for all $j \neq i$. Because all the entries in a diagonal matrix off the main diagonal are zero, we need only specify those entries on the main diagonal, using the notation $\mathbf{D} = \text{diag}(d_1, d_2, \ldots, d_n)$. If the $n \times n$ matrix **A** has an inverse, we say that **A** is *nonsingular* and denote its inverse by $\mathbf{A}^{-1}$; if **A** does not have an inverse, then **A** is *singular*. We often use an $n \times n$ *identity* matrix; **I** is an $n \times n$ identity matrix if $\mathbf{I} = \text{diag}(1, 1, \ldots, 1)$ where there are n ones on the diagonal.

An $n \times n$ *permutation matrix* is obtained from the $n \times n$ identity matrix by interchanging its rows. An $n \times n$ real matrix **P** is called *orthogonal* if $\mathbf{P}^T\mathbf{P} = \mathbf{I}$.

The $n \times n$ matrix **A** is *similar* to the $n \times n$ matrix **B** if there is a nonsingular $n \times n$ matrix **S** such that $\mathbf{B} = \mathbf{S}^{-1}\mathbf{A}\mathbf{S}$. The matrix **A** is called *diagonalizable* if **A** is similar to a diagonal matrix.

A.3 DETERMINANTS

Let $\mathbf{A} = (a_{ij})$ be an $n \times n$ matrix. The *determinant* of **A**, denoted by det(**A**) or

$$\begin{vmatrix} a_{11} & \cdot & \cdot & \cdot & a_{1n} \\ \vdots & \vdots & \vdots & \vdots & \vdots \\ a_{n1} & \cdot & \cdot & \cdot & a_{nn} \end{vmatrix}$$

is defined by

$$\det(\mathbf{A}) = \sum_{\Pi} (-1)^{\sigma(\Pi)} a_{1j_1} a_{2j_2} \cdots a_{nj_n}$$

where $\Pi = \{j_1, j_2, \ldots, j_n\}$ is a permutation of the n numbers $\{1, 2, \ldots, n\}$,

$$\sigma(\Pi) = \begin{cases} 0 & \text{if } \Pi \text{ is an } even \text{ permutation} \\ 1 & \text{if } \Pi \text{ is an } odd \text{ permutation} \end{cases}$$

and the summation is taken over all possible permutations of the n numbers $\{1, 2, \ldots, n\}$.

If $\mathbf{A} = (a_{ij})$ and **B** are $n \times n$ matrices, the determinant function has the following properties:

1. $\det(\mathbf{A}) = \det(\mathbf{A}^T)$.
2. If **B** is obtained from **A** by interchanging two rows (or two columns), then $\det(\mathbf{B}) = -\det(\mathbf{A})$.

3. If **B** is obtained from **A** by multiplying a row (or a column) by the scalar s, then $\det(\mathbf{B}) = s \times \det(\mathbf{A})$.
4. If **B** is obtained from **A** by adding a scalar multiple of one row to another row (or by adding a scalar multiple of one column to another column), then $\det(\mathbf{B}) = \det(\mathbf{A})$.
5. If **A** is an upper triangular matrix, then $\det(\mathbf{A}) = a_{11}a_{22} \ldots a_{nn}$.
6. $\det(\mathbf{AB}) = \det(\mathbf{A}) \det(\mathbf{B})$.
7. **A** is nonsingular if, and only if, $\det(\mathbf{A}) \neq 0$.
8. If **A** is nonsingular, then $\det(\mathbf{A}^{-1}) = [\det(\mathbf{A})]^{-1}$.
9. If **B** is nonsingular, then $\det(\mathbf{B}^{-1}\mathbf{AB}) = \det(\mathbf{A})$.

If $\mathbf{A} = (a_{ij})$ is an $n \times n$ matrix, we can define $\mathbf{A}_{pq}$ as the submatrix obtained from **A** by deleting the pth row and qth column of **A**. The number $\det(\mathbf{A}_{pq})$ is called the *minor* associated with the element a_{pq} of **A**. The signed minor, $(-1)^{p+q}\det(\mathbf{A}_{pq})$, is called the *cofactor* of a_{pq}.

The determinant of $\mathbf{A} = (a_{ij})$ can be calculated by the method of *expansion by minors* (or *cofactor expansion*) because, for fixed indices p and q, we have

$$\det(\mathbf{A}) = \sum_{j=1}^{n} a_{pj}(-1)^{p+j} \det(\mathbf{A}_{pj}) \qquad \text{(expansion by row } p\text{)}$$

$$= \sum_{j=1}^{n} a_{jq}(-1)^{j+q} \det(\mathbf{A}_{jq}) \qquad \text{(expansion by column } q\text{)}$$

The linear algebraic system

$$\mathbf{Ax} = \mathbf{b}$$

is often solved using *Cramer's Rule*, when the $n \times n$ matrix **A** is nonsingular. The solution is written as the ratios of determinants. Let $\mathbf{A}_k$ denote the $n \times n$ matrix obtained from **A** by replacing the kth column of **A** by the vector **b**. The kth component of the solution is

$$x_k = \frac{\det(\mathbf{A}_k)}{\det(\mathbf{A})} \qquad \text{for} \qquad 1 \leq k \leq n.$$

A.4 EIGENVALUES AND EIGENVECTORS

Let **A** be an $n \times n$ matrix. The number λ is an *eigenvalue* of **A** if there is a nonzero vector **x** such that

$$\mathbf{Ax} = \lambda\mathbf{x}$$

The nonzero vector **x** is called an *eigenvector* of **A** associated with the eigenvalue λ. The eigenvalues of **A** satisfy the *characteristic equation* of **A**, which is given by

$$\det(s\mathbf{I} - \mathbf{A}) = 0. \tag{A.4–1}$$

The left-hand side of (A.4–1),

$$\det(s\mathbf{I} - \mathbf{A}) \tag{A.4–2}$$

a degree n polynomial in the variable s is called the *characteristic polynomial* of **A**. Consequently, (A.4–2) has n roots, which may be complex numbers even if the entries in **A** are all real numbers.

If λ is an eigenvalue of **A**, then the eigenvector(s) associated with λ satisfy the equation

$$(\lambda\mathbf{I} - \mathbf{A})\mathbf{x} = \mathbf{0}. \tag{A.4–3}$$

Because the matrix $(\lambda\mathbf{I} - \mathbf{A})$ is singular, (A.4–3) has nonzero solutions **x**.

If λ is a root of multiplicity k of the characteristic polynomial (A.4–2), then λ is said to have *algebraic multiplicity* k. If λ has associated with it a set of q linearly independent eigenvectors, where q is maximal, then the *geometric multiplicity* of λ is q. In general, $q \le k$. If $q < k$ then **A** is said to be *defective*. Let the q linearly independent eigenvectors associated with λ be $\mathbf{v}_1, \mathbf{v}_2, \ldots, \mathbf{v}_q$. We can always find sets of *generalized eigenvectors*, associated with λ,

$$\{\mathbf{x}_{i,1}, \mathbf{x}_{i,2}, \ldots, \mathbf{x}_{i,j_i}\}$$

with $\mathbf{x}_{i,1} = \mathbf{v}_i$, $j_i \ge 1$, for $i = 1, 2, \ldots, q$. The integer j_i depends on the integer i. The generalized eigenvectors satisfy the equations

$$\begin{aligned}
\mathbf{A}\mathbf{x}_{i,1} &= \lambda\mathbf{x}_{i,1} \\
\mathbf{A}\mathbf{x}_{i,2} &= \lambda\mathbf{x}_{i,2} + \mathbf{x}_{i,1} \\
\vdots \quad & \quad \vdots \\
\mathbf{A}\mathbf{x}_{i,j_i} &= \lambda\mathbf{x}_{i,j_i} + \mathbf{x}_{i,j_i-1}
\end{aligned}$$

for $i = 1, 2, \ldots, q$ and

$$\sum_{i=1}^{q} j_i = k - q$$

If the $n \times n$ matrix **A** has a linearly independent set of n eigenvectors, then **A** is similar to a diagonal matrix **D**, which contains the eigenvalues along its main diagonal. Thus, $\mathbf{S}^{-1}\mathbf{A}\mathbf{S} = \mathbf{D}$, where the columns of **S** are the eigenvectors of **A**.

If **A** is an $n \times n$ real symmetric matrix, then **A** has only real eigenvalues. In addition, the eigenvectors of **A** are mutually orthogonal and so can be made into *orthonormal* vectors (orthogonal unit vectors); a real symmetric matrix **A** is orthogonally similar to a diagonal matrix **D**, $\mathbf{S}^T\mathbf{A}\mathbf{S} = \mathbf{D}$ where **S** is an orthogonal matrix. If its eigenvalues are all positive, **A** is called *positive definite*; if all the eigenvalues are negative, **A** is called *negative definite*.

A.5 REFERENCES

G. L. Bradley. *A Primer of Linear Algebra*. Englewood Cliffs, NJ: Prentice-Hall, 1975.
G. Strang. *Linear Algebra and Its Applications*. New York: Academic Press, 1976.

Appendix B

A Review of Pascal

B.1 INTRODUCTION

The object of this appendix is to introduce the uninitiated reader to Pascal. It is recommended that the study of this appendix be followed by reading one of the books listed in the reference section.

Pascal was designed for the teaching of good programming style. Its main merits are transparency (ease of understanding), security (most errors will be detected by the computer, especially while compiling) and efficiency. One of its main advantages is the provision for structuring programs as well as data. The former makes the program more readable and hence, transparent to the user. Data structures allow the programmer to group together related items of data. This makes programming easier and adds to the readability of programs.

As usual with most programming languages, many versions of Pascal have been implemented on different computers. The versions most commonly used on microcomputers are (1) MS (or Microsoft) Pascal, (2) UCSD (University of California, San Diego) Pascal and (3) Turbo Pascal. The programs listed in this book will be based on Turbo Pascal mainly because it is inexpensive and the compiler requires less than 42 kilobytes of memory. These programs can be utilized with other Pascal compilers with only minor modifications.

The best way to get familiar with Pascal is through an example. We shall, therefore, consider a simple program for evaluating the roots of a quadratic using the algorithm presented in Section 3.3.

B.2 A SIMPLE PROGRAM FOR SOLVING QUADRATICS

```
PROGRAM Quadratic;

VAR
  a, b, c, discriminant, root1,
    root2, RealPart, ImaginaryPart, sgn : REAL;

BEGIN
  ClrScr;  (* Turbo Pascal procedure for clearing the screen *)
  WRITELN('This program will calculate the roots of the quadratic');
  WRITELN('x(ax + b) + c = 0, for given a, b and c.');
  WRITE('Press the <enter> key when ready. ');
  READLN;
  ClrScr;
  REPEAT
    WRITELN('Enter a, b and c, separated by spaces. ');
    READLN(a, b, c);
    IF  a = 0.0  THEN
      WRITELN('The coefficient  a  cannot be zero for a quadratic! ')
     (* This avoids division by zero. *)
  UNTIL  a <> 0.0;
    discriminant := Sqr(b) - 4.0*a*c;
  IF discriminant < 0.0 THEN
    BEGIN  (* Complex roots *)
      ImaginaryPart := Sqrt(-discriminant)/(2.0*Abs(a));
      RealPart := -b/(2.0*a);
      WRITELN('The roots are ',RealPart:12:6,' + j ',ImaginaryPart:12:6);
      WRITELN(' and          ',RealPart:12:6,' - j ',ImaginaryPart:12:6)
    END
  ELSE
    BEGIN  (* Real roots *)
      IF  b > 0.0  THEN
        sgn := 1.0
      ELSE
        sgn := -1.0;
     (* Calculate the root with the larger magnitude *)
          root1 := (-b - Sqrt(discriminant)*sgn)/(2.0*a);
    (* Now calculate the other root. *)
          root2 := c/(a*root1);
      WRITELN('The roots are ',root1:12:6,' and ',root2:12:6)
    END;
END.
```

It will be seen that this program has the following components:

1. Heading of the program.
2. Declaration of variables.
3. Program body, including comments.

The heading of the program, which starts with the "reserved word," PROGRAM, is intended to tell us the purpose for which the program has been written.

Some words in Pascal are reserved for specific purposes and cannot be used as names of variables. It is customary to write all reserved words in such a way that one may recognize them easily. We shall follow the practice of using capital letters for reserved words and lower case letters in most other contexts. This will improve the readability of the program. The heading is followed by the declaration of variables. In this program we have several real variables. Note that the names of the variables have been selected so that they are easily recognized by a reader. In Pascal, the variables are of four basic types, integers (INTEGER), real numbers (REAL), alphanumeric characters (CHAR) and logical (BOOLEAN). In addition, the user may define other types of variables, like sets, arrays, and records. All variables used in a program must be predefined in the declaration section preceded by the reserved word VAR.

The body of the program starts with BEGIN and terminates with END and a period. Between these are the various statements, separated by semicolons. In addition, there is a semicolon between the last declaration and the first BEGIN, as well as after the heading of the program. The semicolon serves as a separator between statements. BEGIN and END may also be regarded as "fat" parentheses as they act as separators between compound statements. The symbol := is the assignment operator, which is used instead of the = sign, and it says that the variable to its left is to be assigned the value given by the expression at its right. The symbol = means equality, not assignment. Within the program we also have several comments which appear between (* and *).

In addition to the assignment statement, we have used other Turbo Pascal statements like WRITE, WRITELN, READLN, and ClrScr. All of these are built-in procedures and are called simply by their names. The WRITE procedure causes a display on the screen of the expression inside the parentheses following the statement. Any expression within single quotes inside the parentheses will be printed on the screen as text or a string of characters. The WRITELN procedure is similar, with the difference that it is followed by an end-of-line. The format for displaying real numbers should be noted. If, in the call for WRITE or WRITELN we include the expression $x{:}f{:}d$, then the real number x will be displayed right-justified in a field of f columns and with d digits after the decimal point. It follows that both f and d must be integers. Moreover, if x is an integer then we cannot have a decimal part of it, and hence in this case, our WRITE or WRITELN procedure can only have $x{:}f$. Similar to WRITE and WRITELN, we have the procedures READ and READLN, which are used for reading data or answers to various questions and assigning them to specified variables. In particular if we have data in a line, with blank spaces between them, we may use the READ statement. If we have several lines of such data, we may use the READLN statement. ClrScr is a built-in procedure that clears the screen of the display and causes the cursor to move to the top left-hand corner of the screen.

In addition, we have several conditional statements in Pascal. Two of these have been used in our example. These are the REPEAT . . . UNTIL loop and the IF . . . THEN . . . ELSE loop. In the former, the statements that follow REPEAT and precede UNTIL are repeated until the condition following UNTIL becomes true. Note that in this case we must go through the loop at least once, since the condition to be tested is at the end of the loop. In the latter, the condition to be tested follows immediately after IF and only if this condition is true is the next statement carried out. We may in this case have either a single statement or a compound statement, bracketed by a BEGIN and an END. If the statement following IF is not true then the statement following ELSE is carried out. Pascal has many other conditional statements like FOR . . . DO, WHILE . . . DO and CASE . . . OF.

The reader would have noted that Pascal does not require all statements to begin in a certain column. Indentation of certain lines enables the human reader to see the various loops more clearly, but the compiler ignores the indentation. In most microcomputer implementations of Pascal, the case is also ignored; advantage can be taken of this fact in combining upper and lower case to make the program more readable. We shall follow the practice of using upper case for the reserved words and common built-in procedures and lower case for other statements. In some cases it is convenient to combine upper and lower case in the names of procedures and variables so that they can be read more easily, especially if they combine two or more words.

B.3 A BETTER PROGRAM FOR SOLVING QUADRATICS

Although the program described in the previous section will work, it is not the best way to write a program in Pascal. A better approach is to divide the main task into smaller subtasks, each of which can be reduced to some simple steps. One then writes procedures or functions to perform these subtasks, which are then called in the main program. This is the so-called "top-down" approach. The main advantage is that it is much easier to write these procedures and functions separately with the result that the main program can be made much smaller and easier to understand. Consider the following implementation of our programming example.

```
PROGRAM Quadratic;
(* Program for calculating the roots of a quadratic equation *)

CONST
  F = 14;  (* Fieldwidth  *)
  D = 7;   (* Number of digits after the decimal point *)

VAR
  a, b, c, discriminant : REAL;

PROCEDURE Greeting;
(* This procedure is intended to inform the user about the purpose of
   the program and the format of the input.  *)
BEGIN
  ClrScr;
  WRITELN(' This program calculates the roots of the quadratic ');
```

```
  WRITELN('x(ax + b) + c = 0, for any given a, b and c. ');
  WRITELN;
  WRITE('   Press the <enter> key to start. ') ; READLN
END; (* Greeting *)

FUNCTION Sgn( x : REAL ) : INTEGER;
BEGIN
  IF  x > 0.0  THEN
    Sgn := 1
  ELSE
    Sgn := -1
END; (* Sgn *)

PROCEDURE InputData;
BEGIN
  REPEAT
    WRITELN('Enter a, b and c, separated by spaces. ');
    READLN(a,b,c);
    IF  a = 0.0  THEN
      WRITELN('The coefficient a must not be zero!  Try again. ')
  UNTIL a <> 0.0
END; (* InputData *)

PROCEDURE ComplexRoots;
VAR
  RealPart, ImaginaryPart : REAL;
BEGIN
  ImaginaryPart := Sqrt(-discriminant)/(2.0*Abs(a));
  RealPart := -b/(2.0*a);
  WRITELN('The roots are ',RealPart:F:D,' + j ',ImaginaryPart:F:D);
  WRITELN(' and          ',RealPart:F:D,' - j ',ImaginaryPart:F:D)
END; (* ComplexRoots *)

PROCEDURE RealRoots;
VAR
  Root1, Root2 : REAL;
BEGIN
  Root1 := -(b + Sqrt(discriminant)*sgn(b))/(2.0*a);
  (* The root with the larger magnitude is first calculated *)
  Root2 := c/(a*Root1);
  WRITELN('The roots are ',Root1:F:D,' and ',Root2:F:D)
END; (* RealRoots *)

BEGIN (* main program *)
  Greeting;
  InputData;
  discriminant := Sqr(b) - 4.0*a*c;
  IF  discriminant < 0.0  THEN
    ComplexRoots
  ELSE
    RealRoots
END.
```

Several features of this version of the program are noticeable. First we note that the field-width and the desired number of digits after the decimal point have been declared as constants. This is more convenient since all the numbers can be displayed with different values of these constants by just changing their values at one location. Otherwise, one would have to go to each of the WRITELN statements to make the desired change. Another feature of this program is the division of the main task into a number of smaller tasks. These smaller tasks are written as the procedures "Greeting," "Input," "ComplexRoots" and "RealRoots," as well as the function "Sgn." Each of these is fairly short and simple, so that it can be written easily. The main program, which appears at the end, calls these procedures, as necessary. As a result, it is much easier to read and understand the main program. This is called the "top-down" approach to the design of a modular program.

B.4 SOME COMMENTS ON GOOD SOFTWARE

Often people write a program that "works." Except in some elementary cases, this is not sufficient. The manner in which the program has been designed and written is as important as whether the program works. One must take into account the ease with which the program can be altered to improve performance, satisfy additional specifications, and remove errors which are discovered later. All of these are related to the quality of the design. The programmer who boasts of using clever tricks for reducing the total number of lines of code is usually on the wrong track. Computer memory is relatively cheap, whereas the modification of software can be quite expensive, especially if the program has not been designed with clarity as an important criterion. Hence, if there is a choice between a little less economy of code or greater clarity in the program, the latter is preferable in most practical cases, especially for long programs.

Although the rules of Pascal do impose a certain amount of discipline on the programmer, they cannot force good programming practice. The reader is advised to study the books mentioned in the next section to learn the basic principles of programming in Pascal and study examples of good programs.

B.5 REFERENCES

D. Alcock, *Illustrating Pascal*. New York: Cambridge University Press, 1987.

P. M. Chirlian. *Pascal*. Portland, OR: Matrix Publishers, 1980.

D. L. Matuzsek. *Quick Pascal*. New York: Wiley, 1982.

D. Price. *Pascal: A Considerate Approach*. Englewood Cliffs, NJ: Prentice-Hall, 1982.

R. H. Perrott and D. C. S. Allison. *Pascal for FORTRAN Programmers*. Rockville, MD: Computer Science Press, 1984.

H. F. Ledgard, P. A. Nagin, and J. F. Hueras. *Pascal with Style*. Rochelle Park, NJ: Hayden Book Company, 1979.

Appendix C

Turbo Pascal

C.1 TURBO PASCAL

Turbo Pascal is a fast-compiling version of the Pascal language that was designed and is sold by Borland International. The first version of Turbo Pascal, Turbo Pascal 1, was first sold in 1983. Turbo Pascal 2 appeared in 1984, followed by Turbo Pascal 3 in 1985 and Turbo Pascal 4 in 1987.

There are some differences between Turbo Pascal and other Pascal "dialects." Some of these differences are discussed below. The reader interested in learning more about Turbo Pascal is advised to refer to the books listed in the references.

(1) Program Header Parameters

In some Pascal dialects the Program header statement looks like

```
PROGRAM Name(Input,Output,File1, . . . , FileN);
```

where the files Input, Output, File1, . . . , FileN are used in the program. Input and Output are the standard input and output devices, respectively. Usually, Input is the keyboard and Output is the video terminal. File1, File2, . . . , FileN are other files defined in the program. Turbo Pascal does not require such a parameter list in the program header.

(2) Functions and Procedures cannot be passed as parameters in Turbo Pascal (up to version 4)

Because of this restriction, procedures or functions that require other user-supplied procedures or functions are written using global names. If you are going to use the book procedures or functions with another Pascal compiler that allows passing of procedures and functions as parameters, you can easily modify the Turbo Pascal versions. You can do this by inserting the global names of the user-supplied procedures or functions into the parameter list in accordance with the specifications for the particular Pascal compiler. After this is done, the formerly global Turbo Pascal names are now local names for the procedure or function using them.

(3) Opening of Files for Reading or Writing

Turbo Pascal uses the statement

```
ASSIGN(TextFile,FileName);
```

together with one of the statements

```
RESET(TextFile);
REWRITE(TextFile);
APPEND(TextFile);
```

to open a text file, TextFile, for reading (RESET) or writing (REWRITE or APPEND). Other dialects of Pascal may use different statements.

(4) Closing of Files after Reading or Writing

Turbo Pascal uses the statement

```
CLOSE(TextFile);
```

to close a text file, TextFile, after reading from or writing to the file. Other dialects of Pascal may use a different statement.

C.2 TURBO PASCAL 3

All programs, procedures, and functions in this book were written in Turbo Pascal 3. They will run with and without the 8087 mathematical coprocessor. If your machine has an 8087 coprocessor you may want to use it, because the programs that are floating-point intensive generally run much faster using the coprocessor.

Turbo Pascal 3 has several predefined logical devices that are treated like text files. We have used

(a) **Con** the video terminal;

(b) **Prn** the printer; and

(c) **Kbd** the keyboard, where input is not echoed to the video terminal.

C.3 TURBO PASCAL 4

The Pascal programs, procedures, and functions in the book will run in Turbo Pascal 4 after some minor modifications. These modifications are described below.

Many of the functions and procedures that were built into the Turbo Pascal 3 compiler have been removed in the Turbo Pascal 4 compiler. Some of these procedures and functions are now found in various Turbo Pascal 4 Units. In order to run any book program in Turbo Pascal 4 you need only add the statement

```
Uses Crt, Dos;
```

after the main program's Program header statement. The "Uses" statement tells the compiler that some of the functions or procedures are in the Units Crt or Dos. If you are using the file Response.Pas in Appendix D.1 you should remove the defined function ReadKey; similarly, remove the defined procedure GetTime from the file Time.Pas in Appendix D.8.

If your microcomputer has an 8087 coprocessor and you want to use it, you must inform the Turbo Pascal 4 compiler that there is an 8087 coprocessor. This is done by the compiler directive

```
{$N+}
```

which is placed immediately after the Program header statement, but preceeding the "Uses" statement mentioned above. If this compiler directive is not inserted then the default size of variables of type REAL is taken as 6 bytes (approximately 12 decimal digits of precision which is the same as in Turbo Pascal 3 without the 8087). The arithmetic for these 6-byte floating point numbers is implemented by software, whereas the floating point arithmetic done by the 8087 coprocessor is in hardware.

Turbo Pascal 4 can use the I.E.E.E. floating point standard if there is an 8087 coprocessor. Thus, type REAL variables can be Single, Double or Extended, which are summarized in the following table.

Type	Approximate Precision (decimal digits)
Single	7
Double	16
Extended	19

To duplicate the action of Turbo Pascal 3 with 8087 support, you need to add the following statement to each main program, in addition to the compiler directive {$N+} and the "Uses" statement:

```
TYPE
  REAL = Double;
```

The statement redefines the Pascal type REAL as the I.E.E.E. type Double. This statement can be placed immediately after the "Uses" statement.

Some of the problems and examples in Chapter 1 deal with the precision of computations. If your microcomputer has an 8087 coprocessor you may want to do these problems with (REAL = Single) or (REAL = Double) or (REAL = Extended). It is also possible to mix these types within the same program so that some variables are Single, some are Double, and others are Extended.

For more details on Turbo Pascal 4, see the Reference Manual.

C.4 REFERENCES

Turbo Pascal Version 3.0 Reference Manual. Scotts Valley, CA: Borland International Inc., 1985.

Turbo Pascal Version 4.0 Reference Manual. Scotts Valley, CA: Borland International Inc., 1987.

J. Duntemann. *The Complete Turbo Pascal.* Glenview, IL: Scott, Foresman and Company, 1986.

S. Wood. *Using Turbo Pascal.* Berkeley, CA: Osborne McGraw-Hill, 1986.

Appendix D

Pascal Utility Procedures and Functions

These functions and procedures are used in many of the book's programs as Turbo Pascal Include files.

D.1 The file Response.Pas

```
FUNCTION ReadKey : CHAR;
(* Turbo Pascal 4 contains a function with the same name.
   If you are using Turbo Pascal 4 remove this function. *)
VAR
  response : CHAR;
BEGIN
  READ(KBD,response);
  ReadKey := response
END; (* ReadKey *)

PROCEDURE Echo( VAR response : CHAR );
BEGIN
  response := ReadKey;
  WRITELN(response)
END; (* Echo *)
```

```
PROCEDURE Echoed( VAR response : CHAR );
BEGIN
  response := ReadKey;
  WRITE(response)
END; (* Echoed *)

PROCEDURE Pause;
VAR
  response : CHAR;
BEGIN
  WRITE('Press <space> to continue ...   ');
  response := ReadKey;
  WRITELN
END; (* Pause *)
```

D.2 The file Files.Pas

The following must be declared in the calling program

TYPE
  String14 = String[*k*];

The value of *k* that is used in the actual type declaration should be at least 14. Files.Pas uses the file Response.Pas, which must be declared before Files.Pas.

```
FUNCTION FileExists( VAR SourceFile : TEXT ; FileName : String14 ) : BOOLEAN;
(*   This function is used to test if the file FileName exits. *)
BEGIN
  ASSIGN(SourceFile,FileName) ; (*$I- *) RESET(SourceFile) ; (*$I+ *)
  IF  IOresult = 0  THEN
    BEGIN
      FileExists := TRUE ; CLOSE(SourceFile)
    END
  ELSE
    FileExists := FALSE
END; (* FileExists *)

FUNCTION GetFile( VAR SourceFile : TEXT ) : BOOLEAN;
(*  This function is used to open the file SourceFile, if it exits. *)
VAR
  answer : CHAR;
  found, abort : BOOLEAN;
  FileName : String14;
BEGIN
  REPEAT
    WRITE('Enter the name of the file:  ') ; READLN(FileName) ;
    IF  FileExists(SourceFile,FileName)  THEN
      BEGIN
        found := TRUE ; abort := FALSE
      END
    ELSE
      BEGIN
        found := FALSE;
        WRITELN(' *** The file ',FileName,' cannot be found. ***');
        WRITE('Do you want to try a different file name?  (Y or N)  ');
```

```
        Echo(answer);
        IF  answer IN ['Y','y']  THEN
          abort := FALSE
        ELSE
          abort := TRUE
      END
  UNTIL  found  OR  abort;
  IF  found  THEN
    BEGIN
      ASSIGN(SourceFile,FileName) ; RESET(SourceFile)
    END;
  GetFile := found
END; (* GetFile *)

PROCEDURE CreateFile( VAR OutFile : TEXT ; VAR FileName : String14 );
(*   This procedure is used to create the new file FileName.  *)
VAR
  answer : CHAR;
  Create : BOOLEAN;
BEGIN
  REPEAT
    WRITE('Enter the name of the file:  ') ; READLN(FileName) ;
    IF  FileExists(OutFile,FileName)  THEN
      BEGIN
        WRITELN(' *** The file ',FileName,' already exists. ***');
        WRITE('Do you want to overwrite it?  (Y or N)  ');
        Echo(answer);
        IF  answer  IN  ['Y','y']  THEN
          Create := TRUE
        ELSE
          Create := FALSE
      END
    ELSE
      Create := TRUE
  UNTIL  Create;
  ASSIGN(OutFile,FileName) ; REWRITE(OutFile)
END; (* CreateFile *)
```

D.3 The file ReadNum.Pas

These Pascal procedures and functions are designed to provide fail-safe input of INTEGER and REAL variables from the keyboard.

The following must be declared in the calling program

TYPE
  String80 = String[80];

```
PROCEDURE ErrorHandler( x, y, offset : INTEGER ; message : String80 );
BEGIN
  HighVideo;
  GoToXY(x + offset,y) ; ClrEol ;
  WRITE(' ',message);
  Delay(1500) ; GoToXY(x,y) ; ClrEol ;
  LowVideo
END; (* ErrorHandler *)
```

```
PROCEDURE ReadNumber( VAR m : INTEGER ; VAR t : REAL ; num : BOOLEAN );
(* If  num = TRUE, then we are reading the integer variable  m.
   If  num = FALSE, then we are reading the real variable  t. *)
VAR
  c : String80;
  x, y, code : INTEGER;
BEGIN
  x := WhereX ; LowVideo ;
  REPEAT
    READLN(c) ; y := WhereY ;
    IF  c = ''  THEN
      code := 1
    ELSE
      BEGIN
        Val(c,t,code);  (* check for a valid real *)
        IF  (code = 0) AND num  THEN
          BEGIN
            IF  Abs(t) <= MaxInt  THEN
              Val(c,m,code)  (* check for a valid integer *)
            ELSE
              code := 1
          END
      END;
    IF  code <> 0  THEN
      ErrorHandler(x,Pred(y),Length(c),'Invalid number')
  UNTIL  code = 0
END; (* ReadNumber *)

FUNCTION ReadReal : REAL;
VAR
  m : INTEGER;
  t : REAL;
BEGIN
  ReadNumber(m,t,FALSE);
  ReadReal := t
END; (* ReadReal *)

FUNCTION ReadInteger : INTEGER;
VAR
  m : INTEGER;
  t : REAL;
BEGIN
  ReadNumber(m,t,TRUE);
  ReadInteger := m
END; (* ReadInteger *)
```

D.4 The file MatrixIO.Pas

The Pascal procedures and functions in this file are designed for input and output of matrices.

The following must be declared in the calling program

```
TYPE
  MatrixNM = ARRAY[ 1..MaxN , 1..MaxM ] OF REAL;
  String80 = String[80];
  String14 = String[ k ];   where  k ≥ 14
```

```
These procedures use the files Response.Pas, Files.Pas, and
ReadNum.Pas which must be declared before MatrixIO.Pas.

PROCEDURE GetMatrix( VAR a : MatrixNM ; VAR n, m : INTEGER ;
                     square : BOOLEAN ; MaxN, MaxM, Field, Decimal : INTEGER );
(*  This procedure can be used to get a matrix, from the keyboard or a file,
    modify it if desired, and store the matrix in a file, if desired.
         a    The matrix to be obtained
         n    Number of rows in the matrix      ( n <= MaxN )
         m    Number of columns in the matrix   ( m <= MaxM )
    square    If set as TRUE, then n = m is assumed for the input matrix
              If set as FALSE, then n and m need not be equal
     Field  : the output field width.  The actual width depends on  Decimal.
   Decimal  : the number of decimal places of the entries.  If  Decimal <= 0
               the exponential output form is used with a field width of at
               least 10  *)
VAR
  KeyBoard, change : BOOLEAN;
  answer : CHAR;

PROCEDURE RowAndColumnDimensions;
VAR
  x, y : INTEGER;
  ok : BOOLEAN;
BEGIN
  LowVideo;
  IF  square  THEN
    BEGIN
      WRITE('The dimension of the matrix:    ') ; x := WhereX ;
      REPEAT
        n := ReadInteger ; m := n ; y := WhereY ;
        ok := (1 <= n) AND (n <= MaxN) AND (m <= MaxM);
        IF  NOT ok  THEN
          BEGIN
            ErrorHandler(1,y,0,'Invalid dimensions for the matrix');
            GoToXY(x,Pred(y)) ; ClrEol
          END
      UNTIL  ok
    END
  ELSE
    BEGIN
      WRITELN('The dimensions of the matrix:');
      REPEAT
        WRITE(' Number of rows =    ') ; n := ReadInteger ;
        WRITE(' Number of columns =    ') ; m := ReadInteger ;
        y := Pred(WhereY);
        ok := (1 <= m) AND (m <= MaxM) AND (1 <= n) AND (n <= MaxN);
        IF  NOT ok  THEN
          BEGIN
            ErrorHandler(1,Succ(y),0,'Invalid dimensions for the matrix');
            GoToXY(1,y) ; ClrEol ;
            GoToXY(1,Pred(y)) ; ClrEol
          END
      UNTIL  ok
    END
END; (* RowAndColumnDimensions *)

PROCEDURE MatrixIn; (* Type in a matrix *)
VAR
  i, j : INTEGER;
BEGIN
  FOR  i := 1  TO  n  DO
```

```
    BEGIN
      WRITELN('Row ',i);
      FOR  j := 1  TO  m  DO
        BEGIN
          WRITE(' ',j,':    ') ; a[i,j] := ReadReal
        END
    END
END; (* MatrixIn *)


PROCEDURE Display; (* Display a matrix on the console *)
VAR
  i, j, NumberPerLine : INTEGER;
BEGIN
  NumberPerLine := 80 DIV (Field + 1);
  FOR  i := 1  TO  n  DO
    FOR  j := 1  TO  m  DO
      BEGIN
        IF  Decimal <= 0  THEN
          WRITE(a[i,j]:Field,' ')
        ELSE
          WRITE(a[i,j]:Field:Decimal,' ');
        IF  ((j MOD NumberPerLine) = 0) OR (j = m)  THEN
          WRITELN
      END;
  WRITELN
END; (* Display *)


PROCEDURE Verify; (* the matrix that was typed or read in *)
VAR
  i, j, x, y : INTEGER;
  answer : CHAR;
BEGIN
  change := FALSE;
  REPEAT
    ClrScr ; LowVideo ; Display ;
    WRITE('Are the entries correct?  (Y or N)  ') ; Echo(answer) ;
    IF  NOT (answer IN ['Y','y'])  THEN
      BEGIN
        change := TRUE;
        WRITELN('Which entry is wrong?');
        WRITE('    Row = ');
        REPEAT
          x := WhereX ; i := ReadInteger ; y := WhereY ;
          IF  (i < 1) OR (i > n)  THEN
            ErrorHandler(x,Pred(y),7,'Invalid row number')
        UNTIL (1 <= i) AND (i <= n);
        WRITE(' Column = ');
        REPEAT
          x := WhereX ; j := ReadInteger ; y := WhereY ;
          IF  (j < 1) OR (j > m)  THEN
            ErrorHandler(x,Pred(y),7,'Invalid column number')
        UNTIL  (1 <= j) AND (j <= m);
        WRITE('entry (',i,',',j,') =  ') ; a[i,j] := ReadReal
      END
  UNTIL  answer IN ['Y','y']
END; (* Verify *)
```

```
PROCEDURE WriteMatrix; (* to a file *)
VAR
  i, j, NumberPerLine : INTEGER;
  SourceFile : TEXT;
  FileName : String14;
BEGIN
  WRITE('The output file.   ');
  CreateFile(SourceFile,FileName);
  NumberPerLine := 80 DIV (Field + 1);
  WRITELN(SourceFile,n,'  ',m);
  FOR  i := 1  TO  n  DO
    FOR  j := 1  TO  m  DO
      BEGIN
        IF  Decimal <= 0  THEN
          WRITE(SourceFile,a[i,j]:Field,' ')
        ELSE
          WRITE(SourceFile,a[i,j]:Field:Decimal,' ');
        IF  ((j MOD NumberPerLine) = 0) OR (j = m)  THEN
          WRITELN(SourceFile)
      END;
  CLOSE(SourceFile)
END; (* WriteMatrix *)



PROCEDURE ReadMatrix; (* from a file *)
VAR
  i, j : INTEGER;
  SourceFile : TEXT;
BEGIN
  WRITELN ; WRITE('The input file.   ') ;
  IF  GetFile(SourceFile)  THEN
    BEGIN
      READLN(SourceFile,n,m);
      FOR  i := 1  TO  n  DO
        FOR  j :=  1  TO  m  DO
          BEGIN
            READ(SourceFile,a[i,j]);
            IF  EOLN(SourceFile) OR (j = m)  THEN
              READLN(SourceFile)
          END;
      CLOSE(SourceFile)
    END
  ELSE
    KeyBoard := TRUE
END; (* ReadMatrix *)

PROCEDURE SelectInputMode;
VAR
  answer : CHAR;
BEGIN
  WRITELN;
  WRITE('Do you want to input the matrix from the keyboard or from');
  WRITE(' a file?  (K or F)  ') ; Echo(answer) ;
  IF  answer  IN  ['F','f']  THEN
    KeyBoard := FALSE
  ELSE
    KeyBoard := TRUE
END; (* SelectInputMode *)
```

```
BEGIN (* GetMatrix *)
  SelectInputMode;
  IF  (Decimal <= 0) AND (Field < 10)  THEN
    Field := 10;
  IF  NOT KeyBoard  THEN
    BEGIN
      ReadMatrix;
      IF  NOT KeyBoard  THEN
        Verify
    END;
  IF  KeyBoard  THEN
    BEGIN
      RowAndColumnDimensions;
      MatrixIn;
      Verify;
      change := TRUE
    END;
  IF  change  THEN
    BEGIN
      WRITELN;
      WRITE('Do you want to write the matrix into a file?  (Y or N)  ');
      Echo(answer);
      IF  answer  IN  ['Y','y']  THEN
        WriteMatrix
    END
END; (* GetMatrix *)



PROCEDURE PutMatrix( VAR a : MatrixNM ; n, m, Field, Decimal : INTEGER ;
                         Title : String80 ; place : CHAR );
(* This procedure can be used to write a matrix to the console, the printer,
   or a file.
      a      The matrix to be written to the output device selected
      n      The number of rows of the matrix
      m      The number of columns of the matrix
    Title    An identifier label that will be written to the selected output
              device, either the printer or the console, along with the matrix.
    Field    The output field width.  The actual width depends on  Decimal.
   Decimal   The number of decimal places of the entries.  If  Decimal <= 0 the
              exponential output form is used with a field width of at least 10
    place    A character used to select the output device
                 C   The console
                 P   The printer
                 F   A new disk file
                 A   Append the matrix to an existing disk file   *)
VAR
  i, j, NumberPerLine : INTEGER;
  FileName : String14;
  SourceFile : TEXT;
BEGIN
  IF  (Decimal <= 0) AND (Field < 10)  THEN
    Field := 10;
  CASE    place   OF
        'C','c'  : BEGIN
                     ASSIGN(SourceFile,'Con') ; REWRITE(SourceFile) ;
                     WRITELN(SourceFile) ; WRITELN(SourceFile,Title) ;
                     WRITELN(SourceFile)
                   END;
```

```
          'P','p'  : BEGIN
                       ASSIGN(SourceFile,'Prn') ; REWRITE(SourceFile) ;
                       WRITELN(SourceFile) ; WRITELN(SourceFile,Title) ;
                       WRITELN(SourceFile)
                     END;
          'F','f'  : BEGIN
                       WRITE('The output file.   ');
                       CreateFile(SourceFile,FileName);
                       WRITELN(SourceFile,n,'  ',m)
                     END;
          'A','a'  : BEGIN
                       REPEAT
                         WRITE('The output file.   ');
                       UNTIL  GetFile(SourceFile);
                       APPEND(SourceFile);
                       WRITELN(SourceFile,n,'  ',m)
                     END
   END;
  NumberPerLine := 80 DIV (Field + 1);
  FOR  i := 1  TO  n  DO
    FOR  j := 1  TO  m  DO
      BEGIN
        IF  Decimal <= 0  THEN
          WRITE(SourceFile,a[i,j]:Field,' ')
        ELSE
          WRITE(SourceFile,a[i,j]:Field:Decimal,' ');
        IF  ((j MOD NumberPerLine) = 0) OR (j = m)  THEN
          WRITELN(SourceFile)
      END;
  CLOSE(SourceFile)
END; (* PutMatrix *)
```

D.5 The file MatrixUt.Pas

The Pascal procedures and functions in this file provide various matrix and vector operations.

The following must be declared in the calling program

```
TYPE
  MatrixNM = ARRAY[ 1..MaxN , 1..MaxM ] OF REAL;
  VectorN = ARRAY[ 1..MaxN ] OF REAL;
```

```
PROCEDURE VectorMultiply( row, column : INTEGER ; VAR A : MatrixNM ;
                                                  VAR x, y : VectorN );
(*  y = A*x  *)
VAR
  i, j : INTEGER;
  t : REAL;
BEGIN
  FOR  i := 1  TO  row  DO
    BEGIN
      t := A[i,1]*x[1];
      FOR  j := 2  TO  column  DO
        t := t + A[i,j]*x[j];
      y[i] := t
    END
END; (* VectorMultiply *)
```

```
FUNCTION DotProduct( n : INTEGER ; VAR x, y : VectorN ) : REAL;
(*  Form the inner product of the vectors  x  and  y  *)
VAR
  i : INTEGER;
  t : REAL;
BEGIN
  t := x[1]*y[1];
  FOR  i := 2  TO  n  DO
    t := t + x[i]*y[i];
  DotProduct := t
END; (* DotProduct *)

PROCEDURE Transpose( row, column : INTEGER ; VAR A, Atr : MatrixNM );
(*  Form the transpose of the matrix  A  *)
VAR
  i, j : INTEGER;
BEGIN
  FOR  i := 1  TO  row  DO
    FOR  j := 1  TO  column  DO
      Atr[j,i] := A[i,j]
END; (* Transpose *)

PROCEDURE MatrixMultiply( rowA, columnA, columnB : INTEGER ;
                                             VAR A, B, C : MatrixNM );
(*  Form the matrix product  A*B = C.  It is assumed that the number of rows
    of B = number of columns of A, denoted by columnA  *)
VAR
  t : REAL;
  i, j, k : INTEGER;
BEGIN
  FOR  i := 1  TO  rowA  DO
    FOR  j := 1  TO  columnB  DO
      BEGIN
        t := A[i,1]*B[1,j];
        FOR  k := 2  TO  columnA  DO
          t := t + A[i,k]*B[k,j];
        C[i,j] := t
      END
END; (* MatrixMultiply *)

PROCEDURE OuterProduct( row, column : INTEGER ; VAR x, y : VectorN ;
                                                    VAR A : MatrixNM );
(*  Form the outer product of the vectors  x  and  y  *)
VAR
  t : REAL;
  i, j : INTEGER;
BEGIN
  FOR  i := 1  TO  row  DO
    BEGIN
      t := x[i];
      FOR  j := 1  TO  column  DO
        A[i,j] := t*y[j]
    END
END; (* OuterProduct *)

FUNCTION Trace( n : INTEGER ; VAR A : MatrixNM ) : REAL;
(* Find the trace of the matrix  A  *)
VAR
  i : INTEGER;
  t : REAL;
```

```
BEGIN
  t := A[1,1];
  FOR  i := 2  TO  n  DO
    t := t + A[i,i];
  Trace := t
END; (* Trace *)

PROCEDURE IdentityMatrix( n : INTEGER ; VAR M : MatrixNM );
(* Set the matrix  M  to be the  n by n identity matrix *)
VAR
  i, j : INTEGER;
BEGIN
  FOR  i := 1  TO  n  DO
    BEGIN
      FOR  j := i+1  TO  n  DO
        BEGIN
          M[i,j] := 0.0 ; M[j,i] := 0.0
        END;
      M[i,i] := 1.0
    END
END; (* IdentityMatrix *)
```

D.6 User's Manual for the Plotter Package, which is in the file PrPlot.Pas

This package does not require a special graphics card to be used. The package can be used to obtain point plots of data. The package consists of the three Pascal procedures

```
SetPlot( VAR Data : GraphData )
StorePoint( x, y : REAL ; c : CHAR ; VAR Data : GraphData )
PlotGraph( VAR Data : GraphData )
```

and requires the following to be declared in the calling program

```
TYPE
  Vector1000 = ARRAY[ 1..1000 ] OF REAL;
  GraphData = RECORD
                n, rows, columns : INTEGER;
                start : BOOLEAN;
                x, y : Vector1000;
                z : ARRAY[ 1..1000 ] OF CHAR;
                grid : ARRAY[ 1..21 , 1..61 ] OF CHAR
              END;
```

The user must also define a variable of type GraphData, called Data here. PrPlot.Pas uses the file Response.Pas which must be declared before PrPlot.Pas.

(1) The plot is initiated by calling SetPlot(Data).

(2) StorePoint(x,y,c,Data)

This procedure is called once for each point to be plotted, where (x,y) is a point to be plotted and c is a character that will be printed on the plot to represent (x,y). Note that x and y are of type REAL and c is CHAR. If more than 1000 calls are made to the procedure StorePoint a warning message is issued and then all points in excess of 1000 are ignored. If two different user characters intersect, then the procedure prints an asterisk, *, in that position. More than one graph can be plotted on the same axis system by using different plotting characters c.

(3) PlotGraph(Data)

After all points to be plotted are stored by StorePoint, a call to PlotGraph does the plotting to the screen. After the graph appears on the screen, the user is asked if the graph should be printed.

The procedures SetPlot, StorePoint, PlotGraph are based on the Pascal procedures setplt, stopnt, plot which appeared in

V.A. Dyck, J.D. Lawson, J.A. Smith, and R.J. Beach,
Computing: An Introduction to Structured Problem Solving using Pascal
Reston Publishing, Reston, Virginia, 1982.

D.7 Sample Program showing how the print plotter package is used

```
PROGRAM Plot;

CONST
  alpha = 0.5 ; scale = 50.0 ; points = 300 ;
  Pi = 3.14159265358979323846264 3;

TYPE
  Vector1000 = ARRAY[ 1..1000 ] OF REAL;
  GraphData = RECORD
                n, rows, columns : INTEGER;
                start : BOOLEAN;
                x, y : Vector1000;
                z : ARRAY[ 1..1000 ] OF CHAR;
                grid : ARRAY[ 1..21 , 1..61 ] OF CHAR
              END;

VAR
  Data : GraphData;
  x, y, r, theta, dTheta : REAL;
  k : INTEGER;

(*$I Response.Pas *)
(*$I PrPlot.Pas *)

BEGIN
  dTheta := 2.0*Pi/points ; theta := 0.0 ;
  SetPlot(Data);
  FOR k := 0 TO points DO
    BEGIN
      y := sin(theta) ; r := scale*(alpha + y) ;
      x := r*cos(theta) ; y := r*y ;
      StorePoint(x,y,'x',Data);
      theta := theta + dTheta
    END;
  dTheta := Pi/points ; theta := 0.0 ;
  FOR k := 0 TO points DO
```

```
    BEGIN
      y := sin(theta) ; r := scale*y ;
      x := r*cos(theta) ; y := r*y ;
      StorePoint(x,y,'.',Data);
      theta := theta + dTheta
    END;
  PlotGraph(Data)
END.
```

D.8 The file Time.Pas

The function ElapsedTime and the procedures SetTime and WriteTime can be used to obtain timings. Typically, one may want to obtain the amount of time required for a certain portion of a program to be executed. The procedure SetTime must be called immediately before the first statement of the section of the program to be timed; this initializes the elapsed time counter. Immediately following the last statement of the section of program being timed, the procedure WriteTime can be called to write out the elapsed time to the screen, or the user may want to call the function ElapsedTime.

```
FUNCTION ElapsedTime( VAR time : REAL ) : REAL;
(* If you are using Turbo 4, remove the following type declaration *)
TYPE
  WORD = INTEGER;
VAR
  h, m, s, f : WORD;
  TimeInSeconds : REAL;

PROCEDURE GetTime( VAR h, m, s, f : WORD );
(* If you are using Turbo 4, remove this procedure because Turbo 4 has
   a GetTime procedure in its Dos unit *)
VAR
  regs : RECORD
           ax,bx,cx,dx,bp,si,di,ds,es,flags : INTEGER
         END;
BEGIN
  regs.ax := $2C00 ; MsDos(regs) ;
  h := Hi(regs.cx) ; m := Lo(regs.cx) ; s := Hi(regs.dx) ; f := Lo(regs.dx)
END; (* GetTime *)

BEGIN (* ElapsedTime *)
  GetTime(h,m,s,f); (* This gets the clock time in Turbo 3 and Turbo 4 *)
  TimeInSeconds := 60.0*(60.0*h + m) + s + f/100.0;
  ElapsedTime := TimeInSeconds - time;
  time := TimeInSeconds
END; (* ElapsedTime *)

PROCEDURE SetTime( VAR time : REAL ); (* Initializes  time *)
BEGIN
  time := 0.0;
  time := ElapsedTime(time)
END;

PROCEDURE WriteTime( VAR time : REAL );
BEGIN
  WRITELN('Elapsed time = ',ElapsedTime(time):8:2,' sec.')
END;
```

Appendix E

Some FORTRAN Equivalent or Substitute Programs

In order to assist persons who may not have access to a Pascal compiler, or who may want to use FORTRAN instead of Pascal, we include the following brief list of some of our program files and possible FORTRAN equivalent or substitute programs from the library of programs published by IMSL Inc. (IMSL Library Version 9.2).

Chapter	File Name	IMSL Equivalent or Substitute
2	LinearSy.Pas	LEQT2F
3	Roots.Pas Jenkins.Pas	ZPOLR ZRPOLY
4	ZeroIn.Pas Powell.Pas	ZBRENT ZSPOW
5	Spline.Pas SVD.Pas	ICSCCU, ICSEVU LSVDF
6	QUANC8.Pas	DCADRE

Chapter	File Name	IMSL Equivalent or Substitute
7	RKF45.Pas	DVERK
	ODE.Pas	DGEAR
8	EigenR.Pas	EIGRF
	EigenRS.Pas	EIGRS
9	Golden.Pas	ZXGSN
	Brent.Pas	ZXLSF
	QNM.Pas	ZXMIN

INDEX